T0320627

# Lectures on Quantum Mechanics

## Second Edition

Nobel Laureate Steven Weinberg combines exceptional physical insight with his gift for clear exposition, to provide a concise introduction to modern quantum mechanics, in this fully updated second edition of his successful textbook. Now including six brand new sections covering key topics such as the rigid rotator and quantum key distribution, as well as major additions to existing topics throughout, this revised edition is ideally suited to a one-year graduate course or as a reference for researchers. Beginning with a review of the history of quantum mechanics and an account of classic solutions of the Schrödinger equation, before quantum mechanics is developed in a modern Hilbert space approach, Weinberg uses his remarkable expertise to elucidate topics such as Bloch waves and band structure, the Wigner–Eckart theorem, magic numbers, isospin symmetry, and general scattering theory. Problems are included at the ends of chapters, with solutions available for instructors at www.cambridge.org/9781107111660.

STEVEN WEINBERG is a member of the Physics and Astronomy Departments at the University of Texas at Austin. His research has covered a broad range of topics in quantum field theory, elementary particle physics, and cosmology, and he has been honored with numerous awards, including the Nobel Prize in Physics, the National Medal of Science, and the Heinemann Prize in Mathematical Physics. He is a member of the US National Academy of Sciences, Britain's Royal Society, and other academies in the USA and abroad. The American Philosophical Society awarded him the Benjamin Franklin medal, with a citation that said he is "considered by many to be the preeminent theoretical physicist alive in the world today." His books for physicists include *Gravitation and Cosmology*, the three-volume work *The Quantum Theory of Fields*, and, most recently, *Cosmology*. Educated at Cornell, Copenhagen, and Princeton, he also holds honorary degrees from sixteen other universities. He taught at Columbia, Berkeley, M.I.T., and Harvard, where he was Higgins Professor of Physics, before coming to Texas in 1982.

"Steven Weinberg, a Nobel Laureate in physics, has written an exceptionally clear and coherent graduate-level textbook on modern quantum mechanics. This book presents the physical and mathematical formulations of the theory in a concise and rigorous manner. The equations are all explained step-by-step, and every term is defined. He presents a fresh, integrated approach to teaching this subject with an emphasis on symmetry principles. Weinberg demonstrates his finesse as an excellent teacher and author."

Barry R. Masters, *Optics and Photonics News*

"... *Lectures on Quantum Mechanics* must be considered among the very best books on the subject for those who have had a good undergraduate introduction. The integration of clearly explained formalism with cogent physical examples is masterful, and the depth of knowledge and insight that Weinberg shares with readers is compelling."

Mark Srednicki, *Physics Today*

"Perhaps what distinguishes this book from the competition is its logical coherence and depth, and the care with which it has been crafted. Hardly a word is misplaced and Weinberg's deep understanding of the subject matter means that he leaves no stone unturned: we are asked to accept very little on faith ... it is for the reader to follow Weinberg in discovering the joys of quantum mechanics through a deeper level of understanding: I loved it!"

Jeff Forshaw, *CERN Courier*

"An instant classic ... clear, beautifully structured and replete with insights. This confirms [Weinberg's] reputation as not only one of the greatest theoreticians of the past 50 years, but also one of the most lucid expositors. Pure joy."

*The Times Higher Education Supplement*

# Lectures on Quantum Mechanics

## Second Edition

Steven Weinberg

*The University of Texas at Austin*

# CAMBRIDGE
## UNIVERSITY PRESS

University Printing House, Cambridge CB2 8BS, United Kingdom

One Liberty Plaza, 20th Floor, New York, NY 10006, USA

477 Williamstown Road, Port Melbourne, VIC 3207, Australia

314-321, 3rd Floor, Plot 3, Splendor Forum, Jasola District Centre, New Delhi - 110025, India

79 Anson Road, #06-04/06, Singapore 079906

Cambridge University Press is part of the University of Cambridge.

It furthers the University's mission by disseminating knowledge in the pursuit of
education, learning and research at the highest international levels of excellence.

www.cambridge.org
Information on this title: www.cambridge.org/9781107111660

© Cambridge University Press 2015

First published 2015
3rd printing 2018

*A catalogue record for this publication is available from the British Library*

*Library of Congress Cataloging in Publication data*
Weinberg, Steven, 1933– author.
Lectures on quantum mechanics / Steven Weinberg, The University
of Texas at Austin. – Second edition.
pages   cm
Includes indexes.
ISBN 978-1-107-11166-0 (hbk.)
1. Quantum theory.   I. Title.
QC174.125.W45   2015
530.12–dc23
2015021123

ISBN 978-1-107-11166-0 Hardback

Additional resources for this publication at www.cambridge.org/9781107111660

*For Louise, Elizabeth, and Gabrielle*

# Contents

# Preface

## Preface to First Edition

The development of quantum mechanics in the 1920s was the greatest advance in physical science since the work of Isaac Newton. It was not easy; the ideas of quantum mechanics present a profound departure from ordinary human intuition. Quantum mechanics has won acceptance through its success. It is essential to modern atomic, molecular, nuclear, and elementary particle physics, and to a great deal of chemistry and condensed matter physics as well.

There are many fine books on quantum mechanics, including those by Dirac and Schiff from which I learned the subject a long time ago. Still, when I have taught the subject as a one-year graduate course, I found that none of these books quite fit what I wanted to cover. For one thing, I like to give a much greater emphasis than usual to principles of symmetry, including their role in motivating commutation rules. (With this approach the canonical formalism is not needed for most purposes, so a systematic treatment of this formalism is delayed until Chapter 9.) Also, I cover some modern topics that of course could not have been treated in the books of long ago, including numerous examples from elementary particle physics, alternatives to the Copenhagen interpretation, and a brief (very brief) introduction to the theory and experimental tests of entanglement and its application in quantum computation. In addition, I go into some topics that are often omitted in books on quantum mechanics: Bloch waves, time-reversal invariance, the Wigner–Eckart theorem, magic numbers, isotopic spin symmetry, "in" and "out" states, the "in–in" formalism, the Berry phase, Dirac's theory of constrained canonical systems, Levinson's theorem, the general optical theorem, the general theory of resonant scattering, applications of functional analysis, photoionization, Landau levels, multipole radiation, etc.

The chapters of the book are divided into sections, which on average approximately represent a single seventy-five minute lecture. The material of this book just about fits into a one-year course, which means that much else has had to be skipped. Every book on quantum mechanics represents an exercise in selectivity – I can't say that my selections are better than those of other authors, but at least they worked well for me when I taught the course.

There is one topic I was not sorry to skip: the relativistic wave equation of Dirac. It seems to me that the way this is usually presented in books on quantum mechanics is profoundly misleading. Dirac thought that his equation is a relativistic generalization of the non-relativistic time-dependent Schrödinger equation that governs the probability amplitude for a point particle in an external electromagnetic field. For some time after, it was considered to be a good thing that Dirac's approach works only for particles of spin one half, in agreement with the known spin of the electron, and that it entails negative-energy states, states that when empty can be identified with the electron's antiparticle. Today we know that there are particles like the $W^\pm$ that are every bit as elementary as the electron, and that have distinct antiparticles, and yet have spin one, not spin one half. The right way to combine relativity and quantum mechanics is through the quantum theory of fields, in which the Dirac wave function appears as the matrix element of a quantum field between a one-particle state and the vacuum, and not as a probability amplitude.

I have tried in this book to avoid an overlap with the treatment of the quantum theory of fields that I presented in earlier volumes.[1] Aside from the quantization of the electromagnetic field in Chapter 11, the present book does not go into relativistic quantum mechanics. But there are some topics that were included in *The Quantum Theory of Fields* because they generally are not included in courses on quantum mechanics, and I think they should be. These subjects are included here, especially in Chapter 8 on general scattering theory, despite some overlap with my earlier volumes.

The viewpoint of this book is that physical states are represented by vectors in Hilbert space, with the wave functions of Schrödinger just the scalar products of these states with basis states of definite position. This is essentially the approach of Dirac's "transformation theory." I do not use Dirac's bra–ket notation, because for some purposes it is awkward, but in Section 3.1 I explain how it is related to the notation used in this book. In any notation, the Hilbert space approach may seem to the beginner to be rather abstract, so to give the reader a greater sense of the physical significance of this formalism I go back to its historic roots. Chapter 1 is a review of the development of quantum mechanics from the Planck black-body formula to the matrix and wave mechanics of Heisenberg and Schrödinger and Born's probabilistic interpretation. In Chapter 2 the Schrödinger wave equation is used to solve the classic bound state problems of the hydrogen atom and harmonic oscillator. The Hilbert-space formalism is introduced in Chapter 3, and used from then on.

---

[1]  S. Weinberg, *The Quantum Theory of Fields* (Cambridge University Press, 1995; 1996; 2000).

## Addendum for the Second Edition

Since the publication of the first edition, I have come to think that several additional topics needed to be included in this book. I have therefore added six new sections: Section 4.9 on the rigid rotator; Section 5.9 on van der Waals forces; Section 6.8 on Rabi oscillations and Ramsey interferometers; Section 6.9 on open systems, including a derivation of the Lindblad equation; Section 8.9 on time reversal of scattering processes, including a proof of the Watson–Fermi theorem; and Section 11.8 on quantum key distribution. There have also been many additions within the sections of the first edition, including discussions of the universality of black-body radiation in Section 1.1, lasers in Section 1.2, unentangled systems in Section 3.3, the groups $O(3)$ and $SO(3)$ in Section 4.1, $3j$ symbols and the addition theorem for spherical harmonics in Section 4.3, the application of the eikonal approximation to scattering by long-range forces in Section 7.10, and error-correcting codes in Section 12.3. I have also taken the opportunity to correct many minor errors, as well as a major error in the formulation of degenerate perturbation theory in Sections 5.1 and 5.4.

In Section 3.7 of the first edition I reviewed various interpretations of quantum mechanics, and explained why none of them seem to me entirely satisfactory. I have now reorganized and expanded this discussion, with no change in its conclusion.

$$* \; * \; * \; * \; *$$

I am grateful to Raphael Flauger and Joel Meyers, who as graduate students assisted me when I taught courses on quantum mechanics at the University of Texas, and suggested numerous changes and corrections to the lecture notes on which the first edition of this book was based. I am also indebted to Robert Griffiths, James Hartle, Allan Macdonald, and John Preskill, who gave me advice on various specific topics that proved helpful in preparing the first edition, and to Scott Aaronson, Jeremy Bernstein, Jacques Distler, Ed Fry, Christopher Fuchs, James Hartle, Jay Lawrence, David Mermin, Sonia Paban, Philip Pearle, and Mark Raizen who helped with the coverage of various topics in the second edition. Thanks are due to many readers who pointed out errors in the first edition, especially Andrea Bernasconi, Lu Quanhui, Mark Weitzman, and Yu Shi. Cumrun Vafa used the first half of the first edition as a textbook for a one-term graduate course on quantum mechanics that he gave at Harvard, and was able to make many valuable suggestions of points that should be included or better explained. Of course, only I am responsible for any errors that may remain in this book. Thanks are also due to Terry Riley, Abel Ephraim, and Josh Perlman for finding countless books and articles, and to Jan Duffy for her helps of many sorts. I am grateful to Lindsay Barnes and Roisin Munnelly of Cambridge University Press for helping to ready this book for publication, to Dr. Steven Holt for his careful and sensitive copy editing, and especially to my editor, Simon Capelin, for his encouragement and good advice.

STEVEN WEINBERG

# Notation

Latin indices $i, j, k$, and so on generally run over the three spatial coordinate labels, usually taken as 1, 2, 3.

The summation convention is not used; repeated indices are summed only where explicitly indicated.

Spatial three-vectors are indicated by symbols in boldface. In particular, $\nabla$ is the gradient operator.

$\nabla^2$ is the Laplacian $\sum_i \partial^2 / \partial x^i \, \partial x^i$.

The three-dimensional 'Levi-Civita tensor' $\epsilon_{ijk}$ is defined as the totally antisymmetric quantity with $\epsilon_{123} = +1$. That is,

$$\epsilon_{ijk} \equiv \begin{cases} +1, & ijk = 123, \ 231, \ 312, \\ -1, & ijk = 132, \ 213, \ 321, \\ 0, & \text{otherwise.} \end{cases}$$

The Kronecker delta is

$$\delta_{nm} = \begin{cases} 1, & n = m, \\ 0, & n \neq m. \end{cases}$$

A hat over any vector indicates the corresponding unit vector: Thus, $\hat{\mathbf{v}} \equiv \mathbf{v}/|\mathbf{v}|$.

A dot over any quantity denotes the time-derivative of that quantity.

The step function $\theta(s)$ has the value $+1$ for $s > 0$ and $0$ for $s < 0$.

The complex conjugate, transpose, and Hermitian adjoint of a matrix $A$ are denoted $A^*$, $A^{\mathrm{T}}$, and $A^\dagger = A^{*\mathrm{T}}$, respectively. The Hermitian adjoint of an operator $O$ is denoted $O^\dagger$. $+$H.c. or $+$c.c. at the end of an equation indicates the addition of the Hermitian adjoint or complex conjugate of the foregoing terms.

Where it is necessary to distinguish operators and their eigenvalues, upper case letters are used for operators and lower case letters for their eigenvalues. This convention is not always used where the distinction between operators and eigenvalues is obvious from the context.

Factors of the speed of light $c$, the Boltzmann constant $k_B$, and Planck's constant $h$ or $\hbar \equiv h/2\pi$ are shown explicitly.

Unrationalized electrostatic units are used for electromagnetic fields and electric charges and currents, so that $e_1 e_2 / r$ is the Coulomb potential of a pair of charges $e_1$ and $e_2$ separated by a distance $r$. Throughout, $-e$ is the unrationalized charge of the electron, so that the fine structure constant is $\alpha \equiv e^2/\hbar c \simeq 1/137$.

Numbers in parenthesis at the end of quoted numerical data give the uncertainty in the last digits of the quoted figure. Where not otherwise indicated, experimental data are taken from K. Nakamura *et al.* (Particle Data Group), "Review of Particle Properties," *J. Phys. G* **37**, 075021 (2010).

# 1

# Historical Introduction

The principles of quantum mechanics are so contrary to ordinary intuition that
they can best be motivated by taking a look at their prehistory. In this chapter
we will consider the problems confronted by physicists in the first years of the
twentieth century that ultimately led to modern quantum mechanics.

## 1.1 Photons

Quantum mechanics had its beginning in the study of black-body radiation.
The universality of the frequency distribution of this radiation was established
on thermodynamic grounds in 1859–1862 by Gustav Robert Kirchhoff (1824–
1887), who also gave black-body radiation its name. Consider an enclosure
whose walls are kept at a temperature $T$, and suppose that the energy per volume of radiation within this enclosure in a frequency interval between $\nu$ and
$\nu + d\nu$ is some function $\rho(\nu, T)$ times $d\nu$. Kirchhoff calculated the energy per
time of the radiation in any frequency interval that strikes a small patch of area
$A$. He reasoned that, from a point in the enclosure with polar coordinates $r, \theta, \phi$
(with $r$ the distance to the patch, and $\theta$ measured from the normal to the patch),
the patch will subtend a solid angle $A \cos \theta / r^2$, so the fraction of the energy at
that point that is aimed at the patch will be $A \cos \theta / 4\pi r^2$. The total energy in
a frequency interval between $\nu$ and $\nu + d\nu$ that strikes the patch in a time $t$ is
then the integral of $A \cos \theta / 4\pi r^2 \times \rho(\nu, T) d\nu$ over a hemisphere with radius
$ct$, where $c$ is the speed of light:

$$
2\pi \int_0^{ct} dr \int_0^{\pi/2} d\theta \, r^2 \sin \theta \times \frac{A \cos \theta \, \rho(\nu, T) \, d\nu}{4\pi r^2} = \frac{ct A \, \rho(\nu, T) \, d\nu}{4}.
$$

If a fraction $f(\nu, T)$ of this energy is absorbed by the walls of the enclosure,
then the total energy per area and per time absorbed by the walls in a frequency
interval between $\nu$ and $\nu + d\nu$ is

$$
E(\nu, T) \, d\nu = \frac{c}{4} f(\nu, T) \, \rho(\nu, T) \, d\nu.
$$

1

In order to be in equilibrium, this must also equal the energy per area and per time emitted by the walls in the same frequency interval. The walls cannot absorb more radiation than they receive, so the absorption fraction $f(v, T)$ is at most equal to one. Any material for which $f(v, T) = 1$ is called *black*. The function $\rho(v, T)$ must be universal, for in order for it to be affected when some change is made in the enclosure, keeping it all at temperature $T$, energy at some frequencies would have to flow from the radiation to the walls or vice versa, which is impossible for materials at the same temperature.

Physicists in the last decades of the nineteenth century were greatly concerned to understand the distribution function $\rho(v, T)$. It had been measured, chiefly at a Berlin research institute, the *Physikalisch-Technische Reichsanstalt*, but how could one understand the measured values?

An answer was attempted using the statistical mechanics of the late nineteenth century, without quantum ideas, in a series of papers[1] in 1900 and 1905 by John William Strutt (1842–1919), more usually known as Lord Rayleigh, and by James Jeans (1877–1946). It was familiar that one can think of the radiation field in a box as a Fourier sum over normal modes. For instance, for a cubical box of width $L$, whatever boundary condition is satisfied on one face of the box must be satisfied on the opposite face, so the phase of the radiation field must change by an integer multiple of $2\pi$ in a distance $L$. That is, the radiation field is the sum of terms proportional to $\exp(i\mathbf{q} \cdot \mathbf{x})$, with

$$\mathbf{q} = 2\pi\mathbf{n}/L, \tag{1.1.1}$$

where the vector $\mathbf{n}$ has integer components. (For instance, to maintain translational invariance, it is convenient to impose periodic boundary conditions: each component of the electromagnetic field is assumed to be the same on opposite faces of the box.) Each normal mode is thus characterized by a triplet of integers $n_1, n_2, n_3$ and a polarization state, which can be taken as either left- or right-circular polarization. The wavelength of a normal mode is $\lambda = 2\pi/|\mathbf{q}|$, so its frequency is given by

$$v = \frac{c}{\lambda} = \frac{|\mathbf{q}|c}{2\pi} = \frac{|\mathbf{n}|c}{L}. \tag{1.1.2}$$

Each normal mode occupies a cell of unit volume in the space of the vectors $\mathbf{n}$, so the number of normal modes $N(v)\,dv$ in the range of frequencies between $v$ and $v + dv$ is twice the volume of the corresponding shell in this space:

$$N(v)\,dv = 2 \times 4\pi|\mathbf{n}|^2\,d|\mathbf{n}| = 8\pi(L/c)^3 v^2\,dv, \tag{1.1.3}$$

the extra factor of 2 taking account of the two possible polarizations for each wave number. In classical statistical mechanics, in any system that can be regarded as a collection of harmonic oscillators, the mean energy of each

---

[1]  Lord Rayleigh, *Phil. Mag.* **49**, 539 (1900); *Nature* **72**, 54 (1905); J. Jeans, *Phil. Mag.* **10**, 91 (1905).

oscillator $\bar{E}(T)$ is simply proportional to the temperature, a relation written as $\bar{E}(T) = k_B T$, where $k_B$ is a fundamental constant, known as Boltzmann's constant. (The derivation is given below.) If this applied to radiation, the energy density in the radiation between frequencies $\nu$ and $\nu + d\nu$ would then be given by what has come to be called the Rayleigh–Jeans formula

$$\rho(\nu, T)\, d\nu = \frac{\bar{E}(T)\, N(\nu)\, d\nu}{L^3} = \frac{8\pi k_B T \nu^2\, d\nu}{c^3}. \tag{1.1.4}$$

The prediction that $\rho(\nu, T)$ is proportional to $T\nu^2$ was actually in agreement with observation for small values of $\nu/T$, but failed badly for larger values. Indeed, if it held for all frequencies at a given temperature, then the total energy density $\int \rho(\nu, T)\, d\nu$ would be infinite. This became known as the *ultraviolet catastrophe*.

To be a bit more specific about who did what when, Rayleigh in 1900 showed in effect that $\rho(\nu, T)$ is proportional for low frequency to $T\nu^2$, but he did not attempt to calculate the constant of proportionality in Eq. (1.1.3) or in $\bar{E}(T)$, and hence could not give the constant factor in Eq. (1.1.4). To avoid the ultraviolet catastrophe, he also included an *ad hoc* factor that decayed exponentially for large values of $\nu/T$, without attempting to calculate the values of $\nu/T$ at which the decay becomes appreciable. Rayleigh went further in 1905, and calculated the constant factor in Eq. (1.1.3), but obtained a result 8 times too large. The correct result was given a little later by Jeans (in a postscript to his 1905 article), who also correctly gave $\bar{E}(T) = k_B T$, and hence obtained (1.1.4) as a low-frequency limit.

The correct complete result had already been published by Max Planck (1858–1947) in 1900.[2] Planck noted that the data on black-body radiation could be fit with the formula

$$\rho(\nu, T)\, d\nu = \frac{8\pi h}{c^3} \frac{\nu^3\, d\nu}{\exp(h\nu/k_B T) - 1}, \tag{1.1.5}$$

where $h$ was a new constant, known ever after as *Planck's constant*. Comparison with observation gave $k_B \approx 1.4 \times 10^{-16}$ erg/K and[3] $h \approx 6.6 \times 10^{-27}$ erg sec. This formula was at first just guesswork, but a little later Planck gave a derivation of the formula,[4] based on the assumption that the radiation was the same as if it were in equilibrium with a large number of charged oscillators with different frequencies, the energy of any oscillator of frequency $\nu$ being an integer multiple of $h\nu$. Planck's derivation is lengthy and not worth repeating here, since its basis is very different from what soon replaced it.

---

[2] M. Planck, *Verh. deutsch. phys. Ges.* **2**, 202 (1900).

[3] The modern value is $6.62606891(9) \times 10^{-27}$ erg sec; see E. R. Williams, R. L. Steiner, D. B. Newell, and P. T. Olson, *Phys. Rev. Lett.* **81**, 2404 (1998).

[4] M. Planck, *Verh. deutsch. phys. Ges.* **2**, 237 (1900).

Planck's formula agrees with the Rayleigh–Jeans formula (1.1.4) for $\nu/T \ll k_B/h$, but it gives an energy density that falls off exponentially for $\nu/T \gg k_B/h$, yielding a finite total energy density

$$\int_0^\infty \rho(\nu, T)\, d\nu = a_B T^4, \qquad a_B \equiv \frac{8\pi^5 k_B^4}{15 h^3 c^3}. \tag{1.1.6}$$

(Using modern values of constants, this gives $a_B = 7.56577(5) \times 10^{-15}$ erg cm$^{-3}$ K$^{-4}$.) According to the Kirchhoff relation between $\rho(\nu, T)$ and the rate of emission from a black body, the total rate of energy emission per area from a black body is $\sigma T^4$, where $\sigma$ is the Stefan–Boltzmann constant:

$$\sigma = \frac{c a_B}{4} = \frac{2\pi^5 k_B^4}{15 h^3 c^2} = 5.670373(21) \times 10^{-5} \text{ erg cm}^{-2} \text{ sec}^{-1} \text{K}^{-4}.$$

Perhaps the most important immediate consequence of Planck's work was to provide long-sought values for atomic constants. The theory of ideal gases gives the well-known law $pV = nRT$, where $p$ is the pressure of a volume $V$ of $n$ moles of gas at temperature $T$, with the constant $R$ given by $R = k_B N_A$, where $N_A$ is Avogadro's number, the number of molecules in one mole of gas. Measurements of gas properties had long given values for $R$, so with $k_B$ known it was possible for Planck to infer a value for $N_A$, the reciprocal of the mass of a hypothetical atom with unit atomic weight (close to the mass of a hydrogen atom). This was in good agreement with estimates of $N_A$ from properties of non-ideal gases that depend on number density and not just mass density, such as viscosity. Knowing the mass of individual atoms, and assuming that atoms in solids are closely packed so that the mass to volume ratio of an atom is similar to the measured density of macroscopic solid samples of that element, one could estimate the sizes of atoms. Similarly, measurements of the amount of various elements produced by electrolysis had given a value for the faraday, $F = e N_A$, where $e$ is the electric charge transferred in producing one atom of unit valence, so with $N_A$ known, $e$ could be calculated. It could be assumed that $e$ is the charge of the electron, which had been discovered in 1897 by Joseph John Thomson (1856–1940), so this amounted to a measurement of the charge of the electron, a measurement much more precise than any direct measurement that could be carried out at the time. Thomson had measured the ratio of $e$ to the mass of the electron, by observing the bending of cathode rays in electric and magnetic fields, so this also gave a value for the electron mass.

It is ironic that all this could have been done by Rayleigh in 1900, without introducing quantum ideas, if he had obtained the correct Rayleigh–Jeans formula (1.1.4) then. He would only have had to compare this formula with experimental data at small values of $\nu/T$, where the formula works, and use the result to find $k_B$ – for this, $h$ is not needed.

Planck's quantization assumption applied to the matter that emits and absorbs radiation, not to radiation itself. As George Gamow later remarked, Planck thought that radiation was like butter; butter itself comes in any quantity, but it can be bought and sold only in multiples of one quarter pound. It was Albert Einstein (1879–1955) who in 1905 proposed that the energy of radiation of frequency $\nu$ was itself an integer multiple of $h\nu$.[5] He used this to predict that in the photoelectric effect no electrons are emitted when light shines on a metal surface unless the frequency of the light exceeds a minimum value $\nu_{min}$, where $h\nu_{min}$ is the energy required to remove a single electron from the metal (the "work function"). The electrons then have energy $h(\nu - \nu_{min})$. Experiments[6] by Robert Millikan (1868–1953) in 1914–1916 verified this formula, and gave a value for $h$ in agreement with that derived from black-body radiation.

The connection between Einstein's hypothesis and the Planck black-body formula is best explained in a derivation of the black-body formula by Hendrik Lorentz (1853–1928) in 1910.[7] Lorentz made use of the fundamental result of statistical mechanics due to J. Willard Gibbs (1839–1903),[8] that in a system containing a large number of identical systems in thermal equilibrium at a given temperature $T$ (like light quanta in a black-body cavity), the probability that one of these systems has an energy $E$ is proportional to $\exp(-E/k_B T)$, with an energy-independent constant of proportionality. If the energies of light quanta were continuously distributed, this would give a mean energy

$$\bar{E} = \frac{\int_0^\infty \exp(-E/k_B T)\, E\, dE}{\int_0^\infty \exp(-E/k_B T)\, dE} = k_B T,$$

the assumption used in deriving the Rayleigh–Jeans formula (1.1.4). But if the energies are instead integer multiples of $h\nu$, then the mean energy is

$$\bar{E} = \frac{\sum_{n=0}^\infty \exp(-nh\nu/k_B T)\, nh\nu}{\sum_{n=0}^\infty \exp(-nh\nu/k_B T)} = \frac{h\nu}{\exp(h\nu/k_B T) - 1}. \tag{1.1.7}$$

The energy density in radiation between frequencies $\nu$ and $\nu + d\nu$ is again given by $\rho\, d\nu = \bar{E} N\, d\nu/L^3$, which now with Eqs. (1.1.3) and (1.1.7) yields the Planck formula (1.1.5).

Even after Millikan's experiments had verified Einstein's prediction for the energies of photoelectrons, there remained considerable skepticism about the reality of light quanta. This was largely dispelled by experiments on the scattering of X-rays by Arthur Compton (1892–1962) in 1922–23.[9] The energy of X-rays is sufficiently high that it is possible to ignore the much smaller binding energy of the electron in a light atom, treating the electron as a free particle. Special relativity says that if a quantum of light has energy $E = h\nu$, then it

[5] A. Einstein, *Ann. Physik* **17**, 132 (1905).

[6] R. A. Millikan, *Phys. Rev.* **7**, 355 (1916).

[7] H. A. Lorentz, *Phys. Z.* **11**, 1234 (1910).

[8] J. W. Gibbs, *Elementary Principles in Statistical Mechanics* (Charles Scribner's Sons, New York, 1902).

[9] A. H. Compton, *Phys. Rev.* **21**, 207 (1923).

has momentum $p = h\nu/c$, in order to have $m_\gamma^2 c^4 = E^2 - p^2 c^2 = 0$. If, for instance, a light quantum striking an electron at rest is scattered backwards, then the scattered quantum has frequency $\nu'$ and the electron scattered forward has momentum $h\nu/c + h\nu'/c$, where $\nu'$ is given by the energy conservation condition:

$$h\nu + m_e c^2 = h\nu' + \sqrt{m_e^2 c^4 + (h\nu/c + h\nu'/c)^2 c^2}$$

(where $m_e$ is the electron mass), so

$$\nu' = \frac{\nu m_e c^2}{2h\nu + m_e c^2}.$$

This is conventionally written as a formula relating the wavelengths $\lambda = c/\nu$ and $\lambda' = c/\nu'$:

$$\lambda' = \lambda + 2h/m_e c. \tag{1.1.8}$$

The length $h/m_e c = 2.425 \times 10^{-10}$ cm is known as the Compton wavelength of the electron. (For scattering at an angle $\theta$ to the forward direction, the factor 2 in Eq. (1.1.8) is replaced with $1 - \cos\theta$.) Verification of such relations convinced physicists of the existence of these quanta. A little later the chemist G. N. Lewis[10] gave the quantum of light the name by which it has been known ever since, the *photon*.

## 1.2 Atomic Spectra

Another problem confronted physicists throughout the nineteenth and early twentieth centuries. In 1802 William Hyde Wollaston (1766–1828) discovered dark lines in the spectrum of the Sun, but these lines were not studied in detail until around 1814, when they were re-discovered by Joseph von Fraunhofer (1787–1826). Later it was realized that hot atomic gases emit and absorb light only at certain definite frequencies, the pattern of frequencies, or spectrum, depending on the element in question. The dark lines discovered by Wollaston and Fraunhofer are caused by the absorption of light as it rises through the cooler outer layers of the Sun's photosphere. The study of bright and dark spectral lines became a useful tool for chemical analysis, for astronomy, and for the discovery of new elements, such as helium, discovered in the spectrum of the Sun. But, like writing in a forgotten language, these atomic spectra provided no intelligible message.

No progress could be made in understanding atomic spectra without knowing something about the structure of atoms. After Thomson's discovery of the electron in 1897, it was widely believed that atoms were like puddings, with

---

[10]  G. N. Lewis, *Nature*, **118**, 874 (1926).

negatively charged electrons stuck in like raisins in a smooth background of positive charge. This picture was radically changed by experiments carried out in the laboratory of Ernest Rutherford (1871–1937) at the University of Manchester in 1909–1911. In these experiments a post-doc, Hans Geiger (1882–1945), and an undergraduate, Ernest Marsden (1889–1970), let a collimated beam of alpha particles ($^4$He nuclei) from a radium source strike a thin gold foil. The alpha particles passing through the foil were detected by flashes of light when they struck a sheet of zinc sulfide. As expected, the beam was found to be slightly spread out by scattering of alpha particles by the gold atoms. Then for some reason Rutherford had the idea of asking Geiger and Marsden to check whether any alpha particles were scattered at large angles. This would not be expected if the alpha particle hit a much lighter particle like the electron. If a particle of mass $M$ with velocity $v$ hits a particle of mass $m$ that is at rest, and continues along the same line with velocity $v'$, giving the target particle a velocity $u$, the equations of momentum and energy conservation give

$$Mv = mu + Mv', \quad \frac{1}{2}Mv^2 = \frac{1}{2}Mv'^2 + \frac{1}{2}mu^2. \tag{1.2.1}$$

(In the notation used here, a positive velocity is in the same direction as the original velocity of the alpha particle, while a negative velocity is in the opposite direction.) Eliminating $u$, we obtain a quadratic equation for $v'/v$:

$$0 = (1 + M/m)(v'/v)^2 - 2(M/m)(v'/v) - 1 + M/m.$$

This has two solutions. One solution is $v' = v$. This solution is one for which nothing happens – the incident particle just continues with the velocity it had at the beginning. The interesting solution is the other one:

$$v' = -v\left(\frac{m - M}{m + M}\right). \tag{1.2.2}$$

But this has a negative value (that is, a recoil backwards) only if $m > M$. (Somewhat weaker limits on $m$ can be inferred from scattering at any large angle.)

Nevertheless, alpha particles were observed to be scattered at large angles. As Rutherford later explained, "It was quite the most incredible event that has ever happened to me in my life. It was almost as incredible as if you fired a 15-inch shell at a piece of tissue paper, and it came back and hit you."[11]

So the alpha particle must have been hitting something in the gold atom much heavier than an electron, whose mass is only about 1/7300 the mass of an alpha particle. Furthermore, the target particle must be quite small to stop the alpha particle by the Coulomb repulsion of positive charges. If the charge of the target

---

[11] Quoted by E. N. da Costa Andrade, *Rutherford and the Nature of the Atom* (Doubleday, Garden City, NY, 1964).

particle is $+Ze$, then in order to stop the alpha particle with charge $+2e$ at a distance $r$ from the target particle, the kinetic energy $Mv^2/2$ must be converted into a potential energy $(2e)(Ze)/r$, so $r = 4Ze^2/Mv^2$. The velocity of the alpha particles emitted from radium is $2.09 \times 10^9$ cm/sec, so the distance at which they would be stopped by a heavy target particle was $3Z \times 10^{-14}$ cm, which for any reasonable $Z$ (even $Z \approx 100$) is much smaller than the size of the gold atom, a few times $10^{-8}$ cm.

Rutherford concluded[12] then that the positive charge of the atom is concentrated in a small heavy nucleus, around which the much lighter negatively charged electrons circulate in orbits, like planets around the Sun. But this only heightened the mystery surrounding atomic spectra. A charged particle like the electron circulating in orbit would be expected to radiate light, with the same frequency as the orbital motion. The frequencies of these orbital motions could be anything. Worse, as the electron lost energy to radiation it would spiral down into the atomic nucleus. How could atoms remain stable?

In 1913 an answer was offered by a young visitor to Rutherford's Manchester laboratory, Niels Bohr (1885–1962). Bohr proposed in the first place that the energies of atoms are quantized, in the sense that the atom exists in only a discrete set of states, with energies (in increasing order) $E_1$, $E_2$, .... The frequency of a photon emitted in a transition $m \to n$ or absorbed in a transition $n \to m$ is given by Einstein's formula $E = h\nu$ and energy conservation by

$$\nu = (E_m - E_n)/h. \qquad (1.2.3)$$

A bright or dark spectral line is formed by atoms emitting or absorbing photons in a transition from a higher to a lower energy state, or vice versa. This explained a rule, known as the Ritz combination principle, that had been noticed experimentally by Walther Ritz (1878–1909) in 1908[13] (but without explaining it), that the spectrum of any atom could be described more compactly by a set of so-called "terms," the frequencies of the spectrum being all given by differences of the terms. These terms, according to Bohr, were just the energies $E_n$, divided by $h$.

Bohr also offered a method for calculating the energies $E_n$, at least for electrons in a Coulomb field, as in hydrogen, singly ionized helium, etc. Bohr noted that Planck's constant $h$ has the same dimensions as angular momentum, and he guessed that the angular momentum $m_e v r$ of an electron of velocity $v$ in a circular atomic orbit of radius $r$ is an integer multiple of some constant $\hbar$,[14] presumably of the same order of magnitude as $h$:

$$m_e v r = n\hbar, \quad n = 1, 2, .... \qquad (1.2.4)$$

[12] E. Rutherford, *Phil. Mag.* **21**, 669 (1911).

[13] W. Ritz, *Phys. Z.* **9**, 521 (1908).

[14] N. Bohr, *Phil. Mag.* **26**, 1, 476, 857 (1913); *Nature* **92**, 231 (1913).

(Bohr did not use the symbol $\hbar$. Readers who know how $\hbar$ is related to $h$ should temporarily forget that information; for the present $\hbar$ is just another constant.) Bohr combined this with the equation for the equilibrium of the orbit,

$$\frac{m_e v^2}{r} = \frac{Ze^2}{r^2}, \tag{1.2.5}$$

and the formula for the electron's energy,

$$E = \frac{m_e v^2}{2} - \frac{Ze^2}{r}. \tag{1.2.6}$$

This gives

$$v = \frac{Ze^2}{n\hbar}, \quad r = \frac{n^2\hbar^2}{Zm_e e^2}, \quad E = -\frac{Z^2 e^4 m_e}{2n^2 \hbar^2}. \tag{1.2.7}$$

Using the Einstein relation between energy and frequency, the frequency of a photon emitted in a transition from an orbit with quantum number $n$ to one with quantum number $n' < n$ is

$$v = \frac{\Delta E}{h} = \frac{Z^2 e^4 m_e}{2h\hbar^2}\left(\frac{1}{n'^2} - \frac{1}{n^2}\right). \tag{1.2.8}$$

To find $\hbar$, Bohr relied on a *correspondence principle*, that the results of classical physics should apply for large orbits – that is, for large $n$. If $n \gg 1$ and $n' = n - 1$, Eq. (1.2.8) gives $v = Z^2 e^4 m_e / h\hbar^2 n^3$. This may be compared with the frequency of the electron in its orbit, $v/2\pi r = Z^2 e^4 m_e / 2\pi n^3 \hbar^3$. According to classical electrodynamics these two frequencies should be equal, so Bohr could conclude that $\hbar = h/2\pi$. Using the value of $h$ obtained by matching observations of black-body radiation with Planck's formula, Bohr was able to derive numerical values for the velocity, radial coordinate, and energy of the electron:

$$v = \frac{Ze^2}{n\hbar} \simeq \frac{Zc}{137n}, \tag{1.2.9}$$

$$r = \frac{n^2\hbar^2}{Zm_e e^2} \simeq n^2 \times 0.529 Z^{-1} \times 10^{-8} \text{ cm}, \tag{1.2.10}$$

$$E = -\frac{Z^2 e^4 m_e}{2n^2\hbar^2} \simeq -\frac{13.6 Z^2 \text{ eV}}{n^2}. \tag{1.2.11}$$

The striking agreement of Eq. (1.2.11) with the atomic energy levels of hydrogen inferred from the frequencies of spectral lines was a strong indication that Bohr was on the right track. The case for Bohr's theory became even stronger when he pointed out (in the *Nature* article cited in footnote 14) that Eq. (1.2.11) also accounts for the spectrum of singly ionized helium (observed both astronomically and in laboratory experiments), with a small but detectable correction.

Bohr realized that the mass appearing in these formulas should be not precisely the electron mass, but rather the reduced mass $\mu \equiv m_e/(1 + m_e/m_N)$, where $m_N$ is the nuclear mass. (This is discussed in Section 2.4.) Hence the constant of proportionality between $E$ and $1/n^2$ is larger for helium than for hydrogen by a factor that is not simply equal to $Z^2_{He} = 4$, but rather by a factor $4(1 + m_e/m_H)/(1 + m_e/m_{He}) = 4.00163$, in agreement with experiment.

In this derivation Bohr had relied on the old idea of classical radiation theory, that the frequencies of spectral lines should agree with the frequency of the electron's orbital motion, but he had assumed this only for the largest orbits, with large $n$. The light frequencies he calculated for transitions between lower states, such as $n = 2 \to n = 1$, did not at all agree with the orbital frequency of the initial or final state. So Bohr's work represented another large step away from classical physics.

Bohr's formulas could be used not only for single-electron atoms, like hydrogen or singly ionized helium, but also roughly for the innermost orbits in heavier atoms, where the charge of the nucleus is not screened by electrons, and we can take $Ze$ as the actual charge of the nucleus. For $Z \geq 10$, the energy of a photon emitted in a transition from $n = 2$ to $n = 1$ orbits is greater than 1 keV, and hence is in the X-ray spectrum. By measuring these X-ray energies, H. G. J. Moseley (1887–1915) was able to find $Z$ for a range of atoms from calcium to zinc. He discovered that, within experimental uncertainty, $Z$ is an integer, suggesting that the positive charge of atomic nuclei is carried by particles of charge $+e$, much heavier than the electron, to which Rutherford gave the name *protons*. Also, with just a few exceptions, $Z$ increased by one unit in going from any element to the element with the next largest atomic weight $A$ (roughly, the mass of the atom in units of the hydrogen atom mass). But $Z$ turned out to be not equal to $A$. For instance, zinc has $A = 65.38$, and it turned out to have $Z = 30.00$. For some years it was thought that the atomic weight $A$ was approximately equal to the number of protons, with the extra charge canceled by $A - Z$ electrons. The discovery by James Chadwick (1891–1974) in 1935 of the neutron,[15] which was found to have a mass close to that of the hydrogen atom, showed that instead nuclei contain $Z$ protons and approximately $A - Z$ neutrons. (The atomic weight is not precisely equal to the number of protons plus the number of neutrons, both because the neutron mass is not precisely the same as the proton mass, and also because, according to Einstein's formula $E = mc^2$, the energy of interaction of the particles inside a nucleus contributes to the nuclear mass.)

Incidentally, Eqs. (1.2.9)–(1.2.11) also hold roughly for electrons in the outermost orbits in heavy atoms, where most of the charge of the nucleus is screened by inner electrons, and $Z$ can therefore be taken to be of order unity. This is why the sizes of heavy atoms are not very much larger than those of light atoms,

---

15   J. Chadwick, *Nature* **129**, 312 (1932).

and the frequency of light emitted in transitions of electrons in the outer orbits of heavy atoms is comparable to the corresponding energies in hydrogen, and hence in the visible range of the spectrum. Heavy atoms are somewhat larger than light ones, because, for reasons outlined in Section 4.5, the electrons in the outer orbits of heavy atoms have larger values of $n$ than for light atoms.

The Bohr theory applied only to circular orbits, but just as in the solar system, the generic orbit of a particle in a Coulomb field is not a circle, but an ellipse. A generalization of the Bohr quantization condition (1.2.4) was proposed by Arnold Sommerfeld (1868–1951) in 1916,[16] and used by him to calculate the energies of electrons in elliptical orbits. Sommerfeld's condition was that in a system described by a Hamiltonian $H(q, p)$, with several coordinates $q_a$ and canonical conjugates $p_a$ satisfying the equations $\dot{q}_a = \partial H/\partial p_a$ and $\dot{p}_a = -\partial H/\partial q_a$, if all $q$s and $p$s have a periodic time dependence (as for closed orbits), then for each $a$

$$\oint p_a \, dq_a = n_a h \tag{1.2.12}$$

(with $n_a$ an integer), the integral taken over one period of the motion. For instance, for an electron in a circular orbit we can take $q$ as the angle traced out by the line connecting the nucleus and the electron, and $p$ as the angular momentum $m_e v r$, in which case $\oint p \, dq = 2\pi m_e v r$, and (1.2.12) is the same as Bohr's condition (1.2.4). We will not pursue this approach here, because it was soon made obsolete by the advent of wave mechanics.

In 1916 (in his spare time while discovering the general theory of relativity), Einstein returned to the theory of black-body radiation,[17] this time combining it with the Bohr idea of quantized atomic energy states. Einstein defined a quantity $A_m^n$ as the rate at which an atom will spontaneously make a transition from a state $m$ to a state $n$ of lower energy, emitting a photon of energy $E_m - E_n$. He also considered the absorption of photons from radiation (not necessarily black-body radiation) with an energy density $\rho(v) \, dv$ at frequencies between $v$ and $v + dv$. The rate at which an individual atom in such a field makes a transition from a state $n$ to a state $m$ of higher energy is written as $B_n^m \rho(v_{nm})$, where $v_{nm} \equiv (E_m - E_n)/h$ is the frequency of the absorbed photon. Einstein also took into account the possibility that the radiation would stimulate the emission of photons by the atom in transitions from a state $m$ to a state $n$ of lower energy, at a rate written as $B_m^n \rho(v_{nm})$. The coefficients $B_n^m$ and $B_m^n$ like $A_m^n$ are assumed to depend only on the properties of individual atoms, not on the temperature or the radiation.

Now, suppose the radiation is black-body radiation at a temperature $T$, with which the atoms are in equilibrium. The energy density of the radiation will

---

[16]  A. Sommerfeld, *Ann. Physik* **51**, 1 (1916).
[17]  A. Einstein, *Phys. Z.* **18**, 121 (1917).

be the function $\rho(v, T)$, given by Eq. (1.1.5). In equilibrium the rate at which atoms make a transition $m \to n$ from higher to lower energy must equal the rate at which atoms make the reverse transition $n \to m$:

$$N_m\big[A_m^n + B_m^n \rho(v_{nm}, T)\big] = N_n B_n^m \rho(v_{nm}, T),  \qquad (1.2.13)$$

where $N_n$ and $N_m$ are the numbers of atoms in states $n$ and $m$. According to the Boltzmann rule of classical statistical mechanics, the number of atoms in a state of energy $E$ is proportional to $\exp(-E/k_B T)$, so

$$N_m/N_n = \exp(-(E_m - E_n)/k_B T) = \exp(-h v_{nm}/k_B T).  \qquad (1.2.14)$$

(It is important here to take the $N_n$ as the numbers of atoms in individual states $n$, some of which may have precisely the same energy, rather than the numbers of atoms with energies $E_n$.) Putting this together, we have

$$A_m^n = \frac{8\pi h}{c^3} \frac{v_{nm}^3}{\exp(h v_{nm}/k_B T) - 1} \Big( \exp(h v_{nm}/k_B T)\, B_n^m - B_m^n \Big).  \qquad (1.2.15)$$

For this to be possible at all temperatures for temperature-independent $A$ and $B$ coefficients, these coefficients must be related by

$$B_m^n = B_n^m, \qquad A_m^n = \left( \frac{8\pi h v_{nm}^3}{c^3} \right) B_m^n.  \qquad (1.2.16)$$

Hence, knowing the rate at which a classical light wave of a given energy density is absorbed or stimulates emission by an atom, we can calculate the rate at which it spontaneously emits photons.[18] This calculation will be presented in Section 6.5.

The phenomenon of stimulated emission makes possible the amplification of beams of light, in a laser (an acronym for "light amplification by stimulated emission of radiation"). Suppose a beam of light with energy density distribution $\rho(v)$ passes through a medium consisting of $N_n$ atoms at energy level $E_n$. Stimulated emission from the first excited state $n = 2$ to the ground state $n = 1$ adds photons of frequency $v_{12} \equiv (E_2 - E_1)/h$ to the beam at a rate $N_2 \rho(v_{12}) B_2^1$,

---

[18] Einstein in the article cited in footnote 17 actually used this argument to give a new derivation of the Planck formula for $\rho(v, T)$ as well as the relations (1.2.16). He first considered the limit of very large temperature, for which $\rho(v_{nm}, T)$ may be assumed to be very large, and Eq. (1.2.14) gives $N_n$ very close to $N_m$. In this limit Eq. (1.2.13) requires that $B_n^m = B_m^n$, which, since the $B$s are independent of temperature, must be generally true. Using $B_n^m = B_m^n$ in Eq. (1.2.13) for a general temperature then gives $\rho(v_{nm}, T) = (A_m^n/B_m^n)/[\exp(h v_{nm}/k_B T) - 1]$. Einstein then used a thermodynamic relation due to Wilhelm Wien (1884–1928), the *Wien displacement law*, which requires that $\rho(v, T)$ equals $v^3$ times some function of $v/T$. This gave $A_m^n/B_m^n$ proportional to $v_{nm}^3$, and Einstein then found the constant of proportionality by requiring that the Rayleigh–Jeans formula (1.1.4) is satisfied for $h v \ll k_B T$. But Einstein's use of the Wien displacement law was actually unnecessary, because in order for the formula $\rho(v_{nm}, T) = (A_m^n/B_m^n)/[\exp(h v_{nm}/k_B T) - 1]$ to agree with the Rayleigh–Jeans formula for $h v \ll k_B T$, it is necessary that the ratio $A_m^n/B_m^n$ be given by Eq. (1.2.16), and Planck's formula then follows immediately.

but absorption from the ground state removes photons at a rate $N_1 \rho(\nu_{12}) B_1^2$, and since $B_2^1 = B_1^2$, there will be a net addition of photons only in the case $N_2 > N_1$. Unfortunately, such a population inversion cannot be produced by exposing the atoms in their ground state to light at this frequency. The *net* rate of change in the population of the first excited state $n = 2$ due to spontaneous and stimulated emission from the excited state and absorption from the ground state will be

$$\dot{N_2} = -N_2 \rho(\nu_{12}) B_2^1 - N_2 A_2^1 + N_1 \rho(\nu_{12}) B_1^2,$$

or, using the Einstein relations (1.2.16),

$$\dot{N_2} = B_2^1 \left[ -N_2 \left[ \rho(\nu_{12}) + 8\pi \nu_{12}^3 h/c^3 \right] + N_1 \rho(\nu_{12}) \right]. \qquad (1.2.17)$$

If we start with $N_2 = 0$, then $N_2$ increases until it approaches a value $N_1/(1 + \xi)$, where $\xi \equiv 8\pi \nu_{12}^3 h / \rho(\nu_{12}) c^3$, when $N_2$ becomes constant. Not only can this process not produce a population inversion, but also, because of spontaneous emission, it cannot even make $N_2$ as large as $N_1$. A population inversion can be produced in other ways, for instance by optical pumping, in which atoms are excited to some state $n = 3$ by absorption of light with frequency $\nu_{31} = (E_3 - E_1)/h$, and then spontaneously decay to the state $n = 2$.

## 1.3 Wave Mechanics

Ever since Maxwell, light had been understood to be a wave of electric and magnetic fields, but after Einstein and Compton, it became clear that it is also manifested in a particle, the photon. So is it possible that something like the electron, that had always been regarded as a particle, could also be manifested as some sort of wave? This was suggested in 1923 by Louis de Broglie (1892–1987),[19] a doctoral student in Paris. Any kind of wave of frequency $\nu$ and wave number $\mathbf{k}$ has a spacetime dependence $\exp(i\mathbf{k} \cdot \mathbf{x} - i\omega t)$, where $\omega = 2\pi\nu$. Lorentz invariance requires that $(\mathbf{k}, \omega)$ transform as a four-vector, just like the momentum four-vector $(\mathbf{p}, E)$. For light, according to Einstein, the energy of a photon is $E = h\nu = \hbar\omega$, and its momentum has a magnitude $|\mathbf{p}| = E/c = h\nu/c = h/\lambda = \hbar|\mathbf{k}|$, so de Broglie was led to suggest that in general a particle of any mass is associated with a wave having the four-vector $(\mathbf{k}, \omega)$ equal to $1/\hbar$ times the four-vector $(\mathbf{p}, E)$:

$$\mathbf{k} = \mathbf{p}/\hbar, \qquad \omega = E/\hbar. \qquad (1.3.1)$$

This idea gained support from the fact that a wave satisfying (1.3.1) would have a group velocity equal to the ordinary velocity $c^2 \mathbf{p}/E$ of a particle of

---

[19] L. de Broglie, *Comptes Rendus Acad. Sci.* **177**, 507, 548, 630 (1923).

momentum **p** and energy $E$. For a reminder about group velocity, consider a wave packet in one dimension:

$$\psi(x, t) = \int dk \, g(k) \exp\left(ikx - i\omega(k)t\right), \qquad (1.3.2)$$

where $g(k)$ is some smooth function with a peak at an argument $k_0$. Suppose also that the wave $\int dk \, g(k) \exp(ikx)$ at $t = 0$ is peaked at $x = 0$. By expanding $\omega(k)$ around $k_0$, we have

$$\psi(x, t) \simeq \exp\left(-it[\omega(k_0) - k_0\omega'(k_0)]\right) \int dk \, g(k) \exp\left(ik\left[x - \omega'(k_0)t\right]\right),$$

and therefore

$$|\psi(x, t)| \simeq \left|\psi\left([x - \omega'(k_0)t], 0\right)\right|. \qquad (1.3.3)$$

The wave packet that was concentrated at time $t = 0$ near $x = 0$ is evidently concentrated at time $t$ near $x = \omega'(k_0)t$, so it moves with speed

$$v = \frac{d\omega}{dk} = \frac{dE}{dp} = \frac{c^2 p}{E}, \qquad (1.3.4)$$

in agreement with the usual formula for velocity in special relativity.

Just as vibrational waves on a violin string are quantized by the condition that, since the string is clamped at both ends, it must contain an integer number of half-wavelengths, so according to de Broglie, the wave associated with an electron in a circular orbit must have a wavelength that just fits into the orbit a whole number $n$ of times, so $2\pi r = n\lambda$, and therefore

$$p = \hbar k = \hbar \times 2\pi/\lambda = n\hbar/r. \qquad (1.3.5)$$

Using the non-relativistic formula $p = mv$, this is the same as the Bohr quantization condition (1.2.4). More generally, the Sommerfeld condition (1.2.12) could be understood as the requirement that the phase of a wave changes by a whole-number multiple of $2\pi$ when a particle completes one orbit. Thus the success of Bohr and Sommerfeld's wild guesses could be explained in a wave theory, though that too was just a wild guess.

There is a story that in his oral thesis examination, de Broglie was asked what other evidence might be found for a wave theory of the electron, and he suggested that perhaps diffraction phenomena might be observed in the scattering of electrons by crystals. Whatever the truth of this story, it is known that (at the suggestion of Walter Elsasser (1904–1991)) this experiment was carried out at the Bell Telephone Laboratories by Clinton Davisson (1881–1958) and Lester Germer (1896–1971), who in 1927 reported that electrons scattered by a single crystal of nickel showed a pattern of diffraction peaks similar to those seen in the scattering of X-rays by crystals.[20]

---

[20]  C. Davisson and L. Germer, *Phys. Rev.* **30**, 707 (1927).

Of course, an atomic orbit is not a violin string. What was needed was some way of extending the wave idea from free particles, described by waves like (1.3.2), to particles moving in a potential, such as the Coulomb potential in an atom. This was supplied in 1926 by Erwin Schrödinger (1887–1961).[21] Schrödinger presented his idea as an adaptation of the Hamilton–Jacobi formulation of classical mechanics, which would take us too far away from quantum mechanics to go into here. There is a simpler way of understanding Schrödinger's wave mechanics as a natural generalization of what de Broglie had already done.

According to the relations $\mathbf{p} = \hbar \mathbf{k}$ and $E = \hbar \omega$, the wave function $\psi \propto \exp(i\mathbf{k} \cdot \mathbf{x} - i\omega t)$ of a free particle of momentum $\mathbf{p}$ and energy $E$ satisfies the differential equations

$$-i\hbar \nabla \psi(\mathbf{x}, t) = \mathbf{p}\psi(\mathbf{x}, t), \quad i\hbar \frac{\partial}{\partial t}\psi(\mathbf{x}, t) = E\psi(\mathbf{x}, t).$$

For any state of energy $E$, we then have

$$\psi(\mathbf{x}, t) = \exp(-iEt/\hbar)\,\psi(\mathbf{x}), \tag{1.3.6}$$

while for a free particle, in the non-relativistic case, $E = \mathbf{p}^2/2m$, so here $\psi(\mathbf{x})$ is some solution of the equation

$$E\,\psi(\mathbf{x}) = \frac{-\hbar^2}{2m}\nabla^2\psi(\mathbf{x}).$$

More generally, the energy of a particle in a potential $V(\mathbf{x})$ is given by $E = \mathbf{p}^2/2m + V(\mathbf{x})$, which suggests that for such a particle we still have Eq. (1.3.6), but now

$$E\,\psi(\mathbf{x}) = \left[\frac{-\hbar^2}{2m}\nabla^2 + V(\mathbf{x})\right]\psi(\mathbf{x}). \tag{1.3.7}$$

This is the Schrödinger equation for a single particle of energy $E$.

Just like the equations for the frequencies of transverse vibrations of a violin string, this equation has solutions only for certain definite values of $E$. The boundary condition that takes the place here of the condition that a violin string does not vibrate where it is clamped at its ends, is that $\psi(\mathbf{x})$ is single-valued (that is, it returns to the same value if $\mathbf{x}$ goes around a closed curve) and vanishes as $|\mathbf{x}|$ goes to infinity. For instance, Schrödinger was able to show that in a Coulomb potential $V(\mathbf{x}) = -Ze^2/r$, for each $n = 1, 2, \ldots$, Eq. (1.3.7) has $n^2$ different single-valued solutions that vanish for $r \to \infty$ with energies given by Bohr's formula $E_n = -Z^2e^4m_e/2n^2\hbar^2$, and no such solutions for any other energies. (We will carry out this calculation in the next chapter.) As Schrödinger remarked in his first paper on wave mechanics, "The essential thing seems to me

---

[21] E. Schrödinger, *Ann. Physik* **79**, 361, 409 (1926).

to be that the postulation of "whole numbers" no longer enters into the quantum rules mysteriously, but that we have traced the matter a step farther back, and found the 'integralness' to have its origin in the finiteness and single-valuedness of a certain space function."

More than that, Schrödinger's equation had an obvious generalization to general systems. If a system is described by a Hamiltonian $H(\mathbf{x}_1, \ldots ; \mathbf{p}_1 \ldots)$ (where dots indicate coordinates and momenta of additional particles) the Schrödinger equation takes the form

$$H(\mathbf{x}_1, \ldots ; -i\hbar\,\nabla_1 \cdots)\psi_n(\mathbf{x}_1, \ldots) = E_n\psi_n(\mathbf{x}_1, \ldots). \tag{1.3.8}$$

For instance, for $N$ particles of masses $m_r$ with $r = 1, 2, \ldots$, with a general potential $V(\mathbf{x}_1, \ldots, \mathbf{x}_N)$, the Hamiltonian is

$$H = \sum_r \frac{\mathbf{p}_r^2}{2m_r} + V(\mathbf{x}_1, \ldots, \mathbf{x}_N), \tag{1.3.9}$$

and the allowed energies $E$ are those for which there is a single-valued solution $\psi(\mathbf{x}_1, \ldots, \mathbf{x}_N)$, vanishing when any $|\mathbf{x}_r|$ goes to infinity, of the Schrödinger equation

$$E\,\psi(\mathbf{x}_1, \ldots, \mathbf{x}_N) = \left[\sum_{r=1}^{N} \frac{-\hbar^2}{2m_r}\,\nabla_r^2 + V(\mathbf{x}_1, \ldots, \mathbf{x}_N)\right]\psi(\mathbf{x}_1, \ldots, \mathbf{x}_N). \tag{1.3.10}$$

So now it was possible at least in principle to calculate the spectrum not only of hydrogen, but of any other atom, and indeed of any non-relativistic system with a known potential.

## 1.4 Matrix Mechanics

A few years after de Broglie introduced the idea of wave mechanics, and a little before Schrödinger developed his version of the theory, a quite different approach to quantum mechanics was developed by Werner Heisenberg (1901–1976). Heisenberg suffered from hay fever, so in 1925 he escaped the pollen-laden air of Göttingen to go on vacation to the grassless North Sea island of Helgoland. While on vacation he wrestled with the mystery surrounding the quantum conditions of Bohr and de Broglie. When he returned to the University of Göttingen he had a new approach to the quantum conditions, which has come to be called matrix mechanics.[22]

Heisenberg's starting point was the philosophical judgment that a physical theory should not concern itself with things like electron orbits in atoms that can never be observed. This is a risky assumption, but in this case it served

---

[22] W. Heisenberg, Z. *Physik* **33**, 879 (1925).

Heisenberg well. He fastened on the energies $E_n$ of atomic states, and the rates $A_m^n$ at which atoms spontaneously make radiative transitions from one state $m$ to another state $n$, as the observables on which to base a physical theory. In classical electrodynamics, a particle with charge $\pm e$ with a position vector $\mathbf{x}$ that is undergoing non-uniform motion emits a radiation power[23]

$$P = \frac{2e^2}{3c^3}\ddot{\mathbf{x}}^2. \tag{1.4.1}$$

Heisenberg guessed that this formula gives the power emitted in a radiative transition from an atomic state with energy $E_m$ to one with a lower energy $E_n$, if we make the replacement

$$\mathbf{x} \mapsto [\mathbf{x}]_{nm} + [\mathbf{x}]_{nm}^*, \tag{1.4.2}$$

where $[\mathbf{x}]_{nm}$ is a complex vector amplitude characterizing this transition, taken proportional to $\exp(-i\omega_{nm}t)$, and $\omega_{nm}$ is the circular frequency (the frequency times $2\pi$) of the radiation emitted in the transition:

$$\omega_{nm} = (E_m - E_n)/\hbar. \tag{1.4.3}$$

(Heisenberg did not actually write the classical formula (1.4.1), but he did give the electric and magnetic fields far from the accelerated charge, from which Eq. (1.4.1) can be inferred. He also did not explicitly state that he was making the replacement (1.4.2), but it is pretty clear from his subsequent results that this is what he did.) With the replacement (1.4.2), Eq. (1.4.1) becomes a formula for the radiation power emitted in the transition $m \to n$:

$$P(m \to n) = \frac{2e^2\omega_{nm}^4}{3c^3}\left([\mathbf{x}]_{nm}^2 + 2[\mathbf{x}]_{nm}[\mathbf{x}]_{nm}^* + [\mathbf{x}]_{nm}^*[\mathbf{x}]_{nm}^*\right).$$

The first and third terms are proportional respectively to $\exp(-2i\omega_{nm}t)$ and to $\exp(2i\omega_{nm}t)$, and hence make no contribution when we average over a time long compared with $1/\omega_{nm}$. The time average (indicated by a bar over $P$) is therefore given by the cross-term, which is time-independent:

$$\overline{P}(m \to n) = \frac{4e^2\omega_{nm}^4}{3c^3}\left|[\mathbf{x}]_{nm}\right|^2. \tag{1.4.4}$$

That is, the rate of emitting photons carrying energy $\hbar\omega_{nm}$ in the transition $m \to n$ is, in Einstein's notation,

$$A_m^n = \frac{\overline{P}(m \to n)}{\hbar\omega_{nm}} = \frac{4e^2\omega_{nm}^3}{3c^3\hbar}\left|[\mathbf{x}]_{nm}\right|^2, \tag{1.4.5}$$

---

[23] J. Larmor, *Phil. Mag.* S.5, **44**, 503 (1897). (This is the total radiation power that at time $t$ passes through a sphere of radius $r$, with $\mathbf{x}$ evaluated at the retarded time $t - r/c$, under the assumption that $r$ is much greater than the distance of the particle from the center of the sphere.)

and, according to the Einstein relations (1.2.16), this gives the coefficients of $\rho(\nu_{nm})$ in the rates for induced emission and absorption

$$B_n^m = B_m^n = \frac{2\pi e^2}{3\hbar^2}\left|[\mathbf{x}]_{nm}\right|^2. \tag{1.4.6}$$

In Eqs. (1.4.5) and (1.4.6), $[x]_{nm}$ appears only with $E_m > E_n$, but Heisenberg extended the definition of $[\mathbf{x}]_{nm}$ to the case where $E_n > E_m$, by the condition

$$[\mathbf{x}]_{nm} = [\mathbf{x}]_{mn}^* \propto \exp(-i\omega_{nm}t), \tag{1.4.7}$$

so that Eq. (1.4.6) holds whether $E_m > E_n$ or $E_n > E_m$.

Heisenberg limited his calculations to the example of an anharmonic oscillator in one dimension, for which the energy is given classically in terms of position and its rate of change by

$$E = \frac{m_e}{2}\dot{x}^2 + \frac{m_e\omega_0^2}{2}x^2 + \frac{m_e\lambda}{3}x^3, \tag{1.4.8}$$

where $\omega_0$ and $\lambda$ are free real parameters. To calculate the $E_n$ and $[x]_{nm}$, Heisenberg used two relations. The first is a quantum-mechanical interpretation of Eq. (1.4.8):

$$\frac{m_e}{2}[\dot{x}^2]_{nm} + \frac{m_e\omega_0^2}{2}[x^2]_{nm} + \frac{m_e\lambda}{3}[x^3]_{nm} = \begin{cases} E_n, & n = m, \\ 0, & n \neq m, \end{cases} \tag{1.4.9}$$

where $E_n$ is the energy of the quantum state labeled $n$. But what meaning should be attached to $[\dot{x}^2]_{nm}$, $[x^2]_{nm}$, and $[x^3]_{nm}$? Heisenberg found that the "simplest and most natural assumption" was to take

$$[x^2]_{nm} = \sum_l [x]_{nl}[x]_{lm}, \quad [x^3]_{nm} = \sum_{l,k}[x]_{nl}[x]_{lk}[x]_{km} \tag{1.4.10}$$

and likewise

$$[\dot{x}^2]_{nm} = \sum_k [\dot{x}]_{nk}[\dot{x}]_{km} = \sum_k \omega_{nk}\omega_{mk}[x]_{nk}[x]_{km}. \tag{1.4.11}$$

Note that because $[x]_{nm}$ is proportional to $\exp(-i(E_m - E_n)t/\hbar)$ for all $n$ and $m$, each term in Eq. (1.4.9) is time-independent for $n = m$. Also, by virtue of the condition (1.4.7), the first two terms are positive for $n = m$ though the last term might not be.

The second relation is a quantum condition. Here Heisenberg adopted a formula that had been published a little earlier by W. Kuhn[24] and W. Thomas,[25] which Kuhn derived using a model of an electron in a bound state as an ensemble of oscillators vibrating in three dimensions at frequencies $\nu_{nm}$. From the

---

[24] W. Kuhn, *Z. Physik* **33**, 408 (1925).
[25] W. Thomas, *Naturwissenschaften* **13**, 627 (1925).

condition that at very high frequency the scattering of light from such an electron should be the same as if the electron were a free particle, Kuhn derived the purely classical statement[26] that, for any given state $n$,

$$\sum_m B_n^m (E_m - E_n) = \frac{\pi e^2}{m_e}. \qquad (1.4.12)$$

Combining this with Eq. (1.4.6) gives

$$\hbar = \frac{2m_e}{3} \sum_m \left| [\mathbf{x}]_{nm} \right|^2 \omega_{nm}. \qquad (1.4.13)$$

Since in three dimensions there are three terms in $\left| [\mathbf{x}]_{nm} \right|^2$, the factor $1/3$ gives the average of these three terms, so in one dimension we would have

$$\hbar = 2m_e \sum_m \left| [x]_{nm} \right|^2 \omega_{nm}. \qquad (1.4.14)$$

This is the quantum condition used by Heisenberg.

Heisenberg was able to find an exact solution[27] of Eqs. (1.4.9) and (1.4.14) for the harmonic oscillator case $\lambda = 0$. For any integer $n \geq 0$,

$$E_n = \left( n + \frac{1}{2} \right) \hbar\omega_0, \quad [x]_{n+1,n}^* = [x]_{n,n+1} = e^{-i\omega_0 t} \sqrt{\frac{(n+1)\hbar}{2m_e\omega_0}}, \qquad (1.4.15)$$

with $[x]_{nm}$ vanishing unless $n - m = \pm 1$. We will see how to derive these results for $\lambda = 0$ in Section 2.5. Heisenberg was also able to calculate the corresponding results for small non-zero $\lambda$, to first order in $\lambda$.

This was all very obscure. On his return from Helgoland, Heisenberg showed his work to Max Born (1882–1970). Born recognized that the formulas in Eq. (1.4.10) were just special cases of a well-known mathematical procedure, known as *matrix multiplication*. A matrix denoted $[A]_{nm}$ or just $A$ is a square array of numbers (real or complex), with $[A]_{nm}$ the number in the $n$th row and $m$th column. In general, for any two matrices $[A]_{nm}$ and $[B]_{nm}$, the matrix $AB$ is the square array

$$[AB]_{nm} \equiv \sum_l [A]_{nl} [B]_{lm}. \qquad (1.4.16)$$

---

[26] Kuhn actually gave this condition only where $n$ is the ground state, the state of lowest energy, but the argument applies to any state. Where $n$ is not the ground state, the terms in the sum over $m$ are positive if $m$ has higher energy than $n$, but negative if $m$ has lower energy.

[27] Somewhat inconsistently, Heisenberg took the time-dependence factor in $[x]_{nm}$ to be $\cos(\omega_{nm} t)$ rather than $\exp(-i\omega_{nm} t)$. The results here apply to the case where $[x]_{nm} \propto \exp(-i\omega_{nm} t)$; $[x]_{nm}$ is the term in Heisenberg's solution proportional to $\exp(-i\omega_{nm} t)$.

We also note for further use that the sum of two matrices is defined so that

$$[A + B]_{nm} \equiv [A]_{nm} + [B]_{nm}, \tag{1.4.17}$$

and the product of a matrix and a numerical factor $\alpha$ is defined as

$$[\alpha A]_{nm} \equiv \alpha [A]_{nm}. \tag{1.4.18}$$

Matrix multiplication is thus associative, namely $A(BC) = (AB)C$, and distributive, meaning that $A(\alpha_1 B_1 + \alpha_2 B_2) = \alpha_1 AB_1 + \alpha_2 AB_2$ and $(\alpha_1 B_1 + \alpha_2 B_2)A = \alpha_1 B_1 A + \alpha_2 B_2 A]$, but in general it is not commutative ($AB$ and $BA$ are not necessarily equal). As defined by Eq. (1.4.10), $[x^2]$ is the square of the matrix $[x]$, $[x^3]$ is the cube of the matrix $[x]$, and so on.

The quantum condition (1.4.14) can also be given a pretty formulation as a matrix equation. Note that according to Eq. (1.4.7), the matrix for momentum is

$$[p]_{nm} = m_e [\dot{x}]_{nm} = -i m_e \omega_{nm} [x]_{nm},$$

so the matrix products $[px]$ and $[xp]$ have the diagonal components

$$[px]_{nn} = \sum_m [p]_{nm} [x]_{mn} = -i m_e \sum_m \omega_{nm} \left| [x]_{mn} \right|^2,$$

$$[xp]_{nn} = \sum_m [x]_{nm} [p]_{mn} = -i m_e \sum_m \omega_{mn} \left| [x]_{mn} \right|^2.$$

(In both formulas, we have used the relation (1.4.7), which says that $[x]_{mn}$ is what is called an *Hermitian* matrix.) Since $\omega_{nm} = -\omega_{mn}$, the quantum condition (1.4.14) can be written in two ways

$$i\hbar = -2[px]_{nn} = +2[xp]_{nn}. \tag{1.4.19}$$

Of course, the relation can then also be written

$$i\hbar = [xp]_{nn} - [px]_{nn} = [xp - px]_{nn}, \tag{1.4.20}$$

where we have used the definitions (1.4.17) and (1.4.18).

Shortly after the publication of Heisenberg's paper, there appeared two papers that extended Eq. (1.4.20) to a general formula for all elements of the matrix $xp - px$:

$$xp - px = i\hbar \times 1, \tag{1.4.21}$$

where here 1 is the matrix

$$[1]_{nm} \equiv \delta_{nm} \equiv \begin{cases} 1 & n = m, \\ 0 & n \neq m. \end{cases} \tag{1.4.22}$$

That is, in addition to Eq. (1.4.20), we have $[xp - px]_{nm} = 0$ for $n \neq m$. Born and his assistant Pascual Jordan[28] (1902–1984) gave a mathematically fallacious

---

[28]  P. Jordan, Z. *Physik* **34**, 858 (1925).

derivation of this fact, on the basis of the Hamiltonian equations of motion. Paul Dirac[29](1902–1984) simply assumed Eq. (1.4.21), from an analogy with the Poisson brackets of classical mechanics, described in Section 9.4.

Matrix mechanics was now a general scheme for calculating the spectrum of any system described classically by a Hamiltonian $H(q, p)$, given as a function of a number of coordinates $q_r$ and the corresponding "momenta" $p_r$. One looks for some representation of the $q$s and $p$s as matrices satisfying the matrix equation

$$q_r p_s - p_s q_r = i\hbar \delta_{rs} \times 1, \qquad (1.4.23)$$

and such that the matrix $H(q, p)$ is diagonal,

$$[H(q, p)]_{nm} = E_n \delta_{nm}. \qquad (1.4.24)$$

The diagonal elements $E_n$ are the energies of the system, and the matrix elements $[x]_{nm}$ can be used with Eqs. (1.4.5) and (1.4.6) to calculate the rates for spontaneous and stimulated emission and absorption of radiation.

Unfortunately, there are very few physical systems for which this sort of calculation is practicable. One is the harmonic oscillator, already solved by Heisenberg. Another is the hydrogen atom, whose spectrum was obtained using matrix mechanics in a display of mathematical brilliance by Wolfgang Pauli[30] (1900–1958), a student of Sommerfeld. (Pauli's calculation is presented in Section 4.8.) These two problems were soluble because of special features of the Hamiltonians, the same features that make the classical orbits of particles closed curves. It was hopeless to use matrix mechanics to solve more complicated problems, like the hydrogen molecule, so wave mechanics largely superseded matrix mechanics among the tools of theoretical physics.

But it must not be thought that wave mechanics and matrix mechanics are different physical theories. In 1926, Schrödinger showed how the principles of matrix mechanics can be derived from those of wave mechanics.[31] To see how this works, note first that the Hamiltonian is what is called an Hermitian operator, meaning that for any functions $f$ and $g$ that satisfy the conditions of single-valuedness and vanishing at infinity imposed on wave functions, we have

$$\int f^*(Hg) = \int (Hf)^* g, \qquad (1.4.25)$$

the integrals being taken over all coordinates. This is trivial for the term $V$ in Eq. (1.3.9), and it is also true for the Laplacian operator, as can be seen by integrating the identity

$$(\nabla^2 f)^* g - f^*(\nabla^2 g) = \nabla \cdot [(\nabla f)^* g - f^* \nabla g].$$

[29]  P. A. M. Dirac, *Proc. Roy. Soc.* A **109**, 642 (1926).

[30]  W. Pauli, *Z. Physik* **36**, 336 (1926).

[31]  E. Schrödinger, *Ann. Physik* **79**, 734 (1926).

It follows that for solutions $\psi_n$ of the Schrödinger equation with energy $E_n$, we have

$$E_n \int \psi_m^* \psi_n = \int \psi_m^* (H\psi_n) = \int (H\psi_m)^* \psi_n = E_m^* \int \psi_m^* \psi_n. \qquad (1.4.26)$$

Taking $m = n$, we see that $E_n$ is real, and then taking $m \neq n$, we see that $\int \psi_m^* \psi_n = 0$ for $E_n \neq E_m$. It can be shown that if there is more than one solution of the Schrödinger equation with the same energy, the solutions can always be chosen so that $\int \psi_m^* \psi_n = 0$ for $n \neq m$. (This is shown in footnote 3 of Section 3.1 in cases where there is a finite number of solutions of the Schrödinger equation with a given energy.) By multiplying the $\psi_n$ with suitable factors we can also arrange that $\int \psi_n^* \psi_n = 1$, so the $\psi_n$ are *orthonormal*, in the sense that

$$\int \psi_m^* \psi_n = \delta_{nm}. \qquad (1.4.27)$$

Now consider any operators $A$, $B$, etc., defined by their action on wave functions. For instance, for a single particle, the momentum operator $\mathbf{P}$ and position operators $\mathbf{X}$ are defined by

$$[\mathbf{P}\psi](\mathbf{x}) \equiv -i\hbar \nabla \psi(\mathbf{x}), \qquad [\mathbf{X}\psi](\mathbf{x}) \equiv \mathbf{x}\psi(\mathbf{x}). \qquad (1.4.28)$$

For any such operator, we define a matrix

$$[A]_{nm} \equiv \int \psi_n^* [A\psi_m]. \qquad (1.4.29)$$

Note that, as a consequence of Eq. (1.3.6), this has the time-dependence (1.4.7) assumed by Heisenberg

$$[A]_{nm} \propto \exp\left(-i(E_m - E_n)t/\hbar\right).$$

With the definition (1.4.29), we can show that the matrix of a product of operators is the product of the matrices:

$$\int \psi_n^* \Big[ A[B\psi_m] \Big] = \sum_l [A]_{nl} [B]_{lm}. \qquad (1.4.30)$$

To prove this, we assume that the function $B\psi_m$ can be written as an expansion in the wave functions:

$$B\psi_m = \sum_r b_r(m)\psi_r,$$

with some coefficients $b_r(m)$. (To make this literally true, it may be necessary to put the system in a box, like that used in Section 1.1, so that the solutions of the Schrödinger equation form a discrete set, including those corresponding to unbound electrons.) We can find these coefficients by multiplying both sides of the expansion with $\psi_l^*$ and integrating over all coordinates, using the orthonormality property (1.4.27):

$$[B]_{lm} = \int \psi_l^*[B\psi_m] = \sum_r b_r(m)\delta_{rl} = b_l(m).$$

It follows that

$$B\psi_m = \sum_l [B]_{lm}\psi_l. \tag{1.4.31}$$

Repeating the same reasoning, we have

$$A[B\psi_m] = \sum_{l,s}[B]_{lm}[A]_{sl}\psi_s. \tag{1.4.32}$$

Multiplying with $\psi_n^*$, integrating over all coordinates, and again using the orthonormality property (1.4.27) then gives Eq. (1.4.30).

We can now derive the Heisenberg quantization conditions. First, note that the matrix $[H]_{nm}$ is simply

$$[H]_{nm} \equiv \int \psi_n^*[H\psi_m] = E_m \int \psi_n^*\psi_m = E_m\delta_{nm}, \tag{1.4.33}$$

which is the same as Eq. (1.4.24). Next, we can verify the condition (1.4.14) in the generalized form (1.4.21). Note that

$$\frac{\partial}{\partial x}(x\psi) = \psi + x\frac{\partial}{\partial x}\psi,$$

so the operators $P$ and $X$ defined by (1.4.28) satisfy

$$\Big[P[X\psi]\Big] = -i\hbar\psi + \Big[X[P\psi]\Big].$$

Applying the general formula (1.4.30), we have then

$$[xp - px]_{nm} = i\hbar\delta_{nm}, \tag{1.4.34}$$

which is the same as (1.4.21). The same argument can evidently be applied to give the more general condition (1.4.23).

The approach that will be adopted when we come to the general principles of quantum mechanics in Chapter 3 will be neither matrix mechanics nor wave mechanics, but a more abstract formulation, that Dirac called *transformation theory*,[32] from which matrix mechanics and wave mechanics can both be derived.

Although we will not be going into quantum electrodynamics until Chapter 11, I should mention here that in 1926 Born, Heisenberg, and Jordan[33]

---

[32]  P. A. M. Dirac, *Proc. Roy. Soc.* A **113**, 621 (1927). This approach is the basis of Dirac's treatise, *The Principles of Quantum Mechanics*, 4th edn. (rev.) (Oxford University Press, Oxford, 1976).

[33]  M. Born, W. Heisenberg, and P. Jordan, *Z. Physik* **35**, 557 (1926). They ignored the polarization of light, and treated the problem in one dimension, rather than as in the three-dimensional version described here.

applied the ideas of matrix mechanics to the electromagnetic field. They showed that the free field in a cubical box with edges of length $L$ can be written as a sum of terms with wave numbers given by (1.1.1), that is, $\mathbf{q_n} = 2\pi \mathbf{n}/L$ with $\mathbf{n}$ a vector with integer components, each term described by a harmonic oscillator Hamiltonian $H_{\mathbf{n}} = [\dot{\mathbf{a}}_{\mathbf{n}}^2 + \omega_{\mathbf{n}}^2 \mathbf{a}_{\mathbf{n}}^2]/2$ (with $\mathbf{a_n}$ replacing $\sqrt{m}\mathbf{x}$), where $\omega_{\mathbf{n}} = c|\mathbf{q_n}|$. The energy of this field in which the $\mathbf{n}$th oscillator is in the $\mathcal{N}_{\mathbf{n}}$th excited state is the sum of the harmonic oscillator energies (1.4.15)

$$E = \sum_{\mathbf{n}} \left[ \mathcal{N}_{\mathbf{n}} + \frac{1}{2} \right] \hbar \omega_{\mathbf{n}}. \tag{1.4.35}$$

Such a state is interpreted as one containing $\mathcal{N}_{\mathbf{n}}$ photons of wave number $\mathbf{q_n} = 2\pi \mathbf{n}/L$, thus justifying the Einstein assumption that light comes in quanta with energy $h\nu = \hbar\omega$. (The additional "zero-point" energy $\sum_{\mathbf{n}} \hbar\omega_{\mathbf{n}}/2$ is the energy of quantum fluctuations in the vacuum, which has no effect, except on the gravitational field. This is one contribution to the "dark energy" that is currently a major concern of physicists and astronomers.) In 1927 Dirac[34] was able to use this quantum theory of radiation to give a completely quantum mechanical derivation of the formula (1.4.5) for the rate of spontaneous emission of photons, without having to rely on analogies with classical radiation theory. This derivation is presented and generalized in Section 11.7.

## 1.5 Probabilistic Interpretation

At first, Schrödinger and others thought that wave functions represent particles that are spread out, like pressure disturbances in a fluid – most of the particle is where the wave function is large. This interpretation became untenable with the analysis of scattering in quantum mechanics by Max Born.[35] For this purpose, Born used a generalization of de Broglie's assumption (1.3.6) for the time-dependence of the wave function of a free particle. For any system described by a Hamiltonian $H$, the time-dependence of any wave function, whether or not for a state of definite energy, is given by

$$i\hbar \frac{\partial}{\partial t} \psi = H\psi. \tag{1.5.1}$$

For instance, for a particle of mass $m$ moving in a potential $V(\mathbf{x})$, the non-relativistic Hamiltonian of classical mechanics is $H = \mathbf{p}^2/2m + V$, and the wave function satisfies the time-dependent Schrödinger equation

$$i\hbar \frac{\partial}{\partial t} \psi(\mathbf{x}, t) = H(\mathbf{X}, \mathbf{P})\psi(\mathbf{x}, t) = \left[ -\frac{\hbar^2 \nabla^2}{2m} + V(\mathbf{x}) \right] \psi(\mathbf{x}, t), \tag{1.5.2}$$

[34]  P. A. M. Dirac, *Proc. Roy. Soc.* A **114**, 710 (1927).

[35]  M. Born, *Z. Physik* **37**, 863 (1926); **38**, 803 (1926).

with the operators **X** and **P** defined by Eq. (1.4.28). By following the time development of a packet like (1.3.2) that is localized within a small region of space, Born found that when a particle strikes a target like an atom or atomic nucleus, the wave function radiates out in all directions, with a magnitude decreasing as $1/r$, where $r$ is the distance to the target. (This is shown here in Chapter 7.) This seemed to contradict the common experience that though a particle striking a target may indeed be scattered in any direction, it does not break up and go in all directions.

Born proposed that the magnitude of the wave function $\psi(\mathbf{x}, t)$ does not tell us how much of the particle is at position **x** at time $t$, but rather the *probability* that the particle is at or near **x** at time $t$. To be precise, Born proposed that for a system consisting of a single particle, the probability that the particle is in a small volume $d^3x$ centered at **x** at time $t$ is

$$dP = |\psi(\mathbf{x}, t)|^2 \, d^3x. \tag{1.5.3}$$

In order that there be a 100% probability of the particle being somewhere, the wave function must be normalized so that

$$\int |\psi(\mathbf{x}, t)|^2 \, d^3x = 1, \tag{1.5.4}$$

the integral being taken over all space. The condition that the integral has the value unity does not set important constraints on the sort of wave function that is physically allowed, for as long as the integral is a finite constant $N$, we can always make (1.5.4) satisfied by dividing the wave function by $\sqrt{N}$. It *is* important that the integral be finite; this is a stronger version of the condition used by Schrödinger, that the wave function must vanish at infinity.

Note that for a wave function whose time-dependence is described by the Schrödinger equation (1.5.1), the integral (1.5.4) remains constant, so a wave function that is normalized to satisfy (1.5.4) at one time will satisfy it at all times. The rate of change of this integral is given by

$$i\hbar \frac{d}{dt} \int |\psi(\mathbf{x}, t)|^2 \, d^3x = i\hbar \int \psi^*(\mathbf{x}, t) \frac{\partial}{\partial t} \psi(\mathbf{x}, t) \, d^3x$$
$$+ i\hbar \int \left( \frac{\partial}{\partial t} \psi^*(\mathbf{x}, t) \right) \psi(\mathbf{x}, t) \, d^3x$$
$$= \int \psi^*(\mathbf{x}, t) \, ([H\psi](\mathbf{x}, t)) \, d^3x$$
$$- \int ([H\psi](\mathbf{x}, t))^* \psi(\mathbf{x}, t) \, d^3x,$$

and this vanishes because $H$ satisfies the condition (1.4.25), that it is an Hermitian operator. In particular, if $\psi$ satisfies the one-particle Schrödinger equation (1.5.2), then

$$\frac{\partial}{\partial t}|\psi(\mathbf{x}, t)|^2 = \frac{i\hbar}{2m} \nabla \cdot \left( \psi^*(\mathbf{x}, t) \nabla \psi(\mathbf{x}, t) - \psi(\mathbf{x}, t) \nabla \psi^*(\mathbf{x}, t) \right). \quad (1.5.5)$$

This is a conservation law like the conservation of electric charge, but here $|\psi|^2$ is the density of probability rather than charge, and $(i\hbar/2m)\left( \psi^* \nabla \psi - \psi \nabla \psi^* \right)$ is the flux of probability rather than the electric current density. If $\psi(\mathbf{x}, t)$ vanishes for $|\mathbf{x}| \to \infty$, then Eq. (1.5.5) and Gauss's theorem tell us again that the integral of $|\psi|^2$ over all space is time-independent.

It follows immediately from (1.5.3) that the mean value (the "expectation value") of any function $f(\mathbf{x})$ is given by

$$\langle f \rangle = \int f(\mathbf{x})|\psi(\mathbf{x}, t)|^2 \, d^3x. \quad (1.5.6)$$

In other words, if $f(\mathbf{X})$ is the operator that multiplies a wave function $\psi(\mathbf{x})$ by $f(\mathbf{x})$, then

$$\langle f \rangle = \int \psi^*(\mathbf{x})[f(\mathbf{X})\psi](\mathbf{x}) \, d^3x.$$

It is only a short step from this to assume that the average of any observable $A$ is

$$\langle A \rangle = \int \psi^*(\mathbf{x})[A\psi](\mathbf{x}) \, d^3x, \quad (1.5.7)$$

where $A\psi$ is the effect of the operator representing the observable $A$ on the wave function $\psi$. In systems with more than one particle, the wave function depends on the coordinates of all the particles, and the integrals in Eqs. (1.5.4)–(1.5.7) run over all these coordinates.

In 1927 Paul Ehrenfest (1880–1933) used these results to show how the classical equations of motion of a non-relativistic particle in a potential emerge from the time-dependent Schrödinger equation.[36] To derive Ehrenfest's results, we use Eq. (1.5.2), and find the time-derivatives of the expectation values of the position and momentum:

$$\frac{d}{dt}\langle \mathbf{X} \rangle = \frac{1}{i\hbar} \int d^3x \, \psi^*(\mathbf{x}, t)\left( \mathbf{X}H - H\mathbf{X} \right)\psi(\mathbf{x}, t) = \langle \mathbf{P} \rangle/m,$$

$$\frac{d}{dt}\langle \mathbf{P} \rangle = \frac{1}{i\hbar} \int d^3x \, \psi^*(\mathbf{x}, t)\left( \mathbf{P}H - H\mathbf{P} \right)\psi(\mathbf{x}, t) = -\langle \nabla V(\mathbf{X}) \rangle.$$

This is not quite the same as the classical equations, because $\langle V(\mathbf{X}) \rangle$ is not in general the same as $V(\langle \mathbf{X} \rangle)$, but if (as usual in macroscopic systems) the force does not vary much over the range in which the wave function is appreciable, then these equations are very close to the classical equations of motion for $\langle \mathbf{P} \rangle$ as well as for $\langle \mathbf{X} \rangle$. (This is made more precise by the use of the eikonal approximation, described in Section 7.10.)

---

[36] P. Ehrenfest, *Z. Physik* **45**, 455 (1927).

We can now see why it is important for all operators representing observable quantities to be Hermitian. Taking the complex conjugate of Eq. (1.5.7) gives

$$\langle A \rangle^* = \int [A\psi](\mathbf{x})^* \, \psi(\mathbf{x}) \, d^3x = \int \psi(\mathbf{x})^* [A\psi](\mathbf{x}) \, d^3x.$$

In the last step, we have used the definition (1.4.25) of Hermitian operators. The final expression is the expectation value of $A$, so we see that Hermitian operators have real expectation values.

We can also now derive the condition for a wave function to represent a state that has a definite real value $a$ for some observable represented by an Hermitian operator $A$. The expectation value of $(A - a)^2$ is

$$\langle (A - a)^2 \rangle = \int \psi^*(\mathbf{x}) \Big[ (A - a)^2 \psi \Big](\mathbf{x}) \, d^3x$$
$$= \int \Big( \big[ (A - a)\psi \big](\mathbf{x}) \Big)^* \Big[ (A - a)\psi \Big](\mathbf{x}) \, d^3x$$
$$= \int \Big| \big[ (A - a)\psi \big](\mathbf{x}) \Big|^2 \, d^3x. \tag{1.5.8}$$

If the state represented by $\psi(\mathbf{x})$ has a definite value $a$ for $A$, then the expectation value of $(A - a)^2$ must vanish, in which case (1.5.8) shows that $(A - a)\psi$ vanishes everywhere, and so

$$[A\psi](\mathbf{x}) = a\psi(\mathbf{x}). \tag{1.5.9}$$

In this case, $\psi(\mathbf{x})$ is said to be an *eigenfunction* of $A$ with *eigenvalue a*. The Schrödinger equation for the energies and wave functions of states of definite energy is just a special case of this condition, with $A$ the Hamiltonian operator, and $a$ the energy.

We can now easily see that it is impossible for any state to have definite values for any component $x$ of position *and* the corresponding component $p$ of momentum. If there were such a state, its wave function would satisfy both

$$X\psi = x\psi \quad \text{and} \quad P\psi = p\psi, \tag{1.5.10}$$

where $x$ and $p$ are the numerical values of the position and momentum. But then

$$XP\psi = pX\psi = px\psi, \qquad PX\psi = xP\psi = xp\psi,$$

and so

$$(XP - PX)\psi = 0$$

in contradiction with the commutation relation $XP - PX = i\hbar$.

Heisenberg[37] was even able to set a lower limit on the product of the uncertainty in position and in momentum, known as the Heisenberg Uncertainty Principle. Using the commutation relation $XP - PX = i\hbar$, he was able to show that

$$\Delta x \, \Delta p \geq \hbar/2, \tag{1.5.11}$$

where $\Delta x$ and $\Delta p$ are the uncertainties in position and momentum, defined as the root mean square deviation of position and momentum from their expectation values:

$$\Delta x \equiv \left\langle (X - \langle X \rangle)^2 \right\rangle^{1/2}, \quad \Delta p \equiv \left\langle (P - \langle P \rangle)^2 \right\rangle^{1/2}. \tag{1.5.12}$$

The proof will be given in Section 3.3. It should be emphasized that $\Delta x$ is the spread in values found for the position if we make a large number of highly accurate measurements of position, always starting with the same state with the same wave function $\psi$, and likewise for $\Delta p$. These uncertainties depend on the state, not on the method of measurement, which in general will introduce an additional uncertainty in the results obtained for $x$ or $p$, which is not taken into account in the definitions (1.5.12). As defined by Eq. (1.5.12), $\Delta x$ and $\Delta p$ are not the same as the uncertainties encountered if we measure $x$, which modifies the state, and then measure $p$ in the modified state (or vice versa).[38]

Heisenberg also offered a heuristic argument for a relation like Eq. (1.5.11), but a relation with a rather different meaning. He supposed that a particle is observed using light of wavelength $\lambda$, in which case the uncertainty in measured position cannot be much less than $\lambda$, no matter how sharply peaked the wave function is at a given position. Each photon will have momentum $2\pi\hbar/\lambda$, so in a *successive* measurement of momentum, the uncertainty $\Delta p$ associated with the new wave function cannot be much less than $2\pi\hbar/\lambda$, and so the product of the uncertainties cannot be much less than $2\pi\hbar$. In Heisenberg's thought experiment, the lower bound on the uncertainty in position arises from the nature of the measurement, while the lower bound on the uncertainty in momentum arises from the nature of the wave function after the measurement of position.

More generally, it is only possible for a state represented by a wave function $\psi$ to have definite values for both of two observables represented by operators $A$ and $B$ if

$$(AB - BA)\psi = 0. \tag{1.5.13}$$

Of course, this will be true for all wave functions if $AB = BA$, and for no wave functions if $AB - BA$ is a non-zero number like $i\hbar$ times the unit operator. The difference $AB - BA$ is known as the *commutator* of $A$ and $B$, and denoted

[37] W. Heisenberg, *Z. Physik* **43**, 172 (1927); *The Physical Principles of the Quantum Theory* (University of Chicago Press, Chicago, 1930), transl. C. Eckart and F. C. Hoyt, Chapter II, pp. 16–21. The discussion here of Heisenberg's work is based on the latter reference.

[38] On the uncertainties in such successive measurements, see M. Ozawa, *Phys. Rev. A* **67**, 042105 (2003); J. Distler and S. Paban, arXiv:1211.4169.

$$[A, B] \equiv AB - BA. \tag{1.5.14}$$

It is only possible for a state to have definite values for both $A$ and $B$ if the wave function $\psi$ satisfies $[A, B]\psi = 0$. Any two operators for which the commutator vanishes are said to commute.

Born also gave a probabilistic interpretation of wave functions that are not eigenfunctions of the Hamiltonian.[39] Suppose a wave function is given by an expansion in energy eigenfunctions

$$\psi = \sum_n c_n \psi_n, \tag{1.5.15}$$

where $H\psi_n = E_n\psi_n$ and $c_n$ are numerical coefficients. As remarked in Section 1.4, we can choose the $\psi_n$ to satisfy the orthonormality condition (1.4.27), in which case a normalized wave function must have

$$1 = \int |\psi|^2 = \sum_{nm} c_n^* c_m \int \psi_n^* \psi_m = \sum_n |c_n|^2. \tag{1.5.16}$$

The expectation value of any function $f(H)$ of the Hamiltonian is

$$\langle f(H) \rangle = \sum_{nm} c_n^* c_m \int \psi_n^* f(H) \psi_m = \sum_{nm} f(E_n) c_n^* c_m \int \psi_n^* \psi_m$$
$$= \sum_n |c_n|^2 f(E_n). \tag{1.5.17}$$

For this to be true for all functions, we must interpret $|c_n|^2$ as the probability that in a measurement of the energy (and, in the case of degeneracy, of other observables that distinguish the individual states), the system will be found to be in the state described by $\psi_n$. This rule was soon extended to general operators, not just the Hamiltonian.

As we saw in Section 1.4, the coefficient $c_n$ can be calculated by multiplying Eq. (1.5.15) with $\psi_m^*$, integrating over coordinates, and using the orthonormality condition (1.4.27), which gives $c_m = \int \psi_m^* \psi$. Thus if a system is in a state represented by a wave function $\psi$, and we make a measurement that puts the system in any one of a set of states represented by orthonormal wave functions $\psi_n$ (which may or may not be energy eigenfunctions) then the probability that the system will be found to be in a particular state represented by the wave function $\psi_m$ is

$$P(\psi \to \psi_m) = \left| \int \psi_m^* \psi \right|^2. \tag{1.5.18}$$

This is known as the *Born rule*, and can be taken as the fundamental interpretive postulate of quantum mechanics.

---

[39] M. Born, *Nature* **119**, 354 (1927).

The probabilistic interpretation of quantum mechanics was controversial from the beginning. In one way or another it was opposed by such leaders of theoretical physics as Schrödinger and Einstein. Debates about this aspect of quantum mechanics continued for years, most notably at the Solvay Conferences in Brussels in 1927 and later years. To the present, there continues to be a tension between the probabilistic interpretation and the deterministic evolution of the wave function, described by Eq. (1.5.1). If physical states, including observers and their instruments, evolve deterministically, where do the probabilities come from? These issues will be discussed in Section 3.7.

## Historical Bibliography

The works listed below contain convenient collections of original articles (in English, or English translation) from the early days of quantum mechanics and atomic physics:

1. *The Question of the Atom – From the Karlsruhe Congress to the First Solvay Conference, 1860–1911*, ed. M. J. Nye (Tomash Publishers, Los Angeles/San Francisco, CA, 1986).

2. *The Collected Papers of Lord Rutherford of Nelson O.M., FRS*, ed. J. Chadwick (Interscience, New York, 1963).

3. *Sources of Quantum Mechanics*, ed. B. L. van der Waerden (North-Holland, Amsterdam, 1967).

4. E. Schrödinger, *Collected Papers on Wave Mechanics*, Third English Edition (Chelsea Publishing, New York, 1982).

5. G. Bacciagaluppi and A. Valentini, *Quantum Theory at the Crossroads – Reconsidering the 1927 Solvay Conference* (Cambridge University Press, Cambridge, 2009).

## Problems

1. Consider a non-relativistic particle of mass $M$ in one dimension, confined in a potential that vanishes for $-a \leq x \leq a$, and becomes infinite at $x = \pm a$, so that the wave function must vanish at $x = \pm a$.

   - Find the energy values of states with definite energy, and the corresponding normalized wave functions
   - Suppose that the particle is placed in a state with a wave function proportional to $a^2 - x^2$. If the energy of the particle is measured, what is the probability that the particle will be found in the state of lowest energy?

2. Consider a non-relativistic particle of mass $M$ in three dimensions, described by a Hamiltonian

$$H = \frac{\mathbf{P}^2}{2M} + \frac{M\omega_0^2}{2}\mathbf{X}^2.$$

- Find the energy values of states with definite energy, and the number of states for each energy.
- Suppose the particle has charge $e$. Find the rate at which a state of next-to-lowest energy decays by photon emission into the state of lowest energy.

Hint: you can express the Hamiltonian as a sum of three Hamiltonians for one-dimensional oscillators, and use the results given in Section 1.4 for the energy levels and $x$-matrix elements for one-dimensional oscillators.

3. Suppose the photon had three polarization states rather than two. What difference would that make in the relations between Einstein's $A$ and $B$ coefficients?

# 2

# Particle States in a Central Potential

Before going on to lay out the general principles of quantum mechanics in the next chapter, we will first in this chapter illustrate the meaning of the Schrödinger equation by solving some important physical problems by the methods of wave mechanics. To start, we will consider a single particle moving in three space dimensions under the influence of a general central potential. Later we will specialize to the case of a Coulomb potential, and work out the spectrum of hydrogen. One other classic problem, the harmonic oscillator, will be treated at the end of this chapter.

## 2.1 Schrödinger Equation for a Central Potential

We consider a particle of mass[1] $\mu$ moving in a *central potential* $V(r)$, which depends only on $r \equiv \sqrt{\mathbf{x}^2}$. The Hamiltonian in this case is[2]

$$H = \frac{\mathbf{p}^2}{2\mu} + V(r) = -\frac{\hbar^2}{2\mu}\nabla^2 + V(r), \qquad (2.1.1)$$

where $\nabla^2$ is the Laplacian operator

$$\nabla^2 \equiv \frac{\partial^2}{\partial x_1^2} + \frac{\partial^2}{\partial x_2^2} + \frac{\partial^2}{\partial x_3^2}. \qquad (2.1.2)$$

The Schrödinger equation for a wave function $\psi(\mathbf{x})$ representing a state of definite energy $E$ is then

---

[1] We are using $\mu$ for the mass here to avoid confusion with an index $m$ that is conventionally used in describing the angular dependence of the wave function. We will see in Section 2.4 that the same Schrödinger equation applies to a problem of two particles with masses $m_1$ and $m_2$, with a potential that depends only on the particle separation, if $\mu$ is taken as the reduced mass $m_1 m_2 / (m_1 + m_2)$.

[2] In this chapter, and in most of the following chapters, we will be using $\mathbf{x}$ both as the argument of the wave function (with $r \equiv |\mathbf{x}|$) and as the operator that multiplies the wave function by its argument, denoted $\mathbf{X}$ in the previous chapter. The context should make it clear which is meant. Also, here $\mathbf{p}$ is the operator $-i\hbar\nabla$, denoted $\mathbf{P}$ in the previous chapter.

$$E\psi = H\psi = -\frac{\hbar^2}{2\mu}\nabla^2\psi + V(r)\psi. \tag{2.1.3}$$

Like any wave function for a state of definite energy $E$, this $\psi(\mathbf{x})$ will have a simple time-dependence contained in a factor $\exp(-iEt/\hbar)$, which we will not generally show explicitly.

It is a good idea when confronted with a problem like this to consider what observables along with the energy may be used to characterize physical states. As explained in Section 1.5, these are operators that commute with the Hamiltonian. One such observable is the angular momentum $\mathbf{L} = \mathbf{x} \times \mathbf{p}$. Making the usual substitution of $\mathbf{p}$ with $-i\hbar\nabla$, this suggests that in quantum mechanics we should define an angular momentum operator

$$\mathbf{L} \equiv -i\hbar\mathbf{x} \times \nabla, \tag{2.1.4}$$

where $\mathbf{x}$ is the operator (called $\mathbf{X}$ in Chapter 1) that multiplies a wave function with its argument. Written in terms of Cartesian components, this operator is

$$L_i = -i\hbar \sum_{jk} \epsilon_{ijk} x_j \frac{\partial}{\partial x_k}, \tag{2.1.5}$$

where $i$, $j$, $k$ each run over the three directions 1, 2, 3, and $\epsilon$ is a totally antisymmetric coefficient, defined by

$$\epsilon_{ijk} \equiv \begin{cases} +1, & i, j, k \text{ even permutation of 1, 2, 3,} \\ -1, & i, j, k \text{ odd permutation of 1, 2, 3,} \\ 0, & \text{otherwise.} \end{cases} \tag{2.1.6}$$

To show that $\mathbf{L}$ commutes with the Hamiltonian, first consider the commutator of $L_i$ with either $x_j$ or $\partial/\partial x_j$. Recall that

$$\frac{\partial}{\partial x_k}(x_j\psi) - x_j\frac{\partial}{\partial x_k}\psi = \delta_{jk}\psi,$$

so

$$\left[\frac{\partial}{\partial x_k}, x_j\right] = \delta_{kj}. \tag{2.1.7}$$

Since the components of $\mathbf{x}$ commute with each other, by changing $j$ in Eq. (2.1.5) with a running index $m$ we find

$$[L_i, x_j] = -i\hbar \sum_m \epsilon_{imj} x_m = +i\hbar \sum_k \epsilon_{ijk} x_k. \tag{2.1.8}$$

To evaluate the commutator of $\mathbf{L}$ with the gradient operator, we need only rewrite Eq. (2.1.7) as

$$\left[x_m, \frac{\partial}{\partial x_j}\right] = -\delta_{jm}$$

so that, since the components of the gradient commute with each other,

$$\left[ L_i, \frac{\partial}{\partial x_j} \right] = +i\hbar \sum_k \epsilon_{ijk} \frac{\partial}{\partial x_k}. \tag{2.1.9}$$

Both Eqs. (2.1.8) and (2.1.9) can be written in the form

$$[L_i, v_j] = i\hbar \sum_k \epsilon_k v_k, \tag{2.1.10}$$

where $v_i$ is either $x_i$ or $\partial/\partial x_i$. It can be shown that Eq. (2.1.10) is true of any vector $\mathbf{v}$ that is constructed from $\mathbf{x}$ or $\nabla$. In particular, it is true of $\mathbf{L}$ itself:

$$[L_i, L_j] = i\hbar \sum_k \epsilon_{ijk} L_k. \tag{2.1.11}$$

This is obviously the case if $i$ and $j$ are equal, because $\epsilon_{ijk}$ vanishes if any two of its indices are equal. To check Eq. (2.1.11) when $i$ and $j$ are not equal, consider the case $i = 1$ and $j = 2$. Here

$$\begin{aligned}
[L_1, L_2] &= -i\hbar \left[ L_1, \left( x_3 \frac{\partial}{\partial x_1} - x_1 \frac{\partial}{\partial x_3} \right) \right] \\
&= -i\hbar \left( -i\hbar x_2 \frac{\partial}{\partial x_1} + i\hbar x_1 \frac{\partial}{\partial x_2} \right) \\
&= i\hbar L_3 = i\hbar \sum_k \epsilon_{12k} L_k,
\end{aligned}$$

and likewise for $[L_2, L_3]$ and $[L_3, L_1]$.

To show that the $L_i$ commute with the Hamiltonian, we note that if $v_i$ is any vector satisfying Eq. (2.1.10), we have

$$[L_i, \mathbf{v}^2] = \sum_j [L_i, v_j] v_j + \sum_j v_j [L_i, v_j] = i\hbar \sum_{jk} \epsilon_{ijk} (v_k v_j + v_k v_j),$$

so, because $\epsilon_{ijk}$ is antisymmetric in $j$ and $k$,

$$[L_i, \mathbf{v}^2] = 0. \tag{2.1.12}$$

(Note that this works even if the components of $\mathbf{v}$ do not commute with each other, as will be the case for some vector operators other than the position and gradient vectors.) In particular, $L_i$ commutes with $\mathbf{x}^2$, and therefore with any function of $r \equiv [\mathbf{x}^2]^{1/2}$, and it commutes with the Laplacian $\nabla^2$, so it commutes with the Hamiltonian (2.1.1). It is the rotational symmetry of the Hamiltonian that ensures that it commutes with $\mathbf{L}$; if the Hamiltonian depended on the direction of $\mathbf{x}$ or $\mathbf{p}$ instead of just their magnitudes, it would not commute with $\mathbf{L}$.

Because $L_j$ is itself a vector $v_j$ that satisfies Eq. (2.1.10), it also follows that $L_i$ commutes with $\mathbf{L}^2$. Furthermore, since $L_i$ commutes with the Hamiltonian, so does $\mathbf{L}^2$. Therefore we can characterize physical states by the eigenvalues of the operators $H$, of $\mathbf{L}^2$, and of any one component of $\mathbf{L}$, all of which operators commute with each other. Note that we can only do this for *one* component of $\mathbf{L}$, because according to Eq. (2.1.11) the three different components do not commute with each other. It is conventional to choose this component as $L_3$, so physical wave functions will be characterized by the eigenvalues of $H$, $\mathbf{L}^2$, and $L_3$.

Since each $L_i$ commutes with $r$, it must act only on the direction of the argument $\mathbf{x}$, not its length. That is, in polar coordinates defined by

$$x_1 = r \sin\theta \, \cos\phi, \quad x_2 = r \sin\theta \, \sin\phi, \quad x_3 = r \cos\theta, \qquad (2.1.13)$$

the operators $L_i$ act only on $\theta$ and $\phi$. From the definition (2.1.5) of these operators, we can work out their explicit form in polar coordinates:

$$L_1 = i\hbar \left( \sin\phi \, \frac{\partial}{\partial\theta} + \cot\theta \, \cos\phi \, \frac{\partial}{\partial\phi} \right)$$

$$L_2 = i\hbar \left( -\cos\phi \, \frac{\partial}{\partial\theta} + \cot\theta \, \sin\phi \, \frac{\partial}{\partial\phi} \right) \qquad (2.1.14)$$

$$L_3 = -i\hbar \frac{\partial}{\partial\phi}.$$

Also, in polar coordinates,

$$\mathbf{L}^2 = -\hbar^2 \left[ \frac{1}{\sin\theta} \frac{\partial}{\partial\theta} \left( \sin\theta \, \frac{\partial}{\partial\theta} \right) + \frac{1}{\sin^2\theta} \frac{\partial^2}{\partial\phi^2} \right]. \qquad (2.1.15)$$

As an example of how these are derived, let us calculate $L_3$, which will be of special importance for us. Note that

$$\frac{\partial}{\partial\phi} = \sum_i \frac{\partial x_i}{\partial\phi} \frac{\partial}{\partial x_i}$$

$$= -r \sin\theta \, \sin\phi \, \frac{\partial}{\partial x_1} + r \sin\theta \, \cos\phi \, \frac{\partial}{\partial x_2} = -x_2 \frac{\partial}{\partial x_1} + x_1 \frac{\partial}{\partial x_2}$$

$$= \frac{i}{\hbar} L_3,$$

justifying the formula in (2.1.14) for $L_3$.

It should be noted that each component of $\mathbf{L}$ is an Hermitian operator, because $x_j$ and $p_k$ are Hermitian operators, and commute with each other for $j \neq k$. This is a special case of a general rule: if $A$ and $B$ are Hermitian and commute, then

$$\int \psi^*(AB\psi) = \int (A\psi)^* B\psi = \int (BA\psi)^* \psi = \int (AB\psi)^* \psi,$$

so $AB$ is Hermitian. Also, since each component of $\mathbf{L}$ is Hermitian and commutes with itself, its square is Hermitian, and so their sum $\mathbf{L}^2$ is Hermitian.

What does this have to do with the Schrödinger equation? To see this, let's calculate the operator $\mathbf{L}^2$ in a different way. According to Eq. (2.1.5), this is

$$\mathbf{L}^2 = \sum_i L_i L_i = -\hbar^2 \sum_{ijklm} \epsilon_{ijk}\epsilon_{ilm} x_j \left(\frac{\partial}{\partial x_k}\right) x_l \left(\frac{\partial}{\partial x_m}\right).$$

The sum over $i$ gives

$$\sum_i \epsilon_{ijk}\epsilon_{ilm} = \delta_{jl}\delta_{km} - \delta_{jm}\delta_{kl}.$$

(This holds because for each $i$, $\epsilon_{ijk}$ will vanish unless $j$ and $k$ are the two directions other than $i$, and $\epsilon_{ilm}$ will vanish unless $l$ and $m$ are the two directions other than $i$, so the product $\epsilon_{ijk}\epsilon_{ilm}$ vanishes unless either $j = l$ and $k = m$, or $j = m$ and $k = \ell$. In the first case we have the product of two $\epsilon$s with indices in the same order, which gives $+1$, and in the second case we have the product of two $\epsilon$s differing by a permutation of the second and third indices, which gives $-1$.) Thus

$$\mathbf{L}^2 = -\hbar^2 \sum_{jk} \left[ x_j \left(\frac{\partial}{\partial x_k}\right) x_j \left(\frac{\partial}{\partial x_k}\right) - x_j \left(\frac{\partial}{\partial x_k}\right) x_k \left(\frac{\partial}{\partial x_j}\right) \right].$$

(As usual in these operator expressions, the partial derivatives here act on everything to the right, including whatever function $\mathbf{L}^2$ acts on.) Moving the second $x_j$ in the first term in square brackets to the left and using the commutation relation (2.1.7) gives

$$\sum_{jk} x_j \left(\frac{\partial}{\partial x_k}\right) x_j \left(\frac{\partial}{\partial x_k}\right) = r^2 \nabla^2 + \sum_j x_j \left(\frac{\partial}{\partial x_j}\right).$$

In the same way, interchanging the $x_j$ and $x_k$ in the second term and using the same commutation relation gives

$$\sum_{jk} x_j \left(\frac{\partial}{\partial x_k}\right) x_k \left(\frac{\partial}{\partial x_j}\right) = \sum_{jk} x_k \left(\frac{\partial}{\partial x_k}\right) x_j \left(\frac{\partial}{\partial x_j}\right) + 3 \sum_j x_j \left(\frac{\partial}{\partial x_j}\right)$$
$$- \sum_j x_j \left(\frac{\partial}{\partial x_j}\right).$$

Putting this together and recalling that $\sum_j x_j \, \partial/\partial x_j = r \, \partial/\partial r$, we have

$$\mathbf{L}^2 = -\hbar^2 \left[ r^2 \nabla^2 - r\frac{\partial}{\partial r} r \frac{\partial}{\partial r} - r\frac{\partial}{\partial r} \right] = -\hbar^2 \left[ r^2 \nabla^2 - \frac{\partial}{\partial r} r^2 \frac{\partial}{\partial r} \right],$$

or in other words

$$\nabla^2 = \frac{1}{r^2}\frac{\partial}{\partial r} r^2 \frac{\partial}{\partial r} - \frac{\mathbf{L}^2}{\hbar^2 r^2}. \qquad (2.1.16)$$

The Schrödinger equation (2.1.3) then takes the form

$$E\psi(\mathbf{x}) = -\frac{\hbar^2}{2\mu r^2}\frac{\partial}{\partial r}\left(r^2\frac{\partial\psi(\mathbf{x})}{\partial r}\right) + \frac{1}{2\mu r^2}\mathbf{L}^2\psi(\mathbf{x}) + V(r)\psi(\mathbf{x}). \quad (2.1.17)$$

Now let us consider the spectrum of the operator $\mathbf{L}^2$. As long as $V(r)$ is not extremely singular at $r = 0$, the wave function $\psi$ must be a smooth function of the Cartesian components $x_i$ near $\mathbf{x} = 0$, in the sense that it can be expressed as a power series in these components. Suppose that, for some specific wave function, the terms in this power series with the smallest total number of factors of $x_1$, $x_2$, and $x_3$ have $\ell$ such factors. Here $\ell$ can be 0, 1, 2, etc. The sum of all these terms forms what is called a homogeneous polynomial of order $\ell$ in $\mathbf{x}$. (For instance, a homogeneous polynomial of order 0 is a constant; a homogeneous polynomial of order 1 is a linear combination of $x_1$, $x_2$, and $x_3$; a homogeneous polynomial of order 2 is a linear combination of $x_1^2$, $x_2^2$, $x_3^2$, $x_1 x_2$, $x_2 x_3$, and $x_3 x_1$; and so on.) When written in polar coordinates, a homogeneous polynomial of order $\ell$ is $r^\ell$ times a function of $\theta$ and $\phi$. Thus in the limit $r \to 0$, $\psi(\mathbf{x})$ will take the form

$$\psi(\mathbf{x}) \to r^\ell Y(\theta, \phi), \quad (2.1.18)$$

with $Y(\theta, \phi)$ a homogeneous polynomial of order $\ell$ in the unit vector

$$\hat{x} \equiv \mathbf{x}/r = (\sin\theta \cos\phi, \ \sin\theta \sin\phi, \ \cos\theta). \quad (2.1.19)$$

Equation (2.1.17) may be written

$$\mathbf{L}^2\psi(\mathbf{x}) = \hbar^2\frac{\partial}{\partial r}\left(r^2\frac{\partial\psi(\mathbf{x})}{\partial r}\right) + 2\mu r^2\Big[E - V(r)\Big]\psi(\mathbf{x}).$$

In the limit $r \to 0$ the first term on the right-hand side is $\hbar^2\ell(\ell+1)\psi$ while as long as the potential is less singular than $1/r^2$ the second term on the right-hand side vanishes as $r \to 0$ more rapidly than $\psi$, so Eq. (2.1.19) requires, for $r \to 0$, that $\psi$ satisfy the eigenvalue equation

$$\mathbf{L}^2\psi \to \hbar^2\ell(\ell+1)\psi. \quad (2.1.20)$$

Hence, if $\psi$ is an eigenfunction of $\mathbf{L}^2$ and $H$, the eigenvalue of $\mathbf{L}^2$ can only be $\hbar^2\ell(\ell+1)$, with $\ell \geq 0$ an integer. We will give a much more general derivation of this result in Section 4.2.

If we choose the wave functions (as we can) to be eigenfunctions of $\mathbf{L}^2$ as well as of $H$, then according to Eq. (2.1.20) the eigenvalues can only be $\hbar^2\ell(\ell+1)$, so Eq. (2.1.20) must apply not only for $r \to 0$, but for all $r$. Since $\mathbf{L}^2$ acts only on angles, such a wave function must be proportional to a function only of angles, with a coefficient of proportionality $R$ that can depend only on $r$. That is, for all $r$,

$$\psi(\mathbf{x}) = R(r)\,Y(\theta, \phi), \quad (2.1.21)$$

where $R(r)$ is a function of $r$ satisfying

$$R(r) \propto r^{\ell} \quad \text{for } r \to 0 \tag{2.1.22}$$

and $Y(\theta, \phi)$ is a function of $\theta$ and $\phi$ satisfying

$$\mathbf{L}^2 Y = \hbar^2 \ell(\ell + 1) Y. \tag{2.1.23}$$

If we also require $\psi$ to be an eigenfunction of $L_3$ with eigenvalue denoted $\hbar m$, then

$$L_3 Y = \hbar m \, Y. \tag{2.1.24}$$

Equation (2.1.14) shows that $Y(\theta, \phi)$ must then have a $\phi$-dependence

$$Y(\theta, \phi) = e^{im\phi} \times \text{function of } \theta. \tag{2.1.25}$$

The condition that $Y(\theta, \phi)$ must have the same value at $\phi = 0$ and $\phi = 2\pi$ requires that $m$ be an integer. We will see in the next section that $|m| \leq \ell$.

Using Eq. (2.1.21) in Eq. (2.1.17), the Schrödinger equation becomes an ordinary differential equation[3] for $R(r)$:

$$E \, R(r) = -\frac{\hbar^2}{2\mu r^2} \frac{d}{dr} \left( r^2 \frac{d R(r)}{dr} \right) + \frac{\hbar^2 \ell(\ell + 1)}{2\mu r^2} R(r) + V(r) R(r). \tag{2.1.26}$$

To these conditions we must add the requirement that $R(r)$ vanishes sufficiently rapidly as $r \to \infty$ that $\int |\psi|^2 \, d^3x$ converges, and hence

$$\int_0^{\infty} |R(r)|^2 r^2 \, dr < \infty. \tag{2.1.27}$$

For a potential that approaches the value zero sufficiently rapidly for $r \to \infty$, the general solution of Eq. (2.1.26) for $E < 0$ will be a linear combination of an exponentially growing and an exponentially decaying solution, and Eq. (2.1.27) requires that we choose the exponentially decaying solution.

Equation (2.1.26) can be made to look more like the Schrödinger equation in one dimension by defining a new radial wave function

$$u(r) \equiv r R(r). \tag{2.1.28}$$

Multiplying Eq. (2.1.26) with $r$, the Schrödinger equation then takes the form

$$-\frac{\hbar^2}{2\mu} \frac{d^2 u(r)}{dr^2} + \left[ V(r) + \frac{\ell(\ell + 1)\hbar^2}{2\mu r^2} \right] u(r) = E \, u(r), \tag{2.1.29}$$

---

[3] Often in attempting to solve a partial differential equation like the Schrödinger equation (2.1.3), one tries a solution that factorizes into functions, each function depending on some subset of the coordinates, as in Eq. (2.1.21). The treatment of the Schrödinger equation presented here shows that the success of this procedure follows from the rotational symmetry of the equation to be solved. This is the general rule: factorizable solutions of partial differential equations can generally be found if the equations are subject to suitable symmetry conditions.

with the normalization condition

$$\int_0^\infty |u(r)|^2 \, dr < \infty. \qquad (2.1.30)$$

This is almost the same as the one-dimensional Schrödinger equation, but with two important differences. One is the extra term $\ell(\ell + 1)\hbar^2/2\mu r^2$ added to the potential, which may be understood as the effect of centrifugal forces. The other is the presence of a boundary at $r = 0$, where $u(r)$ is required to go as $r^{\ell+1}$.

## 2.2 Spherical Harmonics

As already remarked in the previous section, we use the eigenvalue of $L_3$ as well as the eigenvalues of $H$ and $\mathbf{L}^2$ to classify the wave functions of definite energy. The angular part of the wave function will therefore be labeled with $\ell$ and $m$, as $Y_\ell^m(\theta, \phi)$, with

$$\mathbf{L}^2 Y_\ell^m = \hbar^2 \ell(\ell + 1) Y_\ell^m, \qquad (2.2.1)$$

and

$$L_3 Y_\ell^m = \hbar m \, Y_\ell^m. \qquad (2.2.2)$$

We will now consider what values of $m$ are allowed for a given $\ell$, and show how to calculate the $Y_\ell^m$.

We can rewrite the eigenvalue condition (2.2.1) in a more convenient form, by using expression (2.1.16) for the Laplacian. Acting on $r^\ell Y_\ell^m$, the first term on the right-hand side of Eq. (2.1.16) is $\ell(\ell + 1)r^{\ell-2}Y_\ell^m$, which according to Eq. (2.2.1) is canceled by the second term, so

$$\nabla^2 \left( r^\ell Y_\ell^m \right) = 0. \qquad (2.2.3)$$

Finally, recall that $r^\ell Y_\ell^m(\theta, \phi)$ is a homogeneous polynomial of order $\ell$ in the Cartesian components of the coordinate vector $\mathbf{x}$. Equivalently, it can be written as a homogeneous polynomial of order $\ell$ in[4]

$$x_\pm \equiv x_1 \pm i x_2 = r \sin\theta \, e^{\pm i\phi} \text{ and } x_3 = r \cos\theta. \qquad (2.2.4)$$

Thus Eq. (2.2.2) tells us that $Y_\ell^m$ must contain numbers $\nu_\pm$ of factors of $x_\pm$ such that

$$m = \nu_+ - \nu_-. \qquad (2.2.5)$$

Since the total number of factors of $x_+$, $x_-$, and $x_3$ is $\ell$, the index $m$ is a positive or negative integer, with a maximum value $\ell$, reached when $\nu_+ = \ell$ and $\nu_- = 0$, and a minimum value $-\ell$, reached when $\nu_- = \ell$ and $\nu_+ = 0$. In Section 4.2 we

---

[4] We sometimes write spherical harmonics as functions of the unit vector $\hat{x} \equiv \mathbf{x}/r$ rather than of $\theta$ and $\phi$, the two sets of variables being related by Eq. (2.2.4).

will see how to use the commutation relations (2.1.11) to give a purely algebraic derivation of this result for the spectrum of $L_3$, and also of Eq. (2.2.1) for the spectrum of $\mathbf{L}^2$.

We must now ask whether $Y_\ell^m$ is uniquely determined (of course, up to a constant factor) by the values of $\ell$ and $m$. For a given $\ell$, the index $m$ can have any integer value from $m = -\ell$ to $m = +\ell$, so it takes $2\ell + 1$ values. On the other hand, a homogeneous polynomial of order $\ell$ in $x_\pm$ and $x_3$ is a linear combination of terms that contain $\nu_+$ factors of $x_+$, with $0 \le \nu_+ \le \ell$, plus $\nu_-$ factors of $x_-$, with $0 \le \nu_- \le \ell - \nu_+$, plus $\ell - \nu_+ - \nu_-$ factors of $x_3$, so the total number of independent homogeneous polynomials of order $\ell$ in these three coordinates is

$$N_\ell = \sum_{\nu_+=0}^{\ell} \sum_{\nu_-=0}^{\ell-\nu_+} 1 = \sum_{\nu_+=0}^{\ell} (\ell - \nu_+ + 1) = \frac{1}{2}(\ell+1)(\ell+2). \tag{2.2.6}$$

The Laplacian of a homogeneous polynomial of order $\ell$ is a homogeneous polynomial of order $\ell - 2$, so Eq. (2.2.3) imposes $N_{\ell-2}$ independent conditions on $Y$, and therefore the number of independent $Y$s for a given $\ell$ is

$$N_\ell - N_{\ell-2} = 2\ell + 1. \tag{2.2.7}$$

Since this is also the number of values taken by $m$ for a given $\ell$, we conclude that there is just one independent polynomial for each $\ell$ and $m$. These functions, denoted $Y_\ell^m(\theta, \phi)$, with $-\ell \le m \le +\ell$, are known as *spherical harmonics*. These functions may be written

$$Y_\ell^m(\theta, \phi) \propto P_\ell^{|m|}(\theta) e^{im\phi}, \tag{2.2.8}$$

with $P_\ell^{|m|}$ satisfying the differential equation (see Eq. (2.1.15))

$$-\frac{1}{\sin\theta} \frac{d}{d\theta} \left( \sin\theta \frac{d P_\ell^{|m|}}{d\theta} \right) + \frac{m^2}{\sin^2\theta} P_\ell^{|m|} = \ell(\ell+1) P_\ell^{|m|}. \tag{2.2.9}$$

The solutions of this equation are known as *associated Legendre functions*. They are polynomials in $\cos\theta$ and $\sin\theta$.

By simply enumerating all the independent homogeneous polynomials in $\mathbf{x}$ of order 0, 1, and 2, and imposing the condition $\nabla^2(r^\ell Y) = 0$, we easily see that the spherical harmonics for $\ell \le 2$ are

$$Y_0^0 = \sqrt{\frac{1}{4\pi}},$$

$$Y_1^1 = -\sqrt{\frac{3}{8\pi}} (\hat{x}_1 + i\hat{x}_2) = -\sqrt{\frac{3}{8\pi}} \sin\theta \, e^{i\phi},$$

$$Y_1^0 = \sqrt{\frac{3}{4\pi}} \hat{x}_3 = \sqrt{\frac{3}{4\pi}} \cos\theta,$$

$$Y_1^{-1} = \sqrt{\frac{3}{8\pi}} \left( \hat{x}_1 - i\hat{x}_2 \right) = \sqrt{\frac{3}{8\pi}} \sin\theta \, e^{-i\phi},$$

$$Y_2^2 = \sqrt{\frac{15}{32\pi}} \left( \hat{x}_1 + i\hat{x}_2 \right)^2 = \sqrt{\frac{15}{32\pi}} (\sin\theta)^2 e^{2i\phi},$$

$$Y_2^1 = -\sqrt{\frac{15}{8\pi}} \left( \hat{x}_1 + i\hat{x}_2 \right)\hat{x}_3 = -\sqrt{\frac{15}{8\pi}} \sin\theta \, \cos\theta \, e^{i\phi},$$

$$Y_2^0 = \sqrt{\frac{5}{16\pi}} \left( 2\hat{x}_3^2 - \hat{x}_1^2 - \hat{x}_2^2 \right) = \sqrt{\frac{5}{16\pi}} \left( 3(\cos\theta)^2 - 1 \right),$$

$$Y_2^{-1} = \sqrt{\frac{15}{8\pi}} \left( \hat{x}_1 - i\hat{x}_2 \right)\hat{x}_3 = \sqrt{\frac{15}{8\pi}} \sin\theta \, \cos\theta \, e^{-i\phi},$$

$$Y_2^{-2} = \sqrt{\frac{15}{32\pi}} \left( \hat{x}_1 - i\hat{x}_2 \right)^2 = \sqrt{\frac{15}{32\pi}} (\sin\theta)^2 e^{-2i\phi}.$$

For instance, $Y_0^0$ and each $Y_1^m$ contain respectively zero and one factor of $\hat{x}_\pm$ or $\hat{x}_3$, so $Y_0^0$ must be a constant, and $Y_1^{+1}$, $Y_1^0$, and $Y_1^{-1}$ must be proportional to $\hat{x}_+$, $\hat{x}_3$, and $\hat{x}_-$ respectively in order to have the right dependence on $\phi$. Similarly, each $Y_2^m$ contains just two factors of $\hat{x}_\pm$ and/or $\hat{x}_3$, so in order to have the right dependence on $\phi$, $Y_2^{\pm 2}$ must be proportional to $\hat{x}_\pm^2$ and $Y_2^{\pm 1}$ must be proportional to $\hat{x}_\pm \hat{x}_3$. The case of $Y_2^0$ is a little more complicated, for both $\hat{x}_+\hat{x}_-$ and $\hat{x}_3^2$ have the right dependence on $\phi$. If we take $Y_2^0$ to be equal to $A\hat{x}_+\hat{x}_- + B\hat{x}_3^2$, then $r^2 Y_2^0$ is equal to $Ax_+x_- + Bx_3^2 = A(x_1^2 + x_2^2) + Bx_3^2$, so $\nabla^2(r^2 Y_2^0) = 4A + 2B$, and hence Eq. (2.2.3) requires that $B = -2A$. Thus $Y_2^0$ is proportional to $\hat{x}_+\hat{x}_- - 2\hat{x}_3^2 = 1 - 3\cos^2\theta$. The numerical factors are chosen here so that the $Y$s are normalized

$$\int d^2\Omega \left| Y_\ell^m(\theta, \phi) \right|^2 \equiv \int_0^\pi \sin\theta \, d\theta \int_0^{2\pi} d\phi \left| Y_\ell^m(\theta, \phi) \right|^2 = 1, \qquad (2.2.10)$$

where $d^2\Omega$ is the solid angle differential $\sin\theta \, d\theta \, d\phi$. This leaves only the phases arbitrary. The reason for the phases chosen here will be made clear when we come to the general theory of angular momentum in Chapter 4.

The spherical harmonics for different $\ell$s and/or $m$s are orthogonal, because they are eigenfunctions of the Hermitian operators $\mathbf{L}^2$ and $L_3$ with different eigenvalues. To check the orthogonality, note first that

$$\int d^2\Omega \, Y_\ell^m(\theta, \phi)^* Y_{\ell'}^{m'}(\theta, \phi) \propto \int_0^{2\pi} \exp(i(m' - m)\phi) \, d\phi \propto \delta_{m'm}. \qquad (2.2.11)$$

Next, considering the case $m' = m$,

$$\int d^2\Omega \, Y_\ell^m(\theta, \phi)^* Y_{\ell'}^m(\theta, \phi) \propto \int_0^\pi P_\ell^{|m|}(\theta) P_{\ell'}^{|m|}(\theta) \sin\theta \, d\theta. \qquad (2.2.12)$$

Multiplying Eq. (2.2.9) with $P_{\ell'}^{|m|}(\theta) \sin \theta$ and subtracting the same expression with $\ell$ and $\ell'$ interchanged gives

$$\left[ \ell(\ell+1) - \ell'(\ell'+1) \right] P_{\ell'}^{|m|}(\theta) P_{\ell}^{|m|}(\theta) \sin \theta$$

$$= \frac{d}{d\theta} \left[ \sin \theta \, P_{\ell}^{|m|}(\theta) \frac{d}{d\theta} P_{\ell'}^{|m|}(\theta) - \sin \theta \, P_{\ell'}^{|m|}(\theta) \frac{d}{d\theta} P_{\ell}^{|m|}(\theta) \right]. \quad (2.2.13)$$

The quantity in square brackets on the right-hand side vanishes at $\theta = 0$ and $\theta = \pi$, so

$$\left[ \ell(\ell+1) - \ell'(\ell'+1) \right] \int_0^\pi P_{\ell'}^{|m|}(\theta) P_{\ell}^{|m|}(\theta) \sin \theta \, d\theta = 0. \quad (2.2.14)$$

It is only possible to have $\ell(\ell+1) = \ell'(\ell'+1)$ with $\ell$ and $\ell'$ positive if $\ell = \ell'$, so

$$\int_0^\pi P_{\ell'}^{|m|}(\theta) P_{\ell}^{|m|}(\theta) \sin \theta \, d\theta = 0 \text{ for } \ell \neq \ell'. \quad (2.2.15)$$

Putting together Eqs. (2.2.10), (2.2.11), and (2.2.15) gives our orthonormality relation

$$\int d^2\Omega \, Y_\ell^m(\theta, \phi)^* Y_{\ell'}^{m'}(\theta, \phi) = \delta_{\ell\ell'} \delta_{mm'}. \quad (2.2.16)$$

We also note the space-inversion (or "parity") property of the wave function. Since the $Y_\ell^m$ are homogeneous polynomials of order $\ell$ in the unit vector $\hat{x}$, it follows that under the transformation $\hat{x} \to -\hat{x}$, the spherical harmonics change by just a sign factor $(-1)^\ell$:

$$Y_\ell^m(\pi - \theta, \pi + \phi) = (-1)^\ell Y_\ell^m(\theta, \phi). \quad (2.2.17)$$

The spherical harmonics for $m = 0$ are conventionally written in terms of *Legendre polynomials* $P_\ell(\cos \theta)$ as

$$Y_\ell^0(\theta) = \sqrt{\frac{2\ell+1}{4\pi}} P_\ell(\cos \theta). \quad (2.2.18)$$

To see that $Y_\ell^0(\theta)$ is a polynomial in $\cos \theta$, recall that it is a polynomial in the components of the unit vector $\hat{x}$, and since it is invariant under rotations around the 3 axis, it must be a polynomial in $\hat{x}_3 = \cos \theta$ and $\hat{x}_+ \hat{x}_- = \sin^2 \theta = 1 - \cos^2 \theta$. (The numerical factor in Eq. (2.2.18) is chosen so that $P_\ell(1) = 1$.) For instance, referring back to the spherical harmonics listed above, Eq. (2.2.18) gives

$$P_0(\cos \theta) = 1, \quad P_1(\cos \theta) = \cos \theta, \quad P_2(\cos \theta) = \frac{1}{2}\left(3 \cos^2 \theta - 1\right), \quad (2.2.19)$$

and so on.

## 2.3 The Hydrogen Atom

At last we come to a realistic three-dimensional system, consisting of a single electron moving in a Coulomb potential

$$V(r) = -\frac{Ze^2}{r}, \qquad (2.3.1)$$

where $-e$ is the electron charge in unrationalized electrostatic units (for which $e^2/\hbar c \simeq 1/137$). We wish here to solve the Schrödinger equation for bound states, which have energy $E < 0$.

The radial Schrödinger equation (2.1.29) (with $\psi(\mathbf{x}) \propto u(r)Y_\ell^m(\theta, \phi)/r$) is then

$$-\frac{\hbar^2}{2m_e}\frac{d^2u(r)}{dr^2} + \left[-\frac{Ze^2}{r} + \frac{\ell(\ell+1)\hbar^2}{2m_e r^2}\right]u(r) = Eu(r),$$

or in other words

$$-\frac{d^2u(r)}{dr^2} + \left[-\frac{2m_e Ze^2}{r\hbar^2} + \frac{\ell(\ell+1)}{r^2}\right]u(r) = -\kappa^2 u(r), \qquad (2.3.2)$$

where $\kappa$ is defined by

$$E = -\frac{\hbar^2\kappa^2}{2m_e}, \qquad \kappa > 0 \qquad (2.3.3)$$

and $m_e$ is the electron mass. We will write this in dimensionless form by introducing

$$\rho \equiv \kappa r. \qquad (2.3.4)$$

After dividing by $\kappa^2$, Eq. (2.3.2) becomes

$$-\frac{d^2u}{d\rho^2} + \left[-\frac{\xi}{\rho} + \frac{\ell(\ell+1)}{\rho^2}\right]u = -u, \qquad (2.3.5)$$

where

$$\xi \equiv \frac{2m_e Ze^2}{\kappa\hbar^2}. \qquad (2.3.6)$$

We must look for a solution that decreases as $\rho^{\ell+1}$ for $\rho \to 0$, and (more or less) like $\exp(-\rho)$ for $\rho \to \infty$, so let's replace $u$ with a new function $F(\rho)$, defined by

$$u = \rho^{\ell+1}\exp(-\rho)F(\rho). \qquad (2.3.7)$$

Then

$$\frac{du}{d\rho} = \rho^{\ell+1}\exp(-\rho)\left[\left(\frac{\ell+1}{\rho} - 1\right)F + \frac{dF}{d\rho}\right]$$

and

$$\frac{d^2 u}{d\rho^2} = \rho^{\ell+1} \exp(-\rho) \left[ \left( 1 - \frac{2(\ell+1)}{\rho} + \frac{\ell(\ell+1)}{\rho^2} \right) F \right.$$
$$\left. + \left( -2 + \frac{2(\ell+1)}{\rho} \right) \frac{dF}{d\rho} + \frac{d^2 F}{d\rho^2} \right].$$

The radial wave equation (2.3.5) thus becomes

$$\frac{d^2 F}{d\rho^2} - 2 \left( 1 - \frac{\ell+1}{\rho} \right) \frac{dF}{d\rho} + \left( \frac{\xi - 2\ell - 2}{\rho} \right) F = 0. \tag{2.3.8}$$

Let's try a power-series solution

$$F = \sum_{s=0}^{\infty} a_s \rho^s, \tag{2.3.9}$$

where $a_0 \neq 0$, because we define $\ell$ so that $u(r) \propto r^{\ell+1}$ for $r \to 0$. Then Eq. (2.3.8) becomes

$$\sum_{s=0}^{\infty} a_s \left[ s(s-1)\rho^{s-2} - 2s\rho^{s-1} + 2s(\ell+1)\rho^{s-2} + (\xi - 2\ell - 2)\rho^{s-1} \right] = 0. \tag{2.3.10}$$

In order to derive a relation between the coefficients in the power series, let us replace the summation variable $s$ with $s+1$ in all terms that go as $\rho^{s-2}$ rather than $\rho^{s-1}$. (The factors $s$ in the first and third terms in Eq. (2.3.10) make the sums over these terms start with $s = 1$, so after redefining $s$ as $s+1$ all the sums start with $s = 0$.) Equation (2.3.10) then becomes

$$\sum_{s=0}^{\infty} \rho^{s-1} \left[ s(s+1)a_{s+1} - 2sa_s + 2(s+1)(\ell+1)a_{s+1} + (\xi - 2\ell - 2)a_s \right] = 0. \tag{2.3.11}$$

This must hold for all $\rho > 0$, so the coefficient of each power of $\rho$ must vanish, which gives a recursion relation

$$(s + 2\ell + 2)(s+1)a_{s+1} = (-\xi + 2s + 2\ell + 2)a_s. \tag{2.3.12}$$

The quantity $(s + 2\ell + 2)(s+1)$ does not vanish for any $s \geq 0$, so this gives all the coefficients $a_s$ in terms of an arbitrary normalization coefficient $a_0$.

Let us consider the asymptotic behavior of this power series for large $\rho$. Equation (2.3.12) shows that, for $s \to \infty$,

$$a_{s+1}/a_s \to 2/s. \tag{2.3.13}$$

Since all the $a_s$ for large $s$ have the same sign, the asymptotic behavior of the power series is dominated by the high powers of $\rho$, for which Eq. (2.3.12) gives

$$a_s \approx C \, 2^s / (s + B)!, \tag{2.3.14}$$

with unknown constants $C$ and $B$. (If $B$ is not an integer the factorial here is a gamma function, but this makes little difference when $s \gg B$.) Thus we expect that asymptotically

$$F(\rho) \approx C \sum_{s=0}^{\infty} \frac{(2\rho)^s}{(s+B)!} \rightarrow C(2\rho)^{-B} e^{2\rho}. \tag{2.3.15}$$

Aside from constants and powers of $\rho$, the function (2.3.7) generically then goes as

$$u \approx e^{\rho}. \tag{2.3.16}$$

This is no surprise, because for generic values of $\xi$ the solution that goes as $\rho^{\ell+1}$ for $\rho \rightarrow 0$ will approach a linear combination of terms proportional to $e^{\rho}$ or $e^{-\rho}$ for $\rho \rightarrow \infty$, which will be dominated in this limit by the term proportional to $e^{\rho}$. But an asymptotic behavior like Eq. (2.3.16) is clearly inconsistent with the condition (2.1.30) that the wave function be normalizable.

The only way to avoid this is to require that the power series terminates, so that $F(\rho)$ goes as some power of $\rho$, rather than as $e^{2\rho}$. The recursion relation (2.3.12) shows that in order for the series to terminate, it is necessary for $\xi$ to be equal to some positive even integer $2n$ with $n \geq \ell + 1$, in which case the series terminates with power $\rho^{n-\ell-1}$. The functions $F(\rho)$ are then polynomials of order $n - \ell - 1$, known as *Laguerre polynomials*, and conventionally written $L_{n-\ell-1}^{2\ell+1}(2\rho)$. The first few examples (aside from normalization constants) are

$$F = \begin{cases} 1, & \text{for } n = \ell + 1, \\ 1 - \rho/(\ell+1), & \text{for } n = \ell + 2. \end{cases} \tag{2.3.17}$$

Although the wave functions depend on $\ell$ and $n$, the energies only depend on $n$. With $\xi = 2n$, Eq. (2.3.6) gives

$$\kappa_n = \frac{2m_e Z e^2}{\xi \hbar^2} = \frac{1}{na}, \tag{2.3.18}$$

where $a$ is the *Bohr radius*:

$$a = \frac{\hbar^2}{m_e Z e^2} = 0.529177249(24) \times 10^{-8} Z^{-1} \text{ cm}. \tag{2.3.19}$$

Since the radial wave function $R(r) \equiv u(r)/r$ decreases at large distances like $\rho^{n-1} \exp(-\rho) \propto r^{n-1} \exp(-r/na)$, the electron is pretty well localized within a radius $na$. Finally, using Eqs. (2.3.18) and (2.3.19) in Eq. (2.3.3) gives the bound-state energies as

$$E_n = -\frac{\hbar^2 \kappa_n^2}{2m_e} = -\frac{\hbar^2}{2m_e a^2 n^2} = -\frac{m_e Z^2 e^4}{2\hbar^2 n^2} = -\frac{13.6056981(40) Z^2 \text{ eV}}{n^2}. \tag{2.3.20}$$

As we saw in Section 1.2, this is the famous formula guessed at by Bohr in 1913. It is an excellent approximation (neglecting magnetic and relativistic effects) for single-electron atoms, such as hydrogen with $Z = 1$, singly ionized helium with $Z = 2$, doubly ionized lithium with $Z = 3$, and so on. As mentioned in Section 1.2, it is also a fair approximation for the states of the outermost electron in neutral atoms of alkali metals such as lithium, sodium, and potassium, for which the charge $Ze$ of the nucleus is partially shielded by the $Z - 1$ inner electrons, so that $Z$ in Eq. (2.3.20) can be taken as effectively of order unity.

Incidentally, note that the energy required to excite a hydrogen atom in the $n = 1$ state to the $n = 2$ state is 10.2 eV, so to excite hydrogen atoms from the ground state to any higher energy state in atomic collisions requires temperatures of at least about $10\,\text{eV}/k_{\text{B}} \simeq 10^5$ K. Hot gases in astrophysics typically cool by emission of radiation from atoms excited in atomic collisions, so a gas of hot hydrogen finds it very difficult to cool below about $10^5$ K. On the other hand, for reasons discussed in Section 4.5, the outer electrons in heavy atoms all have larger values of $n$, so it takes much less energy to excite these atoms to the next higher state, and even small quantities of heavy elements make a large difference in the cooling rate.

For each $n$ we have $\ell$ values running from 0 to $n - 1$, and for each $\ell$ we have $2\ell + 1$ values of $m$, so the total number of states with energy $E_n$ is

$$\sum_{\ell=0}^{n-1}(2\ell + 1) = 2\frac{n(n-1)}{2} + n = n^2. \tag{2.3.21}$$

We will see in Section 4.5 that this formula plays an essential role in explaining the periodic table. In multi-electron atoms the energies of these states are actually separated from each other by departures of the effective electrostatic potential from a strict proportionality to $1/r$, due to the nucleus and other electrons, as well as by relativistic effects and by magnetic fields within the atom, and may be further split by external fields.

There is a standard nomenclature for these states. In general, one-electron atomic states with $\ell = 0,\ 1,\ 2,\ 3$ are labeled $s$, $p$, $d$, $f$. (The letters stand for "sharp," "principal," "diffuse," etc., for reasons having to do with the appearance of spectral lines.) In hydrogen, or hydrogen-like atoms, this letter is preceded by a number giving the energy level. Thus the lowest energy state of hydrogen is $1s$, the next lowest $2s$ and $2p$, the next lowest $3s$, $3p$, and $3d$, and so on.

As discussed in Section 1.4, in the approximation that the wavelength of light emitted in an atomic transition is much larger than the Bohr radius, the rate at which a state represented by a wave function $\psi$ decays by single-photon emission into a state represented by a wave function $\psi'$ is proportional to $|\int \psi'^*\mathbf{x}\psi|^2$. If we change the variable of integration from $\mathbf{x}$ to $-\mathbf{x}$, then as mentioned in Section 2.2, the wave functions $\psi$ and $\psi'$ change by factors $(-1)^\ell$ and $(-1)^{\ell'}$, and so the whole integrand changes by a factor

$$(-1)^{\ell+\ell'+1}.$$

Thus the transition rate vanishes (in this approximation) unless the signs $(-1)^\ell$ and $(-1)^{\ell'}$ are opposite. (There are other selection rules, which will be described in Section 4.4.) For instance, the $2p$ state can emit a photon and decay into the $1s$ state (this is known as Lyman-$\alpha$ radiation), but the $2s$ state cannot. This selection rule actually helps the recombination of hydrogen ions and electrons in hot gases, such as in the early universe at a temperature of about 3000 K. Emission of a Lyman-$\alpha$ photon may not provide an effective way for hydrogen to reach the lowest-energy state (the "ground state"), because that photon just excites another hydrogen atom in the $1s$ state to the $2p$ state.[5] On the other hand, the $2s$ state can only decay to the $1s$ state by emitting *two* photons, neither of which has enough energy to excite another hydrogen atom from the ground state.

## 2.4 The Two-Body Problem

So far, we have considered the quantum mechanics of a single particle in a fixed potential. Of course, real one-electron atoms consist of two particles, a nucleus and an electron, with a potential that depends on the difference of their coordinate vectors. It is well known in classical mechanics that the latter two-body problem is equivalent to a one-body problem, with the electron mass replaced with a reduced mass:

$$\mu = \frac{m_e m_N}{m_e + m_N}, \tag{2.4.1}$$

where $m_N$ is the nuclear mass. We will now see that the same is true in quantum mechanics.

In both classical and quantum mechanics, the Hamiltonian for a one-electron atom is

$$H = \frac{\mathbf{p}_e^2}{2m_e} + \frac{\mathbf{p}_N^2}{2m_N} + V(\mathbf{x}_e - \mathbf{x}_N), \tag{2.4.2}$$

where $\mathbf{p}_e$ and $\mathbf{p}_N$ are the electron and nuclear momenta. (To a good approximation the potential only depends on $|\mathbf{x}_e - \mathbf{x}_N|$, but for the purposes of the present section it is just as easy to deal with the more general case.) Also, in both classical and quantum mechanics, we introduce a relative coordinate $\mathbf{x}$ and a center-of-mass coordinate $\mathbf{X}$ by

$$\mathbf{x} \equiv \mathbf{x}_e - \mathbf{x}_N, \qquad \mathbf{X} \equiv \frac{m_e \mathbf{x}_e + m_N \mathbf{x}_N}{m_e + m_N}, \tag{2.4.3}$$

---

[5] There is an exception to this. In cosmology, a Lyman-$\alpha$ photon that survives long enough will lose energy through the cosmological expansion, to the point where it can no longer excite a hydrogen atom from the ground state to any higher state. This also contributes to hydrogen recombination.

and a relative momentum $\mathbf{p}$ and a total momentum $\mathbf{P}$ by

$$\mathbf{p} \equiv \mu \left( \frac{\mathbf{p}_e}{m_e} - \frac{\mathbf{p}_N}{m_N} \right), \quad \mathbf{P} \equiv \mathbf{p}_e + \mathbf{p}_N. \tag{2.4.4}$$

It is easy to see then that the Hamiltonian (2.4.2) may be written

$$H = \frac{\mathbf{p}^2}{2\mu} + \frac{\mathbf{P}^2}{2(m_e + m_N)} + V(\mathbf{x}) \tag{2.4.5}$$

and this too is true in both classical and quantum mechanics.

In quantum mechanics we identify the momenta as the operators

$$\mathbf{p}_e = -i\hbar \nabla_e, \quad \mathbf{p}_N = -i\hbar \nabla_N. \tag{2.4.6}$$

It is then elementary to calculate that the momenta (2.4.4) are

$$\mathbf{p} = -i\hbar \nabla_\mathbf{x}, \quad \mathbf{P} = -i\hbar \nabla_\mathbf{X}. \tag{2.4.7}$$

So the momenta (2.4.4) and the coordinates (2.4.3) satisfy the commutation relations

$$[x_i, p_j] = [X_i, P_j] = i\hbar\delta_{ij}, \quad [x_i, P_j] = [X_i, p_j] = 0. \tag{2.4.8}$$

It is obvious then that the Hamiltonian (2.4.2) commutes with all components of $\mathbf{P}$, which also commute with each other, so the wave functions representing physical states of definite energy can also be taken to have definite total momentum.

Such a wave function will have the form

$$\psi(\mathbf{x}, \mathbf{X}) = e^{i\mathbf{P}\cdot\mathbf{X}/\hbar} \psi(\mathbf{x}), \tag{2.4.9}$$

where $\mathbf{P}$ is now a c-number eigenvalue, and $\psi(\mathbf{x})$ is a wave function for an internal energy $\mathcal{E}$, satisfying the one-particle Schrödinger equation

$$-\frac{\hbar^2 \nabla_x^2 \psi(\mathbf{x})}{2\mu} + V(\mathbf{x})\psi(\mathbf{x}) = \mathcal{E}\psi(\mathbf{x}). \tag{2.4.10}$$

For example, in single-electron atoms the internal energy $\mathcal{E}$ is given by Eq. (2.3.20), with $m_e$ replaced with $\mu$. The total energy is just the internal energy $\mathcal{E}$ of the atom, plus the kinetic energy of its overall motion:

$$E = \mathcal{E} + \frac{\mathbf{P}^2}{2(m_e + m_N)}. \tag{2.4.11}$$

The most important aspect of the replacement of the electron mass with the reduced mass (2.4.1) is that internal energies then depend very slightly on the mass of the nucleus. There are two stable isotopes of the hydrogen nucleus, the proton with mass $1836m_e$, and the deuteron with mass $3670m_e$, giving reduced masses

$$\mu_{pe} = 0.99945m_e, \quad \mu_{de} = 0.99973m_e. \tag{2.4.12}$$

This tiny difference is enough to produce a detectable split in the frequencies of light emitted from a mixture of ordinary hydrogen and deuterium. The relative intensity of the observed hydrogen and deuterium spectral lines is used by astronomers to measure the relative abundance of hydrogen and deuterium in the interstellar medium, which in turn reveals conditions in the early universe when a tiny fraction of matter was formed into deuterons. Also, as mentioned in Section 1.2, the experimental confirmation of the predicted differences between the energy levels of different one-electron atoms such as hydrogen and ionized helium helped to confirm the Bohr theory of these atoms.

## 2.5 The Harmonic Oscillator

As a final bound-state problem in three dimensions, let's consider a particle of mass $M$ in a potential

$$V(r) = \frac{1}{2} M \omega^2 r^2, \tag{2.5.1}$$

where $\omega$ is a constant with the dimensions of frequency. Of course, this is not the potential felt by electrons in atoms, but it is worth considering for at least four reasons. One is its historical importance. As we saw in Section 1.4, this is the problem (though in one dimension) studied by Heisenberg in his ground-breaking 1925 paper introducing matrix mechanics. Another reason is that this theory provides a nice illustration of how we can find energy levels and radiative transition amplitudes by algebraic methods (the methods used by Heisenberg), without having to solve second-order differential equations. Third, the harmonic oscillator potential *is* used in models of atomic nuclei, which, as we will see in Section 4.5, lead to the idea of "magic numbers" of neutrons or protons for which nuclei are particularly stable. Finally, the methods described here for dealing with the harmonic oscillator will turn out to be useful in Section 10.3 for dealing with the energy levels of electrons in magnetic fields, and in Sections 11.5 and 11.6 for calculating the properties of photons.

The Schrödinger equation (2.1.3) is here

$$E\psi = -\frac{\hbar^2}{2M} \nabla^2 \psi + \frac{1}{2} M \omega^2 r^2 \psi. \tag{2.5.2}$$

Both the Laplacian and $r^2 = \mathbf{x}^2$ may be written as sums over the three coordinate directions, so that the Schrödinger equation may be written

$$\left( \frac{-\hbar^2}{2M} \frac{\partial^2 \psi}{\partial x_1^2} + \frac{M\omega^2 x_1^2 \psi}{2} \right) + \left( \frac{-\hbar^2}{2M} \frac{\partial^2 \psi}{\partial x_2^2} + \frac{M\omega^2 x_2^2 \psi}{2} \right)$$
$$+ \left( \frac{-\hbar^2}{2M} \frac{\partial^2 \psi}{\partial x_3^2} + \frac{M\omega^2 x_3^2 \psi}{2} \right) = E\psi. \tag{2.5.3}$$

This has separable solutions, of the form

$$\psi(\mathbf{x}) = \psi_{n_1}(x_1)\psi_{n_2}(x_2)\psi_{n_3}(x_3), \tag{2.5.4}$$

where $\psi_n(x)$ is a solution of the one-dimensional Schrödinger equation

$$\frac{-\hbar^2}{2M}\frac{\partial^2\psi_n(x)}{\partial x^2} + \frac{M\omega^2 x^2\psi_n(x)}{2} = E_n\psi_n(x). \tag{2.5.5}$$

The energy is the sum of the energies of three one-dimensional harmonic oscillators in the $n_1$th, $n_2$th and $n_3$th energy states:

$$E = E_{n_1} + E_{n_2} + E_{n_2}. \tag{2.5.6}$$

So our problem has been reduced to the one considered by Heisenberg in 1925, the one-dimensional harmonic oscillator.

To solve this problem, we introduce so-called lowering and raising operators

$$a_i \equiv \frac{1}{\sqrt{2M\hbar\omega}}\left(-i\hbar\frac{\partial}{\partial x_i} - iM\omega x_i\right),$$

$$a_i^\dagger \equiv \frac{1}{\sqrt{2M\hbar\omega}}\left(-i\hbar\frac{\partial}{\partial x_i} + iM\omega x_i\right), \tag{2.5.7}$$

with $i = 1, 2,$ and 3. These operators obey the commutation relations

$$\left[a_i, a_j^\dagger\right] = \delta_{ij} \tag{2.5.8}$$

and

$$\left[a_i, a_j\right] = \left[a_i^\dagger, a_j^\dagger\right] = 0. \tag{2.5.9}$$

Also, the one-dimensional Hamiltonian here is

$$H_i \equiv -\frac{\hbar^2}{2M}\nabla_i^2 + \frac{M\omega^2 x_i^2}{2} = \hbar\omega\left[a_i^\dagger a_i + \frac{1}{2}\right]. \tag{2.5.10}$$

(The summation convention, that repeated indices are summed, is not being used here.) Now, it follows from Eqs. (2.5.8)–(2.5.10) that

$$[H_i, a_i] = -\hbar\omega a_i, \quad [H_i, a_i^\dagger] = +\hbar\omega a_i^\dagger. \tag{2.5.11}$$

Hence if $\psi$ represents a state with energy $E$, then $a_i\psi$ represents a state with energy $E - \hbar\omega$, and $a_i^\dagger\psi$ represents a state with energy $E + \hbar\omega$, provided of course that $a_i\psi$ and $a_i^\dagger\psi$ respectively do not vanish. There is a wave function $\psi_0(x_i)$ for which $a_i\psi_0 = 0$; it is

$$\psi_0(x_i) \propto \exp(-M\omega x_i^2/2\hbar), \tag{2.5.12}$$

so this represents a state for which the energy $E_{n_i}$ is $\hbar\omega/2$, and no wave function representing a state with a lower value of $E_{n_i}$ can be formed by operating on this wave function with $a_i$. On the other hand, there is no wave function $\psi(x_i)$ for which $a_i^\dagger \psi$ vanishes, because the solution of the differential equation $a_i^\dagger \psi = 0$ is $\psi \propto \exp(M\omega x_i^2/2\hbar)$, and this is not normalizable. In consequence, there is no upper bound to the energies of states represented by wave functions formed by operating any number of times with $a_i^\dagger$ on $\psi_0$. These wave functions take the form

$$\psi_{n_i}(x_i) \propto a_i^{\dagger n_i} \psi_0(x_i) \propto H_{n_i}(x_i) \exp(-M\omega x_i^2/2\hbar), \qquad (2.5.13)$$

where $H_n(x)$ is a polynomial of order $n$ in $x$. (It is proportional to the Hermite polynomial $He_n(z)$ of order $n$ and argument $z = x\sqrt{2M\omega/\hbar}$.) For instance, $H_0(x) \propto 1$, $H_1(x) \propto x$, $H_2(x) \propto 1 - 2M\omega x^2/\hbar$, and so on. These polynomials satisfy the parity condition

$$H_n(-x) = (-1)^n H_n(x). \qquad (2.5.14)$$

Using Eq. (2.5.10) and the commutation relations shows that Eq. (2.5.13) is an eigenfunction of $H_i$ with eigenvalue $\hbar\omega(n_i + 1/2)$. The general wave function representing a state of definite energy is therefore

$$\psi_{n_1 n_2 n_3}(\mathbf{x}) \propto a_1^{\dagger n_1} a_2^{\dagger n_2} a_3^{\dagger n_3} \psi_0(r)$$

$$\propto H_{n_1}(x_1) H_{n_2}(x_2) H_{n_3}(x_3) \exp(-M\omega r^2/2\hbar), \qquad (2.5.15)$$

and the state has energy

$$E_{n_1 n_2 n_3} = \hbar\omega \left[ N + \frac{3}{2} \right] \qquad (2.5.16)$$

and the parity property

$$\psi_{n_1 n_2 n_3}(-x) = (-1)^N \psi_{n_1 n_2 n_3}(x), \qquad (2.5.17)$$

where

$$N = n_1 + n_2 + n_3. \qquad (2.5.18)$$

All but the lowest of these energy levels have a great deal of degeneracy. For a fixed value of $N = n_1 + n_2 + n_3$ there is just one possible value of $n_3$ for a given $n_1$ and $n_2$, so the number of ways of writing a positive integer $N$ as the sum of three positive (perhaps zero) integers $n_1$, $n_2$, and $n_3$ is

$$\mathcal{N}_N = \sum_{n_1=0}^{N} \sum_{n_2=0}^{N-n_1} 1 = \sum_{n_1=0}^{N} (N - n_1 + 1) = (N+1)^2 - \frac{N(N+1)}{2}$$

$$= \frac{(N+1)(N+2)}{2}. \qquad (2.5.19)$$

Since the potential (2.5.1) is spherically symmetric, these wave functions can also be written as sums of the spherical harmonics $Y_\ell^m(\theta, \phi)$, times $m$-independent radial wave functions $R_{N\ell}(r)$, with numerical coefficients that may depend on $N$, $\ell$, and $m$. The wave function (2.5.15) is a polynomial of order

$N = n_1 + n_2 + n_3$ in the $x_i$ times a function of $r$, so the maximum value of $\ell$ is $N$. Also, according to Eq. (2.5.17) the wave function (2.5.15) is even or odd in $\mathbf{x}$ depending on whether $N$ is even or odd. Thus this wave function is at most a sum of terms proportional to $Y_\ell^m(\theta, \phi)$, with $\ell = N$, $N - 2$, and so on down to $\ell = 1$ or $\ell = 0$. For instance, $H_1(x) \propto x$, so the three wave functions of the form (2.5.15) with $N = 1$ take the form $x_1 \exp(-M\omega r^2/2\hbar)$, $x_2 \exp(-M\omega r^2/2\hbar)$, and $x_3 \exp(-M\omega r^2/2\hbar)$, which can be written as linear combinations of the $\ell = 1$ terms $rY_1^m(\theta, \phi)\exp(-M\omega r^2/2\hbar)$ with $m = +1$, $m = 0$, and $m = -1$.

It turns out that for higher values of $N$ there *are* independent wave functions proportional to $Y_\ell^m(\theta, \phi)$, with $\ell = N$, $N - 2$, and so on down to $\ell = 1$ or $\ell = 0$, with just the usual $2\ell + 1$ wave functions for each such $\ell$. To check this, note that this gives the total degeneracy as

$$\mathcal{N}_N = \sum_{\ell=N, N-2, \dots} (2\ell + 1). \tag{2.5.20}$$

For instance, if $N$ is even we can set $\ell = 2k$, and find a degeneracy

$$\mathcal{N}_N = \sum_{k=0}^{N/2}(4k + 1) = 4\frac{(N/2)(N/2 + 1)}{2} + N/2 + 1 = \frac{(N + 1)(N + 2)}{2},$$

in agreement with Eq. (2.5.19). The same result holds for $N$ odd.

The degeneracy of the energy eigenstates, and in particular the existence of states with different values of $\ell$ but the same energy, is a peculiar feature of the Coulomb and harmonic oscillator potentials, that is not expected to occur for generic potentials. In both cases this degeneracy arises from the existence of operators that commute with the Hamiltonian, and which therefore when operating on a wave function with definite energy give another wave function with the same energy. Some of these operators do not commute with $\mathbf{L}^2$, and when acting on a wave function with a given orbital angular momentum give a wave function with a different orbital angular momentum, though with the same energy. What these operators are for the Coulomb potential will be explained in Section 4.8. For the harmonic oscillator potential, they are the nine operators $a_j^\dagger a_k$, with $j$ and $k$ running over the coordinate indices 1, 2, 3, which can easily be seen to commute with the three-dimensional Hamiltonian given by the sum of the one-dimensional Hamiltonians (2.5.10):

$$H = \hbar\omega\left[\sum_i a_i^\dagger a_i + \frac{3}{2}\right].$$

As we will see in Section 4.6, the fact that these operators commute with the Hamiltonian is related to a symmetry of this Hamiltonian and of the commutation rules. Incidentally, both for the Coulomb potential and for the harmonic oscillator potential, the existence of operators that commute with the Hamiltonian is also related to the peculiar property that classical orbits in these two potentials form closed curves.

In order to calculate mean values and radiation transition probabilities, it is necessary to construct properly normalized wave functions. This can most easily be done using the raising and lowering operators (2.5.7). First, in order that the ground-state wave function $\psi_0$ for one-dimensional oscillators be normalized, we must take it as

$$\psi_0(x) = \left[\frac{M\omega}{\pi\hbar}\right]^{1/4} \exp(-M\omega x^2/2\hbar), \qquad (2.5.21)$$

so that

$$\int_{-\infty}^{+\infty} |\psi_0(x)|^2 \, dx = 1. \qquad (2.5.22)$$

Also, note that $a_i^\dagger$ is the *adjoint* of the operator $a_i$, in the sense that for any two normalizable functions $f$ and $g$, we have

$$\int_{-\infty}^{+\infty} f^*(x_i) a_i g(x_i) \, dx_i = \int_{-\infty}^{+\infty} \left(a_i^\dagger f(x_i)\right)^* g(x_i) \, dx_i. \qquad (2.5.23)$$

It follows that

$$\int_{-\infty}^{\infty} |a_i^{\dagger n_i} \psi_0(x_i)|^2 \, dx_i = \int_{-\infty}^{\infty} \left(a_i^{\dagger(n_i-1)} \psi_0(x_i)\right)^* a_i a_i^{\dagger n_i} \psi_0(x_i) \, dx_i.$$

The commutation relations (2.5.8) and (2.5.9) give

$$a_i a_i^{\dagger n_i} = a_i^{\dagger n_i} a_i + n_i a_i^{\dagger(n_i-1)},$$

and since $a_i$ annihilates $\psi_0(x_i)$, we have

$$\int_{-\infty}^{\infty} \left|a_i^{\dagger n_i} \psi_0(x_i)\right|^2 \, dx_i = n_i \int_{-\infty}^{\infty} \left|a_i^{\dagger(n_i-1)} \psi_0(x_i)\right|^2,$$

and so

$$\int_{-\infty}^{\infty} \left|a_i^{\dagger n_i} \psi_0(x_i)\right|^2 \, dx_i = n_i!. \qquad (2.5.24)$$

The properly normalized wave functions are then

$$\psi_{n_1 n_2 n_3}(\mathbf{x}) = \frac{1}{\sqrt{n_1! n_2! n_3!}} \left[\frac{M\omega}{\pi\hbar}\right]^{3/4} a_i^{\dagger n_1} a_2^{\dagger n_2} a_3^{\dagger n_3} \exp(-M\omega r^2/2\hbar). \qquad (2.5.25)$$

To calculate the matrix element of one of the components of $\mathbf{x}$, say $x_1$, we note that according to Eq. (2.5.7)

$$x_1 = \frac{i\sqrt{\hbar}}{\sqrt{2M\omega}} \left(a_1 - a_1^\dagger\right).$$

Since $a_1$ and $a_1^\dagger$ respectively lower and raise the index $n_1$ by one unit, $[x_1]_{nm}$ must vanish unless $n - m = \pm 1$. Also,

$$[x_1]_{n+1,n} \equiv \int \psi_{n+1}^*(x_1) x_1 \psi_n(x_1) \, dx_1$$

$$= \frac{1}{\sqrt{n!}\sqrt{(n+1)!}} \int \left(a_1^{\dagger(n+1)}\psi_0\right)^* \left(\frac{-ia_1^\dagger\sqrt{\hbar}}{\sqrt{2M\omega}}\right)\left(a_1^{\dagger n}\psi_0\right) dx_1$$

$$= -i\sqrt{\frac{(n+1)\hbar}{2M\omega}}. \tag{2.5.26}$$

If we had included the time-dependence factors $\exp(-iEt/\hbar)$ in the wave functions, this would be the same as Heisenberg's result (1.4.15), except for a conventional constant phase factor, which of course has no effect on $|\mathbf{x}_{nm}|^2$, and hence no effect on radiative transition rates.

## Problems

1. Use the method described in Section 2.2 to calculate the spherical harmonics (aside from constant factors) for $\ell = 3$.

2. Derive a formula for the rate of single-photon emission from the $2p$ to the $1s$ state of hydrogen.

3. Calculate the expectation values of the kinetic and potential energies in the $1s$ state of hydrogen.

4. Calculate the expectation values of the kinetic and potential energies in the lowest-energy state of the three-dimensional harmonic oscillator, using the algebraic methods that were used in Section 2.5 to find the energy levels in this system.

5. Derive the formula for the energy levels of the three-dimensional harmonic oscillator by using the power-series method (with suitable modifications) that was used in Section 2.3 for the hydrogen atom.

6. Find the difference between the energies of the Lyman-$\alpha$ transitions in hydrogen and deuterium.

7. Calculate the wave function (aside from normalization) of the $3s$ state of the hydrogen atom.

**Hint:** in problems 2 and 3, don't forget to use properly normalized wave functions.

# 3

# General Principles of Quantum Mechanics

We have seen in the previous chapter how useful wave mechanics can be in solving physical problems. But wave mechanics has several limitations. It describes physical states by means of wave functions, which are functions of the positions of the particles of the system, but why should we single out position as the fundamental physical observable? For instance, we might want to describe states in terms of probability amplitudes for particles to have certain values of the momentum or energy rather than the position. A more fundamental limitation is that there are attributes of physical systems that cannot be described at all in terms of the positions and momenta of a set of particles. One of these attributes is *spin*, which will be a chief subject of Chapter 4. Another is the value of the electric or magnetic field at some point in space, treated in Chapter 11. This chapter will describe the principles of quantum mechanics in a formalism which is essentially the "transformation theory" of Dirac, mentioned briefly in Section 1.4. This formalism generalizes both the wave mechanics of Schrödinger and the matrix mechanics of Heisenberg, and is sufficiently comprehensive to apply to any sort of physical system.

## 3.1 States

The first postulate of quantum mechanics is that physical states can be represented as vectors in a sort of abstract space known as *Hilbert space.*

Before getting into Hilbert space, I need to say a bit about vectors in general. In kindergarten we learn that vectors are quantities with both magnitude and direction. Later, when we study analytic geometry, we learn instead to describe a vector in $d$ dimensions as a string of $d$ numbers, the components of the vector. The latter approach lends itself well to calculation, but in some respects the kindergarten version is better, because it allows us to describe relations among vectors without specifying a coordinate system. For instance, a statement that one vector is parallel to a second vector, or perpendicular to a third, has nothing to do with how we choose our coordinate system.

Here we will formulate what we mean by vector spaces in general, and Hilbert space in particular, in a way that is independent of the coordinates we use to describe directions in these spaces. From this point of view, the wave functions that we have been using to describe physical states in wave mechanics should be considered as the set of *components* $\psi(\mathbf{x})$ of an abstract vector $\Psi$, known as the *state vector*, in an infinite-dimensional space in which we happen to choose coordinate axes that are labeled by all the values that can be taken by the position $\mathbf{x}$. The same state vector could be described instead by a wave function $\tilde{\psi}(\mathbf{p})$ in momentum space, defined as the coefficient of $\exp\left(i\mathbf{p}\cdot\mathbf{x}/\hbar\right)$ in a wave packet like (1.3.2).[1]

$$\psi(\mathbf{x}) = (2\pi\hbar)^{-3/2} \int d^3p \,\exp\left(i\mathbf{p}\cdot\mathbf{x}/\hbar\right) \tilde{\psi}(\mathbf{p}).$$

In this case, $\tilde{\psi}(\mathbf{p})$ is regarded as the component of the same state vector $\Psi$ along the direction corresponding to a definite value $\mathbf{p}$ of the momentum. This is not conceptually very different from switching to a description of position vectors in terms of latitude, longitude, and altitude to some other set of three coordinates. Or, as in Eq. (1.5.15), we could write $\psi(\mathbf{x})$ as an expansion in wave functions $\psi_n(\mathbf{x})$ of definite energy,

$$\psi(\mathbf{x}) = \sum_n c_n \psi_n(\mathbf{x}),$$

and regard the coefficients $c_n$ as the components of the same state vector along directions characterized by different values of the energy. These are just examples; our discussion of Hilbert space will not depend on any particular choice of coordinates.

Hilbert space is a certain kind of *normed complex vector space*. In general, any sort of vector space consists of quantities $\Psi$, $\Psi'$, etc., with the following properties.

- If $\Psi$ and $\Psi'$ are vectors, then so is $\Psi + \Psi'$. The operation of addition is associative and commutative:

$$\Psi + (\Psi' + \Psi'') = (\Psi + \Psi') + \Psi'', \tag{3.1.1}$$

$$\Psi + \Psi' = \Psi' + \Psi. \tag{3.1.2}$$

- If $\Psi$ is a vector, then so is $\alpha\Psi$, where $\alpha$ is any number. A *real vector space* is one in which these numbers are restricted to be real. In a *complex vector space*, like the Hilbert space of quantum mechanics, the numbers like $\alpha$ can be complex. For either real or complex vector spaces, multiplication by a number is taken to be associative and distributive:

$$\alpha(\alpha'\Psi) = (\alpha\alpha')\Psi, \tag{3.1.3}$$

---

[1] This definition is framed so that the momentum operator $-i\hbar\nabla$ acting on $\psi(\mathbf{x})$ has the effect of multiplying $\tilde{\psi}(\mathbf{p})$ with $\mathbf{p}$. The factor $(2\pi\hbar)^{-3/2}$ is included so that, for a wave function normalized to have $\int |\psi(\mathbf{x})|^2 \, d^3x = 1$, by a theorem of Fourier analysis we have $\int |\tilde{\psi}(\mathbf{p})|^2 \, d^3p = 1$.

$$\alpha(\Psi + \Psi') = \alpha\Psi + \alpha\Psi', \tag{3.1.4}$$

$$(\alpha + \alpha')\Psi = \alpha\Psi + \alpha'\Psi. \tag{3.1.5}$$

- There is a single zero vector[2] **o**, with the obvious properties that, for any vector $\Psi$ and number $\alpha$,

$$\mathbf{o} + \Psi = \Psi, \quad 0\Psi = \mathbf{o}, \quad \alpha\mathbf{o} = \mathbf{o}. \tag{3.1.6}$$

A *normed vector space* is a vector space in which for any two vectors $\Psi$ and $\Psi'$ there is a number, the scalar product $(\Psi, \Psi')$, with the properties of *linearity*,

$$\left(\Psi'', [\alpha\Psi + \alpha'\Psi']\right) = \alpha\left(\Psi'', \Psi\right) + \alpha'\left(\Psi'', \Psi'\right), \tag{3.1.7}$$

*symmetry*,

$$\left(\Psi', \Psi\right)^* = \left(\Psi, \Psi'\right), \tag{3.1.8}$$

and *positivity*, which requires that the scalar product of a vector with itself is a real number with

$$(\Psi, \Psi) > 0 \quad \text{for} \quad \Psi \neq \mathbf{o}. \tag{3.1.9}$$

(Note that $(\Psi, \mathbf{o}) = 0$ for any $\Psi$, and in particular for $\Psi = \mathbf{o}$, because for any number $\alpha$ and vector $\Psi$ we have $\alpha(\Psi, \mathbf{o}) = (\Psi, \alpha\mathbf{o}) = (\Psi, \mathbf{o})$, which is only possible if $(\Psi, \mathbf{o}) = 0$.) For real vector spaces the scalar products $(\Psi, \Psi')$ are all taken to be real, and the complex conjugation in Eq. (3.1.8) has no effect; for complex vector spaces the scalar products must be allowed to be complex. From Eqs. (3.1.7) and (3.1.8) it follows that

$$\left([\alpha\Psi + \alpha'\Psi'], \Psi''\right) = \alpha^*\left(\Psi, \Psi''\right) + \alpha'^*\left(\Psi', \Psi''\right). \tag{3.1.10}$$

In addition to being a normed complex vector space, a Hilbert space is either finite-dimensional, or satisfies certain technical assumptions of continuity that allow it to be treated in some respects as if it were finite-dimensional. To explain this, it is necessary first to say something about sets of vectors that are independent, or complete, and how this allows us to define the dimensionality of a vector space.

A set of vectors $\Psi_1$, $\Psi_2$, etc., is said to be *independent* if no non-trivial linear combination of these vectors can vanish. That is, if $\Psi_1$, $\Psi_2$, etc. are independent, and if for some set of numbers $\alpha_1$, $\alpha_2$, etc. we have $\alpha_1\Psi_1 + \alpha_2\Psi_2 + \cdots = \mathbf{o}$, then it follows that $\alpha_1 = \alpha_2 = \cdots = 0$. Equivalently, no one of a set of independent vectors can be expressed as a linear combination of the others. In particular, vectors $\Psi_1$, $\Psi_2$, etc. are independent if they are orthogonal; that is, if $(\Psi_i, \Psi_j) = 0$ for $i \neq j$, for if such a set of orthogonal vectors satisfies a

---

[2] In future chapters, where no confusion can arise, we will not bother to use the special symbol **o** for the zero state vector, and will instead just use the familiar zero 0.

relation $\alpha_1 \Psi_1 + \alpha_2 \Psi_2 + \cdots = \mathbf{0}$, then by taking the scalar product with any of the $\Psi$s we have $\alpha_i (\Psi_i, \Psi_i) = 0$, so $\alpha_i = 0$ for all $i$. The converse does not hold – the vectors of an independent set do not have to be orthogonal – but if a set $\Psi_i$ of vectors with $1 \leq i \leq n$ are all independent, then we can always find $n$ linear combinations $\Phi_i$ of these vectors that are not only independent but also orthogonal.[3]

A set of vectors $\Psi_1$, $\Psi_2$, ..., $\Psi_n$, is said to be *complete* if any vector $\Psi$ can be expressed as a linear combination of the $\Psi_i$:

$$\Psi = \alpha_1 \Psi_1 + \alpha_2 \Psi_2 + \cdots + \alpha_n \Psi_n.$$

The vectors of a complete set do not have to be independent, but if they are not, then we can always find a subset that is both complete and independent, by deleting in turn any vectors of the set that can be written as linear combinations of the others. Given a complete independent set of vectors $\Psi_i$, by the method described earlier we can find a set of vectors $\Phi_i$ that are orthogonal as well as independent, and since according to this construction every $\Psi_i$ is a linear combination of the $\Phi_i$, the $\Phi_i$ are also complete. A complete set of orthogonal vectors is said to form a *basis* for the Hilbert space.

A vector space is said to have a finite dimensionality $d$ if the largest possible number of independent vectors is $d$. In such a space, *any* set of $d$ independent vectors $\Phi_i$ is also complete, because if there were a vector $\Psi$ that could not be written as $\sum_{i=1}^{d} \alpha_i \Phi_i$, then there would be $d+1$ independent vectors: namely, $\Psi$ and the $\Phi_i$. Also, no set of fewer than $d$ vectors $\Upsilon_j$ could be complete, because if it were then each vector $\Phi_i$ of the $d$ independent vectors could be written as $\Phi_i = \sum_{j=1}^{d-1} c_{ij} \Upsilon_j$, and for any $(d \times (d-1))$-dimensional matrix $c_{ij}$ there is always a $d$-component quantity $u_i$ such that $\sum_{i=1}^{d} u_i c_{ij} = 0$, contradicting the assumption that the $\Phi_i$ are independent.

For our present purposes, a Hilbert space can be defined as a normed complex vector space that is either of finite dimensionality, or in which there exists an infinite set of independent orthogonal vectors $\Phi_i$, that are complete in the sense that for any vector $\Psi$ we can find a set of numbers $\alpha_i$ such that the sum $\sum_{i=1}^{\infty} \alpha_i \Phi_i$ converges to $\Psi$. (By this, we mean that $(\Omega_N, \Omega_N) \to 0$ for $N \to \infty$,

---

[3]  In this case we can construct a vector

$$\Phi_n \equiv \Psi_n - \sum_{i,j=1}^{n-1} (\omega^{-1})_{ji} \Psi_j (\Psi_i, \Psi_n)$$

that is orthogonal to all the $\Psi_i$ with $1 \leq i \leq n - 1$, where $\omega_{ij} \equiv (\Psi_i, \Psi_j)$. (We know that $\omega_{ij}$ has an inverse, because if there were a non-zero vector $v_j$ for which $\sum_j \omega_{ij} v_j = 0$ then the vector $\Omega \equiv \sum_i v_i \Psi_i$ would have norm $(\Omega, \Omega) = \sum_{ij} v_i^* \omega_{ij} v_j = 0$, and would therefore have to vanish, which since the $\Psi_i$ are independent is only possible if all $v_i$ vanish.) Also, we know that $\Phi_n$ does not vanish, because that would contradict the independence of the $\Psi_i$. Continuing along the same lines, we can also construct a non-zero vector $\Phi_{n-1}$ that is orthogonal to all $\Psi_i$ with $1 \leq i \leq n - 2$ and also to $\Phi_n$, and so on, until we have a set of $n$ orthogonal vectors $\Phi_i$.

where $\Omega_N \equiv \Psi - \sum_{i=1}^{N} \alpha_i \Phi_i$.) The latter condition allows us to apply some of the same mathematical methods as if the Hilbert space were finite-dimensional.

The *components* of a state vector $\Psi$ in a basis provided by a complete orthogonal set of vectors $\Phi_i$ are just the numbers $\alpha_i$ in the expression $\Psi = \sum_i \alpha_i \Phi_i$. They are unique, because if $\Psi$ could be written in this way with two different sets of $\alpha_i$, then the difference of the sums would vanish, contradicting the assumption that the $\Phi_i$ are independent. In fact, by taking the scalar product of the sum $\sum_i \alpha_i \Phi_i$ with $\Phi_j$, we see that we can write these components as

$$\alpha_j = \frac{(\Phi_j, \Psi)}{(\Phi_j, \Phi_j)}$$

so that any vector $\Psi$ is expressed in terms of a complete set of orthogonal vectors $\Phi_i$ by

$$\Psi = \sum_j \frac{(\Phi_j, \Psi)}{(\Phi_j, \Phi_j)} \Phi_j. \tag{3.1.11}$$

This allows a concrete realization of the scalar product of any two vectors $\Psi$ and $\Psi'$:

$$(\Psi, \Psi') = \sum_{i,j} \frac{(\Phi_j, \Psi)^* (\Phi_i, \Psi')}{(\Phi_j, \Phi_j)(\Phi_i, \Phi_i)} (\Phi_j, \Phi_i),$$

or, since the $\Phi_i$ are orthogonal,

$$(\Psi, \Psi') = \sum_i \frac{(\Phi_i, \Psi)^*(\Phi_i, \Psi')}{(\Phi_i, \Phi_i)}, \tag{3.1.12}$$

(At this point, we are limiting ourselves to a complete set of basis vectors $\Phi_i$ that is denumerable. The case of a continuum of basis vectors will be considered in the next section.)

Now at last we can put some flesh on these bones, and state the interpretation of scalar products in terms of probabilities. The first interpretive postulate of quantum mechanics is that any complete orthogonal set of states $\Phi_i$ are in one-to-one correspondence with all the possible results of some sort of measurement (what sort will be considered in Section 3.3), and that if the system before the measurement is in a state $\Psi$, then the probability that the measurement will yield a result corresponding to the state $\Phi_i$ is

$$P(\Psi \mapsto \Phi_i) = \frac{\left|\left(\Phi_i, \Psi\right)\right|^2}{\left(\Psi, \Psi\right)\left(\Phi_i, \Phi_i\right)}. \tag{3.1.13}$$

It is important to note that the probabilities given by this formula have the fundamental properties that must be possessed by any probabilities. First, they are obviously all positive. Also, since the $\Phi_i$ are a complete orthogonal set, Eq. (3.1.12) gives

$$(\Psi, \Psi) = \sum_i \frac{|(\Phi_i, \Psi)|^2}{(\Phi_i, \Phi_i)}$$

so the probabilities (3.1.13) add up to one.

The probabilities (3.1.13) are unchanged if we multiply $\Psi$ with a constant $\alpha$, or multiply the $\Phi_i$ with constants $\beta_i$. In quantum mechanics state vectors that differ by a constant factor are regarded as representing the same physical state. (But $\Psi + \Psi'$ and $\alpha\Psi + \Psi'$ do *not* generally represent the same state.) We can if we like multiply the state vectors $\Psi$ and $\Phi_i$ with constants chosen so that

$$(\Psi, \Psi) = (\Phi_i, \Phi_i) = 1, \tag{3.1.14}$$

in which case the probabilities (3.1.13) are

$$P(\Psi \mapsto \Phi_i) = \left|\left(\Phi_i, \Psi\right)\right|^2. \tag{3.1.15}$$

This is essentially the Born rule mentioned in Section 1.5.

A set of vectors $\Phi_i$ that are orthogonal and also normalized so that $(\Phi_i, \Phi_i) = 1$ is said to be *orthonormal*. For a complete orthonormal set of basis vectors $\Phi_i$, Eqs. (3.1.11) and (3.1.12) become

$$\Psi = \sum_j (\Phi_j, \Psi)\, \Phi_j, \tag{3.1.16}$$

and

$$(\Psi, \Psi') = \sum_i (\Phi_i, \Psi)^*(\Phi_i, \Psi'). \tag{3.1.17}$$

Even after choosing $\Psi$ and $\Phi_i$ to satisfy Eq. (3.1.14), we can still multiply the state vectors with complex numbers of magnitude unity (that is, phase factors), with no change in Eqs. (3.1.14) and (3.1.15). Thus physical states in quantum mechanics are in one-to-one correspondence with *rays* in the Hilbert space, each ray consisting of a set of state vectors of unit norm that differ only by multiplication with phase factors.

This is a good place to mention the "bra–ket" notation used by Dirac. In Dirac's notation, a state vector $\Psi$ is denoted $|\Psi\rangle$, and the scalar product $(\Phi, \Psi)$ of two state vectors is written $\langle\Phi|\Psi\rangle$. The symbol $\langle\Phi|$ is called a "bra," and $|\Psi\rangle$ is called a "ket," so that $\langle\Phi|\Psi\rangle$ is a bra–ket, or bracket (not to be confused with the entirely different Dirac bracket described in Section 9.5). In the special cases where $\Psi$ is identified as a state with a definite value $a$ for some observable $A$, the corresponding ket in Dirac's notation is frequently written as $|a\rangle$.

The notation we use, with scalar products denoted $(\Phi, \Psi)$, is commonly used by mathematicians, while the Dirac notation with scalar products denoted $\langle\Phi|\Psi\rangle$ is more common among physicists. In Section 3.3 I will explain how for some purposes the Dirac notation is particularly convenient, and in some cases inconvenient.

## 3.2  Continuum States

Before going on to the next interpretive postulate of quantum mechanics, it is necessary to explain how the description of physical states given in the previous section is modified when we consider a system for which the complete orthogonal states form a continuum. Suppose that instead of being labeled as $\Phi_i$ with a discrete index $i$, they are labeled $\Phi_\xi$, where $\xi$ is a continuous variable, like position. (The mathematical condition that defines a state with a definite value of position or any other observable is discussed in the next section.) We can adapt the results of the previous section by treating such systems approximately, letting $\xi$ take a very large number $\rho(\xi)\,d\xi$ of discrete values of $\xi$ in any small interval from $\xi$ to $\xi+d\xi$. (For instance, if $\xi$ is the $x$-coordinate of some particle, we might replace the $x$-axis with a large number of discrete points, with successive points separated by a small distance $1/\rho(x)$.) It is convenient in such cases when introducing a complete orthogonal set of basis vectors $\Phi_\xi$ to normalize them so that

$$(\Phi_{\xi'}, \Phi_\xi) = \rho(\xi)\delta_{\xi',\xi}. \tag{3.2.1}$$

Then according to Eq. (3.1.11), an arbitrary state can be expressed as a linear combination of basis states

$$\Psi = \sum_\xi \frac{(\Phi_\xi, \Psi)}{\rho(\xi)}\Phi_\xi. \tag{3.2.2}$$

In the limit as the points $\xi$ become increasingly close together, any sum over $\xi$ of a smooth function $f(\xi)$ can be expressed as an integral

$$\sum_\xi f(\xi) \mapsto \int f(\xi)\rho(\xi)\,d\xi. \tag{3.2.3}$$

(The sum over all values of $\xi$, in an interval $d\xi$ that is small enough that within this interval $f(\xi)$ and $\rho(\xi)$ are essentially constant, equals the number $\rho(\xi)\,d\xi$ of allowed values of $\xi$ in this interval, times $f(\xi)$. Summing this over intervals gives the integral.) Hence in this limit Eq. (3.2.2) may be written

$$\Psi = \int (\Phi_\xi, \Psi)\Phi_\xi\,d\xi, \tag{3.2.4}$$

the factors $\rho(\xi)$ here canceling. Similarly, the scalar product (3.1.12) of two such states may be written

$$(\Psi, \Psi') = \sum_\xi \frac{(\Phi_\xi, \Psi)^*(\Phi_\xi, \Psi')}{\rho(\xi)} = \int (\Phi_\xi, \Psi)^*(\Phi_\xi, \Psi')\,d\xi. \tag{3.2.5}$$

In particular, the condition for a state $\Psi$ to have unit norm is that

$$1 = \int |(\Phi_\xi, \Psi)|^2 \, d\xi. \tag{3.2.6}$$

If a system is initially in a state represented by a vector $\Psi$ of unit norm, and we perform an experiment whose possible outcomes are represented by a complete set of states $\Phi_\xi$, then the differential probability $dP(\Psi \mapsto \Phi_\xi)$ that the outcome will be in an interval from $\xi$ to $\xi + d\xi$ will equal the probability of finding an individual state with a label near $\xi$, given by Eq. (3.1.13), times the number of states in this interval:

$$dP(\Psi \mapsto \Phi_\xi) = \frac{|(\Phi_\xi, \Psi)|^2}{(\Phi_\xi, \Phi_\xi)} \times \rho(\xi) \, d\xi = |(\Phi_\xi, \Psi)|^2 \, d\xi. \tag{3.2.7}$$

According to Eq. (3.2.6), this satisfies the essential condition that the total probability of any result should be unity:

$$\int dP(\Psi \mapsto \Phi_\xi) = 1. \tag{3.2.8}$$

For instance, we might take $\Phi_x$ to represent states in which a particle has definite values $x$ for its position in one dimension. As mentioned at the beginning of this chapter, the wave function of Schrödinger's wave mechanics is nothing but the scalar product

$$\psi(x) = (\Phi_x, \Psi). \tag{3.2.9}$$

Equation (3.2.5) shows that the scalar product of two state vectors $\Psi_1$ and $\Psi_2$ is

$$(\Psi_1, \Psi_2) = \int \psi_1^*(x)\psi_2(x) \, dx. \tag{3.2.10}$$

In particular, the condition (3.2.6) for a state vector of unit norm now reads

$$1 = \int |\psi(x)|^2 \, dx, \tag{3.2.11}$$

and for states satisfying this condition, Eq. (3.2.7) gives the probability that the particle is located between $x$ and $x + dx$:

$$dP = |\psi(x)|^2 \, dx \tag{3.2.12}$$

as Born guessed in 1926. (See Section 1.5.)

We will occasionally use a "delta function" notation due to Dirac.[4] Let us define

$$\delta(\xi - \xi') \equiv \rho(\xi)\delta_{\xi,\xi'} \tag{3.2.13}$$

---

[4]  P. A. M. Dirac, *Principles of Quantum Mechanics*, 4th edn. (Clarendon Press, Oxford, 1958).

so that the normalization condition (3.2.1) for continuum states reads

$$(\Phi_\xi, \Phi_{\xi'}) = \delta(\xi - \xi'). \tag{3.2.14}$$

According to Eq. (3.2.3), the integral over $\xi'$ of this function times any smooth function $f(\xi')$ is

$$\int \delta(\xi - \xi') f(\xi') \, d\xi' = \sum_{\xi'} \frac{\delta(\xi - \xi') f(\xi')}{\rho(\xi')} = f(\xi). \tag{3.2.15}$$

That is, the function (3.2.13) vanishes except at $\xi' = \xi$, but is so large there that its integral over $\xi'$ is unity, so that in an integral like Eq. (3.2.15) it picks out the value of the function where $\xi' = \xi$.

Sometimes it is convenient to represent the delta function as a smooth function that is negligible away from zero argument, but so strongly peaked there that its integral is unity. For instance, we might define

$$\delta(\xi - \xi') \equiv \frac{1}{\epsilon \sqrt{\pi}} \exp\left(-(\xi - \xi')^2/\epsilon^2\right), \tag{3.2.16}$$

where $\epsilon$ is allowed to go to zero through positive values. Or we might give up continuity, and define

$$\delta(\xi - \xi') \equiv \begin{cases} 1/2\epsilon, & |\xi - \xi'| < \epsilon, \\ 0, & |\xi - \xi'| \geq \epsilon. \end{cases} \tag{3.2.17}$$

Another representation is suggested by the fundamental theorem of Fourier analysis. According to this theorem, if $g(k)$ is a sufficiently smooth function which is sufficiently well-behaved as $k \to \pm\infty$, and we define

$$f(x) \equiv \frac{1}{\sqrt{2\pi}} \int_{-\infty}^{\infty} g(k) e^{ikx} \, dk, \tag{3.2.18}$$

then

$$g(k) = \frac{1}{\sqrt{2\pi}} \int_{-\infty}^{\infty} f(x) e^{-ikx} \, dx. \tag{3.2.19}$$

If we use Eq. (3.2.19) in the integrand of Eq. (3.2.18), then we have, at least formally,

$$f(x) = \frac{1}{2\pi} \int_{-\infty}^{\infty} dx' \, f(x') \int_{-\infty}^{\infty} dk \, e^{ik(x-x')}, \tag{3.2.20}$$

so we can take

$$\delta(x - x') = \frac{1}{2\pi} \int_{-\infty}^{\infty} dk \, e^{ik(x-x')}. \tag{3.2.21}$$

The reader can check that if we give meaning to this integral by inserting a convergence factor $\exp(-\epsilon^2 k^2/4)$ in the integrand, with $\epsilon$ infinitesimal, then Eq. (3.2.21) becomes the same as the representation (3.2.16).

There is a rigorous approach to the delta function known as the *theory of distributions*, due to the mathematician Laurent Schwartz[5] (1915–2002), in which we give up the idea of representing the delta function itself as an actual function, and instead only define integrals involving the delta function by Eq. (3.2.15). In the same way, the derivative of the delta function is defined by the statement that

$$\int \delta'(\xi - \xi') f(\xi') \, d\xi' = -f'(\xi),  \tag{3.2.22}$$

as obtained from (3.2.15) by a formal integration by parts.

## 3.3 Observables

Now we come to the second postulate of quantum mechanics. This postulate requires that observable physical quantities like position, momentum, energy, etc., are represented as Hermitian operators on Hilbert space, in a sense to be explained below. An Hermitian operator is one that is linear and self-adjoint, so before we spell out what this postulate means, we need to consider what is meant by operators in general, by linear operators in particular, and by the adjoint of an operator.

An operator is any mapping of the Hilbert space on itself. That is, an operator $A$ takes any vector $\Psi$ in the Hilbert space into another vector in the Hilbert space, denoted $A\Psi$. This leads to natural definitions of products of operators with each other and with numbers, and of sums of operators. The product $AB$ of two operators is defined as the operator that operates on an arbitrary state vector $\Psi$ first with $B$ and then with $A$. That is,

$$(AB)\Psi \equiv A(B\Psi).  \tag{3.3.1}$$

An ordinary complex number $\alpha$ can also be regarded as the operator that multiplies any state vector with that number, so according to Eq. (3.3.1), the product $\alpha A$ of a number $\alpha$ with an operator $A$ is the operator that operates on an arbitrary state vector $\Psi$ first with $A$ and then multiplies the result with $\alpha$:

$$(\alpha A)\Psi \equiv \alpha(A\Psi).  \tag{3.3.2}$$

The sum of two operators $A$ and $B$ is defined as the operator that, acting on an arbitrary state vector $\Psi$, gives the sum of the state vectors produced by acting on $\Psi$ with $A$ and $B$ individually:

$$(A + B)\Psi \equiv A\Psi + B\Psi.  \tag{3.3.3}$$

---

[5] L. Schwartz, *Théorie des distributions* (Hermann et Cie, Paris, 1966).

We can define a zero operator $\mathbf{0}$ that, acting on any state vector $\Psi$, gives the zero state vector $\mathbf{o}$:

$$\mathbf{0}\Psi \equiv \mathbf{o}. \tag{3.3.4}$$

It follows then that, for an arbitrary operator $A$ and number $\alpha$,

$$\mathbf{0}A = \mathbf{0}, \quad \mathbf{0} + A = A, \quad \alpha\mathbf{0} = \mathbf{0}\alpha = \mathbf{0}. \tag{3.3.5}$$

We also define a unit operator $\mathbf{1}$ that, acting on any state vector $\Psi$, gives the same state vector:

$$\mathbf{1}\Psi \equiv \Psi. \tag{3.3.6}$$

For an arbitrary operator $A$, we then have

$$\mathbf{1}A = A\mathbf{1} = A. \tag{3.3.7}$$

A *linear* operator $A$ is one for which

$$A(\Psi + \Psi') = A\Psi + A\Psi', \quad A(\alpha\Psi) = \alpha A\Psi, \tag{3.3.8}$$

for arbitrary state vectors $\Psi$ and $\Psi'$ and arbitrary numbers $\alpha$. It is easy to see that if $A$ and $B$ are linear, then so are $AB$ and $\alpha A + \beta B$ for any numbers $\alpha$ and $\beta$. Also, both $\mathbf{0}$ and $\mathbf{1}$ are linear.

The *adjoint* $A^\dagger$ of any operator $A$ (linear or not) is defined as that operator (if there is one) for which[6]

$$(\Psi', A^\dagger\Psi) = (A\Psi', \Psi), \tag{3.3.9}$$

or equivalently

$$(\Psi', A^\dagger\Psi) = (\Psi, A\Psi')^*,$$

for any two state vectors $\Psi$ and $\Psi'$. It is elementary to show the following general properties of adjoints:

$$(AB)^\dagger = B^\dagger A^\dagger, \quad (A^\dagger)^\dagger = A, \quad (\alpha A)^\dagger = \alpha^* A^\dagger, \quad (A + B)^\dagger = A^\dagger + B^\dagger. \tag{3.3.10}$$

Both $\mathbf{0}$ and $\mathbf{1}$ are their own adjoints.

If we introduce a complete orthonormal set of basis vectors $\Phi_i$, we can represent any linear operator $A$ by a matrix $A_{ij}$, given by

$$A_{ij} \equiv (\Phi_i, A\Phi_j). \tag{3.3.11}$$

Using Eq. (3.1.16), we see that the matrix representing any operator product $AB$ is the product of the matrices

$$(AB)_{ij} = (\Phi_i, AB\Phi_j) = \sum_k (\Phi_i, A\Phi_k)(\Phi_k, B\Phi_j) = \sum_k A_{ik}B_{kj}. \tag{3.3.12}$$

---

[6] Equation (3.3.9) is awkward to express in Dirac's bra–ket notation, since in $\langle\Psi'|B|\Psi\rangle$ the operator $B$ is always presumed to act to the right. Instead of Eq. (3.3.9), one must write $\langle\Psi'|A^\dagger|\Psi\rangle = \langle\Psi|A|\Psi'\rangle^*$.

The adjoint of an operator is represented by the transposed complex conjugate of the matrix representing the operator:

$$(A^\dagger)_{ij} = A^*_{ji}. \tag{3.3.13}$$

As discussed in the previous section, we frequently encounter complete sets of state vectors $\Phi_\xi$, labeled with a continuum variable $\xi$ instead of a discrete label $i$, and orthonormal in the sense that

$$(\Phi_{\xi'}, \Phi_\xi) = \delta(\xi' - \xi). \tag{3.3.14}$$

In this case, we define

$$A_{\xi'\xi} \equiv (\Phi_{\xi'}, A\Phi_\xi), \tag{3.3.15}$$

and instead of Eq. (3.3.12), we have

$$(AB)_{\xi'\xi} = \int d\xi''\, A_{\xi'\xi''} B_{\xi''\xi}. \tag{3.3.16}$$

The second postulate of quantum mechanics holds that a state has a definite value $a$ for an observable represented by a linear Hermitian operator $A$ if and only if the state vector $\Psi$ is an eigenstate of $A$ with eigenvalue $a$, in the sense that

$$A\Psi = a\Psi. \tag{3.3.17}$$

If also $A\Psi' = a'\Psi'$, then because $A$ is Hermitian,

$$a(\Psi', \Psi) = (\Psi', A\Psi) = (A\Psi', \Psi) = a'^*(\Psi', \Psi).$$

In the case $\Psi = \Psi' \neq \mathbf{0}$ and $a' = a$ this gives $a^* = a$, while for $a \neq a'$ we have $(\Psi', \Psi) = 0$. That is, the allowed values of observables are real, and state vectors with different values for any observable are orthogonal. In terms of the matrices (3.3.11) or (3.3.15), the condition (3.3.17) may be written

$$\sum_j A_{ij}(\Phi_j, \Psi) = a(\Phi_i, \Psi), \tag{3.3.18}$$

or else

$$\int d\xi\, A_{\xi'\xi}(\Phi_\xi, \Psi) = a(\Phi_{\xi'}, \Psi). \tag{3.3.19}$$

If a state vector $\Psi$ has a definite value $a$ for an observable represented by $A$ and also a definite value $b$ for an observable represented by $B$, then

$$AB\Psi = bA\Psi = ba\Psi = ab\Psi = aB\Psi = BA\Psi,$$

so $\Psi$ has the definite value zero for the commutator $[A, B] \equiv AB - BA$. In particular, it is impossible for there to be a state with definite values for a pair of observables, if their commutator does not have a zero eigenvalue, as is the case for instance if the commutator is a non-zero number times the unit operator. This obstacle to the existence of states in which $A$ and $B$ each have definite values does not arise if the operators *commute*, in the sense that the commutator $[A, B]$ vanishes.

The Hermitian operators representing observables are assumed to have the important property that their eigenvectors form complete sets, which can be taken to be orthonormal. This is automatic for Hermitian operators acting in spaces of finite dimensionality.[7] It is more difficult to show that a given Hermitian operator in an infinite-dimensional space has this property, especially when its eigenvalues form a continuum, and we will simply assume that this is the case.

This is often referred to as the *diagonalization* of the matrix $A$, because we can regard the $i$th component of the $r$th orthonormal eigenvector $u_r$ of $A$ as the $ir$ component of a matrix $U_{ir}$, so that the eigenvalue condition can be written $AU = UD$, where $D_{rs} = a_r \delta_{rs}$ is a diagonal matrix. The condition that the eigenvectors are orthonormal tells us that $U^\dagger U = 1$, so $U$ has an inverse equal to $U^\dagger$, and $U^{-1}AU = D$.

To see what goes wrong when an operator is not Hermitian, consider the $2 \times 2$ matrix

$$M = \begin{pmatrix} a & c \\ 0 & b \end{pmatrix},$$

which is not Hermitian if $c \neq 0$, whatever the values of $a$ and $b$. It has eigenvalues $a$ and $b$, with respective eigenvectors

$$\begin{pmatrix} 1 \\ 0 \end{pmatrix}, \qquad \begin{pmatrix} c \\ b - a \end{pmatrix}.$$

These eigenvectors form a complete set in this two-dimensional space, except in the case $a = b$, where for $c \neq 0$ both eigenvalues are the same and both eigenvectors are in the same direction, and so are not a complete set. On the other hand, in the Hermitian case with $c = 0$ the two eigenvectors can be taken to be the complete set $(1, 0)$ and $(0, 1)$, irrespective of whether or not $b = a$.

---

[7] Here is the proof. It follows from the theory of determinants that a matrix $A_{ij}$ in a finite number $d$ of dimensions will have an eigenvalue $a$ if and only if the determinant of $A - a\mathbf{1}$ vanishes. This determinant is a polynomial in $a$ of order $d$, and therefore by a fundamental theorem of algebra, there is always at least one value of $a$ where it vanishes, and hence at least one eigenvector $u$ for which $Au = au$. Consider the space of vectors $v$ that are orthogonal to $u$ – that is, for which $(v, u) = 0$. If $A$ is Hermitian, this space is invariant under $A$, for if $(v, u) = 0$ then $(Av, u) = (v, Au) = a(v, u) = 0$. According to the argument given in footnote 3 of Section 3.1, we can introduce a complete orthonormal basis of vectors $v_i$ in this space, so that $Av_i$ is a linear combination $\sum_j A_{ji} v_j$ of these basis vectors. Because $A_{ji} = (v_j, Av_i) = (Av_j, v_i) = A^*_{ij}$, the coefficients $A_{ij}$ form an Hermitian matrix, but now in $d - 1$ dimensions. We then apply the same argument as before to show that there is some linear combination of the $v_i$ orthogonal to $u$ that is also an eigenvector of $A$. Then by considering the action of $A$ on the $(d - 2)$-dimensional space of vectors orthogonal to both $u$ and $v$, we can find an eigenvector of $A$ in this space. We can continue in this way to construct $d$ orthogonal eigenvectors of $A$. Since they are orthogonal, they are independent, and since there are $d$ of them, they form a complete set.

These results can be generalized to the case of several commuting Hermitian operators. Suppose that $A$ and $B$ are Hermitian and satisfy $[A, B] = 0$. As remarked above, we can find a complete set of vectors $u_r$ satisfying the eigenvalue condition $A u_r = a_r u_r$. Let us make a small change in notation, using $r$ to label different values of the eigenvalue $a_r$, and using an index $s$ to distinguish different eigenvectors $u_{rs}$ of $A$, all with eigenvalue $a_r$. For fixed $r$, the space of linear combinations $u$ of the $u_{rs}$ with different values of $s$ is invariant under $B$, because if $Au = a_r u$ then $A(Bu) = BAu = a_r Bu$. Hence by the same argument as for $A$, in this space we can find a complete orthonormal set of eigenvectors of $B$. That is, we can choose the orthonormal vectors $u_{rs}$ so that $A u_{rs} = a_r u_{rs}$ and $B u_{rs} = b_s u_{rs}$. Hence in the same sense as before, we can choose a basis in which $A$ and $B$ are both represented by diagonal matrices.

The second postulate of quantum mechanics leads to a simple formula for the expectation value of any observable. Let $\Psi_r$ be a complete orthonormal set of state vectors that for some self-adjoint linear operator $A$ represent states with values $a_r$ for the observable represented by $A$, and so for which $A\Psi_r = a_r \Psi_r$. The expectation value of this observable in a state represented by a normalized vector $\Psi$ is the sum over allowed values, weighted by the probability (3.1.15) of each:

$$\langle A \rangle_\Psi = \sum_r a_r |(\Psi_r, \Psi)|^2 = \sum_r (\Psi, A\Psi_r)(\Psi_r, \Psi) = (\Psi, A\Psi). \qquad (3.3.20)$$

It is easy to see that if the state represented by $\Psi$ has a definite value $a$ for an observable represented by an operator $A$, then $A^n \Psi = a^n \Psi$, and so it has a definite value $p(a)$ for the observable represented by any power series $p(A)$ in the operator $A$. More generally, we can define functions $f(A)$ of Hermitian operators by specifying that for an arbitrary linear combination $\sum_r c_r \Psi_r$ of a complete independent set of eigenvectors $\Psi_r$ of $A$ with eigenvalues $a_r$, we have

$$f(A) \sum_r c_r \Psi_r \equiv \sum_r c_r f(a_r) \Psi_r.$$

In general, the expectation value of a function of an operator is not equal to that function of the expectation value. That is, $\langle f(A) \rangle_\Psi \neq f(\langle A \rangle_\Psi)$. In fact, for Hermitian operators, $\langle A^2 \rangle_\Psi \geq \langle A \rangle_\Psi^2$, with equality if and only if $\Psi$ is an eigenvector of $A$. To see this, we note that the expectation value of the square of any Hermitian operator $B$ is

$$\langle B^2 \rangle_\Psi = (B\Psi, B\Psi),$$

so the expectation value is always positive, and vanishes only if $B$ annihilates the state vector $\Psi$. Thus in particular

$$0 \leq \left\langle (A - \langle A \rangle_\Psi)^2 \right\rangle_\Psi = \langle A^2 \rangle_\Psi - 2\langle A \rangle_\Psi^2 + \langle A \rangle_\Psi^2 = \langle A^2 \rangle_\Psi - \langle A \rangle_\Psi^2. \qquad (3.3.21)$$

As this shows, $\langle A \rangle^2_\Psi$ is at most equal to $\langle A^2 \rangle_\Psi$, and equals it only if $\Psi$ is an eigenstate of $A$.

We are now in a position to prove a generalized version of the Heisenberg uncertainty principle. For this purpose, we will need a general inequality, known as the *Schwarz inequality*, which states that for any two state vectors $\Psi$ and $\Psi'$, we have

$$|(\Psi', \Psi)|^2 \le (\Psi', \Psi')(\Psi, \Psi). \tag{3.3.22}$$

(This is a generalization of the familiar fact that $\cos^2 \theta \le 1$.) The Schwarz inequality is proved by introducing

$$\Psi'' \equiv \Psi - \Psi'(\Psi', \Psi)/(\Psi', \Psi')$$

and noting that

$$0 \le (\Psi'', \Psi'')(\Psi', \Psi') = (\Psi, \Psi)(\Psi', \Psi') - 2(\Psi, \Psi')(\Psi', \Psi) + |(\Psi', \Psi)|^2$$
$$= (\Psi, \Psi)(\Psi', \Psi') - |(\Psi', \Psi)|^2.$$

To give a precise statement of the uncertainty principle, we may define the root mean square deviation of an Hermitian operator $A$ from its expectation value in a state represented by $\Psi$ as

$$\Delta_\Psi A \equiv \sqrt{\left\langle \left( A - \langle A \rangle_\Psi \right)^2 \right\rangle_\Psi}. \tag{3.3.23}$$

For our purposes, it is convenient to rewrite this as

$$\Delta_\Psi A = \sqrt{(\Psi_A, \Psi_A)},$$

where

$$\Psi_A \equiv (A - \langle A \rangle_\Psi)\Psi/\sqrt{(\Psi, \Psi)}.$$

For any pair of Hermitian operators $A$ and $B$, the Schwarz inequality (3.3.22) then gives

$$\Delta_\Psi A \, \Delta_\Psi B \ge |(\Psi_A, \Psi_B)|.$$

The scalar product on the right-hand side may be expressed as

$$(\Psi_A, \Psi_B) = \frac{(\Psi, [A - \langle A \rangle_\Psi][B - \langle B \rangle_\Psi]\Psi)}{(\Psi, \Psi)} = \frac{(\Psi, [AB - \langle A \rangle_\Psi \langle B \rangle_\Psi]\Psi)}{(\Psi, \Psi)}.$$

In particular, since for Hermitian operators $(\Psi, AB\Psi)^* = (\Psi, BA\Psi)$, the imaginary part of this scalar product is

$$\mathrm{Im}(\Psi_A, \Psi_B) = \frac{(\Psi, [A, B]\Psi)}{2i(\Psi, \Psi)} = \langle [A, B] \rangle_\Psi /2i.$$

The absolute value of any complex number is equal to or greater than the absolute value of its imaginary part, so at last

$$\Delta_\Psi A \,\Delta_\Psi B \geq \frac{1}{2}|\langle [A, B]\rangle_\Psi|. \tag{3.3.24}$$

For example, if we have a pair of operators $X$ and $P$ for which $[X, P] = i\hbar$, then in any state $\Psi$,

$$\Delta_\Psi X \,\Delta_\Psi P \geq \frac{\hbar}{2}. \tag{3.3.25}$$

This is the Heisenberg uncertainty relation, discussed in Section 1.5. It is not possible to derive an improved general lower bound on $\Delta_\Psi X \,\Delta_\Psi P$, because for a Gaussian wave packet this product actually equals $\hbar/2$.

For some operators $A$, we may define a number called the *trace*, written Tr $A$. The trace is defined by introducing a complete orthonormal set of basis vectors $\Psi_i$, and writing

$$\text{Tr } A \equiv \sum_i (\Psi_i, A\Psi_i). \tag{3.3.26}$$

This definition is useful because the trace, where it exists, is independent of the choice of basis vectors. According to Eq. (3.1.16), for any other complete orthonormal set of basis vectors $\Phi_i$, we have

$$A\Psi_i = \sum_j (\Phi_j, A\Psi_i)\Phi_j,$$

so Eqs. (3.3.26) and (3.1.17) give

$$\text{Tr } A = \sum_{ij} (\Phi_j, A\Psi_i)(\Psi_i, \Phi_j) = \sum_j (\Phi_j, A\Phi_j).$$

The trace has some obvious properties:

$$\text{Tr}(\alpha A + \beta B) = \alpha \,\text{Tr}A + \beta \,\text{Tr } B, \quad \text{Tr } A^\dagger = (\text{Tr } A)^*. \tag{3.3.27}$$

Also,

$$\text{Tr}(AB) = \sum_i (\Psi_i, AB\Psi_i) = \sum_{ij} (\Psi_i, A\Psi_j)(\Psi_j, B\Psi_i)$$

$$= \sum_{ij} (\Psi_j, B\Psi_i)(\Psi_i, A\Psi_j)$$

$$= \text{Tr}(BA). \tag{3.3.28}$$

But not all operators have traces. The trace of the unit operator $\mathbf{1}$ is just $\sum_i 1$, which is the dimensionality of the Hilbert space, and hence is not defined in Hilbert spaces of infinite dimensionality. Note in particular that in a space of finite dimensionality the trace of the commutation relation $[X, P] = i\hbar\mathbf{1}$ would give the contradictory result $0 = i\hbar \operatorname{Tr} \mathbf{1}$, so this commutation relation can only be realized in Hilbert spaces of infinite dimensionality, where the traces do not exist.

Operators can be constructed from state vectors. For any two state vectors $\Psi$ and $\Omega$, we may define a linear operator $\left[\Psi\Omega^\dagger\right]$ known as a *dyad*, by the statement that, acting on an arbitrary state vector $\Phi$, this operator gives[8]

$$\left[\Psi\Omega^\dagger\right]\Phi \equiv \Psi(\Omega, \Phi). \tag{3.3.29}$$

The adjoint of this dyad is $\left[\Psi\Omega^\dagger\right]^\dagger = \left[\Omega\Psi^\dagger\right]$. The result of operating on an arbitrary state vector $\Phi$ with a product of such dyads is

$$\left[\Psi_1\Omega_1^\dagger\right]\left[\Psi_2\Omega_2^\dagger\right]\Phi = \left(\Omega_2, \Phi\right)\left[\Psi_1\Omega_1^\dagger\right]\Psi_2 = \left(\Omega_2, \Phi\right)\left(\Omega_1, \Psi_2\right)\Psi_1,$$

so the product is a numerical factor times another dyad:

$$\left[\Psi_1\Omega_1^\dagger\right]\left[\Psi_2\Omega_2^\dagger\right] = \left(\Omega_1, \Psi_2\right)\left[\Psi_1\Omega_2^\dagger\right]. \tag{3.3.30}$$

(For any given state vector $\Omega$ we can if we like introduce an operator $\Omega^\dagger$, which operating on any state vector $\Phi$ yields the number $(\Omega, \Phi)$, but in this book we will not have occasion to employ the symbol $\Omega^\dagger$ except as an ingredient in the symbols for dyads like $\left[\Psi\Omega^\dagger\right]$.)

In particular, if $\Phi$ is a normalized state vector, then the dyad $\left[\Phi\Phi^\dagger\right]$ is an Hermitian operator equal to its own square:

$$[\Phi\Phi^\dagger]^2 = [\Phi\Phi^\dagger]. \tag{3.3.31}$$

Such operators are called *projection operators*. From Eq. (3.3.31) it follows that the eigenvalues $\lambda$ of projection operators satisfy $\lambda^2 = \lambda$, and therefore are all either one or zero. The projection operator $[\Phi\Phi^\dagger]$ represents an observable, that takes the value one in the state represented by $\Phi$, and the value zero in any state represented by a vector orthogonal to $\Phi$. For a complete orthonormal set of state vectors $\Phi_i$, the relation (3.1.17) may be expressed as a statement about the sum of the corresponding projection operators

$$\sum_i \left[\Phi_i\Phi_i^\dagger\right] = \mathbf{1}. \tag{3.3.32}$$

An Hermitian operator $A$ with eigenvalues $a_i$ and a complete set of orthonormal eigenvectors $\Phi_i$ can be expressed as a sum of projection operators with coefficients equal to the eigenvalues:

---

[8] Here the Dirac bra–ket notation is particularly convenient. The dyad $\left[\Psi\Omega^\dagger\right]$ is written in this notation as $|\Psi\rangle\langle\Omega|$, which immediately suggests that $(|\Psi\rangle\langle\Omega|)|\Phi\rangle = |\Psi\rangle(\langle\Omega|\Phi\rangle)$, which is the same as Eq. (3.3.29).

$$A = \sum_i a_i \left[ \Phi_i \Phi_i^\dagger \right]. \tag{3.3.33}$$

(To see this, it is only necessary to check that the operator $A - \sum_i a_i \left[ \Phi_i \Phi_i^\dagger \right]$ annihilates any of the $\Phi_i$; since the $\Phi_i$ form a complete set, this operator therefore vanishes.)

From Eq. (3.3.33) it is easy to see that for any polynomial function $P(A)$ of an Hermitian operator $A$, we have

$$P(A) = \sum_i P(a_i) \left[ \Phi_i \Phi_i^\dagger \right].$$

We extend this to a definition of general functions of operators: for any function $f(a)$ that is finite at the eigenvalues $a_i$, we define

$$f(A) \equiv \sum_i f(a_i) \left[ \Phi_i \Phi_i^\dagger \right]. \tag{3.3.34}$$

Probabilities can enter in quantum mechanics not only because of the probabilistic nature of state vectors, but also because (just as in classical mechanics) we may not know the state of a system. A system may be in any one of a number of states, represented by state vectors $\Psi_n$ that are normalized *but not necessarily orthogonal*, with probabilities $P_n$ satisfying $\sum_n P_n = 1$. (For instance, an atomic state with $\ell = 1$ may have a 20% chance of being in a state with $L_z = \hbar$, a 30% chance of having $L_x = 0$, and a 50% chance of having $(L_x + L_y)/\sqrt{2} = \hbar$.) In such cases, it is often convenient to define a *density matrix* (actually an operator, not a matrix) as a sum of projection operators, with coefficients equal to the corresponding probabilities

$$\rho \equiv \sum_n P_n \left[ \Psi_n \Psi_n^\dagger \right]. \tag{3.3.35}$$

We note that the expectation value of the observable represented by an arbitrary Hermitian operator $A$ is the sum of the expectation values in the individual states $\Psi_n$, weighted with the probabilities of these states:

$$\langle A \rangle = \sum_n P_n \left( \Psi_n, A\Psi_n \right) = \mathrm{Tr}\{A\rho\}. \tag{3.3.36}$$

So in quantum mechanics the physical properties of a statistical ensemble of possible states are completely characterized by the density matrix of the ensemble. This is remarkable, because the same density matrix can be written in different ways as sums over various sets of states with various probabilities. In particular, because the density matrix (3.3.35) is Hermitian, it has a complete set of orthonormal eigenvectors $\Phi_i$ with eigenvalues $p_i$, so it can also be written

$$\rho = \sum_i p_i \left[ \Phi_i \Phi_i^\dagger \right]. \tag{3.3.37}$$

Also, $\rho$ is a positive operator, in the sense that any of its expectation values is a positive number, so all $p_i$ have $p_i \geq 0$. Finally, using Eq. (3.1.17), we can see that the operator (3.3.35) has unit trace

$$\mathrm{Tr}\,\rho = \sum_n P_n = 1,$$

so applying this to the representation (3.3.37), we also have $\sum_i p_i = 1$. As far as calculating expectation values is concerned, we can equally well say that the system is in any of the states represented by possibly non-orthogonal state vectors $\Psi_n$, with probabilities $P_n$, or in any of the states represented by the orthogonal state vectors $\Phi_i$, with probabilities $p_i$. It is a special feature of quantum mechanics that our knowledge of the same system can be expressed in different ways, as different sets of probabilities that the system is in different sets of states. As we shall see in Section 12.1, it is this feature of quantum mechanics that prevents the instantaneous transmission of information between distant isolated observers.

It is sometimes convenient to express the degree to which the state of a system differs from a single pure state by the *von Neumann entropy*:

$$S[\rho] \equiv -k_{\mathrm{B}}\,\mathrm{Tr}\Big(\rho \ln \rho\Big) = -k_{\mathrm{B}} \sum_i p_i \ln p_i, \qquad (3.3.38)$$

where $k_{\mathrm{B}}$ (often omitted) is the Boltzmann constant. For a pure state, with one $p_i$ equal to unity and all others equal to zero, the von Neumann entropy vanishes, while in all other cases we have $S > 0$.

We often encounter systems that are composed of two subsystems, so that we label states with compound indices $ma$, $nb$, etc.: $\Psi_{ma}$ would be a vector representing a state in which subsystem $I$ is in state $m$ and subsystem $II$ is in state $a$. These two subsystems might be just two atoms, or subsystem $I$ might be some microscopic system of interest while subsystem $II$ is its environment. If an observable is represented by an operator $A$ that acts non-trivially only on the states of subsystem $I$, that is,

$$A_{ma,nb} = A^I_{mn}\delta_{ab}, \qquad (3.3.39)$$

then its mean value in an ensemble of states with density matrix $\rho_{ma,nb}$ is

$$\langle A \rangle = \mathrm{Tr}(A\rho) = \sum_{manb} A_{ma,nb}\rho_{nb,ma} = \sum_{mn} A^I_{mn}\rho^I_{nm}, \qquad (3.3.40)$$

where

$$\rho^I_{mn} \equiv \sum_a \rho_{ma,na}. \qquad (3.3.41)$$

We can thus think of $\rho^I_{mn}$ as the density matrix for subsystem $I$, relevant to the case in which nothing is being done to probe subsystem $II$. Note that like any density matrix, $\rho^I$ is Hermitian, positive, and has unit trace. In the same sense, $\rho^{II}_{ab} \equiv \sum_m \rho_{ma,nb}$ can be regarded as the density matrix of subsystem $II$.

Where there is no correlation between the two subsystems, the density matrix of the whole system is the direct product of the density matrices of the subsystems: $\rho = \rho^I \otimes \rho^{II}$, or more explicitly

$$\rho_{ma,nb} = \rho_{mn}^I \, \rho_{ab}^{II}. \tag{3.3.42}$$

In this case, each eigenvalue of $\rho$ is the product of an eigenvalue $p_i^I$ of $\rho^I$ and an eigenvalue $p_r^{II}$ of $\rho^{II}$, and the von Neumann entropy (3.3.38) is therefore simply additive:

$$S[\rho] = -k_{\rm B} \sum_{ir} p_i^I p_r^{II} \ln[p_i^I p_r^{II}] = -- k_{\rm B} \sum_{ir} p_i^I p_r^{II} \left( \ln[p_i^I] + \ln[p_r^{II}] \right)$$

$$= S[\rho^I] + S[\rho^{II}]. \tag{3.3.43}$$

The case of *entanglement*, in which neither Eq. (3.3.42) nor Eq. (3.3.43) holds, is the subject of Chapter 12.

## 3.4 Symmetries

Historically, it was classical mechanics that provided quantum mechanics with a menu of observable quantities and with their properties. But much of this can be learned from fundamental principles of symmetry, without recourse to classical mechanics.

A symmetry principle is a statement that, when we change our point of view in certain ways, the laws of nature do not change. For instance, moving or rotating our laboratory should not change the laws of nature observed in the laboratory. Such special ways of changing our point of view are called symmetry transformations. This definition does not mean that a symmetry transformation does not change physical states, but only that the new states after a symmetry transformation will be observed to satisfy the same laws of nature as the old states.

In particular, symmetry transformations must not change transition probabilities. Recall that if a system is in a state represented by a normalized Hilbert space vector $\Psi$, and we perform a measurement (say, of a set of observables represented by commuting Hermitian operators) which puts the system in any one of a complete set of states represented by orthonormal state vectors $\Phi_i$, then the probability of finding the system in a state represented by a particular $\Phi_i$ is given by Eq. (3.1.15):

$$P(\Psi \mapsto \Phi_i) = \left| \left( \Phi_i, \Psi \right) \right|^2. \tag{3.4.1}$$

Thus symmetry transformations must leave all $\left| \left( \Phi, \Psi \right) \right|^2$ invariant. One way to satisfy this condition is to suppose that a symmetry transformation takes general state vectors $\Psi$ into other state vectors $U\Psi$, where $U$ is a linear operator satisfying the condition of *unitarity*, namely that for any two state vectors $\Phi$ and $\Psi$, we have

$$\left( U\Phi, U\Psi \right) = \left( \Phi, \Psi \right). \tag{3.4.2}$$

Recall that the adjoint of an operator $U$ is defined so that

$$\left( U\Phi, U\Psi \right) = \left( \Phi, U^\dagger U\Psi \right),$$

so the condition of unitarity may also be expressed as an operator relation:

$$U^\dagger U = 1. \tag{3.4.3}$$

We limit ourselves to symmetry transformations that, like rotations and translations, have inverses, which undo the effect of the transformation. (For instance, the symmetry transformation of rotating around some axis by an angle $\theta$ has an inverse symmetry transformation, in which one rotates around the same axis by an angle $-\theta$.) If a symmetry transformation is represented by a linear unitary operator that takes any $\Psi$ into $U\Psi$, then its inverse must be represented by a left-inverse operator $U^{-1}$ that takes $U\Psi$ into $\Psi$, so that

$$U^{-1}U = 1. \tag{3.4.4}$$

The same must be true for $U^{-1}$ itself, so it has a left-inverse $(U^{-1})^{-1}$ for which $(U^{-1})^{-1}U^{-1} = 1$. Multiplying this on the right with $U$ and using Eq. (3.4.4) then gives

$$(U^{-1})^{-1} = U, \tag{3.4.5}$$

so by applying Eq. (3.4.4) to $U^{-1}$, we see that the left-inverse of $U$ is also a right-inverse:

$$UU^{-1} = 1. \tag{3.4.6}$$

Acting on Eq. (3.4.3) on the right with $U^{-1}$, we see that the inverse of a unitary operator is its adjoint:

$$U^\dagger = U^{-1}. \tag{3.4.7}$$

Now, is this the only way that symmetry transformations can act on physical states? In formulating the mathematical conditions for symmetry principles in quantum mechanics, we immediately run into a complication. As discussed in Section 3.1, in quantum mechanics a physical state is not represented by a specific individual normalized vector in Hilbert space, but by a ray, the whole class of normalized state vectors that differ from one another only by phase factors, numerical factors with modulus unity. We have no right simply to assume that a symmetry transformation must map an arbitrary vector in Hilbert space into some other definite vector. We are only entitled to require that symmetry transformations map rays into rays – that is, a symmetry transformation acting on the normalized state vectors differing by phase factors that represent a given physical state will yield some other class of normalized state vectors differing only by phase factors that represent some other physical state. To represent a symmetry, such a transformation of rays must preserve transition probabilities – that is, if $\Psi$ and $\Phi$ are state vectors belonging to the rays representing two different physical states, and a symmetry transformation takes these two rays into two other rays containing the state vectors $\Psi'$ and $\Phi'$, then we must have

$$|(\Phi', \Psi')|^2 = |(\Phi, \Psi)|^2. \tag{3.4.8}$$

Notice that this is only a condition on rays – if it is satisfied by a given set of state vectors, then it is satisfied by any other set of state vectors that differ from the first set only by arbitrary phases.

There is a fundamental theorem due to Eugene Wigner[9] (1902–1995), which says that there are just two ways that this condition can be satisfied for all $\Psi$ and $\Phi$. One is the way we have already discussed: phases can be chosen so that the effect of a symmetry transformation on any state vector $\Psi$ is a transformation $\Psi \rightarrow U\Psi$, with $U$ a linear unitary operator satisfying the condition (3.4.2). The other possibility is that $U$ is antilinear and antiunitary, by which it is meant that

$$U(\alpha\Psi + \alpha'\Psi') = \alpha^* U\Psi + \alpha'^* U\Psi' \tag{3.4.9}$$

and

$$(U\Phi, U\Psi) = (\Phi, \Psi)^*. \tag{3.4.10}$$

(Note that an antiunitary operator cannot be linear, because if it were then we would have $\alpha(U\Phi, U\Psi) = (U\Phi, U\alpha\Psi) = (\Phi, \alpha\Psi)^* = \alpha^*(U\Phi, U\Psi)$, which is not true for complex $\alpha$.) For antiunitary operators the definition of the adjoint is changed to

$$(U^\dagger\Phi, \Psi) = (\Phi, U\Psi)^*,$$

so Eq. (3.4.3) applies to antiunitary as well as to unitary operators. We will see in Section 3.6 that symmetries represented by antilinear antiunitary operators all involve a change in the direction of time's flow. We will mostly be concerned with symmetries represented by linear unitary operators.

The operator **1** represents a trivial symmetry, that does nothing to state vectors. It is of course unitary as well as linear. If $U_1$ and $U_2$ both represent symmetry transformations, then so does $U_1 U_2$. This property, together with the existence of inverses and a trivial transformation **1**, means that the set of all operators representing symmetry transformations forms a *group*.

There is a special class of symmetries represented by linear unitary operators – those for which $U$ can be arbitrarily close to **1**. Any such symmetry operator can conveniently be written

---

[9] E. P. Wigner, *Ann. Math.* **40**. 149 (1939). Some missing steps are provided by S. Weinberg, *The Quantum Theory of Fields*, Vol. 1 (Cambridge University Press, Cambridge, 1995), pp. 91–96.

$$U_\epsilon = 1 + i\epsilon T + O(\epsilon^2), \tag{3.4.11}$$

where $\epsilon$ is an arbitrary real infinitesimal number, and $T$ is some $\epsilon$-independent operator. The unitarity condition is

$$\left(1 - i\epsilon T^\dagger + O(\epsilon^2)\right)\left(1 + i\epsilon T + O(\epsilon^2)\right) = 1,$$

or, to first order in $\epsilon$,

$$T = T^\dagger. \tag{3.4.12}$$

Thus Hermitian operators arise naturally in the presence of infinitesimal symmetries. If we take $\epsilon = \theta/N$, where $\theta$ is some finite $N$-independent parameter, and then carry out the symmetry transformation $N$ times and let $N$ go to infinity, we find a transformation represented by the operator

$$\left[1 + i\theta T/N\right]^N \to \exp(i\theta T) = U(\theta). \tag{3.4.13}$$

(To see that this is true for Hermitian operators $T$, note that it is true when both sides of the equation act on any eigenvector of $T$, where $T$ can be replaced with the eigenvalue, and since these eigenvectors form a complete set, it is true in general.) The operator $T$ appearing in Eq. (3.4.11) is known as the *generator* of the symmetry. As we shall see, *many if not all of the operators representing observables in quantum mechanics are the generators of symmetries.* For instance, the total momentum is the generator of translations of spatial coordinates (Section 3.5); the Hamiltonian is the generator of translations of the time (Section 3.6); and the total angular momentum is the generator of spatial rotations (Section 4.1).

Under a symmetry transformation $\Psi \mapsto U\Psi$, the expectation value of any observable $A$ is subjected to the transformation

$$(\Psi, A\Psi) \mapsto (U\Psi, AU\Psi) = (\Psi, U^{-1}AU\Psi), \tag{3.4.14}$$

so we can find the transformation properties of expectation values (or any other matrix elements) by subjecting observables to the transformation

$$A \mapsto U^{-1}AU. \tag{3.4.15}$$

Transformations of this type are called *similarity transformations*. Note that similarity transformations preserve algebraic relations:

$$U^{-1}AU \times U^{-1}BU = U^{-1}(AB)U, \quad U^{-1}AU + U^{-1}BU = U^{-1}(A+B)U.$$

Also, similarity transformations do not change the eigenvalues of operators; if $\Psi$ is an eigenvector of $A$ with eigenvalue $a$, then $U^{-1}\Psi$ is an eigenvector of $U^{-1}AU$ with the same eigenvalue. Where $U$ takes the form (3.4.11) with $\epsilon$ infinitesimal, an arbitrary operator $A$ is transformed into

$$A \mapsto A - i\epsilon[T, A] + O(\epsilon^2). \tag{3.4.16}$$

Thus the effect of infinitesimal symmetry transformations on any operator is expressed in the commutation relations of the symmetry generator with that operator. This is in particular true when the operator $A$ is itself a symmetry generator; as we will see in several examples, in that case the commutation relations reflect the nature of the symmetry group.

## 3.5 Space Translation

As an example of a symmetry transformation of great physical importance, let us consider the symmetry under spatial translation: the laws of nature should not change if we shift the origin of our spatial coordinate system, so that any particle coordinate $\mathbf{X}_n$ (where $n$ labels the individual particles) is transformed to $\mathbf{X}_n + \boldsymbol{a}$, where $\boldsymbol{a}$ is an arbitrary three-vector. It follows that there must exist a unitary operator[10] $U(\mathbf{a})$ such that

$$U^{-1}(\mathbf{a})\mathbf{X}_n U(\mathbf{a}) = \mathbf{X}_n + \mathbf{a}. \tag{3.5.1}$$

In particular, for $\mathbf{a}$ infinitesimal, $U$ must take a form like (3.4.11), which in this case we will write with an Hermitian three-vector operator $-\mathbf{P}/\hbar$ in place of $T$:

$$U(\mathbf{a}) = 1 - i\mathbf{P} \cdot \mathbf{a}/\hbar + O(\mathbf{a}^2). \tag{3.5.2}$$

The condition (3.5.1) then requires that, for any infinitesimal three-vector $\mathbf{a}$,

$$i[\mathbf{P} \cdot \mathbf{a}, \mathbf{X}_n]/\hbar = \mathbf{a},$$

and therefore

$$[X_{ni}, P_j] = i\hbar\delta_{ij}. \tag{3.5.3}$$

The presence of $\hbar$ in this familiar commutation relation arises because we conventionally express the generator of spatial translations in units of mass times velocity, rather than in natural units of inverse length. Equation (3.5.2) can simply be taken as the definition of what we mean by momentum, leaving it to experience to justify the identification of this symmetry generator with what is called momentum in classical mechanics.

It should be noted that the operator $\mathbf{P}$ introduced here has the same commutation relation (3.5.3) with the coordinate vector of any particle, so $\mathbf{P}$ must be interpreted as the *total* momentum of any system. In a system containing a number of different particles labeled $n$, the total momentum usually takes the form

$$\mathbf{P} = \sum_n \mathbf{P}_n, \tag{3.5.4}$$

---

[10] We will generally not bother to label such unitary operators with the nature of the symmetry they represent, leaving this to be indicated by the argument of the unitary operator.

where the operator $\mathbf{P}_n$ acts only on the $n$th particle, and therefore

$$[\mathbf{P}_n, \mathbf{X}_m] = 0 \text{ for } n \neq m. \tag{3.5.5}$$

It follows then from Eq. (3.5.3) that

$$[X_{ni}, P_{mj}] = i\hbar \delta_{ij} \delta_{nm}. \tag{3.5.6}$$

Of course, the individual momentum operators $\mathbf{P}_n$ are not the generators of any symmetry of nature.

A translation by a vector $\mathbf{a}$ followed by a translation by a vector $\mathbf{b}$ gives the same change of coordinates as a translation by a vector $\mathbf{b}$ followed by a translation by a vector $\mathbf{a}$, so

$$U(\mathbf{b})U(\mathbf{a}) = U(\mathbf{a})U(\mathbf{b}).$$

The terms in this relation proportional to $a_i b_j$ tell us that the components of momentum commute with each other:

$$[P_i, P_j] = 0. \tag{3.5.7}$$

Because they commute, we can find a complete set of eigenvectors of all three components of momentum, so by the same argument we used earlier in deriving Eq. (3.4.13), for finite translations we have

$$U(\mathbf{a}) = \exp\left(-i\mathbf{P} \cdot \mathbf{a}/\hbar\right). \tag{3.5.8}$$

This is a very simple example of the derivation of commutation relations from the structure of a transformation group. It isn't always so easy. The effect of two rotations around different axes depends on the order in which the rotations are carried out, so, as we shall see in the next chapter, the different components of the generator of rotations, the angular momentum vector, do not commute with each other.

If $\Phi_0$ is a one-particle state with a definite position at the origin (that is, an eigenstate of the position operator $\mathbf{X}$ with eigenvalue zero), then according to Eq. (3.5.1), we can form a state with definite position $\mathbf{x}$:

$$\Phi_{\mathbf{x}} \equiv U(\mathbf{x})\Phi_0, \tag{3.5.9}$$

in the sense that

$$\mathbf{X}\Phi_{\mathbf{x}} = \mathbf{x}\Phi_{\mathbf{x}}. \tag{3.5.10}$$

From Eq. (3.5.6) we can infer that

$$P_j \Phi_{\mathbf{x}} = i\hbar \frac{\partial}{\partial x_j} \Phi_{\mathbf{x}}, \tag{3.5.11}$$

so the scalar product of this state with a state $\Psi_{\mathbf{p}}$ of definite momentum is

$$\left(\Psi_{\mathbf{p}}, \Phi_{\mathbf{x}}\right) = \exp\left(-i\mathbf{p} \cdot \mathbf{x}/\hbar\right)\left(\Psi_{\mathbf{p}}, \Phi_0\right).$$

It is convenient to normalize these states so that

$$\left(\Psi_{\mathbf{p}}, \Phi_{\mathbf{x}}\right) = (2\pi\hbar)^{-3/2} \exp\left(-i\mathbf{p} \cdot \mathbf{x}/\hbar\right).$$

The complex conjugate gives the usual plane wave formula for the coordinate-space wave function of a particle of definite momentum

$$\psi_{\mathbf{p}}(\mathbf{x}) \equiv \left(\Phi_{\mathbf{x}}, \Psi_{\mathbf{p}}\right) = (2\pi\hbar)^{-3/2} \exp\left(i\mathbf{p} \cdot \mathbf{x}/\hbar\right). \qquad (3.5.12)$$

This normalization has the virtue that, if the states $\Phi_{\mathbf{x}}$ satisfy the usual normalization condition for continuum states

$$\left(\Phi_{\mathbf{x}'}, \Phi_{\mathbf{x}}\right) = \delta^3(\mathbf{x} - \mathbf{x}'),$$

then so do the states $\Psi_{\mathbf{p}}$. That is, the scalar product of these states is

$$\left(\Psi_{\mathbf{p}'}, \Psi_{\mathbf{p}}\right) = \int d^3x \, \psi_{\mathbf{p}'}^*(\mathbf{x})\psi_{\mathbf{p}}(\mathbf{x}) = \int d^3x \, (2\pi\hbar)^{-3} \exp\left(i(\mathbf{p} - \mathbf{p}') \cdot \mathbf{x}/\hbar\right).$$

We recognize this integral as the product of the representations (3.2.21) of the delta function (with $k_i = p_i/\hbar$) for each coordinate direction, so

$$\left(\Psi_{\mathbf{p}'}, \Psi_{\mathbf{p}}\right) = \delta^3(\mathbf{p} - \mathbf{p}'), \qquad (3.5.13)$$

as required by Eq. (3.2.14).

$$* \ * \ * \ * \ *$$

In some external environments, the Hamiltonian is not invariant under all translations, but only under a subgroup of the translation group. In a three-dimensional crystal, the Hamiltonian is invariant under spatial translations

$$\mathbf{x} \mapsto \mathbf{x} + \mathbf{L}_r, \qquad r = 1, 2, 3, \qquad (3.5.14)$$

as well as any combinations of these. The $\mathbf{L}_r$ are the three independent translation vectors that take any atom to the neighboring atom with an identical crystal environment. (Of course, $\mathbf{L}_r$ are three independent vectors, not the three components of a single vector.) For instance, in a cubic lattice like sodium chloride the three $L_r$ are orthogonal vectors of equal length, but in general they do not need to be either orthogonal or equal in length.

Because of this symmetry, if $\psi(\mathbf{x})$ is a solution of the time-independent Schrödinger equation for an electron in the crystal, then each of $\psi(\mathbf{x} + \mathbf{L}_r)$ with $r = 1, 2, 3$ is also a solution with the same energy. Assuming no degeneracy,[11]

---

[11]  The conclusion (3.5.15) applies also in the case of degeneracy, but a few more words are needed in the argument. In the case of an $N$-fold degeneracy, in place of the factors $\exp(i\theta_r)$ in Eq. (3.5.15) we have three $N \times N$ unitary matrices. Because translations commute, these three unitary matrices commute with each other, and hence we can choose a basis for the $N$ degenerate wave functions in which the unitary matrices are diagonal: they have phase factors $\exp(i\theta_{rv})$ on the main diagonal, with $v = 1, 2, \ldots, N$,

this requires that $\psi(\mathbf{x} + \mathbf{L}_r)$ is simply proportional to $\psi(\mathbf{x})$, with a proportionality constant that is required by the normalization of the wave function to be a phase factor:

$$\psi(\mathbf{x} + \mathbf{L}_r) = e^{i\theta_r}\psi(\mathbf{x}), \tag{3.5.15}$$

where $\theta_r$ are three real angles. In the language of group theory, the wave function provides a one-dimensional representation of the group of translations that consists of all combinations of the three fundamental translations (3.5.14). Without loss of generality, we can limit each of the $\theta_r$ by

$$0 \le \theta_r < 2\pi, \qquad r = 1,\, 2,\, 3. \tag{3.5.16}$$

We will define a wave vector $\mathbf{q}$ by the three conditions

$$\mathbf{q} \cdot \mathbf{L}_r = \theta_r, \qquad r = 1,\, 2,\, 3. \tag{3.5.17}$$

In the special case of a cubic lattice, this directly gives the Cartesian components of $\mathbf{q}$. More generally, it is necessary to solve these three linear equations to find the three components of $\mathbf{q}$. In any case, it follows from Eqs. (3.5.15) and (3.5.17) that the function $e^{-i\mathbf{q}\cdot\mathbf{x}}\psi(\mathbf{x})$ is periodic, the factors arising from the change in the exponential canceling the factors $e^{i\theta_r}$ in Eq. (3.5.15). Hence we may write

$$\psi(\mathbf{x}) = e^{i\mathbf{q}\cdot\mathbf{x}}\varphi(\mathbf{x}), \tag{3.5.18}$$

where $\varphi(\mathbf{x})$ is periodic, in the sense that

$$\varphi(\mathbf{x} + \mathbf{L}_r) = \varphi(\mathbf{x}), \qquad r = 1,\, 2,\, 3. \tag{3.5.19}$$

Such solutions of the Schrödinger equation are known as *Bloch waves.*[12]

If $\psi(\mathbf{x})$ satisfies a Schrödinger equation of the form

$$H(\nabla, \mathbf{x})\psi(\mathbf{x}) = E\psi(\mathbf{x}), \tag{3.5.20}$$

then $\varphi(\mathbf{x})$ satisfies a $\mathbf{q}$-dependent equation

$$H(\nabla + i\mathbf{q}, \mathbf{x})\varphi(\mathbf{x}) = E\varphi(\mathbf{x}). \tag{3.5.21}$$

Just as in the case of free particles in a box with periodic boundary conditions, the periodicity conditions (3.5.19) make the spectrum of eigenvalues for each $\mathbf{q}$ appearing in the differential equation (3.5.21) a discrete set $E_n(\mathbf{q})$. Of course, $\mathbf{q}$ is a continuous variable, but according to Eqs. (3.5.16) and (3.5.17) it varies only over a finite range, defined by[13]

---

and zero everywhere else. In this basis Eq. (3.5.15) applies to the $\nu$th degenerate wave function, with a phase $\theta_{r\nu}$ in place of $\theta_r$.

[12] F. Bloch, *Z. Physik* **52**, 555 (1928).

[13] This is known as the first Brillouin zone, identified by L. Brillouin, *Comptes Rendus* **191**, 292 (1930). If we had adopted a convention for the angles $\theta_r$ in Eq. (3.5.15) other than Eq. (3.5.16), then the wave vector $\mathbf{q}$ would lie in one of various other finite regions, known as the second, third, etc. Brillouin zones. This would just amount to a re-definition of the periodic function $\varphi(\mathbf{x})$, with no change in physical results.

$$|\mathbf{q} \cdot \mathbf{L}_r| < 2\pi, \qquad r = 1, 2, 3. \tag{3.5.22}$$

Hence for each $n$ the energies $E_n(\mathbf{q})$ occupy a finite band. As will briefly be described in Section 4.5, many of the properties of crystalline solids depend on the occupancy of these bands.

## 3.6 Time Translation and Inversion

One of the fundamental symmetries of nature is time-translation invariance – the laws of nature should not depend on how we set our clocks. Thus whatever time-dependence a physical state vector $\Psi(t)$ may have, the results $\Psi(t + \tau)$ of a time translation by an arbitrary amount $\tau$ should be physically equivalent, so there must be some linear unitary operator $U(\tau)$ such that the state of a system at time $t$ is transformed to

$$U(\tau)\Psi(t) = \Psi(t + \tau). \tag{3.6.1}$$

Because $\tau$ is a continuous variable, it must be possible to express $U(\tau)$ in a form like (3.4.13). For time translation in place of the general Hermitian operator $T$ in Eq. (3.4.13), we introduce an Hermitian operator $-H/\hbar$, so that

$$U(\tau) = \exp\left(-i H\tau/\hbar\right). \tag{3.6.2}$$

This can be taken as the definition of the Hamiltonian $H$.

It follows, by setting $t = 0$ in Eq. (3.6.1) and then replacing $\tau$ with $t$, that the time-dependence of any physical state vector is given by

$$\Psi(t) = \exp\left(-i H t/\hbar\right) \Psi(0). \tag{3.6.3}$$

Like any symmetry transformation represented by linear unitary operators, this leaves scalar products invariant:

$$\left(\Phi(t), \Psi(t)\right) = \left(\Phi(0), \Psi(0)\right). \tag{3.6.4}$$

From Eq. (3.6.3) we can easily derive a differential equation for the time-dependence of the state vector:

$$i\hbar\dot{\Psi}(t) = H\Psi(t). \tag{3.6.5}$$

This is the general version of the time-dependent Schrödinger equation.

This formalism, in which we ascribe time-dependence to physical states (and hence to wave functions), is known as the *Schrödinger picture*. There is a completely equivalent formalism, in which we keep the state vectors fixed, by describing any state in terms of its appearance at a fixed time such as $t = 0$, and instead ascribe time-dependence to operators representing observables. In order that the time-dependence of expectation values should be the same in both pictures, we must define operators in the Heisenberg picture by

$$A_H(t) = \exp\left(+iHt/\hbar\right) A \exp\left(-iHt/\hbar\right). \qquad (3.6.6)$$

Note that, since $H$ commutes with itself,

$$\exp\left(+iHt/\hbar\right) H \exp\left(-iHt/\hbar\right) = H,$$

so the Hamiltonian is the same in the Heisenberg and Schrödinger pictures. The time-dependence of any operator in the Heisenberg picture is given by

$$\dot{A}_H(t) = i[H, A_H(t)]/\hbar, \qquad (3.6.7)$$

provided that the definition of $A$ does not refer explicitly to time. The Hamiltonian thus determines the time-dependence of most physical quantities. Any operator $A$ that commutes with the Hamiltonian and that does not depend explicitly on time is *conserved*, in the sense that $\dot{A}_H(t) = 0$, which means that expectation values of this observable are time-independent, irrespective of whether we use the Heisenberg picture or the Schrödinger picture.

Symmetry principles provide a natural reason why physical theories should involve conserved quantities. If an observer sees a state $\Psi(t)$ evolving according to Eq. (3.6.3), then another observer for whom the laws of nature are the same must see the state $U\Psi(t)$ evolving according to the same equation

$$U\Psi(t) = \exp\left(-iHt/\hbar\right) U\Psi(0). \qquad (3.6.8)$$

In order for this to be consistent with Eq. (3.6.3) for all states, we must have

$$\exp\left(-iHt/\hbar\right) U = U \exp\left(-iHt/\hbar\right), \qquad (3.6.9)$$

and therefore, provided $U$ is a linear operator,

$$U^{-1}HU = H. \qquad (3.6.10)$$

That is, the Hamiltonian must be invariant under the symmetry transformation. For an infinitesimal symmetry transformation with $U$ given by Eq. (3.4.11), this tells us that

$$[H, T] = 0, \qquad (3.6.11)$$

so *observables represented by the generators of symmetries of the Hamiltonian commute with the Hamiltonian*. It is invariance under space and time translation that is responsible for the conservation of momentum and energy.

Note that this would not work if $U$ were antilinear. In that case, because of the $i$ in the exponent in Eq. (3.6.9), in place of Eq. (3.6.10) we would find $U^{-1}HU = -H$. This would imply that for every eigenstate $\Phi$ of the Hamiltonian with energy $E$, there would be another eigenstate $U\Phi$ with energy $-E$, which is clearly in conflict with observation and with the stability of matter.[14] The only way to avoid this conclusion for symmetries represented by antilinear operators is to suppose that, instead of Eq. (3.6.8), such symmetries reverse the direction of time:

$$U\Psi(t) = \exp\left(iHt/\hbar\right) U\Psi(0). \tag{3.6.12}$$

Then in place of Eq. (3.6.9), consistency with Eq. (3.6.3) would require that

$$\exp\left(iHt/\hbar\right) U = U \exp\left(-iHt/\hbar\right). \tag{3.6.13}$$

With $U$ antilinear, this again yields the result that $U$ commutes with $H$, avoiding the disaster of negative energies. So we see that symmetries represented by antilinear operators are possible, but they necessarily involve a reversal of the direction of time.

It used to be thought that nature respects a symmetry under a transformation $t \to -t$ with everything else left unchanged. As discussed in Section 4.7, it is now known that this symmetry is violated by the weak interactions, although it is a good approximation even there. The application of time-reversal symmetry to scattering processes is described in Section 8.9. There is also a transformation that reverses both the direction of time and of space, and also interchanges matter and antimatter, which is believed to be an exact symmetry of all interactions. This is discussed further in Section 4.7.

Not all symmetries are represented by operators that commute with the Hamiltonian. The leading example of a different sort of symmetry is invariance under Galilean transformations, which take the spatial coordinate $\mathbf{x}$ into $\mathbf{x} + \mathbf{v}t$ (where $\mathbf{v}$ is a constant velocity) while leaving the time coordinate unchanged. In quantum mechanics this symmetry requires there to be a unitary linear operator $U(\mathbf{v})$ such that

$$U^{-1}(\mathbf{v})\mathbf{X}_H(t)U(\mathbf{v}) = \mathbf{X}_H(t) + \mathbf{v}t, \tag{3.6.14}$$

where $\mathbf{X}_H(t)$ is the Heisenberg-picture operator representing the spatial coordinate of any particle. Taking the time-derivative of Eq. (3.6.14) and using Eq. (3.6.7) gives

$$iU^{-1}(\mathbf{v})[H, \mathbf{X}_H(t)]U(\mathbf{v}) = i[H, \mathbf{X}_H(t)] + \hbar\mathbf{v},$$

and therefore, setting $t = 0$,

$$i\left[U^{-1}(\mathbf{v})HU(\mathbf{v}), U^{-1}(\mathbf{v})\mathbf{X}U(\mathbf{v})\right] = i[H, \mathbf{X}] + \hbar\mathbf{v}.$$

---

[14] Negative-energy states were encountered by Dirac, not as a consequence of time reversal symmetry, but as negative-energy solutions of his relativistic wave equation. Dirac supposed that matter is stable because all or almost all of these negative-energy states are filled. (See P. A. M. Dirac, *Proc. Roy. Soc.* A **126**, 360 (1930).) Dirac's interpretation of negative-energy states is untenable, for reasons indicated in Section 4.6.

For $t = 0$ Eq. (3.6.14) tells us that $U(\mathbf{v})$ commutes with the Schrödinger-picture operator $\mathbf{X}$, so this gives

$$i\left[U^{-1}(\mathbf{v})HU(\mathbf{v}), \mathbf{X}\right] = i[H, \mathbf{X}] + \hbar\mathbf{v}. \tag{3.6.15}$$

This requires that

$$U^{-1}(\mathbf{v})HU(\mathbf{v}) = H + \mathbf{P} \cdot \mathbf{v}, \tag{3.6.16}$$

where $\mathbf{P}$ is an operator satisfying the familiar commutation relation $[X_i, P_j] = i\hbar\delta_{ij}$ with *every* particle coordinate – that is, $\mathbf{P}$ is the total momentum vector.

For $\mathbf{v}$ infinitesimal we can write

$$U(\mathbf{v}) = 1 - i\mathbf{v} \cdot \mathbf{K} + O(\mathbf{v}^2), \tag{3.6.17}$$

with $\mathbf{K}$ some Hermitian operator, known as the *boost generator*. Since the transformations (3.6.14) are additive, we have $U(\mathbf{v})U(\mathbf{v}') = U(\mathbf{v} + \mathbf{v}')$, and hence

$$[K_i, K_j] = 0. \tag{3.6.18}$$

Also, letting $\mathbf{v}$ in Eq. (3.6.16) become infinitesimal, we find

$$[\mathbf{K}, H] = -i\mathbf{P}. \tag{3.6.19}$$

It is because $\mathbf{K}$ does not commute with the Hamiltonian that we do not use its eigenvalues to classify physical states of definite energy. The boost generator is an exception to the general rule that the generators of symmetries commute with the Hamiltonian. This exception arises because $\mathbf{K}$ is associated with a symmetry transformation (3.6.14) that depends explicitly on time.

Since Eq. (3.6.14) applies to the coordinate $\mathbf{X}_n$ of any particle (now labeling individual particles with a subscript $n$), by taking the time-derivative and multiplying with the particle mass $m_n$, we have

$$U^{-1}(\mathbf{v})\mathbf{P}_{nH}(t)U(\mathbf{v}) = \mathbf{P}_{nH}(t) + m_n\mathbf{v}, \tag{3.6.20}$$

where $\mathbf{P}_{nH} \equiv m_n\dot{\mathbf{X}}_{nH}$ is the momentum of the $n$th particle in the Heisenberg picture. Setting $t = 0$ and specializing to the infinitesimal Galilean transformations (3.6.17), this gives

$$[K_i, P_{nj}] = -im_n\delta_{ij}. \tag{3.6.21}$$

Note that then Eq. (3.6.19) is satisfied by the usual Hamiltonian for a multi-particle system

$$H = \sum_n \frac{\mathbf{P}_n^2}{2m_n} + V, \tag{3.6.22}$$

provided the potential $V$ depends only on the differences of the particle coordinate vectors. Indeed, from a point of view that regards symmetries as fundamental, we can say that Galilean invariance is the reason why Hamiltonians for non-relativistic particles take this form.

Note that the operators $\mathbf{K}$, $H$, and the total momentum $\mathbf{P} = \sum_n \mathbf{P}_n$ form a closed Lie algebra, in the sense that the commutators of these generators are linear combinations of the same generators. But there is a complication: the commutator of $K_i$ and $P_j$ is proportional to the total mass $\sum_n m_n$. Quantities like the total mass that appear in commutation relations but commute with all the operators in these relations are known as *central charges*. In theories that obey Lorentz invariance rather than Galilean invariance, there are again symmetries generated by the total momentum $\mathbf{P}$, the Hamiltonian $H$, and a boost generator $\mathbf{K}$, but the commutation relations are different: the commutator of $\mathbf{K}$ with $\mathbf{P}$ is proportional to $H$, not to the total mass; there are no central charges; and the commutators $[K_i, K_j]$ do not vanish, but are proportional to the total angular momentum operator.

$$* * * * *$$

It is sometimes useful to follow the time-dependence of the density matrix. Suppose that at time $t = 0$ the probabilities that a system is in various states represented by independent normalized (but not necessarily orthogonal) state vectors $\Psi_n$ are the positive quantities $P_n$, with $\sum_n P_n = 1$. Then, as discussed in Section 3.3, the density matrix at $t = 0$ is

$$\rho(0) = \sum_n P_n \left[\Psi_n \, \Psi_n^\dagger\right]. \tag{3.6.23}$$

At a later time $t$ the state vectors $\Psi_n$ turn into $\exp(-iHt/\hbar)\Psi_n$, and the density matrix becomes

$$\rho(t) = \sum_n P_n \exp(-iHt/\hbar) \left[\Psi_n \, \Psi_n^\dagger\right] \exp(+iHt/\hbar)$$
$$= \exp(-iHt/\hbar) \, \rho(0) \exp(+iHt/\hbar). \tag{3.6.24}$$

This is a unitary transformation, so $\rho(t)$ is Hermitian, and has the same eigenvalues as $\rho(0)$, and therefore is positive, has unit trace, and has the same von Neumann entropy as $\rho(0)$.

## 3.7 Interpretations of Quantum Mechanics

The discussion of probabilities in Section 3.1 was implicitly based on what is called the Copenhagen interpretation of quantum mechanics, formulated under the leadership of Niels Bohr.[15] According to Bohr,[16] "The essentially new

---

[15] N. Bohr , *Nature* **121**, 580 (1928), reprinted in *Quantum Theory and Measurement*, eds. J. A. Wheeler and W. H. Zurek (Princeton University Press, Princeton, NJ, 1983); *Essays 1958–1962 on Atomic Physics and Human Knowledge* (Interscience Publishers, New York, 1963).

[16] N. Bohr, "Quantum Mechanics and Philosophy – Causality and Complementarity," in *Philosophy in the Mid-Century*, ed. R. Klibansky (La Nuova Italia Editrice, Florence, 1958), reprinted in N. Bohr, *Essays 1958–1962 on Atomic Physics and Human Knowledge* (Interscience Publishers, New York, 1963).

feature of the analysis of quantum phenomena is ... the introduction of a *fundamental distinction between the measuring apparatus and the objects under investigation.* This is a direct consequence of the necessity of accounting for the functions of the measuring apparatus in purely classical terms, excluding in principle any regard to the quantum of action."

As Bohr acknowledged, in the Copenhagen interpretation a measurement changes the state of a system in a way that cannot itself be described by quantum mechanics.[17] This can be seen from the interpretive rules of the theory. If we measure an observable represented by an Hermitian operator $A$, and the system is initially in a normalized superposition $\sum_r c_r \Psi_r$ of orthonormal eigenvectors $\Psi_r$ of $A$ with eigenvalues $a_r$, then the state is supposed to collapse during the measurement to a state in which the observable has a definite one of the values $a_r$, and the probability of finding the value $a_r$ is given by what is known as the Born rule, as $|c_r|^2$. This interpretation of quantum mechanics entails a departure during measurement from the dynamical assumptions of quantum mechanics. In quantum mechanics the evolution of the state vector described by the time-dependent Schrödinger equation is deterministic. If the time-dependent Schrödinger equation described the measurement process, then whatever the details of the process, the end result would be some definite pure state, not a number of possibilities with different probabilities.

We can see this more concretely by considering the effect of a measurement on the density matrix. For a system that can be in various possible states $\Psi_r$ with probabilities $P_r$, the density matrix is

$$\rho = \sum_r \Lambda_r P_r, \tag{3.7.1}$$

where $\Lambda_r \equiv [\Psi_r \Psi_r^\dagger]$ is the projection operator on the normalized state vector $\Psi_r$. If the system is in a state $\Psi_r$ and we make a measurement of some quantity or quantities that have definite values in a complete orthonormal set of state vectors $\Phi_\alpha$, then the probability that we will find the values characteristic of some particular state $\Phi_\alpha$ is $|(\Phi_\alpha, \Psi_r)|^2$, so the density matrix after the measurement is

$$\rho' = \sum_\alpha \Lambda_\alpha \sum_r P_r |(\Phi_\alpha, \Psi_r)|^2 = \sum_\alpha \Lambda_\alpha \operatorname{Tr}\left(\rho \Lambda_\alpha\right) = \sum_\alpha \Lambda_\alpha \rho \Lambda_\alpha, \tag{3.7.2}$$

where $\Lambda_\alpha \equiv [\Phi_\alpha \Phi_\alpha^\dagger]$ is the projection operator on state vector $\Phi_\alpha$. On the other hand, for the familiar deterministic evolution of state vectors in quantum mechanics, a system that is in state $\Psi_r$ at time $t$ will at time $t'$ be in a state $\Psi_r' = \exp(-iH(t'-t)/\hbar)\Psi_r$, so the density matrix at time $t'$ will be

---

[17] There are variants of the Copenhagen interpretation sharing this feature, some of them described by B. S. DeWitt, *Physics Today*, September, p. 30 (1970).

$$\rho' = \sum_r P_r \exp(-iH(t'-t)/\hbar) \, \Lambda_r \exp(+iH(t'-t)/\hbar)$$

$$= \exp(-iH(t'-t)/\hbar) \, \rho \, \exp(+iH(t'-t)/\hbar). \qquad (3.7.3)$$

There is no possible Hamiltonian for which for all initial density matrices $\rho$ the final density matrices (3.7.3) would take the form (3.7.2).

This is clearly unsatisfactory. If quantum mechanics applies to everything, then it must apply to a physicist's measurement apparatus, and to physicists themselves. On the other hand, if quantum mechanics does not apply to everything, then we need to know where to draw the boundary of its area of validity. Does it apply only to systems that are not too large? Does it apply if a measurement is made by some automatic apparatus, and no human reads the result? Also, for Bohr, classical mechanics was not merely an approximation to quantum mechanics – it was an essential part of the world, necessary for the interpretation of quantum mechanics. Even if we reject this as absurd, the Copenhagen interpretation still leaves us with the question, what *does* lie beyond the boundary of validity of quantum mechanics?

This puzzle has led some physicists to propose ways to replace quantum mechanics with a more satisfactory theory. One possibility is to add "hidden variables" to the theory. The probabilities encountered in quantum mechanics would then reflect our ignorance of these variables, rather than any intrinsic indeterminacy in nature.[18] Another possibility, which goes in the opposite direction, is to introduce intrinsically random terms into the equation for the evolution of the state vector, with no hidden variables, so that superpositions spontaneously collapse in an unpredictable way into the sorts of states familiar in classical physics, too slowly for it to be observed for microscopic systems like atoms or photons, but much more quickly for macroscopic systems such as measuring instruments.[19] In this section we will limit ourselves to interpretations of quantum mechanics that do not entail any change in its dynamical foundations – no hidden variables, and no modifications to the time-dependent Schrödinger equation.

There has emerged in recent years a clearer picture of what actually happens in a measurement. This has been largely due to the attention given to the phenomenon of decoherence.[20] But as I will try to show, even with this clarification, there still seems to be something important missing in our present understanding of quantum mechanics.

From the beginning, it was clear that the first requirement in a measurement is an evolution of the state vector in the Schrödinger picture, which establishes

---

[18] The best known theory of this sort is that of D. Bohm, *Phys. Rev.* D **85**, 166, 180 (1952).

[19] The leading theory of this type is that of G. C. Ghirardi, A. Rimini, and T. Weber, *Phys. Rev.* D **34**, 470 (1986). For a review, see A. Bassi and G. C. Ghirardi, *Phys. Rep.* **379**, 257 (2003).

[20] For a review of decoherence, see W. H. Zurek, *Rev. Mod. Phys.* **75**, 715 (2003).

a correlation between the system under study (which I will call the microscopic system, though in principle it need not be small), such as an atom's angular momentum or a radioactive nucleus, and a macroscopic apparatus, such as a detector that determines the atom's trajectory, or a cat. Suppose that the microscopic system can be in various states labeled with an index $n$, while the apparatus can be in states labeled with an index $a$, so that the states of the combined system can be expressed in terms of a complete orthonormal basis of state vectors denoted $\Psi_{na}$. (There must be at least as many apparatus states $a$ as system states $n$, though there may be many more.) The apparatus is placed at $t = 0$ in a suitable known initial state denoted $a = 0$, with the microscopic system in a general superposition of states, so that the combined system has an initial state vector

$$\Psi(0) = \sum_n c_n \Psi_{n0}. \tag{3.7.4}$$

We then turn on an interaction between the microscopic system and the measuring apparatus, so that the system evolves in a time $t$ to $U\Psi(0)$, where $U$ is the unitary operator $U = \exp(-itH/\hbar)$. We suppose that we are free to choose the Hamiltonian $H$ to be anything we like, so that $U$ is whatever unitary transformation we need. For an ideal measurement, what we need is that the basis states $\Psi_{n0}$ should evolve into states $U\Psi_{n0} = \Psi_{na_n}$, with $n$ unchanged,[21] and with $a_n$ labeling some definite state of the apparatus in a unique correspondence with the state of the microscopic system, so that $a_n \neq a_{n'}$ if $n \neq n'$. That is, we need[22]

$$U_{n'a',n0} = \delta_{n'n}\delta_{a'a_n}. \tag{3.7.5}$$

---

[21] Measurements that are ideal in this sense, with the state of the microscopic system unchanged, were called by J. A. Wheeler (1911–2008) "quantum non-demolition" measurements. In some cases measurements that change the state of the microscopic system are also useful.

[22] We can always choose the other elements of $U_{n'a',na}$, those with $a \neq 0$, to make the whole matrix unitary. For instance, for $a \neq 0$, we can take

$$U_{n'a',na} = \begin{cases} \delta_{n'n}\mathcal{U}^{(n)}_{a'a}, & a' \neq a_{n'}, \\ 0, & a' = a_{n'}, \end{cases}$$

where the submatrix $\mathcal{U}^{(n)}$ is constrained by the condition that, for all $a \neq 0$ and $\bar{a} \neq 0$,

$$\delta_{a\bar{a}} = \sum_{a' \neq a_n} \mathcal{U}^{(n)*}_{a'\bar{a}}\mathcal{U}^{(n)}_{a'a}.$$

The submatrices $\mathcal{U}^{(n)}_{a'a}$ are square, because $a'$ runs over all apparatus states except $a' = a_n$, and $a$ runs over all apparatus states except $a = 0$. These conditions thus simply require that these submatrices are unitary, and since they are subject to no other constraints, we can find any number of matrices that satisfy this condition. The reader can check that these conditions make the whole matrix $U_{n'a',na}$ unitary.

After the microscopic system and the measuring apparatus have interacted, the combined system is in a state $U\Psi(0)$, which according to Eqs. (3.7.4) and (3.7.5) is a superposition of apparatus states:[23]

$$U\Psi(0) = \sum_n c_n \Psi_{na_n}.  \tag{3.7.6}$$

This is not yet a measurement, because the system is still in a pure state, a definite superposition of the basis states $\Psi_{n,a_n}$. Somehow the system must make a transition to one or other of these states, with probabilities given by the Born rule as $|c_n|^2$.

Even before we consider how this happens, we face a problem. Ordinary experience shows that there are severe limitations on the states produced in measurements. We may observe the pointer on a meter in any one of a number of definite directions on the dial, but in practice we never see it in a superposition of directions. We will refer to the favored states produced by measurement as *classical states*. (These states were identified by Zurek,[24] with the name of "pointer states.") Quantum mechanics itself does not indicate anything special about the classical states. As far as our discussion so far is concerned, we could have taken the $\Psi_{na}$ to be any orthonormal basis we like. The solution turns out to involve the phenomenon of decoherence. To illustrate this, let's look at two classic examples of measurement, which will also be useful later as illustrations in dealing with deeper problems.

The first example is the 1922 Stern–Gerlach experiment, which will be considered in detail (more detail than we need here) in Section 4.2. In this sort of experiment a beam of atoms is sent into a magnetic field, with a homogeneous term in, say, the $z$-direction, and a smaller inhomogeneous term, which puts the atoms on different trajectories according to the value of the $z$-component $J_z$ of the total angular momentum of the atom. If the atom is initially in a state that is a linear combination of eigenstates of $J_z$ with different eigenvalues, then the

---

[23] A frequently quoted example was given by John von Neumann (1903–1957), in *Mathematical Foundations of Quantum Mechanics*, transl. R. T. Beyer (Princeton University Press, Princeton, NJ, 1955). Instead of discrete indices $n$ and $a$, the states of the microscopic system and the apparatus are characterized by the position coordinate $x$ of a particle and the coordinate $X$ of a pointer. The Hamiltonian is taken as $H = \omega x P$, where $\omega$ is some constant and $P$ is the pointer momentum operator, satisfying the usual commutation relation $[X, P] = i\hbar$ (and with $X$ and $P$ commuting with $x$ and its associated momentum $p$). If at $t = 0$ the coordinate-space wave function is $\psi(x, X, 0) = f(x - \xi)g(X)$, then at a later time $t$ the wave function in this case will be

$$\psi(x, X, t) = f(x - \xi)g(X - x\omega t).$$

If both $f$ and $g$ are sharply peaked at zero values of their arguments, then observation of the pointer position $X$ will tell us the position $\xi$ of the particle, with an uncertainty that can be made as small as we like by choosing the peaks in $f$ and $g$ to be sufficiently sharp. But if we start with the particle described by a broad wave packet $f$, then no matter how sharply peaked we take the function $g$, the pointer will be left in a superposition of states with a broad range of different positions $X$.

[24] W. H. Zurek, *Phys. Rev.* D **24**, 1516 (1981).

state vector evolves to become a superposition of terms in which the atoms are following different trajectories. So why do we always see the particle on one definite trajectory, corresponding to a definite value of $J_z$?

The answer has to do with the phenomenon of decoherence. This occurs because any real macroscopic apparatus will always be subject to tiny perturbations from the external environment, if only from the black-body photons that are present at any temperature above absolute zero.[25] Joos and Zeh[26] have considered an experiment in which electrons can classically follow either one of two possible trajectories, and shown how room temperature radiation will in one second introduce large random phases in the state vectors of trajectories separated by only 1 mm. These perturbations cannot normally change one classical state into another. For instance, exposure to low-temperature black-body photons will not cause a particle on one trajectory in a Stern–Gerlach experiment to switch to an entirely different trajectory. So if we choose the basis states $\Psi_{na}$ to be classical states, such as the states in a Stern–Gerlach experiment in which the particle has definite values of $J_z$ and travels on definite trajectories, then the effect of decoherence can only be to convert Eq. (3.7.6) to

$$\sum_n \exp(i\varphi_n)\, c_n \Psi_{na_n}, \qquad (3.7.7)$$

where the $\varphi_n$ are randomly fluctuating phases.[27] In consequence, when we calculate expectation values the interferences between different terms in this superposition average to zero, and the observed expectation value of any Hermitian operator $A$ (not necessarily one for which the $\Psi_{na_n}$ are eigenstates) will be

$$\overline{\langle A \rangle} = \sum_n |c_n|^2 \left( \Psi_{na_n}, A\Psi_{na_n} \right), \qquad (3.7.8)$$

with the bar over the expectation value indicating that it is averaged over the phases $\varphi_n$. This is commonly interpreted as meaning that the probability of the system under study and the apparatus being in the state $\Psi_{na_n}$ is $|c_n|^2$, but, as discussed below, this interpretation is far from clear.

A more melodramatic example of measurement in quantum mechanics was offered in 1935 by Schrödinger.[28] A cat is placed in a closed chamber with a radioactive nucleus, a Geiger counter that can detect the nuclear decay, and a capsule of poison that is released when the counter records that the decay

---

[25] The possibility of suppressing decoherence so that superpositions of classical states can be observed is discussed by A. J. Leggett, *Contemp. Phys.* **25**, 583 (1984).

[26] E. Joos and H. D. Zeh, *Z. Phys. B: Condensed Matter* **59**, 223 (1985).

[27] The classical states $\Psi_{na}$ of the sort discussed above are here assumed to form a complete orthonormal basis. In simple cases such as a Stern–Gerlach experiment, the classical states do form a complete orthonormal set. This is not necessarily true in more complicated cases.

[28] E. Schrödinger, *Naturwissenschaften* **48**, 52 (1935).

has occurred. After one half-life, the state vector of the combined system is a superposition of terms with equal magnitude: in one term, the nucleus has not yet decayed and the cat is still alive; in the other term the decay has occurred and the cat has been killed by the poison. Just looking at the cat perturbs the state, but it cannot change a dead cat into one that is alive, or vice versa. But these perturbations can and do rapidly change the *phase* of classical states, in which the cat is definitely alive or dead. These rapid and random phase changes almost immediately change any superposition of classical states to other superpositions. A feline superposition $c_{\text{alive}}\Psi_{\text{alive}} + c_{\text{dead}}\Psi_{\text{dead}}$ will become $e^{i\alpha}c_{\text{alive}}\Psi_{\text{alive}} + e^{i\delta}c_{\text{dead}}\Psi_{\text{dead}}$, with $\alpha$ and $\delta$ randomly fluctuating phases. Again, the expectation value of an operator $A$ that represents an observable in such a superposition when averaged over phases will become the average of the expectation values of $A$ in the states in which the cat is alive or dead, weighted with $|c_{\text{alive}}|^2$ and $|c_{\text{dead}}|^2$.

There seems to be a wide-spread impression that decoherence removes all obstacles to this class of interpretations of quantum mechanics. But there is still a problem with the Born rule, that tells us that in a state (3.7.8), the probability that an observer sees the system in the state $\Psi_{na_n}$ is $|c_n|^2$. The "derivation" given above, based on Eq. (3.7.8), is clearly circular, because it relies on the formula for expectation values as matrix elements of operators, which is itself derived from the Born rule. So where does the Born rule come from? There are two main approaches to this question, that are often called *instrumentalist* and *realist*, each with its own drawbacks.

### *Instrumentalism*

In instrumentalist approaches, one gives up the idea that the state vector of a closed system gives a complete account of the condition of the system, and instead regards it as just an instrument that provides a prescription for the calculation of probabilities. This point of view can be regarded as a re-interpretation of the Copenhagen version of quantum mechanics: instead of invoking a mysterious collapse of the state of a system during measurement, one simply assumes that in a state with a normalized state vector $\Psi$, the probability that the system will be found to have a value $a_n$ for some quantity represented by an Hermitian operator $A$ (rather than any other value of that quantity) is $p_n = \sum_r |(\Phi_{nr}, \Psi)|^2$, where $\Phi_{nr}$ are all the orthonormal eigenvectors of $A$ with eigenvalue $a_n$. This Born rule would simply be taken as one of the laws of nature. But if these probabilities are taken to be the probabilities of obtaining various results when people make observations, then this approach brings people into the laws of nature.

This is not a problem for those physicists who, as did Bohr, view the laws of nature as no more than a set of methods for ordering and surveying human experience. They are certainly that, but it would be sad to give up the hope that they are something more, that the laws of nature are in some sense "out there" in objective reality, the same laws (aside from language) for whoever studies them, and the same whether or not anyone is studying them.

For some physicists the intrusion of humans into the laws of nature is not unwelcome. David Mermin[29] approvingly cites the approach known as QBism,[30] which "attributes the muddle at the foundations of quantum mechanics to our unacknowledged removal of the scientist from the science."

The problem with instrumentalism is not that, in considering what happens in a measurement, one takes into account the scientist making the measurement. That is unobjectionable, and perhaps inevitable. The problem arises precisely because we want to be able to understand scientists along with everything else scientifically, and for that very reason, we need to keep humans (scientists, observers, or anyone else) out of the laws of nature, which by definition are unexplained. Only if the laws are expressed in impersonal terms, whether particle trajectories or wave functions or something else that does not refer to people making observations, can we hope to come to a scientific understanding of what is going on when people do observe nature or make a measurement.

This has a parallel in the theory of evolution. Before Charles Darwin and Alfred Russel Wallace, those naturalists who accepted the reality of evolution generally explained it in terms of an inherent tendency of life to evolve toward something better, like us. That put humans into the laws of biology, in a way that would rule out a unified view of nature encompassing both life and physics. The great achievement of Darwin and Wallace was to show how species like humans could evolve from earlier species, without invoking any law of nature to that effect. Much of the progress in biology since then would have been impossible without this achievement.

Some physicists who follow the instrumentalist approach claim that the probabilities predicted by the Born rule can be regarded as objective probabilities, not necessarily having anything to do with people making measurements. For instance, it is argued that when we say that the probability that a particle is in a small interval $\Delta x$ around the coordinate $x$ is $|\psi(x)|^2 \Delta x$, this is simply a statement about where the particle actually is likely to be, not necessarily about where we are likely to find it when we look at the particle. I don't find this tenable, because in general the particle has no definite position or momentum until people choose to observe one or the other. It can't have both a definite position $x$ and a definite momentum $p$ (with $\Delta x \, \Delta p < \hbar/2$), because there is no such state.

By not attributing any reality to the state vector, except as a predictor of probabilities, instrumentalism also gives up the classic and classical idea of an objective evolution of physical systems. We can live with the idea that the state of a physical system is described by a vector in Hilbert space rather than by numerical values of the positions and momenta of all the particles in the system,

29 N. D. Mermin, *Nature* **507**, 421 (2014).

30 C. A. Fuchs, N. D. Mermin, and R. Schack, *Am. J. Phys.* **82**, 749 (2014).

but it is hard to live with no description whatever of the evolution of physical states. This objection is met in part by the "decoherent histories" or "consistent histories" approach, due originally to Griffiths,[31] and developed by Omnès[32] and in detail by Gell-Mann and Hartle.[33] In this approach, one defines histories of closed systems (such as the whole universe) to which one can attribute probabilities that are consistent with the usual properties of probability.

A history is characterized first by a normalized initial state $\Psi$, which evolves from the initial time $t_0$ to a time $t_1$ according to the time-dependent Schrödinger equation, At time $t_1$ the system is averaged over its properties, holding fixed only the values $a_{1\eta}$ of a few observables $A_{1\eta}$. This is followed by evolution to a time $t_2$, at which time the system is again averaged over its properties, now holding fixed only values $a_{2\eta}$ of another set of observables $A_{2\eta}$, and so on. That is, the history is defined by $\Psi$, by the times $t_1$, $t_2$, etc., by the choice of the observables $A_{1\eta}$, $A_{2\eta}$, etc. whose values are held fixed in the averaging at each of these times, and by the fixed values $a_{1\eta}$, $a_{2\eta}$, etc. of these observables. This corresponds to what is actually done in observations, say of particle trajectories, in which only a few properties of a system are measured, and other properties such as the surrounding thermal radiation field are ignored.

To simplify our notation, we will suppress the index $\eta$, as if each averaging held fixed the value of just a single observable $A_1$, $A_2$, etc. To each history one assigns a state vector:

$$
\begin{aligned}
\Psi_{a_1 a_2 \ldots a_N} &\equiv \Lambda_N(a_N) \exp\left(-i H(t_N - t_{N-1})/\hbar\right) \ldots \\
&\times \exp\left(-i H(t_3 - t_2)/\hbar\right) \Lambda_2(a_2) \\
&\times \exp\left(-i H(t_2 - t_1)/\hbar\right) \Lambda_1(a_1) \exp\left(-i H(t_1 - t_0)/\hbar\right) \Psi,
\end{aligned}
$$
$$(3.7.9)$$

where $\Lambda_1(a_1)$, $\Lambda_2(a_2)$, etc. are sums of projection operators on all states of the system that are consistent with restrictions labeled by $a_1$, $a_2$, etc. For instance, if the $r$th sum held fixed only the value $a_r$ of a single observable $A_r$, then $\Lambda_r(a_r)$ would be the sum $\sum_i^{(a_r)}[\Phi_i \Phi_i^\dagger]$ of the projection operators on a set of

[31] R. B. Griffiths, *J. Stat. Phys.* **36**, 219 (1984); also see R. B. Griffiths, *Consistent Quantum Theory* (Cambridge University Press, Cambridge, 2002).

[32] R. Omnès, *Rev. Mod. Phys.* **64**, 339 (1992); also see R. Omnès, *The Interpretation of Quantum Mechanics* (Princeton University Press, Princeton, 1994).

[33] M. Gell-Mann and J. B. Hartle, in *Complexity, Entropy, and the Physics of Information*, ed. W. H. Zurek (Addison–Wesley, Reading, MA, 1990); in *Proceedings of the Third International Symposium on the Foundations of Quantum Mechanics in the Light of New Technology*, ed. S. Kobayashi, H. Ezawa, Y. Murayama, and S. Nomura (Physical Society of Japan, 1990); in *Proceedings of the 25th International Conference on High Energy Physics*, Singapore, August 2–8, 1990, ed. K. K. Phua and Y. Yamaguchi (World Scientific, Singapore, 1990); J. B. Hartle, *Directions in Relativity*, Vol. 1, ed. B.-L. Hu, M. P. Ryan, and C. V. Vishveshwars (Cambridge University Press, Cambridge, 1993).

orthonormal states $\Phi_i$ that are complete in the subspace consisting of eigenstates of $A_r$ with eigenvalue $a_r$. (This is called *coarse-graining* by Gell-Mann and Hartle in the texts cited in footnote 33. Projection operators were discussed in Section 3.3.) Equivalently, we have

$$\Psi_{a_1 a_2 \dots a_{\mathcal{N}}} = e^{-i H t_{\mathcal{N}}/\hbar} \Lambda_{\mathcal{N}}(a_{\mathcal{N}}, t_{\mathcal{N}}) \dots \Lambda_2(a_2, t_2) \Lambda_1(a_1, t_1) e^{i H t_0/\hbar} \Psi, \quad (3.7.10)$$

where $\Lambda_r(a_r, t_r)$ are the same sums of projection operators, but in the Heisenberg picture:

$$\Lambda_r(a_r, t_r) = e^{i H t_r/\hbar} \Lambda_r(a_r) e^{-i H t_r/\hbar}. \quad (3.7.11)$$

A positive probability is assumed for each history by a generalization of the Born rule:

$$P(a_1 a_2 \dots) \equiv \left( \Psi_{a_1 a_2 \dots}, \Psi_{a_1 a_2 \dots} \right). \quad (3.7.12)$$

It is necessary to show that Eq. (3.7.12) possesses the usual properties of probabilities, but this is true only for a limited class of possible histories. Specifically, we must show that the sum of these probabilities over all possible values of one of the observables, say $a_r$, equals the probability of the history in which this observable is not held fixed:

$$\sum_{a_r} P(a_1 a_2 \dots a_{r-1} a_r a_{r+1} \dots a_{\mathcal{N}}) = P(a_1 a_2 \dots a_{r-1} a_{r+1} \dots a_{\mathcal{N}}). \quad (3.7.13)$$

This is the case for histories that satisfy the consistency condition, that

$$\left( \Psi_{a'_1 a'_2 \dots a'_{\mathcal{N}}}, \Psi_{a_1 a_2 \dots a_{\mathcal{N}}} \right) = 0 \text{ unless } a'_1 = a_1, \ a'_2 = a_2, \dots. \quad (3.7.14)$$

Here is the proof. According to Eq. (3.7.12), the sum in Eq. (3.7.13) is

$$\sum_{a_r} P(a_1 a_2 \dots a_{r-1} a_r a_{r+1} \dots a_{\mathcal{N}})$$

$$= \sum_{a_r} \left( \Psi_{a_1 a_2 \dots a_{r-1} a_r a_{r+1} \dots a_{\mathcal{N}}}, \Psi_{a_1 a_2 \dots a_{r-1} a_r a_{r+1} \dots a_{\mathcal{N}}} \right).$$

By using the consistency condition (3.7.14), we can write this as

$$\sum_{a_r} P(a_1 a_2 \dots a_{r-1} a_r a_{r+1} \dots a_{\mathcal{N}})$$

$$= \left( \sum_{a'_r} \Psi_{a_1 a_2 \dots a_{r-1} a'_r a_{r+1} \dots a_{\mathcal{N}}}, \sum_{a_r} \Psi_{a_1 a_2 \dots a_{r-1} a_r a_{r+1} \dots a_{\mathcal{N}}} \right).$$

But the completeness relation (3.3.32) gives

$$\sum_{a_r} \Lambda_r(a_r, t_r) = 1,$$

so

$$\sum_{a_r} \Psi_{a_1 a_2 \ldots a_{r-1} a_r a_{r+1} \ldots a_{\mathcal{N}}} = \Psi_{a_1 a_2 \ldots a_{r-1} a_{r+1} \ldots a_{\mathcal{N}}},$$

from which Eq. (3.7.13) follows immediately. This theorem has the important consequence that the sum of probabilities for all histories of a given type (that is, all histories with a given initial state $\Psi$, given times $t_1, \ldots, t_{\mathcal{N}}$, and given observables $A_r$ that are held fixed at each of these times) is unity:

$$\sum_{a_1 a_2 \ldots a_{\mathcal{N}}} P(a_1 a_2 \ldots a_{\mathcal{N}}) = \left( \Psi, \Psi \right) = 1. \tag{3.7.15}$$

The histories that satisfy the consistency condition (3.7.14) are identified by considerations of decoherence. For instance, the history of a planet's motion around the Sun is characterized by a set of projection operators, with labels $a$ that distinguish various cells of finite spatial volume in which the planet might be found. (It is necessary to deal with finite volumes of space, since a precise measurement of position would give the planet an unwanted change in momentum.) In evaluating (3.7.9) or (3.7.10) for any given history, we average over all other variables characterizing perturbations of the planet's orbit, including those that describe solar radiation, interplanetary matter, etc. These perturbations do not move a planet from one cell to another, but they do change the phase of the state vector (3.7.9), and the averaging over perturbations thus destroys the correlations that would invalidate the consistency condition (3.7.14).

Some adherents of the decoherent-histories approach describe the probabilities (3.7.12) as objective properties of the various histories, not necessarily related to anything seen by any observer, and applying even where there are no actual observers, in particular to the early universe. This view seems to me untenable, for reasons like those already described in the case of a single measurement. The requirement that histories have to satisfy the consistency condition (3.7.13) does not uniquely determine the choice of the observables $A_1$, $A_2$, etc. over whose eigenvectors we do not average at times $t_1$, $t_2$, etc. The problem here is not that the choice is not unique, but rather, that it can only be made by people. Of course, the answers to questions depend on what questions we choose to ask, in classical as well as in quantum mechanics, but in classical physics the necessity of choice can be evaded because in principle we can choose to measure everything. It cannot be evaded in this way in quantum mechanics because in general many of these choices are incompatible with each other. For instance, we can choose to leave the eigenvalues of $J_x$ or $J_y$ or $J_z$ unaveraged at a given time, but we can't leave all three unaveraged, because there is no state in which all three have definite non-zero values. So the Born rule in the decoherent-histories approach seems to bring people into the laws of nature, as is apparently inevitable for any instrumentalist approach.

### *Realism*

The drawbacks of the Copenhagen and instrumentalist approaches to quantum mechanics have led some physicists to adopt an approach in which one attributes reality not to classical observables like position and momentum, but instead to the state vector itself. Taking the state vector seriously, as a complete description of the physical condition of the system, we can attempt to understand how probabilities arise from the deterministic evolution of the state vector, without introducing measurements or the people making measurements into the laws of nature.

One trouble with attributing reality to the state vector is that in an entangled state of two systems that are entirely isolated from each other, the state vector of one system can be instantaneously changed by intervention in the other system, We will take this up when we come to entanglement in Section 12.1.

Another aspect of the realist approach, which some physicists find implausible, is that it seems to lead inevitably to the "many-worlds interpretation" of quantum mechanics, presented originally in the 1957 Princeton Ph.D. thesis[34] of Hugh Everett (1930–1982). In this approach, the state vector does not collapse; it continues to be governed by the deterministic time-dependent Schrödinger equation, but different components of the state vector of the system studied become associated with different components of the state vector of the measuring apparatus and observer, so that the history of the world effectively splits into different paths, each characterized by different results of the measurement.

The difference between this interpretation of quantum mechanics and the Copenhagen interpretation can be illustrated by considering the classic examples of the measurement process mentioned earlier. In a Stern–Gerlach experiment, according to the Copenhagen interpretation somehow when the atom interacts with an observer, the system collapses to a state in which the atom has a definite value for the component $J_z$ of the angular momentum in the direction of the homogeneous magnetic field, and is following just one trajectory. According to the many-worlds interpretation, the state vector of the system comprising both the atom and the observer remains a superposition: in one term, the observer sees the atom with one value for $J_z$ and following one definite trajectory; in another term of the state vector, the observer sees the atom with a different value for $J_z$ and following a different trajectory. Either interpretation is in accord with experience, but the Copenhagen interpretation relies on something happening during a measurement that is outside the scope of quantum mechanics, while the many-worlds interpretation strictly follows quantum mechanics, but supposes that the history of the universe is continually splitting into an inconceivably large number of branches.

---

[34] The published version is H. Everett, *Rev. Mod. Phys.* **29**, 454 (1957).

Similarly, in the case of Schrödinger's cat, according to the Copenhagen interpretation, when the cat is observed (perhaps by the cat itself – it is not clear) the state of the nucleus and the cat and the observer collapses, either to a state with the nucleus not yet decayed and the cat still alive, or to a state with the decay having occurred and the cat being dead, each with its own probability. In contrast, according to the many-worlds interpretation, the state vector remains a superposition of terms, one with the cat alive and the observer seeing the cat alive, and the other term with the cat dead and the observer seeing it dead. (Of course, even in the term in the state vector in which the cat is still alive after a single half-life, its future is dim.)

In addition to its other problems, the realist approach faces the challenge of deriving the Born rule. If measurement is really described by quantum mechanics, then we ought to be able to derive such formulas by applying the time-dependent Schrödinger equation to the case of repeated measurement. This is not just a matter of intellectual tidiness, of wanting to reduce the postulates of physical theory to the minimum number needed. If the Born rule cannot be derived from the time-dependent Schrödinger equation, then something else is needed, something outside the scope of quantum mechanics, and in this respect the many-worlds interpretation would share the inadequacies of the instrumentalist and Copenhagen interpretations.[35]

To address this problem, we need to be specific about the circumstances in which probabilities are to be measured. If we regard probability as a matter of the frequencies of things seen by observers, we have to specify when it is that the observer becomes so tangled with the system that we can think of different terms in the state vector as including different conclusions of the observer.

One possibility is that a sequence of experiments is carried out, each one of these experiments starting with the same state vector (3.7.4), and in each case followed by a measurement of the sort described above, with the observer treated as part of the measuring apparatus. In each measurement the history of the world splits into as many branches as there are states $n$, and (as long as none of the $c_n$ vanish) for every possible sequence of experimental results $n_1$, $n_2$, etc. there is one history in which the observer sees those results. For instance, consider a system with only two possible states, which appear in the state vector with coefficients $c_1$ and $c_2$. As long as neither coefficient vanishes, after a single measurement of the observable that distinguishes these states, the state of the world will have two branches, in one of which the observer finds that the system is in state 1, and in the other of which the observer finds that the system is in state 2. After $N$ repeated measurements, the history of the world will have $2^N$ branches, in which there will occur every possible history of results of these experiments. No matter how large or small the ratio $c_1/c_2$ may be, as long

---

[35] For a strong expression of this view, see A. Kent, *Int. J. Mod. Phys.* A **5**, 1745 (1990).

as it is neither zero nor infinity, there is nothing to pick out one sequence of experimental results as being more or less likely than another. There is nothing in this picture that corresponds to the usual assumption of quantum mechanics, that assigns a probability $|c_{n_1}|^2 |c_{n_2}|^2 \ldots$ to a history in which the sequence of results found by the observer is $n_1$, $n_2$, etc.

In a different sort of experiment for the measurement of probabilities, a large number $N$ of copies of the same system is prepared in the same state $\sum_n c_n \Psi_n$, so that the state vector of the combined system is a direct product:

$$\Psi = \sum_{n_1 n_2 \ldots n_N} c_{n_1} c_{n_2} \ldots c_{n_N} \Psi_{n_1 n_2 \ldots n_N}, \qquad (3.7.16)$$

where $\Psi_{n_1 n_2 \ldots n_N}$ is the state in which system copy $s$ is in state $n_s$. If the $\Psi_n$ are suitable classical states, of the sort that survive decoherence, then the effect of the environment will be to multiply each $c_{n_s}$ with a phase factor $\exp(i\varphi_{s,n_s})$, so that Eq. (3.7.16) becomes

$$\Psi = \sum_{n_1 n_2 \ldots n_N} c_{n_1} c_{n_2} \ldots c_{n_N} \exp\left[i\varphi_{1,n_1} + \cdots + i\varphi_{N,n_N}\right] \Psi_{n_1 n_2 \ldots n_N} \qquad (3.7.17)$$

with the phases $\varphi_{s,n_s}$ random and uncorrelated. We take the states of this basis to be orthonormal, in the sense that

$$\left( \Psi_{n_1' n_2' \ldots n_N'}, \Psi_{n_1 n_2 \ldots n_N} \right) = \delta_{n_1' n_1} \delta_{n_2' n_2} \ldots \delta_{n_N' n_N},$$

and the state (3.7.17) is then normalized if $\sum_n |c_n|^2 = 1$. In this scenario, it is only after the microscopic system has been prepared in the state (3.7.17) that, by correlating this state with a measuring apparatus and observer, the observer finds herself in a branch of the history of the world in which each of the copies of the system is in some definite basis state, say in the states $n_1$, $n_2$, $\ldots$, $n_N$. Let's say that she finds $N_n$ copies in each state $n$, of course with $\sum_n N_n = N$. She will conclude that the probability that any one copy is in the state $n$ is $P_n = N_n/N$.

Note that this is pretty much how probabilities are actually measured in practice. For instance, if we want to measure the probability that a nucleus in a given initial state will experience a radioactive decay in a certain time $t$, we assemble a large number $N$ of these nuclei in the same initial state, and count how many have experienced the decay after a time $t$; the decay probability is that number divided by $N$.

Here again, all results are possible. The observer can find any set of results $n_1$, $n_2$, $\ldots$, $n_N$ for the states of the identical subsystems. This is not so different from the situation in classical mechanics. An observer tossing a coin a few times might find that it comes up heads every time, and has to hope that if the number $N$ of repetitions were sufficiently large, the relative frequencies $N_n/N$ would give a good approximation to the actual probability $P_n$.

Even in the limit of large $N$, does this picture lead to the usual assumption of quantum mechanics, that the quantities $P_n$ approach $|c_n|^2$? Of course, state vectors tell us nothing without some sort of interpretive postulate. The one postulate that does not seem to raise problems of consistency with the deterministic dynamics of the Schrödinger equation, and does not drag reference to people into the laws of nature, is the "second postulate of quantum mechanics" described in Section 3.3: if the state vector of a system is an eigenstate of the Hermitian operator $A$ representing some observable, with eigenvalue $a$, then the system definitely has the value $a$ for that observable. The operators that interest us here are frequency operators $P_n$, defined by the conditions that they are linear and act on the basis states of the combined system as

$$P_n \Psi_{n_1 n_2 \dots n_N} \equiv (N_n/N) \Psi_{n_1 n_2 \dots n_N}, \tag{3.7.18}$$

where $N_n$ is the number of the indices $n_1$, $n_2$, $\dots$, $n_N$ equal to $n$. It would solve all our problems if we could show that the state (3.7.17) is an eigenstate of $P_n$ with eigenvalue $|c_n|^2$, but of course this is not true (except in the special cases where $|c_n|$ is zero or one, where $\Psi$ either does not contain any term $\Psi_{n_1 n_2 \dots n_N}$ where any index equals $n$, or is just proportional to a term where all indices equal $n$). What we can show is that this eigenvalue condition is *nearly* true for large $N$. Specifically, for the states (3.7.17) we have[36]

$$||(P_n - |c_n|^2)\Psi||^2 = \frac{|c_n|^2(1 - |c_n|^2)}{N} \leq \frac{1}{4N}, \tag{3.7.19}$$

where for any state $\Phi$, the norm $||\Phi||$ denotes $(\Phi, \Phi)^{1/2}$.

Here is the proof. It is convenient to replace the set of indices $n_1 n_2 \dots n_N$ with a compound index $\nu$, and let $N_{\nu,n}$ be the number of the indices $n_1 n_2 \dots n_N$ that are equal to $n$. Of course, for any $\nu$, we have $\sum_n N_{\nu,n} = N$. The state (3.7.17) can be written in this notation as

$$\Psi = \sum_\nu \left( \prod_n c_n^{N_{\nu,n}} \right) e^{i\varphi_\nu} \Psi_\nu,$$

and Eq. (3.7.18) gives

$$P_n \Psi = \sum_\nu \left( \prod_m c_m^{N_{\nu,m}} \right) e^{i\varphi_\nu} \left( \frac{N_{\nu,n}}{N} \right) \Psi_\nu.$$

Instead of summing over $\nu$, we can sum here independently over $N_1$, $N_2$, etc. The number of $\nu$s with $N_{\nu,n} = N_n$ for some given values of $N_1$, $N_2$, etc. is the binomial coefficient $N!/N_1!N_2!\dots$. Thus we have

36  The proof that $||(P_n - |c_n|^2)\Psi||$ vanishes for large $N$ was given by J. B. Hartle, *Am. J. Phys.* **36**, 704 (1968). Also see B. S. DeWitt, in *Battelle Rencontres, 1967 Lectures in Mathematics and Physics*, eds. C. DeWitt and J. A. Wheeler (W. A. Benjamin, New York, 1968); N. Graham, in *The Many Worlds Interpretation of Quantum Mechanics*, eds. B. S. DeWitt and N. Graham (Princeton University Press, Princeton, NJ, 1973) [who gives Eq. (3.7.19) explicitly]; E. Farhi, J. Goldstone, and S. Gutmann, *Ann. Phys.* **192**, 368 (1989); D. Deutsch, *Proc. Roy. Soc.* A **455**, 3129 (1999).

$$||(P_n - |c_n|^2)\Psi||^2 = \sum_{N_1 N_2 \ldots} \left( \prod_m |c_m|^{2N_m} \right) \left( \frac{N_n}{N} - |c_n|^2 \right)^2 \frac{N!}{N_1! N_2! \ldots},$$

with the sum constrained by $N_1 + N_2 + \cdots = N$. According to the binomial theorem,

$$\sum_{N_1 N_2 \ldots} \left( \prod_m |c_m|^{2N_m} \right) \frac{N!}{N_1! N_2! \ldots} = \left( \sum_m |c_m|^2 \right)^N,$$

so

$$||(P_n - |c_n|^2)\Psi||^2 = \left[ \frac{1}{N^2} \left( |c_n|^2 \frac{\partial}{\partial |c_n|^2} \right)^2 - \frac{2}{N} \left( |c_n|^4 \frac{\partial}{\partial |c_n|^2} \right) + |c_n|^4 \right]$$
$$\times \left( \sum_m |c_m|^2 \right)^N$$
$$= N(N-1) \left( \frac{|c_n|^4}{N^2} \right) \left( \sum_m |c_m|^2 \right)^{N-2}$$
$$+ N \left( \frac{|c_n|^2}{N^2} \right) \left( \sum_m |c_m|^2 \right)^{N-1}$$
$$- 2N \left( \frac{|c_n|^4}{N} \right) \left( \sum_m |c_m|^2 \right)^{N-1} + |c_n|^4 \left( \sum_m |c_m|^2 \right)^N.$$

If we now use the normalization condition $\sum_m |c_m|^2 = 1$, we find Eq. (3.7.19), as was to be proved.

What should we make of this? Eq. (3.7.19) does not show that the states $\Psi_\nu$ approach eigenstates of the frequency operators $P_n$ for $N \to \infty$, because these states do not approach any limit. Indeed, the size of the Hilbert space they inhabit depends on $N$. Hartle and Farhi, Goldstone, and Gutmann in the texts cited in footnote 36 showed how to construct a Hilbert space for the case $N = \infty$,[37] and showed that the operators $P_n$ acting on this space have eigenvalues $|c_n|^2$, but to apply this construction it is necessary to extend the usual interpretive assumption about eigenvalues from the Hilbert space for any finite number $N$ of systems to the Hilbert space for $N = \infty$, which seems a stretch.

We might try introducing a strengthened version of the postulate about eigenstates and eigenvalues, assuming that, if a normalized state vector $\Psi$ is nearly an eigenvector of an Hermitian operator $A$ with eigenvalue $a$, in the sense that the norm $||(A - a)\Psi||$ is small, then in the state represented by $\Psi$, it is almost certain that the value of the observable represented by $A$ is close to $a$. This is hardly

---

[37] For criticisms of this construction, see C. M. Caves and R. Schack, *Ann. Phys.* **315**, 123 (2005).

precise, and in any case, since this assumption refers to something being "almost certain," it re-introduces a postulate regarding probability, without showing how it follows from the dynamical assumptions of quantum mechanics.

Apart from these problems, which are perhaps not so different from those that afflict discussions of probability in classical physics, there is the additional difficulty that the Born rule emerges from this analysis precisely because we use the quantum-mechanical norm $\|\Psi\| \equiv (\Psi, \Psi)^{1/2}$ as a measure of the departure of physical states from being eigenstates of the operator $P_n$ with eigenvalue $|c_n|^2$. The smallness of all $\|(P_n - |c_n|^2)\Psi\|$ for large $N$ does tell us that the scalar product of $\Psi$ with any eigenstate of $P_n$ with an eigenvalue appreciably different from $|c_n|^2$ is small. (Specifically, the sum of $|(\Phi, \Psi)|^2$ over states $\Phi$ for which $P_n$ has an eigenvalue that differs from $|c_n|^2$ by more than terms of order $1/\sqrt{N}$ is at most of order $1/N$.) If we assume the Born rule, then this means that the probability of an observer observing such "wrong" values of $N_n/N$ is small, but of course it is circular to use this reasoning to derive the Born rule.

$$* * * * *$$

My own conclusion is that today there is no interpretation of quantum mechanics that does not have serious flaws. This view is not universally shared. Indeed, many physicists are satisfied with their own interpretation of quantum mechanics. But different physicists are satisfied with different interpretations. In my view, we ought to take seriously the possibility of finding some more satisfactory other theory, to which quantum mechanics is only a good approximation.

## Problems

1. Consider a system with a pair of observable quantities $A$ and $B$, whose commutation relations with the Hamiltonian take the form $[H, A] = iwB$, $[H, B] = -iwA$, where $w$ is some real constant. Suppose that the expectation values of $A$ and $B$ are known at time $t = 0$. Give formulas for the expectation values of $A$ and $B$ as a function of time.

2. Consider a normalized initial state $\Psi$ at $t = 0$ with a spread $\Delta E$ in energy, defined by

$$\Delta E \equiv \sqrt{\left\langle \left( H - \langle H \rangle_\Psi \right)^2 \right\rangle_\Psi}.$$

Calculate the probability $|(\Psi(\delta t), \Psi)|^2$ that after a very short time $\delta t$ the system is still in the state $\Psi$. Express the result in terms of $\Delta E$, $\hbar$, and $\delta t$, to *second* order in $\delta t$.

3. Suppose that the Hamiltonian is a linear operator with

$$H\Psi = g\Phi, \quad H\Phi = g^*\Psi, \quad H\Upsilon_n = 0,$$

where $g$ is an arbitrary constant, $\Psi$ and $\Phi$ are a pair of normalized independent (but not necessarily orthogonal) state vectors, and $\Upsilon_n$ runs over all state vectors orthogonal to both $\Psi$ and $\Phi$. What are the conditions that $\Phi$ and $\Psi$ must satisfy in order for this Hamiltonian to be Hermitian? With these conditions satisfied, find the states with definite energy, and the corresponding energy values.

4. Suppose that a linear operator $A$, though not Hermitian, satisfies the condition that it commutes with its adjoint. What can be said about the relation between the eigenvalues of $A$ and of $A^\dagger$? What can be said about the scalar product of two eigenstates of $A$ with unequal eigenvalues?

5. Suppose the state vectors $\Psi$ and $\Psi'$ are eigenvectors of a unitary operator with eigenvalues $\lambda$ and $\lambda'$, respectively. What relation must $\lambda$ and $\lambda'$ satisfy if $\Psi$ is not orthogonal to $\Psi'$?

6. Show that the product of the uncertainties in position and momentum takes its minimum value $\hbar/2$ for a Gaussian wave packet of free-particle wave functions.

# 4

# Spin *et cetera*

Wave mechanics failed badly in accounting for the multiplicity of atomic energy levels. This was most conspicuous in the case of the alkali metals, lithium, sodium, potassium, and so on. It was known that an atom of any of these elements can be treated as a more-or-less inert core, consisting of the nucleus and $Z - 1$ inner electrons, together with a single outer "valence" electron whose transitions between energy levels are responsible for spectral lines. Since the electrostatic field felt by the outer electron is not a Coulomb field, its energy levels in the absence of external fields depend on the orbital angular momentum quantum number $\ell$ as well as a radial quantum number $n$, but because of the spherical symmetry of the atom, not on the angular momentum $z$ component $\hbar m$. (See Eq. (2.1.29).) For each $n$, $\ell$, and $m$ there should be just one energy level. But observations of atomic spectra showed that in fact all but the $s$-states were doubled. For instance, even a spectroscope of low resolution shows that the D line of sodium, which is produced in a $3p \rightarrow 3s$ transition of the valence electron, is a doublet, with wavelengths 5896 and 5890 Angstroms. Pauli was led to propose that there is a fourth quantum number for electrons in such atoms, in addition to $n$, $\ell$, and $m$, with the fourth quantum number taking just two values in all but $s$-states. But the physical significance of this fourth quantum number was obscure.

Then in 1925 two young physicists, Samuel Goudsmit (1902–1978) and George Uhlenbeck (1900–1988), suggested[1] that the doubling of energy levels was due to an internal angular momentum of the electron, whose component in the direction of **L** (for **L** $\neq$ 0) can only take two values, and whose interaction with the weak magnetic field produced by the orbital motion of the electron therefore splits all but $s$ states into nearly degenerate doublets. Any component of angular momentum $s$ would take $2s + 1$ values, so the quantity $s$ corresponding to $\ell$ for the internal angular momentum would have to have the unusual value $1/2$. This internal angular momentum came to be called the electron's *spin*.

At first this idea was widely disbelieved. As we saw in Section 2.1, orbital angular momentum cannot have the non-integer value $\ell = 1/2$. Another worry was that if a sphere with the mass of the electron and with angular momentum

---

[1] S. Goudsmit and G. Uhlenbeck, *Naturwissenschaften* **13**, 953 (1925); *Nature* **117**, 264 (1926).

$\hbar/2$ has a rotation velocity at its surface less than the speed of light, then its radius must be larger than $\hbar/2m_ec \simeq 2 \times 10^{-11}$ cm, and it was presumed that an electron radius that large would not have escaped observation. Electron spin became more respectable a little later, when several authors[2] showed that the coupling between the electron's spin and its orbital motion accounted for the fine structure of hydrogen – the splitting of states with $\ell \neq 0$ into doublets. (This is discussed in Section 4.2.)

The worries about models of spinning electrons were due to the lingering wish to understand quantum phenomena in classical terms. Instead, we should think of the existence of both spin and orbital angular momenta as consequences of a symmetry principle. We saw in Sections 3.4–3.6 how symmetry principles imply the existence of conserved observables such as energy and momentum. There is another classic symmetry of both non-relativistic and relativistic physics, invariance under spatial rotations. In Section 4.1 we will show how rotational invariance leads in quantum mechanics to the existence of a conserved angular-momentum three-vector **J**. The commutation relations of these operators will be used in Section 4.2 to derive the spectrum of eigenvalues of $\mathbf{J}^2$ and $J_3$, and to find how all three components of **J** act on the corresponding eigenstates. It turns out that the eigenvalues of $J_3$ can be integer *or half-integer* multiples of $\hbar$.

In general the angular momentum **J** of any particle is the sum of its orbital angular momentum, already discussed in Section 2.1, and a spin angular momentum, that can take half-integer as well as integer values. Also, in a multiparticle system, the total angular momentum of the system is the sum of the angular momenta of the individual particles. For both reasons, in Section 4.3 we will consider how the eigenstates of $\mathbf{J}^2$ and $J_3$ for the sum of two angular momenta are constructed from the corresponding eigenstates for the individual angular momenta. In Section 4.4 the rules for angular-momentum addition are applied to derive a formula, known as the Wigner–Eckart theorem, for the matrix elements of operators between multiplets of angular-momentum eigenstates.

It turns out that not only the electron but also the proton and neutron have spin 1/2. It is sometimes said that this value of the spin of the electron and other particles is a consequence of relativity. This is because Dirac in 1928 developed a kind of relativistic wave mechanics,[3] which required that the particles of the theory have spin 1/2. But Dirac's relativistic wave mechanics is not the only way to combine relativity and quantum mechanics. Indeed, in 1934 Pauli and Victor Weisskopf[4] (1908–2002) showed how a relativistic quantum theory could be constructed for particles with no spin. Today we know of particles like the Z and W particles that seem to be every bit as elementary as the electron, and

---

[2] W. Heisenberg and P. Jordan, Z. *Physik* **37**, 263 (1926); C. G. Darwin, *Proc. Roy. Soc.* A **116**, 227 (1927).

[3] P. A. M. Dirac, *Proc. Roy. Soc.* A **117**, 610 (1928).

[4] W. Pauli and V. F. Weisskopf, *Helv. Phys. Acta* **7**, 709 (1934).

that have spins with $j = 1$ rather than $j = 1/2$. There is nothing about spin that requires relativity to be taken into account, and nothing about relativity that requires elementary particles to have spin one-half.

Though it was not known at first, the spin of a particle determines whether the wave function of several particles of the same type is symmetric or anti-symmetric in the particle coordinates (including their spins). This is discussed in Section 4.5, along with some of its implications for atoms, nuclei, gases, and crystals.

Using what we have learned about angular momentum, in Sections 4.6 and 4.7 we will consider two other kinds of symmetry: internal symmetries, such as isotopic spin symmetry, and symmetry under space inversion. Section 4.8 shows that for the Coulomb potential there are two different three-vectors with the properties of angular momentum, and uses the properties of such three-vectors derived in Section 4.2 to give an algebraic calculation of the spectrum of hydrogen. This long chapter ends in Section 4.9 with a discussion of the rigid rotator, whose energy levels can be calculated exactly, and that provides an approximation to the rotational spectra of molecules.

## 4.1 Rotations

A rotation is a real linear transformation $x_i \mapsto \sum_j R_{ij} x_j$ of the Cartesian coordinates $x_i$ that leaves invariant the scalar product $\mathbf{x} \cdot \mathbf{y} = \sum_i x_i y_i$. That is,

$$\sum_i \left( \sum_j R_{ij} x_j \right) \left( \sum_k R_{ik} y_k \right) = \sum_i x_i y_i,$$

with sums over $i$, $j$, $k$, etc. running over the values 1, 2, 3. By equating coefficients of $x_j y_k$ on both sides of the equation, we find the fundamental condition for a rotation:

$$\sum_i R_{ij} R_{ik} = \delta_{jk}, \tag{4.1.1}$$

or in matrix notation

$$R^{\mathrm{T}} R = 1, \tag{4.1.2}$$

where $R^{\mathrm{T}}$ denotes the transpose of a matrix, $[R^{\mathrm{T}}]_{ji} = R_{ij}$, and 1 is here the unit matrix, $[1]_{jk} = \delta_{jk}$. Real matrices satisfying Eq. (4.1.2) are said to be *orthogonal*.

Taking the determinant of Eq. (4.1.2) and using the facts that the determinant of a product of matrices is the product of the determinants, and that the determinant of the transpose of a matrix equals the determinant of the matrix, we see that $[\mathrm{Det}\, R]^2 = 1$, so $\mathrm{Det}\, R$ can only be $+1$ or $-1$. There is a theorem of matrix algebra that tells us that, since Det $R$ does not vanish, $R$ has an inverse $R^{-1}$ such that $R^{-1}R = RR^{-1} = 1$. Multiplying Eq. (4.1.2) on the left with $R^{-1}$ tells us that $R^{-1} = R^T$. Note that this inverse is also an orthogonal matrix, for $(R^{-1})^T R^{-1} = RR^T = 1$.

It should be noted that the transpose of a product of matrices is the product of the transposes in the opposite order:

$$[AB]_{ij}^T = [AB]_{ji} = \sum_k A_{jk}B_{ki} = \sum_k B_{ik}^T A_{kj}^T = [B^T A^T]_{ij}.$$

It follows in particular that the product of orthogonal matrices is orthogonal: if $A^T A = 1$ and $B^T B = 1$ then

$$(AB)^T AB = B^T A^T AB = B^T B = 1.$$

The set of all real orthogonal matrices includes the unit matrix, and these matrices all have inverses that are also real orthogonal matrices, so this set satisfies all the conditions for a group. This group is known as $O(3)$, the group of real orthogonal $3 \times 3$ matrices.

Not all transformations $x_i \mapsto \sum_j R_{ij}x_j$ with $R_{ij}$ satisfying Eq. (4.1.2) are rotations. We have already noted that with $R_{ij}$ satisfying Eq. (4.1.2), the determinant of $R$ can only be $+1$ or $-1$. The transformations with Det $R = -1$ are space-inversions; an example is the simple transformation $\mathbf{x} \mapsto -\mathbf{x}$. These transformations will be considered in Section 4.7. The transformations with Det $R = +1$ are the rotations, which concern us here. The rotations form a group by themselves, since any product of matrices with unit determinant will have unit determinant. This subgroup of $O(3)$ is known as the special orthogonal group in three dimensions, or $SO(3)$, where $O(3)$ again means that these are real orthogonal $3 \times 3$ matrices, and the $S$ stands for "special," meaning that these matrices have unit determinant.

Like other symmetry transformations, a rotation $R$ induces on the Hilbert space of physical states a unitary transformation, in this case $\Psi \mapsto U(R)\Psi$. If we perform a rotation $R_1$ and then a rotation $R_2$, physical states undergo the transformation $\Psi \mapsto U(R_2)U(R_1)\Psi$, but this must be the same as if we had performed a rotation $R_2 R_1$, so[5]

$$U(R_2)U(R_1) = U(R_2 R_1). \tag{4.1.3}$$

Acting on the operator $\mathbf{V}$ representing a vector observable (such as the coordinate vector $\mathbf{X}$ or the momentum vector $\mathbf{P}$), $U(R)$ must induce a rotation

---

[5]  In general it might be possible for a phase factor $\exp[i\alpha(R_1, R_2)]$ to appear on the right-hand side of this relation. But this does not occur for rotations that can be built up from rotations by very small angles, the case that will be of interest here. For a detailed discussion of this point, see S. Weinberg, *The Quantum Theory of Fields*, Vol. I (Cambridge University Press, Cambridge, 1995), pp. 52–53 and Section 2.7.

$$U^{-1}(R)V_iU(R) = \sum_j R_{ij}V_j. \tag{4.1.4}$$

Rotations, unlike inversions, can be infinitesimal. In this case,

$$R_{ij} = \delta_{ij} + \omega_{ij} + O(\omega^2), \tag{4.1.5}$$

with $\omega_{ij}$ infinitesimal. The condition (4.1.2) gives here

$$1 = \left(1 + \omega^{\mathrm{T}} + O(\omega^2)\right)\left(1 + \omega + O(\omega^2)\right) = 1 + \omega^{\mathrm{T}} + \omega + O(\omega^2),$$

so $\omega^{\mathrm{T}} = -\omega$, or in other words

$$\omega_{ji} = -\omega_{ij}. \tag{4.1.6}$$

For such infinitesimal rotations, the unitary operator $U(R)$ must take the form

$$U(1 + \omega) \rightarrow 1 + \frac{i}{2\hbar} \sum_{ij} \omega_{ij} J_{ij} + O(\omega^2), \tag{4.1.7}$$

with $J_{ij} = -J_{ji}$ a set of Hermitian operators. (The factor $1/\hbar$ is inserted in the definition (4.1.7) in order to give $J_{ij}$ the dimensions of $\hbar$, the same as distance times momentum.)

As usual with the generators of symmetry transformations, the transformation property of other observables can be expressed in commutation relations of these observables with the symmetry generators. For instance, by using Eq. (4.1.7) in the transformation rule (4.1.4) for a vector $\mathbf{V}$, we find

$$\frac{i}{\hbar}[V_k, J_{ij}] = \delta_{ik}V_j - \delta_{jk}V_i. \tag{4.1.8}$$

We can also find the transformation rule of the $J_{ij}$s, and their commutators with each other. As an application of Eq. (4.1.3), we have

$$U(R'^{-1})U(1 + \omega)U(R') = U(R'^{-1}(1 + \omega)R') = U(1 + R'^{-1}\omega R'),$$

for any $\omega_{ij} = -\omega_{ji}$ and any rotation $R'$, unrelated to $\omega$. To first order in $\omega$, we then have

$$\sum_{ij} \omega_{ij} U(R'^{-1}) J_{ij} U(R') = \sum_{kl} (R'^{-1}\omega R)_{kl} J_{kl} = \sum_{ijkl} R'_{ik} R'_{jl} \omega_{ij} J_{kl},$$

in which we have used Eq. (4.1.2), which gives $R'^{-1} = R'^{\mathrm{T}}$. Equating the coefficients of $\omega_{ij}$ on both sides of this equation then gives the transformation rule of the operator $J_{ij}$:

$$U(R'^{-1}) J_{ij} U(R') = \sum_{kl} R'_{ik} R'_{jl} J_{kl}. \tag{4.1.9}$$

That is, $J_{ij}$ is a *tensor*. We can take this a step further, and let $R'$ itself be an infinitesimal rotation, of the form $R' \to 1 + \omega'$, with $\omega'_{ij} = -\omega'_{ji}$ infinitesimal. Then, to first order in $\omega'$, Eq. (4.1.9) gives

$$\frac{i}{2\hbar}\left[ J_{ij} , \sum_{kl} \omega'_{kl} J_{kl} \right] = \sum_{kl} \left( \omega'_{ik} \delta_{jl} + \omega'_{jl} \delta_{ik} \right) J_{kl} = \sum_{k} \omega'_{ik} J_{kj} + \sum_{l} \omega'_{jl} J_{il}.$$

Equating the coefficients of $\omega'_{kl}$ on both sides of this equation gives the commutation rule of the $J$s:

$$\frac{i}{\hbar}\left[ J_{ij} , J_{kl} \right] = -\delta_{il} J_{kj} + \delta_{ik} J_{lj} + \delta_{jk} J_{il} - \delta_{jl} J_{ik}. \tag{4.1.10}$$

So far, all this could be applied to rotationally invariant theories in spaces of any dimensionality. In three dimensions it is very convenient to express $J_{ij}$ in terms of a three-component operator $\mathbf{J}$, defined by

$$J_1 \equiv J_{23}, \quad J_2 \equiv J_{31}, \quad J_3 \equiv J_{12},$$

or more compactly,

$$J_k \equiv \frac{1}{2}\sum_{ij} \epsilon_{ijk} J_{ij}, \quad J_{ij} = \sum_k \epsilon_{ijk} J_k, \tag{4.1.11}$$

where $\epsilon_{ijk}$ is a totally antisymmetric quantity, whose only non-vanishing components are $\epsilon_{123} = \epsilon_{231} = \epsilon_{312} = +1$ and $\epsilon_{213} = \epsilon_{321} = \epsilon_{132} = -1$. The unitary operator (4.1.7) for infinitesimal rotations then takes the form

$$U(1 + \omega) \to 1 + \frac{i}{\hbar}\boldsymbol{\omega} \cdot \mathbf{J} + \mathbf{O}(\omega^2), \tag{4.1.12}$$

where $\omega_k \equiv \frac{1}{2}\sum_{ij} \epsilon_{ijk}\omega_{ij}$. The rotation here is by an infinitesimal angle $|\boldsymbol{\omega}|$ around an axis in the direction of $\boldsymbol{\omega}$.

In terms of $\mathbf{J}$, the characteristic property (4.1.8) of a three-vector $\mathbf{V}$ takes the form

$$[J_i, V_j] = i\hbar \sum_k \epsilon_{ijk} V_k. \tag{4.1.13}$$

(For instance, Eq. (4.1.8) gives $[J_1, V_2] = [J_{23}, V_2] = i\hbar V_3$.) Also, the commutation relation (4.1.10) takes the form

$$[J_i, J_j] = i\hbar \sum_k \epsilon_{ijk} J_k. \tag{4.1.14}$$

(For instance, Eq. (4.1.10) gives $[J_1, J_2] = [J_{23}, J_{31}] = -i\hbar J_{21} = i\hbar J_3$.) That is, $\mathbf{J}$ is itself a three-vector. We may recall that Eq. (4.1.14) is the same commutation relation as the commutation relation (2.1.11) satisfied by the orbital angular momentum operator $\mathbf{L}$, but derived here from the assumption of rotational symmetry, with no assumptions regarding coordinates or momenta. This commutation relation will be the basis of our treatment of angular momentum in the following sections.

Incidentally, it should not be surprising that the quantity $\mathbf{J}$ defined by Eq. (4.1.11) should be a vector, because although the components of $\epsilon_{ijk}$ are the same in all coordinate systems, it is a tensor, in the sense that

$$\epsilon_{ijk} = \sum_{i'j'k'} R_{ii'} R_{jj'} R_{kk'} \epsilon_{i'j'k'}. \tag{4.1.15}$$

This is because the right-hand side is totally antisymmetric in $i$, $j$, and $k$, so it must be proportional to $\epsilon_{ijk}$. According to the definition of determinants, the proportionality coefficient is just Det $R$, which for rotations is $+1$. Knowing that $\epsilon_{ijk}$ and $J_{ij}$ are tensors, it becomes obvious from Eq. (4.1.11) that $J_i$ is a three-vector.

Now let's return to the point raised in the introduction to this chapter, that the total angular momentum $\mathbf{J}$ of a particle may be different from its orbital angular momentum $\mathbf{L}$. If $\mathbf{J}$ is the true generator of rotations, then it is $\mathbf{J}$ rather than $\mathbf{L}$ that has the commutator (4.1.13) with *any* vector. As we saw in Section 2.1, direct calculation shows that in the case of a particle in a central potential the operator $\mathbf{L} \equiv \mathbf{X} \times \mathbf{P}$ satisfies the same commutation relation (4.1.14) as $\mathbf{J}$:

$$[L_i, L_j] = i\hbar \sum_k \epsilon_{ijk} L_k, \tag{4.1.16}$$

and since $\mathbf{L}$ is a vector we must have

$$[J_i, L_j] = i\hbar \sum_k \epsilon_{ijk} L_k. \tag{4.1.17}$$

Therefore, if we define an operator $\mathbf{S} \equiv \mathbf{J} - \mathbf{L}$, so that

$$\mathbf{J} = \mathbf{L} + \mathbf{S}, \tag{4.1.18}$$

then by subtracting Eq. (4.1.16) from Eq. (4.1.17), we find

$$[S_i, L_j] = 0. \tag{4.1.19}$$

From Eqs. (4.1.19), (4.1.18), (4.1.16), and (4.1.14) we then have

$$[S_i, S_j] = i\hbar \sum_k \epsilon_{ijk} S_k. \tag{4.1.20}$$

Thus $\mathbf{S}$ acts as a new kind of angular momentum, and may be thought of as an internal property of a particle, called the *spin*. In Section 2.1 we assumed in effect that the particle in question had $\mathbf{S} = 0$, but this is not the case for electrons and various other particles.

The spin operator is not constructed from the particle's position and momentum operators. Indeed, it commutes with them. Direct calculation gives

$$[L_i, X_j] = i\hbar \sum_k \epsilon_{ijk} X_k, \quad [L_i, P_j] = i\hbar \sum_k \epsilon_{ijk} P_k, \tag{4.1.21}$$

while, as a special case of Eq. (4.1.13),

$$[J_i, X_j] = i\hbar \sum_k \epsilon_{ijk} X_k, \quad [J_i, P_j] = i\hbar \sum_k \epsilon_{ijk} P_k. \tag{4.1.22}$$

The difference of Eqs. (4.1.21) and (4.1.22) then gives

$$[S_i, X_j] = [S_i, P_j] = 0. \tag{4.1.23}$$

A system containing several particles has a total angular momentum given by the sum of the orbital angular momenta $\mathbf{L}_n$ and spins $\mathbf{S}_n$ of the individual particles (labeled here with indices $n, n'$)

$$\mathbf{J} = \sum_n \mathbf{L}_n + \sum_n \mathbf{S}_n. \tag{4.1.24}$$

Because they act on different particles, the commutation relations of the contributions to $\mathbf{J}$ take the general form

$$[L_{ni}, L_{n'j}] = i\hbar \delta_{nn'} \sum_k \epsilon_{ijk} L_{nk}, \tag{4.1.25}$$

$$[L_{ni}, S_{n'j}] = 0, \tag{4.1.26}$$

$$[S_{ni}, S_{n'j}] = i\hbar \delta_{nn'} \sum_k \epsilon_{ijk} S_{nk}, \tag{4.1.27}$$

so that $\mathbf{J}$ satisfies Eq. (4.1.14). Also, $\mathbf{L}_n$ acts only on the coordinates of the $n$th particle, so

$$[L_{ni}, X_{n'j}] = i\hbar \delta_{nn'} \sum_k \epsilon_{ijk} X_{nk}, \quad [L_{ni}, P_{n'j}] = i\hbar \delta_{nn'} \sum_k \epsilon_{ijk} P_{nk}, \tag{4.1.28}$$

while

$$[S_{ni}, X_{n'j}] = [S_{ni}, P_{n'j}] = 0. \tag{4.1.29}$$

Without an explicit formula for $\mathbf{S}$ or $\mathbf{J}$, it is important to be able to calculate how angular momentum operators act on physical state vectors in general, using just the commutation relations. We will work this out in the next section for $\mathbf{J}$, but exactly the same analysis applies to $\mathbf{S}$ and $\mathbf{L}$, and to the total or spin or orbital angular momenta of individual particles.

## 4.2 Angular-Momentum Multiplets

We will now work out the eigenvalues of $\mathbf{J}^2$ and $J_3$, and the action of $\mathbf{J}$ on a multiplet of eigenvectors of these operators, for any Hermitian operator $\mathbf{J}$ satisfying the commutation relations (4.1.14).

First, we note that

$$\left[ J_3, \left( J_1 \pm iJ_2 \right) \right] = i\hbar J_2 \pm i(-i\hbar J_1) = \pm \hbar \left( J_1 \pm iJ_2 \right). \qquad (4.2.1)$$

Therefore $J_1 \pm iJ_2$ act as *raising and lowering operators*: for a state vector $\Psi^m$ that satisfies the eigenvalue condition $J_3 \Psi^m = \hbar m \Psi^m$ (with any $m$), we have

$$J_3 \left( J_1 \pm iJ_2 \right) \Psi^m = (m \pm 1)\hbar \left( J_1 \pm iJ_2 \right) \Psi^m,$$

so if $\left( J_1 \pm iJ_2 \right) \Psi^m$ does not vanish, then it is an eigenstate of $J_3$ with eigenvalue $\hbar(m \pm 1)$. Since $\mathbf{J}^2$ commutes with $J_3$, we can choose $\Psi^m$ to be an eigenvector of $\mathbf{J}^2$ as well as $J_3$, and since $\mathbf{J}^2$ commutes with $\left( J_1 \pm iJ_2 \right)$, all the state vectors that are connected with each other by lowering and/or raising operators will have the same eigenvalue for $\mathbf{J}^2$.

Now, there must be a maximum and a minimum to the eigenvalues of $J_3$ that can be reached in this way, because the square of any eigenvalue of $J_3$ is necessarily *less* than the eigenvalue of $\mathbf{J}^2$. This is because in any normalized state $\Psi$ that has an eigenvalue $a$ for $J_3$ and an eigenvalue $b$ for $\mathbf{J}^2$, we have

$$b - a^2 = \left( \Psi, (\mathbf{J}^2 - J_3^2)\Psi \right) = \left( \Psi, (J_1^2 + J_2^2)\Psi \right) \geq 0.$$

It is conventional to define a quantity $j$ as the maximum value of the eigenvalues of $J_3/\hbar$ for a particular set of state vectors that are related by raising and lowering operators. We will also temporarily define $j'$ as the minimum eigenvalue of $J_3/\hbar$ for these state vectors. The state vector $\Psi^j$ for which $J_3$ takes its maximum eigenvalue $\hbar j$ must satisfy

$$\left( J_1 + iJ_2 \right) \Psi^j = 0, \qquad (4.2.2)$$

since otherwise $\left( J_1 + iJ_2 \right) \Psi^j$ would be a state vector with a larger eigenvalue of $J_3$. Likewise, acting on the state vector $\Psi^j$ with $\left( J_1 - iJ_2 \right)$ gives an eigenstate of $J_3$ with eigenvalue $\hbar(j - 1)$, unless of course this state vector vanishes. Continuing in this way, we must eventually get to a state vector $\Psi^{j'}$ with the minimum eigenvalue $\hbar j'$ of $J_3$, which satisfies

$$\left( J_1 - iJ_2 \right) \Psi^{j'} = 0, \qquad (4.2.3)$$

since otherwise $\left(J_1 - iJ_2\right)\Psi^{j'}$ would be a state vector with an even smaller eigenvalue of $J_3$. We get to $\Psi^{j'}$ from $\Psi^j$ by applying the lowering operator $\left(J_1 - iJ_2\right)$ a whole number of times, so $j - j'$ must be a whole number.

To go further, we use the commutation relations of $J_1$ and $J_2$ to show that

$$\left(J_1 - iJ_2\right)\left(J_1 + iJ_2\right) = J_1^2 + J_2^2 + i[J_1, J_2] = \mathbf{J}^2 - J_3^2 - \hbar J_3, \tag{4.2.4}$$

$$\left(J_1 + iJ_2\right)\left(J_1 - iJ_2\right) = J_1^2 + J_2^2 - i[J_1, J_2] = \mathbf{J}^2 - J_3^2 + \hbar J_3. \tag{4.2.5}$$

According to Eq. (4.2.2), the operator (4.2.4) gives zero when acting on $\Psi^j$, so

$$\mathbf{J}^2\Psi^j = \hbar^2 j(j + 1)\Psi^j. \tag{4.2.6}$$

On the other hand, according to Eq. (4.2.3) the operator (4.2.5) gives zero when acting on $\Psi^{j'}$, so

$$\mathbf{J}^2\Psi^{j'} = \hbar^2 j'(j' - 1)\Psi^{j'}. \tag{4.2.7}$$

But all these state vectors are eigenstates of $\mathbf{J}^2$ with the same eigenvalue, so $j'(j'-1) = j(j+1)$. This quadratic equation for $j'$ has two solutions, $j' = j+1$ and $j' = -j$. The first solution is impossible, because $j'$ is the minimum eigenvalue of $J_3/\hbar$, and therefore cannot be greater than the maximum eigenvalue $j$. This leaves us with the other solution

$$j' = -j. \tag{4.2.8}$$

But we saw that $j - j'$ must be an integer, so $j$ *must be an integer or a half-integer.* The eigenvalues of $J_3$ range over the $2j+1$ values of $\hbar m$ with $m$ running by unit steps from $-j$ to $+j$. The corresponding eigenstates will be denoted $\Psi_j^m$, so that

$$J_3\Psi_j^m = \hbar m\Psi_j^m, \qquad m = -j, -j+1, \dots, +j, \tag{4.2.9}$$
$$\mathbf{J}^2\Psi_j^m = \hbar^2 j(j + 1)\Psi_j^m. \tag{4.2.10}$$

These are the same eigenvalues that we found previously in the case of orbital angular momentum, with the one big difference that $j$ and $m$ may be half-integers rather than integers.

The state vectors $\Psi_j^m$ for different values of $m$ are orthogonal, because they are eigenvectors of the Hermitian operator $J_3$ with different eigenvalues, and they can be multiplied with suitable constants to normalize them, so that

$$\left(\Psi_j^{m'}, \Psi_j^m\right) = \delta_{m'm}. \tag{4.2.11}$$

Also, we have noted that $\left(J_1 \pm iJ_2\right)\Psi_j^m$ has eigenvalue $\hbar(m \pm 1)$ for $J_3$, so it must be proportional to $\Psi_j^{m\pm1}$:

$$\left(J_1 \pm iJ_2\right)\Psi_j^m = \alpha^{\pm}(j, m)\Psi_j^{m\pm1}. \tag{4.2.12}$$

It follows then from Eq. (4.2.4) that

$$\alpha^-(j, m+1)\alpha^+(j, m) = \hbar^2[j(j+1) - m^2 - m].  \tag{4.2.13}$$

In order to satisfy the normalization condition (4.2.11), it is necessary that

$$|\alpha^\pm(j, m)|^2 = \left((J_1 \pm iJ_2)\Psi_j^m, (J_1 \pm iJ_2)\Psi_j^m\right)$$
$$= \left(\Psi_j^m, (J_1 \mp iJ_2)(J_1 \pm iJ_2)\Psi_j^m\right),$$

and therefore, according to Eqs. (4.2.4) and (4.2.5),

$$|\alpha^\pm(j, m)|^2 = \hbar^2[j(j+1) - m^2 \mp m].  \tag{4.2.14}$$

We can adjust the phases of the coefficients $\alpha^-(j, m)$ to be anything we want, by multiplying the state vectors $\Psi_j^m$ with phase factors (complex numbers with modulus unity), which do not affect Eq. (4.2.11). (To adjust the phase of $\alpha^-(j, j)$, multiply $\Psi_j^{j-1}$ by a suitable phase factor; then, to adjust the phase of $\alpha^-(j, j-1)$, multiply $\Psi_j^{j-2}$ by a suitable phase factor; and so on.) It is conventional to adjust these phases so that all $\alpha^-(j, m)$ are real and positive, in which case Eq. (4.2.13) requires that all $\alpha^+(j, m)$ are also real and positive. Equation (4.2.14) then gives these factors as

$$\alpha^\pm(j, m) = \hbar\sqrt{j(j+1) - m^2 \mp m},  \tag{4.2.15}$$

so that

$$\left(J_1 \pm iJ_2\right)\Psi_j^m = \hbar\sqrt{j(j+1) - m^2 \mp m}\ \Psi_j^{m\pm1}.  \tag{4.2.16}$$

It can now be revealed that the phases of the spherical harmonics $Y_\ell^m$ were chosen in Section 2.2 so that the same relations apply to them, with $L_i$ and $\ell$ in place of $J_i$ and $j$. Equations (4.2.9) and (4.2.16) provide a complete statement of how the quantum-mechanical operators $J_i$ act on the state vectors $\Psi_j^m$. In group theory, we say that the relations (4.2.9) and (4.2.16) furnish a *representation* of the commutation relations (4.1.14). (Of course, the state vectors $\Psi_j^m$ can depend on any number of other dynamical variables, which are invariant under the action of the symmetry generators $J_i$.)

As an example, consider the case $j = 1/2$. We note that Eq. (4.2.16) here gives

$$(J_1 \pm iJ_2)\Psi_{1/2}^{\mp1/2} = \hbar\Psi_{1/2}^{\pm1/2}, \qquad (J_1 \pm iJ_2)\Psi_{1/2}^{\pm1/2} = 0,$$

and of course

$$J_3\Psi_{1/2}^{\pm1/2} = \pm\frac{\hbar}{2}\Psi_{1/2}^{\pm1/2}.$$

These results can be summarized in the statement that

$$\left(\Psi_{1/2}^{m'}, \mathbf{J}\Psi_{1/2}^m\right) = \frac{\hbar}{2}\boldsymbol{\sigma}_{m'm},  \tag{4.2.17}$$

where $\sigma_i$ are $2 \times 2$ matrices, known as *Pauli matrices*:

$$\sigma_1 = \begin{pmatrix} 0 & 1 \\ 1 & 0 \end{pmatrix}, \quad \sigma_2 = \begin{pmatrix} 0 & -i \\ i & 0 \end{pmatrix}, \quad \sigma_3 = \begin{pmatrix} 1 & 0 \\ 0 & -1 \end{pmatrix}. \qquad (4.2.18)$$

There is a simple application of Eq. (4.2.16) that is useful in many physical calculations. Suppose we know that a system is in a state with normalized state vector $\Psi_j^m$, and we want to know the probability that a certain measurement will put the system in a state with normalized state vector $\Phi_j^m$ (rather than any other of a complete orthonormal set), where the various $\Psi_j^m$ form a multiplet related to each other by Eq. (4.2.16), and likewise for the $\Phi_j^m$. According to the general principles of quantum mechanics, this probability is the absolute value squared of the matrix element[6] $(\Phi_j^m, \Psi_j^m)$. Using Eq. (4.2.16), we can show that this matrix element, and hence the probability, is *independent* of $m$. To see this, we use Eq. (4.2.16) to calculate

$$\hbar\sqrt{j(j+1) - m^2 \mp m} \left( \Phi_j^{m\pm1}, \Psi_j^{m\pm1} \right)$$

$$= \left( \Phi_j^{m\pm1}, (J_1 \pm iJ_2) \Psi_j^m \right)$$

$$= \left( (J_1 \mp iJ_2) \Phi_j^{m\pm1}, \Psi_j^m \right)$$

$$= \hbar\sqrt{j(j+1) - (m\pm1)^2 \pm (m\pm1)} \left( \Phi_j^m, \Psi_j^m \right)$$

$$= \hbar\sqrt{j(j+1) - m^2 \mp m} \left( \Phi_j^m, \Psi_j^m \right),$$

and therefore

$$\left( \Phi_j^{m\pm1}, \Psi_j^{m\pm1} \right) = \left( \Phi_j^m, \Psi_j^m \right). \qquad (4.2.19)$$

This can be repeated, leading to the conclusion that $\left( \Phi_j^m, \Psi_j^m \right)$ is independent of $m$, as was to be proved. By the same reasoning, if $A$ is an operator (such as the Hamiltonian) that commutes with $\mathbf{J}$, then also its matrix elements $\left( \Phi_j^m, A \, \Psi_j^m \right)$ are independent of $m$. This little theorem will be used in Section 4.4 to calculate the $m$-dependence of matrix elements of operators with various transformation properties under rotations.

$$* * * * *$$

As we have seen, the angular momentum of bound-state energy levels determines the multiplicity of these levels. The components of angular momentum can also be measured directly. The classic example of such a measurement is that

---

[6] We consider only the matrix elements in which both state vectors have equal values of $j$ and $m$, because both state vectors are eigenstates of the Hermitian operators $\mathbf{J}^2$ and $J_3$, so the matrix element would vanish unless they both had the same eigenvalues.

of Walter Gerlach (1889–1979) and Otto Stern (1888–1969) in 1922,[7] already briefly mentioned in Section 3.7 in connection with the interpretation of quantum mechanics. In the Stern–Gerlach experiment, a beam of neutral atoms[8] is sent into a slowly varying magnetic field. The magnetic field is of the form

$$\mathbf{B}(\mathbf{x}) = \mathbf{B}_0 + \mathbf{B}_1(\mathbf{x}), \tag{4.2.20}$$

where $\mathbf{B}_0$ is a constant, and the variable term $\mathbf{B}_1(\mathbf{x})$ is much smaller than $\mathbf{B}_0$. As we will see, the direction of $\mathbf{B}_0$ determines what it is that is measured in this experiment. We will take the three-axis to be in this direction. The precise form of $\mathbf{B}_1(\mathbf{x})$ is not very important, though of course it must satisfy the free-field Maxwell equations

$$\nabla \cdot \mathbf{B}_1 = 0, \quad \nabla \times \mathbf{B}_1 = 0. \tag{4.2.21}$$

For instance, we might have $B_{1i} = \sum_j D_{ij} x_j$, with the constant matrix $D_{ij}$ both symmetric and traceless. The atom is supposed to have a total angular momentum $\mathbf{J}$. The Hamiltonian of the atom is then

$$H = \frac{\mathbf{p}^2}{2m} - \left(\frac{\mu}{\hbar j}\right) \left(J_3 |\mathbf{B}_0| + \mathbf{J} \cdot \mathbf{B}_1(\mathbf{x})\right), \tag{4.2.22}$$

where $\mathbf{J}^2 = \hbar^2 j(j+1)$, and $\mu$ is a property of the atom, known as its magnetic moment. In the original Stern–Gerlach experiment, the atoms in question were of silver, with angular momentum $j = 1/2$ arising from the spin of a single electron (though this was not known at the time), but it is just as easy to consider the general case, of arbitrary $j$. According to the arguments of Ehrenfest described in Section 1.5, the expectation values of the position and the momentum will obey the equations of motion

$$\frac{d}{dt}\langle\mathbf{x}\rangle = \langle\mathbf{p}\rangle/m, \quad \frac{d}{dt}\langle\mathbf{p}\rangle = \left(\frac{\mu}{\hbar j}\right)\left\langle\nabla\Big(\mathbf{J}\cdot\mathbf{B}_1(\mathbf{x})\Big)\right\rangle. \tag{4.2.23}$$

For sufficiently large $\mathbf{B}_0$, the time dependence of the component of a state vector having the eigenvalue $\hbar\sigma \neq 0$ for $J_3$ is dominated by a rapidly oscillating factor $\exp(i\sigma\mu|\mathbf{B}_0|t/\hbar j)$. We have seen that the eigenvalues of $J_3$ are $\hbar\sigma$, where $\sigma = -j, -j+1, \ldots, +j$. Also, Eq. (4.2.16) shows that $J_1$ and $J_2$ have matrix elements only between eigenstates of $J_3$ that differ by $\pm\hbar$, so these matrix elements are proportional to $\exp(\pm i\mu|\mathbf{B}_0|t/j)$, and therefore vanish when averaged even over short time intervals. Thus the equations of motion (4.2.23) of a particle for which $J_z = \hbar\sigma$ become effectively

$$\frac{d}{dt}\langle\mathbf{x}\rangle = \langle\mathbf{p}\rangle/m, \quad \frac{d}{dt}\langle\mathbf{p}\rangle = \left(\frac{\mu\sigma}{j}\right)\langle\nabla B_{13}(\mathbf{x})\rangle. \tag{4.2.24}$$

---

[7] W. Gerlach and O. Stern, Z. *Physik* **9**, 353 (1922).

[8] Neutral atoms are used, both to avoid Coulomb forces from incidental electric fields, and to avoid the Lorentz force produced by the motion of a charged particle through a magnetic field.

For instance, in the case discussed above where $B_{1i} = \sum_j D_{ij} x_j$, these two equations can be combined to give a single second-order differential equation for $\langle \mathbf{x} \rangle$:

$$m \frac{d^2}{dt^2} \langle x_i \rangle = \left( \frac{\mu \sigma}{j} \right) D_{3i}.$$

Whatever the form of $\mathbf{B}_1$, there are $2j + 1$ possible trajectories, and observation of the actual trajectory that is followed by the particle tells us the value of $\sigma$.

## 4.3 Addition of Angular Momenta

It often happens that a physical system will contain angular momenta of two or more different types. For instance, in the ground state of the helium atom there are two electrons, each with its own spin, but no orbital angular momentum. In the excited states of the hydrogen atom with $\ell > 0$ there is both an orbital angular momentum and a spin angular momentum. The presence of interactions between the individual angular momenta usually has the effect that they are not separately conserved – that is, the individual angular momenta do not commute with the Hamiltonian. In such cases it is useful to introduce a *total* angular-momentum operator, given by the sum of the individual angular-momentum operators, which *does* commute with the Hamiltonian. The problem is, how to relate the states labeled by values of the total angular momentum to states described in terms of the individual angular momenta?

Suppose we have two angular-momentum operator vectors $\mathbf{J}'$ and $\mathbf{J}''$, which may be spins or orbital angular momenta or the sums of spins and/or angular momenta, with each satisfying the commutation relations (4.1.14):

$$[J_1', J_2'] = i\hbar J_3', \qquad [J_2', J_3'] = i\hbar J_1', \qquad [J_3', J_1'] = i\hbar J_2', \qquad (4.3.1)$$

$$[J_1'', J_2''] = i\hbar J_3'', \qquad [J_2'', J_3''] = i\hbar J_1'', \qquad [J_3'', J_1''] = i\hbar J_2'', \qquad (4.3.2)$$

but commuting with each other,

$$[J_i', J_k''] = 0. \qquad (4.3.3)$$

We consider a set of states having two independent angular momenta $j'$ and $j''$, with $J_3'$ and $J_3''$ taking values $\hbar m'$ and $\hbar m''$, respectively,[9] and with $m'$ and $m''$ running by unit steps from $-j'$ to $j'$ and from $-j''$ to $j''$, respectively. The normalized state vectors $\Psi_{j'j''}^{m'm''}$ of these states satisfy

$$\mathbf{J}'^2 \Psi_{j'j''}^{m'm''} = \hbar^2 j'(j'+1) \Psi_{j'j''}^{m'm''}, \qquad (4.3.4)$$

$$J_3' \Psi_{j'j''}^{m'm''} = \hbar m' \Psi_{j'j''}^{m'm''}, \qquad (4.3.5)$$

---

[9] Of course there is no connection between the $j'$ used here and that introduced temporarily in the previous section.

$$\left(J_1' \pm i J_2'\right)\Psi_{j'j''}^{m'm''} = \hbar\sqrt{j'(j'+1) - m'^2 \mp m'}\; \Psi_{j'j''}^{m'\pm1,\,m''}, \tag{4.3.6}$$

$$\mathbf{J}'^2\,\Psi_{j'j''}^{m'm''} = \hbar^2 j'(j'+1)\,\Psi_{j'j''}^{m'm''}, \tag{4.3.7}$$

$$J_3'\Psi_{j'j''}^{m'm''} = \hbar m'\,\Psi_{j'j''}^{m'm''}, \tag{4.3.8}$$

$$\left(J_1'' \pm i J_2''\right)\Psi_{j'j''}^{m'm''} = \hbar\sqrt{j''(j''+1) - m''^2 \mp m''}\; \Psi_{j'j''}^{m',\,m''\pm1}. \tag{4.3.9}$$

We can then introduce a total angular momentum

$$\mathbf{J} = \mathbf{J}' + \mathbf{J}'', \tag{4.3.10}$$

which also satisfies the commutation relations (4.1.14):

$$[J_1,\,J_2] = i\hbar J_3, \qquad [J_2,\,J_3] = i\hbar J_1, \qquad [J_3,\,J_1] = i\hbar J_2. \tag{4.3.11}$$

Both $\mathbf{J}'^2$ and $\mathbf{J}''^2$ commute with all the components of $\mathbf{J}'$ and $\mathbf{J}''$. On the other hand, the Hamiltonian will in general contain interaction terms that do not commute with either $\mathbf{J}'$ or $\mathbf{J}''$, such as a possible term proportional to $\mathbf{J}' \cdot \mathbf{J}''$. We then have to look for other operators that do commute with such interaction terms.

This usually (though not always!) includes $\mathbf{J}'^2$ and $\mathbf{J}''^2$, since they each commute with both $\mathbf{J}'$ and $\mathbf{J}''$. Also, as we have seen in Section 4.1, the *total* angular momentum $\mathbf{J}$ commutes with all rotationally invariant operators. For instance,

$$\mathbf{J}' \cdot \mathbf{J}'' = \frac{1}{2}\Big[\mathbf{J}^2 - \mathbf{J}'^2 - \mathbf{J}''^2\Big],$$

and each term on the right-hand side commutes with $\mathbf{J}$. Instead of states of definite energy being characterized by the values $\hbar^2 j'(j'+1)$, $\hbar m'$, $\hbar^2 j''(j''+1)$, and $\hbar m''$ of $\mathbf{J}'^2$, $J_3'$, $\mathbf{J}''^2$, and $J_3''$, they will be characterized by the values $\hbar^2 j'(j'+1)$, $\hbar^2 j''(j''+1)$, $\hbar^2 j(j+1)$, and $\hbar m$ of $\mathbf{J}'^2$, $\mathbf{J}''^2$, $\mathbf{J}^2$, and $J_3$, respectively. Our problems are, what values of $j$ occur for a given $j'$ and $j''$, how many states for a given $j'$, $j''$, $j$, and $m$ can be constructed from the states with state vectors $\Psi_{j'j''}^{m'm''}$, and how can we express the state vectors of these states in terms of the $\Psi_{j'j''}^{m'm''}$?

The general rule is, that there is precisely one state for each $j$ and $m$ in the ranges

$$j = |j' - j''|,\; |j' - j''|+1,\; \ldots,\; j'+j'', \qquad m = j,\; j-1,\; \ldots,\; -j. \tag{4.3.12}$$

The normalized state vectors $\Psi_{j'j''j}^{m}$ of these states are then uniquely defined (up to a common phase factor) by

$$\mathbf{J}'^2\,\Psi_{j'j''j}^{m} = \hbar^2 j'(j'+1)\Psi_{j'j''j}^{m}, \tag{4.3.13}$$

$$\mathbf{J}''^2\,\Psi_{j'j''j}^{m} = \hbar^2 j''(j''+1)\Psi_{j'j''j}^{m}, \tag{4.3.14}$$

$$\mathbf{J}^2\,\Psi_{j'j''j}^{m} = \hbar^2 j(j+1)\Psi_{j'j''j}^{m}, \tag{4.3.15}$$

$$J_3\Psi_{j'j''j}^{m} = \hbar m\Psi_{j'j''j}^{m}, \tag{4.3.16}$$

$$\left(J_1 \pm i J_2\right)\Psi^m_{j'j''j} = \hbar\sqrt{j(j+1) - m^2 \mp m}\ \Psi^{m\pm 1}_{j'j''j}. \tag{4.3.17}$$

These state vectors may be expressed as linear combinations

$$\Psi^m_{j'j''j} = \sum_{m'm''} C_{j'j''}(j\,m\,;\,m'\,m'')\Psi^{m'm''}_{j'j''}, \tag{4.3.18}$$

where $C_{j'j''}(j\,m\,;\,m'\,m'')$ are a set of constants known as *Clebsch–Gordan coefficients*. Of course, since $J_3 = J'_3 + J''_3$, the only non-vanishing Clebsch–Gordan coefficients are those for which

$$m = m' + m''. \tag{4.3.19}$$

To verify that the values of $j$ for which the Clebsch–Gordan coefficients do not vanish are limited by Eq. (4.3.12), we note first that the values of $m = m' + m''$ can only lie between $j' + j''$ and $-j' - j''$, so the maximum possible value for $j$ is $j' + j''$. On the other hand, a state vector with $m' = j'$ and $m'' = j''$ has $j \geq |m| = j' + j''$, so it can only have $j = j' + j''$. Furthermore, the only way to have $m = j' + j''$ is to have $m' = j'$ and $m'' = j''$, so there is precisely one state with $j = j' + j''$ and $m = j' + j''$, and hence only one state with $j = j' + j''$ and any $m$ between $j' + j''$ and $-j' - j''$. With an appropriate choice of phase, the state vector for this state is simply

$$\Psi^{j'+j''}_{j'j''\,j'+j''} = \Psi^{j'j''}_{j'j''}. \tag{4.3.20}$$

That is,

$$C_{j'j''}(j\,m\,;\,j'\,j'') = \delta_{j,\,j'+j''}\delta_{m,\,j'+j''}. \tag{4.3.21}$$

Now consider the state vectors $\Psi^{m'm''}_{j'j''}$ with $m = m' + m'' = j' + j'' - 1$. There are generally two such state vectors, one with $m' = j'$ and $m'' = j'' - 1$, and the other with $m' = j' - 1$ and $m'' = j''$. (The only exceptions occur for $j' - 1 < -j'$, or in other words $j' = 0$, in which case $m'$ cannot equal $j' - 1$, or for $j'' - 1 < -j''$, or in other words $j'' = 0$, in which case $m''$ cannot equal $j'' - 1$.) One linear combination of these two state vectors is a state vector with $j = j' + j''$, which is formed by operating with the lowering operator $J_1 - iJ_2$ on the state vector (4.3.20). The factor (4.2.15) here is

$$\sqrt{j(j+1) - j^2 + j} = \sqrt{2j} = \sqrt{2(j' + j'')},$$

so

$$\begin{aligned}
\Psi^{j'+j''-1}_{j'j''\,j'+j''} &= (2(j'+j''))^{-1/2}\left(J_1 - iJ_2\right)\Psi^{j'+j''}_{j'j''\,j'+j''} \\
&= (2(j'+j''))^{-1/2}\left(J'_1 - iJ'_2 + J''_1 - iJ''_2\right)\Psi^{j'j''}_{j'j''} \\
&= (j'+j'')^{-1/2}\left(\sqrt{j'}\,\Psi^{j'-1,\,j''}_{j'j''} + \sqrt{j''}\,\Psi^{j',\,j''-1}_{j'j''}\right). \tag{4.3.22}
\end{aligned}$$

There is no other state vector with $j = j' + j''$ and $m = j' + j'' - 1$, because if there were then there would also have to be two state vectors with $j = j' + j''$ and $m = j' + j''$, and we have seen that there is only one. Therefore the only other state vector with $m = j' + j'' - 1$ must have the only other value of $j$ that is possible for such a state vector, $j = j' + j'' - 1$. The state vector with this value of $j$ must be orthogonal to the state vector (4.3.22), since it is a state vector with a different value of $\mathbf{J}^2$, so (apart from an arbitrary choice of a phase factor) if properly normalized it can only be the state vector

$$\Psi^{j'+j''-1}_{j'j''\,j'+j''-1} = (j'+j'')^{-1/2} \left( \sqrt{j''}\, \Psi^{j'-1,\,j''}_{j'j''} - \sqrt{j'}\, \Psi^{j',\,j''-1}_{j'j''} \right). \quad (4.3.23)$$

That is,

$$C_{j'j''}(jm;\, j'-1\, j'')$$

$$= \delta_{m,\,j'+j''-1} \left[ \sqrt{\frac{j'}{j'+j''}}\,\delta_{j,\,j'+j''} + \sqrt{\frac{j''}{j'+j''}}\,\delta_{j,\,j'+j''-1} \right], \quad (4.3.24)$$

and

$$C_{j'j''}(jm;\, j'\, j''-1)$$

$$= \delta_{m,\,j'+j''-1} \left[ \sqrt{\frac{j''}{j'+j''}}\,\delta_{j,\,j'+j''} - \sqrt{\frac{j'}{j'+j''}}\,\delta_{j,\,j'+j''-1} \right]. \quad (4.3.25)$$

Continuing in this way, we find that at first for each step down in $m$ there is just one new state vector $\Psi^m_{j'j''j}$ that is orthogonal to all the state vectors of this type that are obtained by applying the lowering operator to the state vectors already constructed (which have $j = m + 1, m + 2, \ldots, j' + j''$), and that therefore can only have $j = m$.

This procedure eventually stops, because $m'$ is limited to the range from $-j'$ to $+j'$, and $m''$ is limited to the range from $-j''$ to $+j''$. It follows that for a given $m$, $m' = m - m''$ runs up from the greater of $-j'$ and $m - j''$ to the lesser of $+j'$ and $m + j''$. For $m = j' + j''$ the greater of $-j'$ and $m - j''$ is $m - j'' = j'$ and the lesser of $+j'$ and $m + j''$ is $j'$, so of course the value of $m'$ is unique, $m' = j'$. As long as the greater of $-j'$ and $m - j''$ is $m - j''$ and the lesser of $+j'$ and $m + j''$ is $j'$, each unit step down in $m$ increases the range of $m'$ by one, giving a new value of $j$ one unit lower at each step. But this continues only until either $m - j'' = -j'$ or $m + j'' = j' -$ in other words, until $m$ equals the greater of $j'' - j'$ and $j' - j''$, which is $|j' - j''|$. After that, we get no new values of $j$, which therefore is limited to the range (4.3.12).

As a check, let's count the total number of all these state vectors. Suppose that $j' \geq j''$, so that (4.3.12) allows values of $j$ running from $j' - j''$ to $j' + j''$, each with $2j + 1$ values of $m$. The total number of state vectors $\Psi^m_{j'j''j}$ is then

$$\sum_{j=j'-j''}^{j'+j''} (2j+1) = 2\frac{(j'+j'')(j'+j''+1)}{2} - 2\frac{(j'-j''-1)(j'-j'')}{2}$$

$$+ 2j'' + 1$$
$$= (2j'+1)(2j''+1), \tag{4.3.26}$$

which is just the number of state vectors $\Psi_{j'j''}^{m'm''}$ with $m'$ and $m''$ taking $2j'+1$ and $2j''+1$ values, respectively. Since the result is symmetric in $j'$ and $j''$, the same result applies for $j'' \geq j'$.

With the phase conventions adopted here, the Clebsch–Gordan coefficients are all real. They also have another important property, that follows from their role as the transformation coefficients between two complete sets of orthonormal state vectors. To see this in general, suppose we have two sets of state vectors, $\Phi_n$ and $\Phi'_a$, that satisfy the orthonormality conditions

$$\left(\Phi_n, \Phi_m\right) = \delta_{nm}, \qquad \left(\Phi'_a, \Phi'_b\right) = \delta_{ab},$$

and are related by a set of coefficients $C_{na}$,

$$\Phi_n = \sum_a C_{na} \Phi'_a. \tag{4.3.27}$$

The orthonormality conditions require that

$$\delta_{nm} = \left(\Phi_n, \Phi_m\right) = \sum_{ab} C_{na}^* C_{mb} \left(\Phi'_a, \Phi'_b\right) = \sum_a C_{na}^* C_{ma}. \tag{4.3.28}$$

There is a general theorem of matrix algebra[10] that tells us that when a finite square array of complex numbers $C_{na}$ satisfies this relation, then we also have

$$\sum_n C_{na}^* C_{nb} = \delta_{ab}. \tag{4.3.29}$$

In consequence

$$\Phi'_a = \sum_n C_{na}^* \Phi_n. \tag{4.3.30}$$

For the real Clebsch–Gordan coefficients the conditions (4.3.28) and (4.3.29) read

---

[10] In matrix notation, the relation $\sum_a C_{na}^* C_{ma} = \delta_{nm}$ is written $CC^\dagger = 1$, where the product $AB$ of any two matrices $A$ and $B$ is defined as a matrix with components $(AB)_{mn} \equiv \sum_a A_{ma} B_{an}$, and $C^\dagger$ is the matrix with $C_{an}^\dagger = (C_{na})^*$. Also, 1 is here the unit matrix with $1_{mn} = \delta_{nm}$. The determinant of a product of matrices is the product of the determinants, and the determinant of $C^\dagger$ is the complex conjugate of the determinant of $C$, so here $|\text{Det } C|^2 = 1$. Since $\text{Det } C \neq 0$, $C$ has an inverse, which in this case is $C^\dagger$, so here also $C^\dagger C = 1$. The $ab$ component of this equation tells us that $\sum_n C_{na}^* C_{nb} = \delta_{ab}$.

$$\sum_{jm} C_{j'j''}(jm;\ m'm'')C_{j'j''}(jm;\ \tilde{m}'\tilde{m}'') = \delta_{m'\tilde{m}'}\delta_{m''\tilde{m}''}, \qquad (4.3.31)$$

and

$$\sum_{m'm''} C_{j'j''}(jm;\ m'm'')C_{j'j''}(\tilde{j}\tilde{m};\ m'm'') = \delta_{j\tilde{j}}\delta_{m\tilde{m}}. \qquad (4.3.32)$$

Also, the relation (4.3.18) may be inverted to read

$$\Psi_{j'j''}^{m'm''} = \sum_{jm} C_{j'j''}(jm;\ m'm'')\Psi_{j'j''j}^{m}. \qquad (4.3.33)$$

Values for some Clebsch–Gordan coefficients are given in Table 4.1.

To take a physical example, consider the state vectors of the hydrogen atom, now taking into account the spin $1/2$ of the electron. For $\ell = 0$ the only possible value of $j$ is of course $j = 1/2$, while for $\ell > 0$ there are two values of $j$, that is, $j = \ell + 1/2$ and $j = \ell - 1/2$. In a standard notation, the hydrogen states are written $n\,\ell_j$, with orbital angular momenta $\ell = 0, 1, 2, 3, 4, \ldots$ represented by the letters $s$, $p$, $d$, $f$, $g$, and from then on alphabetically. Recall also that $\ell \leq n - 1$. We saw that the ground state, with $n = 1$, has $\ell = 0$, so this state has a unique $j$ value, $j = 1/2$, and is denoted $1s_{1/2}$. The first excited energy level, with $n = 2$, has $\ell = 0$ and $\ell = 1$. The $n = 2$ state with $\ell = 0$ has $j = 1/2$, and is denoted $2s_{1/2}$. The $n = 2$ state with $\ell = 1$ can be decomposed into states with $j = 1/2$ and $j = 3/2$, denoted $2p_{1/2}$ and $2p_{3/2}$. The hydrogen states are therefore $1s_{1/2}, 2p_{3/2}, 2p_{1/2}, 2s_{1/2}, 3d_{5/2}, 3d_{3/2}, 3p_{3/2}, 3p_{1/2}, 3s_{1/2}$, etc.

If for instance we measure the values $S_3$ and $L_3$ of the 3-component of the electron's spin and orbital angular momentum[11] in the $2p_{3/2}$ state with $m = 1/2$, then we will either get values $1/2$ and $0$, or values $-1/2$ and $+1$, with probabilities equal to the squares of the corresponding Clebsch–Gordan coefficients, which according to Table 4.1 are $2/3$ and $1/3$, respectively.

The spin–orbit interaction proportional to $\mathbf{L} \cdot \mathbf{S}$ splits the states with the same $n$ and $\ell$ but different $j$ from each other by what is known as the *fine structure* of the hydrogen atom. For instance, the energy difference of the $2p_{1/2}$ and $2p_{3/2}$ states is $4.5283 \times 10^{-5}$ eV. These effects would leave states with the same $j$ and $n$ but different $\ell$ with the same energy, but they are split by a smaller energy difference known as the *Lamb shift*, due chiefly to a continual emission and

---

[11] This can be done for example by a Stern–Gerlach experiment, with a strong magnetic field in the 3-direction. As we will see in Section 5.2, $\mathbf{L}$ and $\mathbf{S}$ contribute differently to the magnetic moment of the atom, so the interaction energy of the atom with the magnetic field will be different for different values of $m_\ell$ and $m_s$, even for states with the same value of $m = m_\ell + m_s$. If this interaction energy is large compared with the interaction between the atom's spin and orbital angular momentum, then the matrix elements of the 1 and 2 components of the magnetic moment, which connect states with different values for $m_\ell$ and/or $m_s$, will oscillate rapidly, and will not contribute to the interaction energy. Thus if the magnetic field also has a weak inhomogeneous term with a non-vanishing 3-component, the atom will pursue different trajectories for different values of $m_\ell$ and/or $m_s$.

Table 4.1 The non-vanishing Clebsch–Gordan coefficients for the addition of angular momenta $j'$ and $j''$ with 3-components $m'$ and $m''$ to give angular momentum $j$ with 3-component $m$, for several low values of $j'$ and $j''$

| $j'$ | $j''$ | $j$ | $m$ | $m'$ | $m''$ | $C_{j'j''}(jm;\ m'm'')$ |
|------|-------|-----|-----|------|-------|------------------------|
| $\frac{1}{2}$ | $\frac{1}{2}$ | 1 | $+1$ | $+\frac{1}{2}$ | $+\frac{1}{2}$ | 1 |
| $\frac{1}{2}$ | $\frac{1}{2}$ | 1 | 0 | $\pm\frac{1}{2}$ | $\mp\frac{1}{2}$ | $1/\sqrt{2}$ |
| $\frac{1}{2}$ | $\frac{1}{2}$ | 1 | $-1$ | $-\frac{1}{2}$ | $-\frac{1}{2}$ | 1 |
| $\frac{1}{2}$ | $\frac{1}{2}$ | 0 | 0 | $\pm\frac{1}{2}$ | $\mp\frac{1}{2}$ | $\pm 1\sqrt{2}$ |
| 1 | $\frac{1}{2}$ | $\frac{3}{2}$ | $\pm\frac{3}{2}$ | $\pm 1$ | $\pm\frac{1}{2}$ | 1 |
| 1 | $\frac{1}{2}$ | $\frac{3}{2}$ | $\pm\frac{1}{2}$ | $\pm 1$ | $\mp\frac{1}{2}$ | $\sqrt{1/3}$ |
| 1 | $\frac{1}{2}$ | $\frac{3}{2}$ | $\pm\frac{1}{2}$ | 0 | $\pm\frac{1}{2}$ | $\sqrt{2/3}$ |
| 1 | $\frac{1}{2}$ | $\frac{1}{2}$ | $\pm\frac{1}{2}$ | $\pm 1$ | $\mp\frac{1}{2}$ | $\pm\sqrt{2/3}$ |
| 1 | $\frac{1}{2}$ | $\frac{1}{2}$ | $\pm\frac{1}{2}$ | 0 | $\pm\frac{1}{2}$ | $\mp\sqrt{1/3}$ |
| 1 | 1 | 2 | $\pm 2$ | $\pm 1$ | $\pm 1$ | 1 |
| 1 | 1 | 2 | $\pm 1$ | $\pm 1$ | 0 | $1/\sqrt{2}$ |
| 1 | 1 | 2 | $\pm 1$ | 0 | $\pm 1$ | $1/\sqrt{2}$ |
| 1 | 1 | 2 | 0 | $\pm 1$ | $\mp 1$ | $1/\sqrt{6}$ |
| 1 | 1 | 2 | 0 | 0 | 0 | $\sqrt{2/3}$ |
| 1 | 1 | 1 | $\pm 1$ | $\pm 1$ | 0 | $\pm 1/\sqrt{2}$ |
| 1 | 1 | 1 | $\pm 1$ | 0 | $\pm 1$ | $\mp 1/\sqrt{2}$ |
| 1 | 1 | 0 | 0 | $\pm 1$ | $\mp 1$ | $1/\sqrt{3}$ |
| 1 | 1 | 0 | 0 | 0 | 0 | $-1/\sqrt{3}$ |

reabsorption of photons by the electron. This splitting of the $2p_{1/2}$ and $2s_{1/2}$ states is $4.35152 \times 10^{-6}$ eV.

The above discussion of the hydrogen spectrum ignored the effect of the magnetic moment of the proton. This is very small, because the proton's large mass gives it a much smaller magnetic moment than the electron. The effect of the magnetic field of the nucleus of any atom on the atom's energy levels is called its *hyperfine splitting*. For instance, there are two $1s$ states of hydrogen, with total proton plus electron spin equal to 1 or 0, separated by an energy difference

$5.87 \times 10^{-6}$ eV, comparable to the Lamb shift of the $n = 2$ states. The radiative transition between the states of total spin 1 and 0 is the famous 21-centimeter line in the radio spectrum of hydrogen.

The Clebsch–Gordan coefficients have an important property of symmetry or antisymmetry:

$$C_{j'j''}(jm; m'm'') = (-1)^{j-j'-j''} C_{j''j'}(jm; m''m').  \qquad (4.3.34)$$

To see this, note that the state vectors $\Psi^{jm}_{j'j''}$ and $\Psi^{jm}_{j''j'}$ both represent the same state, one in which angular momenta $\mathbf{J}'$ and $\mathbf{J}''$ combine to form a total angular momentum $\mathbf{J}$ with $\mathbf{J}^2 = \hbar^2 j(j+1)$ and $J_z = \hbar m$, and are therefore equal up to a constant factor. By interchanging $j'$ with $j''$ and then interchanging $j''$ with $j'$ we must get back to the same state vector, so this factor must have unit square, and is therefore just a sign. Further, since all the $\Psi^{jm}_{j'j''}$ with the same $j'$, $j''$, and $j$ and different values of $m$ are related to one another by multiplication with the operators $J_1 + iJ_2$ or $J_1 - iJ_2$, which are symmetric between $\mathbf{J}'$ and $\mathbf{J}''$, these state vectors all have the same symmetry or antisymmetry property, the choice depending only on $j'$, $j''$, and $j$, so

$$C_{j'j''}(jm; m'm'') = (\pm 1)_{jj'j''} C_{j''j'}(jm; m''m').$$

For the case of maximum $j$ and $m$, with $j = m = j' + j''$, Eq. (4.3.21) shows that the sign is $+1$. There are two states with $m' + m'' = j - 1$, one with $j = j' + j''$, which must have a Clebsch–Gordan coefficient that is symmetric under interchange of $j'$ and $j''$, as we see in Eq. (4.3.24), and another state vector with $j = j' + j'' - 1$, which must be orthogonal to the state with $m' + m'' = j - 1$ and $j = j' + j''$, which requires it to have a Clebsch–Gordan coefficient that is antisymmetric under interchange of $j'$ and $j''$, as we see in Eq. (4.3.25). This argument can then be repeated for all lower values of $m$, with the result that for fixed $j'$ and $j''$ the sign $(\pm 1)_{jj'j''}$ changes for each decrease of $j$ by one unit, with the result that $(\pm 1)_{jj'j''} = (-1)^{j-j'-j''}$, as was to be proved.

The result (4.3.34) can be observed in the entries in Table 4.1. For instance, the state consisting of two particles of spin $1/2$ is symmetric or antisymmetric in the spin 3-components of the two particles depending on whether the total spin $s$ is $s = 1$, for which $s - 1/2 - 1/2 = 0$, or $s = 0$, for which $s - 1/2 - 1/2 = -1$.

There is an important special case of the addition of angular momenta: the construction of a rotationally invariant state $\Psi$ with total angular momentum $j = 0$, $m = 0$ from states that have two separate angular momenta $j'$, $m'$ and $j''$, $m''$. According to Eqs. (4.3.12) and (4.3.19), this is only possible if $j' = j''$ and $m' = -m''$, so this rotationally invariant state must take the form

$$\Psi = \sum_{m'} C_{j'm'} \Psi^{m'\,-m'}_{j'\,j'}.$$

Rotational invariance requires this state to be annihilated by the raising operator, so

$$
\begin{aligned}
0 = (J_1 + iJ_2)\Psi &= (J_1' + iJ_2')\Psi + (J_1'' + iJ_2'')\Psi \\
&= \sum_{m'} \Big[ C_{j'm'}\sqrt{(j'-m')(j'+m'+1)}\,\Psi_{j'\,m'+1,\,j'\,-m'} \\
&\qquad + C_{j'm'}\sqrt{(j'+m')(j'-m'+1)}\,\Psi_{j'\,m',\,j'\,-m'+1} \Big].
\end{aligned}
$$

Changing the summation variable in the second term in square brackets from $m'$ to $m'+1$, we see that this is equivalent to the requirement that $C_{j'm'} = -C_{j'\,m'+1}$. We can therefore adjust the overall phase of $C_{j'm'}$ so that $C_{j'm'} = (-1)^{j'-m'} N_{j'}$, with $N_{j'}$ real and positive. The normalization condition (4.3.32) then tells us that $N_{j'} = 1/\sqrt{2j'+1}$. Thus (dropping unnecessary primes) the Clebsch–Gordan coefficient here is

$$
C_{jj}(00;\, m\, -m) = \frac{(-1)^{j-m}}{\sqrt{2j+1}}. \tag{4.3.35}
$$

The reader can check that this is the same, with the same phase conventions, as the results in the fourth line and the last two lines of Table 4.1.

In particular, we can use this result to combine spherical harmonic functions of two different unit vectors $\hat{a}$ and $\hat{b}$ to form a function of $\hat{a}$ and $\hat{b}$ that is rotationally invariant, and hence can only depend on $\hat{a} \cdot \hat{b}$:

$$
F_\ell(\hat{a} \cdot \hat{b}) = \sum_{m=-\ell}^{\ell} (-1)^{\ell-m} Y_\ell^m(\hat{a}) Y_\ell^{-m}(\hat{b}).
$$

We can identify the function $F_\ell$ by looking at the special case where $\hat{b} = \hat{z} \equiv (0,0,1)$ and $\hat{a} = (\sin\theta\cos\phi, \sin\theta\sin\phi, \cos\theta)$. The spherical harmonics $Y_\ell^{-m}(\hat{z})$ vanish except for $m = 0$, and in this case Eq. (2.2.18) gives

$$
Y_\ell^0(\hat{a}) = \sqrt{\frac{2\ell+1}{4\pi}}\, P_\ell(\cos\theta), \quad Y_\ell^0(\hat{b}) = \sqrt{\frac{2\ell+1}{4\pi}}.
$$

It follows that $F_\ell(\cos\theta) = [(2\ell+1)/4\pi]P_\ell(\cos\theta)$, which yields the important *addition theorem for spherical harmonics*:

$$
P_\ell(\hat{a} \cdot \hat{b}) = \frac{4\pi}{2\ell+1} \sum_{m=-\ell}^{\ell} (-1)^{\ell-m} Y_\ell^m(\hat{a}) Y_\ell^{-m}(\hat{b}). \tag{4.3.36}
$$

Instead of using Clebsch–Gordan coefficients to construct states of total angular momentum $j, m$ from states which have two individual angular momenta $j', m'$ and $j'', m''$, we can use these coefficients together with

Eq. (4.3.35) to construct a state $\Psi$ of total angular momentum zero from a state $\Psi^{m\,m'\,m''}_{j\,j'\,j''}$ with *three* individual angular momenta:

$$\Psi = \sum_{mm'm''} \begin{pmatrix} j & j' & j'' \\ m & m' & m'' \end{pmatrix} \Psi^{m\,m'\,m''}_{j\,j'\,j''}, \tag{4.3.37}$$

where the coefficients are

$$\begin{pmatrix} j & j' & j'' \\ m & m' & m'' \end{pmatrix} \equiv \frac{(-1)^{j+m}}{\sqrt{2j+1}} C_{j'j''}(j-m;m'm''), \tag{4.3.38}$$

and are known as $3j$ *symbols*. Because of the symmetric way in which the three angular momenta appear in Eq. (4.3.37), it will not be a surprise that the $3j$ symbols are symmetric or antisymmetric not only under interchange of $j',m'$ with $j'',m''$, as in Eq. (4.3.34), but also under interchange of $j,m$ with $j',m'$ (or $j'',m''$):

$$\begin{pmatrix} j & j' & j'' \\ m & m' & m'' \end{pmatrix} = (-1)^{m'-m''+m} \begin{pmatrix} j' & j & j'' \\ m' & m & m'' \end{pmatrix}. \tag{4.3.39}$$

In other words,

$$C_{jj''}(j'-m';mm'') = (-1)^{j-j'-2m'+m''} \sqrt{\frac{2j'+1}{2j+1}} C_{j'j''}(j-m;m'm''). \tag{4.3.40}$$

(The signs appearing here will play no role in what follows, and we will make no attempt to derive them.) From the orthonormality condition (4.3.32), we then obtain another useful orthonormality condition,

$$\sum_{m'm''} C_{jj''}(j'm';mm'')\, C_{\bar{j}j''}(j'm';\bar{m}m'') = \frac{2j'+1}{2j+1} \delta_{j\bar{j}} \delta_{m\bar{m}}. \tag{4.3.41}$$

$$* \ * \ * \ * \ *$$

There is an alternative description of angular momentum multiplets that is useful in some contexts, and can be extended to other symmetry groups of importance in elementary-particle physics. According to Eqs. (4.2.17) and (4.1.12), the action of an infinitesimal rotation $1 + \omega$ on a spin one-half state vector $\Psi_m$ (with $m = \pm 1/2$) is

$$\Psi_m \rightarrow \sum_{m'=\pm 1/2} \left(1 + \frac{i}{2}\omega \cdot \sigma\right)_{mm'} \Psi_{m'}. \tag{4.3.42}$$

Now, for general real $\omega$,

$$\omega \cdot \sigma = \begin{pmatrix} \omega_3 & \omega_1 - i\omega_2 \\ \omega_1 + i\omega & -\omega_3 \end{pmatrix},$$

which is the most general traceless Hermitian $2 \times 2$ matrix. Hence (4.3.34) is the most general $2 \times 2$ unitary infinitesimal transformation with unit determinant. (Recall that for $M$ infinitesimal, $\mathrm{Det}(1 + M) = 1 + \mathrm{Tr}\, M$.) So, acting on spin one-half indices, the three-dimensional rotation group is the same as the group known as $SU(2)$, the group of $2 \times 2$ unitary matrices that are "special" in the sense of having unit determinant. We see that, at least for rotations that can be built up from infinitesimal rotations, the three-dimensional rotation group $SO(3)$ is the same as the two-dimensional unitary unimodular group $SU(2)$. (There are similar relations in a few higher dimensions, for instance a similar relation between $SO(6)$ and $SU(4)$, but nothing like this occurs in spaces of general dimensionality.)

More generally, a state vector $\Psi_{m_1 \ldots m_N}$ that combines $N$ spin one-half angular momenta, with each $m_i$ equal to $\pm 1/2$, transforms as a tensor under $SU(2)$:

$$\Psi_{m_1 \ldots m_N} \to \sum_{m_1' \ldots m_N'} U_{m_1 m_1'} \cdots U_{m_N m_N'} \, \Psi_{m_1' \ldots m_N'}, \qquad (4.3.43)$$

where $U$ is a unitary $2 \times 2$ matrix with unit determinant. In general, from such a tensor we can derive tensors with fewer indices. Note that the condition that $U$ has unit determinant means that

$$\sum_{m_1' m_2'} U_{m_1' m_1} U_{m_2' m_2} \epsilon_{m_1' m_2'} = \epsilon_{m_1 m_2}, \qquad (4.3.44)$$

where

$$\epsilon_{\frac{1}{2}, -\frac{1}{2}} = -\epsilon_{-\frac{1}{2}, \frac{1}{2}} = 1, \quad \epsilon_{\frac{1}{2}, \frac{1}{2}} = \epsilon_{-\frac{1}{2}, -\frac{1}{2}} = 0. \qquad (4.3.45)$$

It follows that by multiplying a general tensor $\Psi_{m_1 \ldots m_N}$ with $\epsilon_{m_r m_s}$ (where $r$ and $s$ are any two different integers between 1 and $N$) and summing over $m_r$ and $m_s$, we can form a tensor with two fewer indices. The only sort of tensor, which is irreducible in the sense that from it we cannot in this way form non-trivial tensors with fewer indices, is one that is totally symmetric, for which the sum over $m_r$ and $m_s$ would vanish.

To put this in the language of angular momentum, we note that by the rules of angular momentum addition, a state vector $\Psi_{m_1 \ldots m_N}$ can be expressed as a sum of state vectors of various total angular momenta, just one of which will be angular momentum $N/2$. From the fourth line of Table 4.1, we see that the tensor (4.3.37) is essentially just the Clebsch–Gordan coefficient for combining two angular momenta one-half to form angular momentum zero:

$$\epsilon_{m_1 m_2} = \sqrt{2}\, C_{\frac{1}{2}, \frac{1}{2}}(0, 0; m_1 m_2) \qquad (4.3.46)$$

so when we multiply $\Psi_{m_1 \ldots m_N}$ with $\epsilon_{m_r m_s}$ and sum over $m_r$ and $m_s$, we get a state vector that combines $N - 2$ spin one-half angular momenta, which can be expressed as a sum of state vectors of various total angular momenta, all of them less than $N/2$. Thus in order to isolate the part of a state vector $\Psi_{m_1 \dots m_{2j}}$ that contains only the angular momentum $j$, the state vector must be symmetrized in the indices $m_1 \dots m_{2j}$. The independent components of this symmetrized state vector are entirely characterized by the numbers $n$ and $2j - n$ of indices with $m = +1/2$ and $m = -1/2$, so the number of independent components is simply the number of values of $n$ between zero and $2j$, which is $2j + 1$. Thus a spin-$j$ state vector can simply be described as a symmetrized combination of $2j$ spins one-half. For instance, a multiplet with total angular momentum unity consists of the three states

$$\Psi_{\frac{1}{2},\frac{1}{2}}, \quad \Psi_{\frac{1}{2},-\frac{1}{2}} + \Psi_{-\frac{1}{2},\frac{1}{2}}, \quad \Psi_{-\frac{1}{2},-\frac{1}{2}}$$

in agreement (apart from normalization) with the first three lines of Table 4.1.

We can use this alternative formalism to work out rules for the addition of angular momenta. When we combine spins $j_1$ and $j_2$, the state vector in this formalism takes the form $\Psi_{m_1 \dots m_{2j_1}; m'_1 \dots m'_{2j_2}}$, symmetrical in the $m$s and symmetrical in the $m'$s, but with no particular symmetry between the $m$s and $m'$s. From this, by multiplying with $M$ factors $\epsilon_{m_r m'_s}$ and summing over indices, we can form a tensor with $M$ fewer $m$ indices and $M$ fewer $m'$ indices. If we symmetrize with respect to the remaining indices, we have a tensor that describes only angular momentum $2j_1 + 2j_2 - 2M$. Here $M$ can be given any value from zero to the lesser of $2j_1$ and $2j_2$. Hence by combining angular momenta $j_1$ and $j_2$, we can form any angular momentum $j = j_1 + j_2 - M$, with $0 \le M \le \min\{2j_1, 2j_2\}$, or in other words, with $|j_1 - j_2| \le j \le j_1 + j_2$, just as we found earlier by the use of raising and lowering operators.

## 4.4 The Wigner–Eckart Theorem

One of the advantages of the algebraic approach to angular momentum is that we can deduce the form of the matrix elements of various operators if we know their commutation relations with the rotation generators, which follow from the rotation transformation properties of the corresponding observables. A set of $2j + 1$ operators $O_j^m$ with $m = j, j - 1, \dots, -j$ is said to have spin $j$ if the commutators of the rotation generators with these operators have the same form as the formulas (4.2.9) and (4.2.16) for their action on state vectors $\Psi_j^m$ of angular momentum $j$:

$$\left[ J_3, O_j^m \right] = \hbar m O_j^m, \tag{4.4.1}$$

$$\left[ J_1 \pm iJ_2, O_j^m \right] = \hbar \sqrt{j(j+1) - m^2 \mp m}\, O_j^{m\pm 1}. \tag{4.4.2}$$

These conditions can be summarized in the statement that

$$[\mathbf{J}, O_j^m] = \hbar \sum_{m'} \mathbf{J}_{m'm}^{(j)} O_j^{m'}, \tag{4.4.3}$$

where $\mathbf{J}^{(j)}_{m'm}$ is the spin-$j$ representation of the angular-momentum operators

$$[J_3^{(j)}]_{m'm} \equiv m\delta_{m'm}, \quad [J_1^{(j)}]_{m'm} \pm i[J_2^{(j)}]_{m'm} \equiv \sqrt{j(j+1) - m^2 \mp m}\, \delta_{m',m\pm1}. \tag{4.4.4}$$

For instance, a scalar operator $S$ is one that commutes with all components of $\mathbf{J}$, which trivially agrees with Eqs. (4.4.1) and (4.4.2) or equivalently with (4.4.3) if we assign the operator $j = m = 0$, for which $\mathbf{J}^{(0)}_{m'm} = 0$. Also, according to Eq. (4.1.13), a vector operator $\mathbf{V}$ is one that satisfies the commutation relations

$$\left[J_i,\, V_j\right] = i\hbar \sum_k \epsilon_{ijk} V_k. \tag{4.4.5}$$

We can define *spherical components* of this vector as the quantities

$$V^{+1} \equiv -\frac{V_1 + iV_2}{\sqrt{2}}, \quad V^{-1} \equiv \frac{V_1 - iV_2}{\sqrt{2}}, \quad V^0 \equiv V_3. \tag{4.4.6}$$

Then we can use the commutation relations (4.4.5) to show that

$$[J_3,\, V^m] = \hbar m V^m, \tag{4.4.7}$$

and

$$[J_1 \pm iJ_2,\, V^m] = \hbar\sqrt{2 - m^2 \mp m}\, V^{m\pm1}, \tag{4.4.8}$$

so the $V^m$ form an operator $V_1^m$ with $j = 1$. A special case of such an operator $V_1^m$ is provided by the spherical harmonic $Y_1^m(\hat{x})$, with $\hat{x}$ treated as an operator. Indeed, for any vector operator $\mathbf{V}$, the $\ell$th-order polynomials $|\mathbf{V}|^\ell Y_\ell^m(\hat{V})$ are operators of type $O_j^m$ with $j = \ell$.

We will prove a fundamental general result due to Wigner[12] and Carl Eckart[13] (1902–1973), known as the *Wigner–Eckart theorem*, that gives

$$\left(\Phi_{j''}^{m''},\, O_j^m \Psi_{j'}^{m'}\right) = C_{jj'}(j''m'';\, mm')\left(\Phi\|O\|\Psi\right), \tag{4.4.9}$$

where $C_{jj'}(j''m'';\, mm')$ is the Clebsch–Gordan coefficient introduced in Section 4.3, and $\left(\Phi\|O\|\Psi\right)$ is a coefficient known as the *reduced matrix element* that can depend on everything except the 3-components $m$, $m'$, and $m''$.

To prove this result, consider a general operator $O_j^m$ of spin $j$. When multiplied with the angular momentum generators, the state vector $\Omega_{jj'}^{mm'} \equiv O_j^m \Psi_{j'}^{m'}$ becomes

$$J_i\, \Omega_{jj'}^{mm'} = [J_i,\, O_j^m]\Psi_{j'}^{m'} + O_j^m J_i\, \Psi_{j'}^{m'}$$
$$= \hbar \sum_{m''} [J_i^{(j)}]_{m''m} \Omega_{jj'}^{m''m'} + \hbar \sum_{m''} [J_i^{(j')}]_{m''m'} \Omega_{jj'}^{mm''}. \tag{4.4.10}$$

[12] E. P. Wigner, *Gruppentheorie* (Vieweg und Sohn, Braunschweig, 1931).
[13] C. Eckart, *Rev. Mod. Phys.* **2**, 305 (1930).

In other words, $J_i$ acts on $\Omega_{jj'}^{mm'}$ just as if $\Omega_{jj'}^{mm'}$ were a state vector for a system consisting of two particles with spins $j$ and $j'$ and 3-components $m$ and $m'$. Therefore

$$O_j^m \Psi_{j'}^{m'} = \sum_{j''m''} C_{jj'}(j''m''; mm')\Omega_{jj'j''}^{m''}, \qquad (4.4.11)$$

where $\Omega_{jj'j''}^{m''}$ is a state vector of angular momentum $j''$ with 3-component $m''$. Applying Eq. (4.2.19) to the state vectors $\Phi$ and $\Omega$ then gives the desired result, Eq. (4.4.9).

There is an immediate application of this result for vector operators: *the matrix elements of all vector operators for state vectors of definite angular momentum are parallel.* That is, for any pair of vectors **V** and **W**, as long as $(\Phi||W||\Psi)$ does not vanish, we have

$$(\Phi_{j''}^{m''}, V_1^m \Psi_{j'}^{m'}) = \left( \frac{\left(\Phi||V||\Psi\right)}{\left(\Phi||W||\Psi\right)} \right) \left(\Phi_{j''}^{m''}, W_1^m \Psi_{j'}^{m'}\right). \qquad (4.4.12)$$

Since this is true of the spherical components of the vectors, it is also true of the Cartesian components

$$\left(\Phi_{j''}^{m''}, V_i \Psi_{j'}^{m'}\right) = \left( \frac{\left(\Phi||V||\Psi\right)}{\left(\Phi||W||\Psi\right)} \right) \left(\Phi_{j''}^{m''}, W_i \Psi_{j'}^{m'}\right). \qquad (4.4.13)$$

In particular, since **J** is itself a vector, we have

$$\left(\Phi_{j'}^{m''}, V_i \Psi_{j'}^{m'}\right) \propto \left(\Phi_{j'}^{m''}, J_i \Psi_{j'}^{m'}\right). \qquad (4.4.14)$$

We have written this last result only for the case $j'' = j'$ because, since **J** commutes with $\mathbf{J}^2$, the reduced matrix element $(\Phi||J||\Psi)$ would vanish if $\Phi$ and $\Psi$ had different angular momenta. But it should not be thought that vector operators generally have vanishing matrix elements between states of different total angular momentum; this is a general rule only for the angular momentum operator itself.

We will use Eq. (4.4.14) in our treatment of the Zeeman effect in Section 5.2. It is often explained "physically," by arguing that any vector's components orthogonal to the angular momentum vector are averaged out by the rotation of a system around **J**, but without the Wigner–Eckart theorem one might think that this essentially classical explanation leaves open the possibility of quantum corrections.

As a further application of the Wigner–Eckart theorem, we will derive the selection rules obeyed by the most common sort of photon emission transition. As we saw in Section 1.4, Heisenberg made use of the classical formula for radiation by an oscillating charge to guess at a formula, Eq. (1.4.5), for the rate of a transition from one atomic state to another. Generalizing to any number of charged particles with position operators $\mathbf{X}_n$ (relative to the center of mass) and charges $e_n$, this formula gives the rate of transition from initial atomic state $a$ to final atomic state $b$ as

$$\Gamma(a \to b) = \frac{4(E_a - E_b)^3}{c^3 \hbar^4} \left| \left( b|\mathbf{D}|a \right) \right|^2, \tag{4.4.15}$$

where $\mathbf{D}$ is the dipole operator

$$\mathbf{D} = \sum_n e_n \mathbf{X}_n. \tag{4.4.16}$$

We will give a quantum-mechanical derivation of this formula in Section 11.7. As shown there, Eq. (4.4.15) gives the radiative transition rate (with $\left( b|\mathbf{X}_n|a \right)$ defined as the matrix element of the $n$th particle coordinate relative to the center of mass, stripped of its momentum conservation delta function), in the approximation that the wavelength $hc/(E_a - E_b)$ of the emitted photon is much larger than the size of the atom, provided that the matrix element $\left( b|\mathbf{D}|a \right)$ does not vanish. What concern us here are the conditions under which the matrix element may not vanish.

The operator $\mathbf{D}$ is a three-vector, and so, as in Eq. (4.4.6), its components can be written as linear combinations of a $j = 1$ multiplet of operators $D^m$:

$$D_1 = \frac{1}{\sqrt{2}} \left( -D^{+1} + D^{-1} \right), \quad D_2 = \frac{i}{\sqrt{2}} \left( D^{+1} + D^{-1} \right), \quad D_3 = D^0. \tag{4.4.17}$$

The matrix elements of the operators $D^m$ have a dependence on $m$ and on the angular-momentum quantum numbers $j_a, m_a$ and $j_b, m_b$ of the initial and final states given by a Clebsch–Gordan coefficient:

$$\left( b|D^m|a \right) \propto C_{j_a 1}(j_b m_b; m_a m), \tag{4.4.18}$$

with a constant of proportionality independent of $m$, $m_a$, and $m_b$. The transition rate (4.4.15) therefore vanishes unless the angular-momentum quantum numbers satisfy

$$|j_a - j_b| \leq 1, \quad j_a + j_b \geq 1, \quad |m_a - m_b| \leq 1. \tag{4.4.19}$$

There is a further parity selection rule, given in Section 4.7.

Where these selection rules are satisfied, and the transition rate is given to a good approximation by Eq. (4.4.15), this is known as an electric dipole, or E1, transition. Of course, not all possible atomic transitions satisfy these selection rules. Where the selection rules are not satisfied, photon transitions may still be possible, but their rates are suppressed by additional factors of the atomic size divided by the photon wavelength. Such transitions are discussed in Section 11.7.

It frequently happens that an atom or molecule or elementary particle of angular momentum $j'$ is unpolarized, with all values of $m'$ between $-j'$ and $j'$ equally likely, so that in finding the expectation value of an operator $O_j^m$ in a state $\Psi_{j'}^{m'}$ we must average over $m'$. The Wigner–Eckart theorem then gives the expectation value

$$\left\langle O_j^m \right\rangle = \frac{1}{2j'+1} \sum_{m'} \left( \Psi_{j'}^{m'}, O_j^m \, \Psi_{j'}^{m'} \right)$$

$$= \frac{1}{2j'+1} \sum_{m'} C_{jj'}(j'm'; mm') \left( \Psi \|O\| \Psi \right). \tag{4.4.20}$$

By setting $\bar{j} = \bar{m} = 0$ in the orthonormality relation (4.3.41) and using the obvious relation $C_{0j''}(j'm'; 0m'') = \delta_{j'j''}\delta_{m'm''}$ we find

$$\sum_{m'} C_{jj'}(j'm'; mm') = (2j'+1)\delta_{j0}\delta_{m0}. \tag{4.4.21}$$

Hence none of the operators $O_j^m$ have non-vanishing expectation values in unpolarized systems, except for those operators with $j = m = 0$. As we will see in Section 5.9, this has important implications for the long-range forces between electrically neutral atoms and molecules.

## 4.5 Bosons and Fermions

As far as we know, every electron in the universe is identical to every other electron, except for the values taken by their positions (or momenta) and spin 3-components. The same is true of the other known elementary particles: photons, quarks, etc. For such indistinguishable particles, it can make no difference what order we write the position and spin labels on a physical state: we can say that in a state with state vector $\Phi_{\mathbf{x}_1,m_1;\mathbf{x}_2,m_2;\dots}$ there is one electron with position $\mathbf{x}_1$ and spin 3-component $\hbar m_1$, another electron with position $\mathbf{x}_2$ and spin 3-component $\hbar m_2$, and so on, and *not* that the first electron has position $\mathbf{x}_1$ and spin 3-component $\hbar m_1$, that the second electron has position $\mathbf{x}_2$ and spin 3-component $\hbar m_2$, and so on. Thus for instance the state vector $\Phi_{\mathbf{x}_2,m_2;\mathbf{x}_1,m_1;\dots}$ must represent the same physical state as the state vector $\Phi_{\mathbf{x}_1,m_1;\mathbf{x}_2,m_2;\dots}$. This does not mean that these state vectors are equal, only that they are equal up to a constant factor,[14] say $\alpha$:

---

[14]  It is important in deriving Eq. (4.5.3) that $\alpha$ should depend only on the species of particle, not on the particle's momentum or spin. This follows from considerations of spacetime symmetry; a dependence of $\alpha$ on momentum or spin would contradict invariance under rotations of the coordinate system or transformations to moving coordinate systems. In two space dimensions there is an exotic possibility, that $\alpha$ might depend on the paths by which the particles are brought to their positions or momenta, but this is not possible in three or more space dimensions.

$$\Phi_{\mathbf{x}_2,m_2;\mathbf{x}_1,m_1;\dots} = \alpha\,\Phi_{\mathbf{x}_1,m_1;\mathbf{x}_2,m_2;\dots}. \tag{4.5.1}$$

Because $\alpha$ does not depend on momentum or spin, we also have

$$\Phi_{\mathbf{x}_1,m_1;\mathbf{x}_2,m_2;\dots} = \alpha\,\Phi_{\mathbf{x}_2,m_2;\mathbf{x}_1,m_1;\dots}. \tag{4.5.2}$$

Inserting Eq. (4.5.1) in the right-hand side of Eq. (4.5.2), we see that

$$\Phi_{\mathbf{x}_1,m_1;\mathbf{x}_2,m_2;\dots} = \alpha^2\,\Phi_{\mathbf{x}_1,m_1;\mathbf{x}_2,m_2\dots},$$

and therefore

$$\alpha^2 = 1. \tag{4.5.3}$$

This argument applies to particles of any type, elementary or not. Particles with $\alpha = +1$ and $\alpha = -1$ are known as *bosons* and *fermions*, respectively, named after Satyendra Nath Bose (1894–1974) and Enrico Fermi (1901–1954).

One of the most important consequences of special relativity in quantum mechanics is that all particles whose spins are half odd integers are fermions, and all particles whose spins are integers are bosons.[15] Thus electrons and quarks, which have spin 1/2, are fermions. The heavy W and Z particles, which play an essential role in the radioactive process known as beta decay, have spin one, and are therefore bosons. (The definition of spin for a massless particle like the photon requires some care. For our purposes here we note only that the component of spin angular momentum in the direction of a photon's motion can only take the values $\pm\hbar$, corresponding to left- and right-circularly polarized electromagnetic waves, and that photons are bosons.)

When we exchange a pair of identical composite particles, we exchange all of their constituents, so we get a sign factor given by the product of all the sign factors for the individual constituents. It follows that *a composite particle consisting of an even number of fermions and any number of bosons is a boson, and a composite particle consisting of an odd number of fermions and any number of bosons is a fermion.* Thus the proton and neutron, which each consist of three quarks, are fermions. The hydrogen atom, which consists of a proton and an electron, is a boson. Note that this rule is consistent with the feature of angular-momentum addition that the addition of an odd number of half-odd-integer angular momenta and any number of integer angular momenta is a half-odd-integer angular momentum, while the addition of an even number of half-odd-integer angular momenta and any number of integer angular momenta is an integer angular momentum. It would have been impossible for

---

[15] This result was first presented in the context of perturbation theory by M. Fierz, *Helv. Phys. Acta* **12**, 3 (1939) and W. Pauli, *Phys. Rev.* **58**, 716 (1940). Non-perturbative proofs in axiomatic field theory were given by G. Lüders and B. Zumino, *Phys. Rev.* **110**, 1450 (1958) and N. Burgoyne, *Nuovo Cimento* **8**, 807 (1958). Also see R. F. Streater and A. S. Wightman, *PCT, Spin & Statistics, and All That* (Benjamin, New York, 1968).

all integer-spin particles to be fermions, because a composite of an even number of integer-spin particles would have integer spin, but would also be a boson.

The distinction between bosons and fermions is particularly important for systems in which to a good approximation the Hamiltonian acts separately on each particle. That is,

$$H\Phi_{\xi_1\xi_2\ldots} = \int d\xi_1' \, H_{\xi_1',\xi_1}\Phi_{\xi_1'\xi_2\ldots} + \int d\xi_2' \, H_{\xi_2',\xi_2}\Phi_{\xi_1\xi_2'\ldots} + \cdots, \qquad (4.5.4)$$

where $H_{\xi',\xi}$ is the matrix element of an effective one-particle Hamiltonian between one-particle states,

$$H_{\xi',\xi} \equiv \left(\Phi_{\xi'}, H^{\text{eff}}\Phi_\xi\right). \qquad (4.5.5)$$

(We are now using $\xi$ to denote a particle momentum and spin $z$-component, and an integral over $\xi$ is understood to include an integral over the momentum vector and a sum over the spin $z$-component.) In atomic physics, this is called the *Hartree approximation*.[16] It is often a good approximation in many-particle systems, where any one particle can be assumed to respond to the potential created by the other particles, while its response to this potential has negligible reaction back on the potential. When the Hamiltonian takes the form (4.5.4), a state $\Psi$ will be an eigenstate of the Hamiltonian if its wave function is a product of single-particle wave functions:

$$\left(\Phi_{\xi_1,\xi_2,\ldots}, \Psi\right) = \psi_1(\xi_1)\psi_2(\xi_2)\ldots, \qquad (4.5.6)$$

where the $\psi_a$ are eigenfunctions of the one-particle Hamiltonian

$$\int d\xi' \, H_{\xi,\xi'}\psi_a(\xi') = E_a\psi_a(\xi). \qquad (4.5.7)$$

In this case, we have

$$\left(\Phi_{\xi_1,\xi_2,\ldots}, H\Psi\right) = \int d\xi_1' \, H^*_{\xi_1',\xi_1}\psi_1(\xi_1')\psi_2(\xi_2)\ldots$$
$$+ \int d\xi_2' \, H^*_{\xi_2',\xi_2}\psi_1(\xi_1)\psi_2(\xi_2')\ldots + \cdots.$$

Using the Hermiticity of the one-particle Hamiltonian, we have $H^*_{\xi',\xi} = H_{\xi,\xi'}$, so with Eq. (4.5.7) this gives

$$\left(\Phi_{\xi_1,\xi_2,\ldots}, H\Psi\right) = (E_1 + E_2 + \cdots)\left(\Phi_{\xi_1,\xi_2,\ldots}, \Psi\right)$$

and therefore $\Psi$ is an eigenvector of $H$ with energy $E_1 + E_2 + \cdots$:

$$H\Psi = (E_1 + E_2 + \cdots)\Psi. \qquad (4.5.8)$$

---

[16] D. R. Hartree, *Proc. Camb. Phil. Soc.* **24**, 111 (1928).

But for identical particles Eq. (4.5.6) is in conflict with the requirement that $\Phi_{\xi_1, \xi_2, \ldots}$ must be symmetric or antisymmetric in the $\xi$s for bosons or fermions, respectively. In this case, in place of (4.5.6), we must symmetrize or antisymmetrize the wave function:

$$\left( \Phi_{\xi_1, \xi_2, \ldots}, \Psi \right) = \sum_P \delta_P \psi_1(\xi_{P1}) \psi_2(\xi_{P2}) \ldots, \qquad (4.5.9)$$

where the sum is over all permutations $1, 2, \ldots \mapsto P1, P2, \ldots$, and $\delta_P$ for fermions is $+1$ or $-1$ for even or odd permutations, respectively, while for bosons $\delta_P = 1$ for all permutations. The argument given above for the energy of the wave function (4.5.6) applies to each term of this sum, so by the same argument, $\Psi$ is again an eigenvector of $H$ with eigenvalue $E_1 + E_2 + \cdots$.

For instance, for a two-particle state there are just two permutations, the identity $1, 2 \mapsto 1, 2$ and the odd permutation $1, 2 \mapsto 2, 1$, so

$$\left( \Phi_{\xi_1, \xi_2}, \Psi \right) = \psi_1(\xi_1) \psi_2(\xi_2) \pm \psi_1(\xi_2) \psi_2(\xi_1),$$

the sign being plus for bosons and minus for fermions. For fermions, the wave function in the general case is a determinant, known as a *Slater determinant*:[17]

$$\left( \Phi_{\xi_1, \xi_2, \ldots}, \Psi \right) = \begin{vmatrix} \psi_1(\xi_1) & \psi_1(\xi_2) & \psi_1(\xi_3) & \ldots \\ \psi_2(\xi_1) & \psi_2(\xi_2) & \psi_2(\xi_3) & \ldots \\ \psi_3(\xi_1) & \psi_3(\xi_2) & \psi_3(\xi_3) & \ldots \\ \ldots & \ldots & \ldots & \ldots \end{vmatrix}. \qquad (4.5.10)$$

For bosons instead of a determinant the wave function is a *permanent*, which is a determinant but with all minus signs replaced with plus signs.

For fermions it is impossible to form a state vector of the form (4.5.10) in which any of the $\psi_a$ are the same, because then two rows of the determinant would be the same, and the state vector would vanish. This is known as the *Pauli exclusion principle*.[18] In contrast, for bosons we can even have a state in which a macroscopic number of the $\psi_a$ are the same. This is known as a *Bose–Einstein condensation*.[19] The peculiar properties of liquid $^4$He can be interpreted as due to a Bose–Einstein condensation, but in this case the wave function cannot be expressed approximately as a symmetrized sum of products of one-particle wave functions. Only in recent years has a Bose–Einstein condensation been observed for a *gas* of atoms,[20] where this approximation is appropriate.

---

[17] J. C. Slater, *Phys. Rev.* **34**, 1293 (1929).

[18] W. Pauli, *Z. Physik* **31**, 763 (1925).

[19] In a letter to Einstein, Bose described the theory of bosons like photons for which the number of particles is not fixed. Einstein translated it himself from English to German, and had it published, as S. N. Bose, *Z. Physik* **26**, 178 (1924). Einstein then worked out the theory of gases of bosons with a fixed number of particles, published in A. Einstein, *Sitzungsber Preuss. Akad. Wiss.* 3 (1925).

[20] M. H. Anderson, J. R. Ensher, M. R. Matthews, C. E. Wieman, and E. A. Cornell, *Science* **269**, 198 (1995).

The exclusion principle does not apply to bosons, even bosons like the hydrogen atom consisting of pairs of fermions, but it does have implications for ensembles of such bosonic bound states. Consider a boson consisting of a pair of fermions with coordinates $\xi$ and $\eta$ (each including a momentum and spin $z$-component) and wave function $\psi(\xi, \eta)$. A gas of such identical bosons will have a wave function given by a product of bound-state wave functions, but antisymmetrized among fermion variables, and therefore equal to a determinant:

$$
\begin{vmatrix}
\psi(\xi_1, \eta_1) & \psi(\xi_1, \eta_2) & \psi(\xi_1, \eta_3) & \cdots \\
\psi(\xi_2, \eta_1) & \psi(\xi_2, \eta_2) & \psi(\xi_2, \eta_3) & \cdots \\
\psi(\xi_3, \eta_1) & \psi(\xi_3, \eta_2) & \psi(\xi_3, \eta_3) & \cdots \\
\cdots & \cdots & \cdots & \cdots
\end{vmatrix}.
$$

There is no limit to how many of these identical bosons can co-exist.

The first great application of the exclusion principle was in explaining the periodic table of the elements. As has already been mentioned, each electron in a multi-electron atom may be considered approximately to move in a potential $V(r)$ arising from the nucleus and the other electrons. This potential is very close to a central potential, depending only on the distance $r$ from the nucleus, but it is not a simple Coulomb potential proportional to $1/r$. It behaves instead like $-Ze^2/r$ near the nucleus (whose charge is $+Ze$), and like $-e^2/r$ outside the atom, where the nuclear charge is screened by the negative charge of $Z - 1$ electrons. Because the potential is a central potential we can still label the wave functions $\psi_a(\xi)$ of the individual electrons with an orbital angular momentum $\ell$ and a principal quantum number $n$, with $2(2\ell + 1)$ of these states for each $n$ and $\ell$ (the extra factor 2 arising from the electron's spin). The integer $n$ can be defined as $\ell + 1$ plus the number of nodes of the radial wave function, just as for a Coulomb potential. But because the potential is not a Coulomb potential we no longer have precisely equal energies for states of different $\ell$ and the same $n$. Instead, there is a tendency of energy to increase with $\ell$, because the wave function behaves near the origin like $r^\ell$, so that electrons with large $\ell$ spend little time near the nucleus, where $r|V(r)|$ is largest. For atoms with a large number $Z$ of electrons, it even sometimes happens that a one-electron state of large $\ell$ has a higher energy than a state of larger $n$ and smaller $\ell$.

The Pauli exclusion principle tells us that no two electrons can have the same wave function $\psi_a(\xi)$, so, as we consider atoms with more and more electrons, the electrons must be placed in one-electron states of higher and higher energy $E_a$. Of course, with increasing numbers of electrons the potential $V(r)$ changes, so the values of the energies $E_a$ and even their order also change. Detailed calculations show that the one-electron states are filled (with sporadic exceptions) in the order (with energies increasing down the list)

$1s,$

$2s, \ 2p,$

$3s, \ 3p,$

$4s, \ 3d, \ 4p,$

$5s, \ 4d, \ 5p,$

$6s, \ 4f, \ 5d, \ 6p,$

$7s, \ 5f, \ 7p, \ \ldots,$ \hfill (4.5.11)

where $s$, $p$, $d$, and $f$ are the time-honored symbols for $\ell = 0$, $\ell = 1$, $\ell = 2$, and $\ell = 3$. The one-electron states listed on the same line have approximately equal energy, but increasing somewhat from left to right.

Taking spin into account, the total numbers of states for the energy levels listed on each line of Eq. (4.5.11) are $2$, $2 + 6 = 8$, $2 + 6 = 8$, $2 + 6 + 10 = 18$, $2 + 10 + 6 = 18$, $2 + 14 + 10 + 6 = 32$, and so on. The first two elements, hydrogen and helium, with $Z = 1$ and $Z = 2$, have electrons only in the first (deepest) of the energy levels (4.5.11); the next eight elements from lithium to neon have electrons also in the second of these energy levels; the eight elements from sodium to argon have electrons in the third as well as the first and second of these energy levels; and so on.

Now, the chemical properties of an element are generally determined by the number of electrons in its highest energy level, which are least tightly bound. (An important exception is noted below.) An element whose atoms have no electrons outside filled energy levels is particularly stable chemically. Such elements are called *noble gases*: helium with $Z = 2$, neon with $Z = 2 + 8 = 10$, argon with $Z = 2 + 8 + 8 = 18$, krypton with $Z = 2 + 8 + 8 + 18 = 36$, xenon with $Z = 2 + 8 + 8 + 18 + 18 = 54$, and radon with $Z = 2 + 8 + 8 + 18 + 18 + 32 = 86$. For elements with a small number of electrons more or fewer than the number for a noble gas, chemical properties are largely determined by that number, known as the *valence* – positive for extra electrons, negative for missing electrons. Stable compounds that are held together by the Coulomb attractions of atoms that have gained or lost one or more electrons are typically formed from elements whose valences add up to zero. If there is just one electron in the highest energy level then it is easily lost, so the element behaves as a chemically reactive metal with valence $+1$. (Metals are characterized by their property of forming solids in which electrons leave individual atoms and travel freely through the solid. This gives metals their high thermal and electrical conductivity.) Such elements are called *alkali metals*, and include lithium with $Z = 2 + 1 = 3$, sodium with $Z = 2 + 8 + 1 = 11$, potassium with $Z = 2 + 8 + 8 + 1 = 19$, etc. Likewise, if there is just one electron *missing* in the highest energy level, then the atom tends strongly to attract one extra electron, so it is a chemically reactive non-metal, with valence $-1$, which can form particularly stable compounds with the alkali metals. Such elements are called *halogens*, and include fluorine with

$Z = 2 + 8 - 1 = 9$, chlorine with $Z = 2 + 8 + 8 - 1 = 17$, bromine with $Z = 2 + 8 + 8 + 18 - 1 = 35$, and so on. Elements with two electrons more than a noble gas are chemically reactive, though not as reactive as the alkali metals; these are known as the *alkali earths*, with valence $+2$, and include beryllium with $Z = 2 + 2 = 4$, magnesium with $Z = 2 + 8 + 2 = 12$, calcium with $Z = 18 + 2 = 20$, and so on. Similarly, elements with two electrons fewer than a noble gas are chemically reactive, with valence $-2$, though not as reactive as the halogens. These include oxygen with $Z = 10 - 2 = 8$, sulfur with $Z = 18 - 2 = 16$, and so on.

The inclusion of $4f$ states in the sixth energy level and $5f$ states in the seventh energy level produces a striking feature of the periodic table of the elements. Detailed calculations show that the mean radius of the $4f$ orbits is smaller than that of the $6s$ states, and the mean radius of the $5f$ orbits is smaller than that of the $7s$ states, so the numbers of $4f$ or $5f$ electrons have little effect on the chemical properties of the atom, even where these are the highest-energy electrons in the atom. Thus the $2(2 \cdot 3 + 1) = 14$ elements in which the highest-energy electrons are in $4f$ states are quite similar chemically, and likewise for the 14 elements in which the highest-energy electrons are in $5f$ states. The elements of the first set are known as *rare earths* or *lanthanides*, and have $Z$ running from $2 + 8 + 8 + 18 + 18 + 2 + 1 = 57$ (lanthanum)[21] to $2 + 8 + 8 + 18 + 18 + 2 + 14 = 70$ (ytterbium). The second set are known as *actinides*, and have $Z$ running from $2+8+8+18+18+32+2+1 = 89$ (actinium) to $2 + 8 + 8 + 18 + 18 + 32 + 2 + 14 = 102$ (nobelium). Much beyond nobelium the question of chemical behavior becomes moot, because for such large values of $Z$ the Coulomb repulsion among the protons makes the nucleus so unstable that the atoms do not last long enough to participate in chemical reactions.

An analogous shell structure is seen in atomic nuclei.[22] There are certain "magic numbers" of protons or neutrons that form closed shells, as shown by the fact that the nucleus with one additional proton or neutron has anomalously small binding energy. The magic numbers observed in this way are

$$2, \ 8, \ 20, \ 28, \ 50, \ 82, \ 126 \qquad (4.5.12)$$

For instance, $^4$He is doubly magic, since it has two protons and two neutrons, and in consequence there is no stable nucleus with one extra proton or neutron, which is one of the reasons why nuclear reactions in the early universe produced hardly any complex nuclei heavier than $^4$He. Other doubly magic nuclei such as $^{16}$O and $^{40}$Ca do allow the binding of an extra proton or neutron, but with

---

[21] Lanthanum is actually one of the sporadic exceptions to the rule of filling energy levels in the order shown in Eq. (4.5.11). The 57th electron is in a $5d$ rather than a $4f$ state. But in the next rare earth (cerium) there are two electrons in the $4f$ state, and none in the $5d$ state, and this pattern continues for all the other rare earths. Similar exceptions occur for the actinides.

[22] M. Goeppert-Mayer and J. H. D. Jensen, *Elementary Theory of Nuclear Shell Structure* (Wiley, New York, 1955).

substantially less binding energy than neighboring nuclei, and as a result these isotopes of oxygen and calcium are produced in stars more abundantly than neighboring nuclei.

The explanation of magic numbers in nuclei is similar to the explanation of the atomic numbers $Z = 2, \ 10, \ 18,$ etc. of noble gases, but of course with a very different potential. To the extent that nucleons can be supposed to move in a common potential $V(r)$ in nuclei, the potential must be analytic in the three-vector **x** at the origin, since unlike the case of atoms, in nuclei there is nothing special about the origin. Thus, for $r \rightarrow 0$, the potential must go as a constant plus a term of order $r^2$. A simple potential that satisfies this condition is the harmonic oscillator potential, $V(r) \propto V_0 + m_N \omega^2 r^2 / 2$, with $\omega$ some constant frequency. As we saw in Section 2.5, the first few energy levels (with energies relative to the zero-point energy $V_0 + 3\hbar/2$) of a particle in this potential, and the degeneracies of these levels, are as follows:

$$
\begin{array}{ccc}
\text{Energy} & \text{States} & \text{Degeneracy} \\
0 & s & 2 \\
\hbar\omega & p & 6 \\
2\hbar\omega & s \ \& \ d & 12 \\
3\hbar\omega & p \ \& \ f & 20 \\
\cdots & \cdots & \cdots
\end{array}
\tag{4.5.13}
$$

An extra factor 2 has been included in these degeneracies to take account of the two spin states of the nucleon. Protons are fermions, and are all identical to each other, so the number of protons in a nucleus with the lowest energy level filled is 2; with all levels filled up to $\hbar\omega$ it is $2 + 6 = 8$; with all levels filled up to $2\hbar\omega$ it is $2 + 6 + 12 = 20$, and so on. Of course, the same applies to neutrons.

This accounts for the first three magic numbers, but would suggest that the next magic number should be $2 + 6 + 12 + 20 = 40$, which is definitely not the case. For all beyond the lightest nuclei, it is necessary to take into account not only inevitable departures from the simple harmonic potential, but also the spin–orbit coupling, which as discussed in Section 4.3 splits the $2(2\ell + 1)$ states with definite $\ell$ into $2\ell + 2$ states with total one-particle angular momentum $j = \ell + 1/2$ and $2\ell$ states with $j = \ell - 1/2$. It turns out that the spin–orbit coupling depresses the energy of the $f$ state with $j = 7/2$ below the other states in the $3\hbar\omega$ level. The degeneracy of the $f_{7/2}$ state is 8, so the next magic number beyond 20 is $20 + 8 = 28$. Similar considerations explain the higher magic numbers.

The distinction between bosons and fermions has a profound effect on the way we count physical states in statistical mechanics. According to the general principles of statistical mechanics, the probability of any state in thermal equilibrium is proportional to an exponential function of linearly conserved quantities – that is, quantities whose sum over subsystems is conserved when the subsystems interact. These conserved quantities include the total energy[23] $E$, and the number $N$ of particles (strictly speaking, the numbers of certain kinds of particles, such as quarks and electrons, minus the numbers of their antiparticles). This exponential probability distribution is known as a *grand canonical ensemble*. We will consider here a system like a monomolecular gas, for which the total energy is the sum over one-particle states labeled $n$ of the energies $E_n$ of these states times the numbers $N_n$ of identical particles in the $n$th state. The probability of any given set of $N_n$ particles being in thermal equilibrium is then

$$P(N_1, N_2, \dots) \propto \exp\left(-\frac{E}{k_B T} + \frac{\mu N}{k_B T}\right) = \exp\left(-\sum_n N_n(E_n - \mu)/k_B T\right),$$

(4.5.14)

where $N = \sum_n N_n$ and $E = \sum_n N_n E_n$ are the total particle number and energy, $k_B$ is Boltzmann's constant, and $T$ and $\mu$ are parameters describing the state of the system, known respectively as the *temperature* and *chemical potential*.

So far, there is no difference between distinguishable and indistinguishable particles, or for indistinguishable particles between bosons and fermions. The difference enters when we sum over states in calculating thermodynamic averages. For distinguishable particles, we sum over the possible states of each particle. For indistinguishable particles, we instead sum over the number of particles in each one-particle state. For bosons, the mean number of particles in the $n$th state is then

$$\overline{N}_n = \frac{\sum_{N_n=0}^{\infty} N_n \exp(-N_n(E_n - \mu)/k_B T)}{\sum_{N_n=0}^{\infty} \exp(-N_n(E_n - \mu)/k_B T)}$$

$$= \frac{1}{\exp((E_n - \mu)/k_B T) - 1}.$$

(4.5.15)

(The sums over the numbers $N_m$ of particles in states $m \neq n$ other than $n$ cancel between numerator and denominator.) This is the case of *Bose–Einstein statistics*.

For instance, the number of photons is not conserved in radiative processes, so for photons we have to take $\mu = 0$. As we saw in Section 1.1, there are $8\pi v^2 \, dv/c^3$ one-photon states between frequencies $v$ and $v + dv$, each with energy $hv$, so the energy per volume between frequencies $v$ and $v + dv$ is $8\pi hv^3 \overline{N} \, dv/c^3$, which immediately yields the Planck black-body formula (1.1.5).

For fermions the calculation of $\overline{N}_n$ is precisely the same as for bosons, except that in accord with the Pauli exclusion principle, the sum over each $N_n$ runs only over the values zero and one. Hence

---

[23] We usually do not include the total momentum, even though it is linearly conserved, because we can always choose a frame of reference in which the total momentum vanishes.

$$\overline{N}_n = \frac{\exp(-(E_n - \mu)/k_{\mathrm{B}}T)}{1 + \exp(-(E_n - \mu)/k_{\mathrm{B}}T)}$$

$$= \frac{1}{\exp((E_n - \mu)/k_{\mathrm{B}}T) + 1}. \tag{4.5.16}$$

Note that $\overline{N}_n \leq 1$, as of course is required by the Pauli principle. This is the case of *Fermi–Dirac statistics*.

When the temperature is sufficiently small, the mean occupation number (4.5.16) is well approximated by

$$\overline{N}_n = \begin{cases} 1, & E_n < \mu, \\ 0, & E_n > \mu. \end{cases} \tag{4.5.17}$$

The surface $E_n = \mu$ in momentum space provides the boundary of the space of filled states, and is known as the *Fermi surface*. The existence of a Fermi surface plays an important role for electrons in white dwarf stars and for neutrons in neutron stars.

The Pauli principle has important implications also for the dynamics of electrons in crystals. As we saw in Section 3.5, in a crystal the allowed energies of an electron fall in several distinct bands. A crystal in which each band has all its states occupied by electrons or all empty is an insulator; the electron states cannot respond to an electric field because these states are completely fixed by the Pauli principle. A crystal in which some band has both an appreciable number of filled states and an appreciable number of unfilled states is a metal, with good electrical and thermal conductivity, because in this case the Pauli principle does not block the change of electron states to other states in an electric field, and there are plenty of electrons to respond. A crystal in which some band is nearly full or nearly empty, while all other bands are entirely full or empty, is a semi-conductor. At zero temperature a pure semi-conductor is an insulator, but it can be made into a conductor by doping it with impurities that either add electrons to the nearly empty band or remove electrons from the nearly full band.

The distinction between Eq. (4.5.15) for bosons and Eq. (4.5.16) for fermions evidently disappears when the exponential $\exp((E_n - \mu)/k_{\mathrm{B}}T)$ is much larger than unity. In this case, we have simply

$$\overline{N}_n = \exp(-(E_n - \mu)/k_{\mathrm{B}}T), \tag{4.5.18}$$

which is the familiar case of Maxwell–Boltzmann statistics.

## 4.6 Internal Symmetries

So far, we have considered only symmetry transformations that act on spacetime coordinates. There are also important symmetry transformations that act instead on the nature of particles, leaving their spacetime coordinates unaffected. This is a very large subject, to which only a very brief introduction can be given here.

An early example grew out of the 1932 discovery of the neutron. From the beginning it was striking that the neutron mass is nearly equal to the proton mass – they are respectively 939.565 MeV/$c^2$ and 938.272 MeV/$c^2$. This suggested that there should be a "charge symmetry," a symmetry under a transformation that, acting on any state, turns neutrons into protons and protons into neutrons. This would clearly not be an exact symmetry, since neutrons and protons do not have precisely the same masses. It would not be a symmetry of the electromagnetic interactions at all, since protons are charged and neutrons are not. But it was at least plausible that it would be a symmetry of whatever strong nuclear forces hold neutrons and protons together inside atomic nuclei and that presumably also have a large effect on neutron and proton masses.

This charge symmetry has important implications for complex nuclei. For light nuclei, where Coulomb forces are not dominant, each energy level of a nucleus with $Z$ protons and $N$ neutrons should be matched by an energy level of a nucleus with $N$ protons and $Z$ neutrons, with the same energy and spin. This is well borne out by experiment. For instance, the spin-1/2 ground state of $^3$H is so close in energy to the spin-1/2 ground state of $^3$He that the energy difference is just barely enough to allow $^3$H to decay into $^3$He with the emission of an electron and an approximately massless antineutrino. Likewise, the spin-1 ground state of $^{12}$B is matched with the spin 1 ground state of $^{12}$N.

Charge symmetry requires that the strong nuclear force between two neutrons be the same as between two protons, but it says nothing about the force between a proton and a neutron. At first only the neutron–proton force could be measured, both directly by scattering neutrons on hydrogen targets and indirectly by measurement of the properties of the deuteron. The neutron–neutron force could not be directly measured for obvious reasons: there are no neutron targets, and no two-neutron bound states. The proton–proton force could be measured, but at low energies the Coulomb repulsion between protons keeps protons from coming close to each other, so the force is almost purely electromagnetic. By 1936 it had become possible to accelerate protons to sufficiently high energy to measure effects of the nuclear force, and it was found that this force was similar to the proton–neutron force. To be more precise, the energy of the protons in this experiment was still small enough that the scattering state had $\ell = 0$ (the connection between low energy and low $\ell$ is explained in Section 7.6), so because protons are fermions they had to be in an antisymmetric spin state, with total spin zero. It was possible to separate out the force between protons and neutrons in the state with $\ell = 0$ and total spin zero from neutron–proton scattering experiments by subtracting the force in the state with $\ell = 0$ and total spin one, as measured from the properties of the deuteron. It was found that the nuclear forces in the neutron–proton and proton–proton states with $\ell = 0$ and total spin zero were similar in strength and range.[24]

This clearly called for a symmetry between protons and neutrons that goes beyond charge symmetry. The correct symmetry transformations were identified[25] as

$$\begin{pmatrix} p \\ n \end{pmatrix} \mapsto u \begin{pmatrix} p \\ n \end{pmatrix}, \tag{4.6.1}$$

where $u$ is a general $2 \times 2$ unitary matrix with unit determinant. As we saw at the end of Section 4.3, this is the same as the group of rotations in three dimensions, but acting on the labels $p$ and $n$ instead of coordinates or momenta or ordinary spin indices, and with the doublet $(p, n)$ transforming the same way that a spin-$1/2$ doublet of states transforms under ordinary rotations. These are known as *isospin* transformations.

For these transformations to be symmetries of a quantum-mechanical theory, there must exist a unitary operator $U(u)$ for each $2 \times 2$ unitary matrix $u$ with unit determinant. These transformations are generated by Hermitian operators $T_a$ (with $a = 1, 2, 3$), in the sense that for an isospin transformation $u$ close to unity, of the general form

$$u = 1 + \frac{i}{2} \begin{pmatrix} \epsilon_3 & \epsilon_1 - i\epsilon_2 \\ \epsilon_1 + i\epsilon_2 & -\epsilon_3 \end{pmatrix}$$

(with $\epsilon_a$ real and infinitesimal), the operator $U(u)$ takes the form

$$U \rightarrow 1 + i \sum_a \epsilon_a T_a. \tag{4.6.2}$$

Because the structure of the isospin group is the same as the structure of the rotation group, the generators satisfy the same commutation relations (4.1.14) (without the conventional factor $\hbar$) as ordinary angular momentum:

$$[T_a, T_b] = i \sum_c \epsilon_{abc} T_c. \tag{4.6.3}$$

The action of these generators on proton and neutron states can be derived in the same way that we derived Eq. (4.2.17):

$$(T_1 + iT_2)\Psi_p = 0, \quad (T_1 - iT_2)\Psi_p = \Psi_n, \quad T_3\Psi_p = \frac{1}{2}\Psi_p$$

$$(T_1 + iT_2)\Psi_n = \Psi_p, \quad (T_1 - iT_2)\Psi_n = 0, \quad T_3\Psi_n = -\frac{1}{2}\Psi_n. \tag{4.6.4}$$

---

[24] M. A. Tuve, N. Heydenberg, and L. R. Hafstad, *Phys. Rev.* **50**, 806 (1936).

[25] B. Cassen and E. U. Condon, *Phys. Rev.* **50**, 846 (1936); G. Breit and E. Feenberg, *Phys. Rev.* **50**, 850 (1936).

We note that single-nucleon states have electric charge $(1/2 + T_3)e$. Hence states consisting of $A$ nucleons have electric charge

$$Q = \left( \frac{A}{2} + T_3 \right) e, \tag{4.6.5}$$

which shows clearly the violation of isospin invariance by electromagnetic interactions.

Isospin invariance has implications for nuclear structure that go beyond those of charge symmetry. Each energy level in a light nucleus must be part of a multiplet of energy levels in $2t + 1$ nuclei (where $t$ is an integer or half-integer, analogous to $j$), with the same atomic weight $A$ and with $T_3$ running by unit steps from $-t$ to $+t$, and hence with atomic numbers $Z$ running from $A/2 - t$ to $A/2 + t$, all of these nuclear states having the same spin and approximately the same energy. For instance, not only do the ground states of $^{12}$B and $^{12}$N have the same spin ($j = 1$) and approximately the same energy – there is also an excited state of $^{12}$C with the same spin and energy, indicating that these three nuclear energy levels form an isospin multiplet with $t = 1$. (The $t = 1$ state in $^{12}$C is not the ground state, which is 15 MeV/$c^2$ below the $t = 1$ excited state, and has spin $j = 0$ instead of $j = 1$.)

Isospin invariance requires that not only nuclei, but all particles that feel the strong nuclear force, form isospin multiplets. Thus, for instance, in 1947 a pair of unstable charged particles $\pi^{\pm}$ with charges $+e$ and $-e$ were discovered, in reactions like $N + N \rightarrow N + N + \pi$ (where N can be either a neutron or a proton.) These "pions" have nucleon number $A = 0$, so according to Eq. (4.6.5), the $\pi^+$ and $\pi^-$ have $T_3 = +1$ and $T_3 = -1$, respectively. Isospin then requires that the pions must be part of a multiplet of $2t + 1$ approximately equal-mass particles with $t \geq 1$. In particular, there would have to be a neutral particle $\pi^0$ with $T_3 = 0$, and indeed, such a neutral pion was soon discovered. But no doubly charged pions were found, so the pions form a triplet, with $t = 1$.

The decays of these particles are quite different: the $\pi^{\pm}$ decay through weak interactions (similar to those in nuclear beta decay) into a heavy counterpart of the positron and electron, the $\mu^{\pm}$, and a neutrino or antineutrino, while the $\pi^0$ decays through electromagnetic interactions into two photons. But isospin invariance is respected in any process that is dominated by the strong nuclear forces. For instance, there is a multiplet of four unstable states $\Delta^{++}$, $\Delta^+$, $\Delta^0$, and $\Delta^-$ of a nucleon and a pion, all $\Delta$s with spin 3/2 and masses of about 1240 MeV/$c^2$. These states show a large uncertainty in energy, about 120 MeV/$c^2$, so by the uncertainty principle they must decay very rapidly, indicating that the decay is not produced by weak or electromagnetic interactions, but by the strong nuclear force, which respects isospin symmetry. Since the $\Delta$s decay into a state with one nucleon, they have $A = 1$, and hence according to Eq. (4.6.5) have $T_3$ respectively equal to 3/2, 1/2, $-1/2$, and $-3/2$. This is evidently an isospin multiplet with $t = 3/2$. The amplitude $M$ for a $\Delta$ with $T_3 = m$ to decay through strong interactions into a $\pi$ with $T_3 = m'$ and a nucleon with $T_3 = m''$ then has a dependence on charges proportional to a Clebsch–Gordan coefficient:

$$M(m, m', m'') = M_0 C_{1\frac{1}{2}} \left( \frac{3}{2} m; m'm'' \right),$$

where $M_0$ is independent of charges. The decay rates are of course proportional to the squares of these amplitudes. Inspection of the fifth, sixth, and seventh lines of Table 4.1 shows that these decay rates have ratios given by

$$\Gamma(\Delta^{++} \to \pi^+ + p) = \Gamma(\Delta^- \to \pi^- + n) \equiv \Gamma_0,$$

$$\Gamma(\Delta^+ \to \pi^+ + n) = \Gamma(\Delta^0 \to \pi^- + p) = \frac{1}{3}\Gamma_0,$$

$$\Gamma(\Delta^+ \to \pi^0 + p) = \Gamma(\Delta^0 \to \pi^0 + n) = \frac{2}{3}\Gamma_0,$$

all in good agreement with observation.[26]

The discovery in 1947 of new particles forced a significant change in the relation (4.6.5) between electric charge and isospin. For example (using modern names), collisions between nucleons were found to produce a number of spin-1/2 particles called hyperons – a neutral particle $\Lambda^0$ with mass 1115 $GeV/c^2$, and a triplet of particles $\Sigma^+$, $\Sigma^0$, and $\Sigma^-$, with masses 1189 $GeV/c^2$, 1192 $GeV/c^2$, and 1197 $GeV/c^2$. These hyperons were always produced in association with a doublet of spin-zero particles $K^+$ and $K^0$, with masses 494 $GeV/c^2$ and 498 $GeV/c^2$. (Superscripts indicate the electric charge in units of $e$.) It had been thought that the number $A$ of nucleons (minus the number of antinucleons) was absolutely conserved in nature, but hyperons were observed to decay into a nucleon and a pion, so it became necessary to extend this conservation law to a quantity $B$ called baryon number, the number of nucleons and hyperons, minus the number of their antiparticles. But it is not enough just to replace $A$ in Eq. (4.6.5) with $B$. Since the $\Lambda^0$ is not part of an isospin multiplet with other particles, it must have $t = 0$ and hence $T_3 = 0$, but if we replace $A$ in Eq. (4.6.5) with the baryon number $B = 1$, then this formula would give the $\Lambda^0$ charge $e/2$, not zero. Similar problems would arise with the $\Sigma$s and Ks. It was suggested that one should replace Eq. (4.6.5) with[27]

$$Q = \left( \frac{B + S}{2} + T_3 \right) e, \tag{4.6.6}$$

where $S$ is a quantity known as *strangeness*, equal to zero for ordinary particles like nucleons and pions, but equal to $-1$ for the $\Lambda$ and $\Sigma$, and equal to $+1$ for

[26] H. L. Anderson, E. Fermi, R. Martin, and D. E. Nagle, *Phys. Rev.* **91**, 151 (1953); J. Orear, C. H. Tsao, J. J. Lord, and A. B. Weaver, *Phys. Rev.* **95**, 624A (1954).

[27] M. Gell-Mann, *Phys. Rev.* **92**, 833 (1953); T. Nakano and K. Nishijima, *Prog. Theor. Phys. (Kyoto)* **10**, 582 (1953).

the K. These assignments fix the charges: the $\Lambda$ and $\Sigma$s have $B + S = 0$, so $Q = T_3 e$, while the Ks have $B + S = 1$, so $Q = T_3 + 1/2$. The conservation of strangeness in strong interactions requires that in nucleon–nucleon collisions these hyperons must be produced in association with K particles, to keep the total strangeness zero.

Other strange particles were discovered: a doublet $\Xi^0$ and $\Xi^-$, with masses 1315 GeV/$c^2$ and 1322 GeV/$c^2$, and the antiparticles $\overline{K}^-$ and $\overline{K}^0$ of the $K^+$ and $K^0$. To get their charges right the $\Xi$ must be assigned strangeness $-2$, and the anti-K strangeness $-1$. Strangeness is not conserved in the decay of hyperons and Ks and $\bar{K}$s into nucleons and pions, but these decays proceed through a class of interactions much weaker than the strong nuclear forces. (Strange particles typically have lifetimes around $10^{-8}$ to $10^{-10}$ seconds, which is enormously long compared with the typical time scale of strong interactions, $\hbar/(1\ \text{GeV}) = 6.6 \times 10^{-25}$ seconds.) So strangeness is not conserved by the weak interactions responsible for strange particle decays, but it is conserved by the strong (and electromagnetic) interactions.

All of these approximate or exact conservation laws, of charge, baryon number, and strangeness, can also be formulated as symmetry principles. For example, we may construct a unitary operator,

$$U(\alpha) \equiv \exp(i\alpha Q), \tag{4.6.7}$$

where here $Q$ is an Hermitian operator that, acting on any state, gives a factor equal to the total electric charge $q$ of the particles in the state, and $\alpha$ is an arbitrary real number. Acting on any state of charge $q$ the operator $U(\alpha)$ gives a phase factor, $\exp(i\alpha q)$. Transition amplitudes are invariant under this symmetry if and only if charge is conserved – that is, if and only if the Hamiltonian $H$ satisfies

$$U^{-1}(\alpha) H U(\alpha) = H. \tag{4.6.8}$$

The symmetry group here is $U(1)$, the group of multiplication by $1 \times 1$ unitary matrices, which of course are just phase factors. The conservation of baryon number and strangeness can likewise be expressed as invariance under other $U(1)$ symmetry groups.

These $U(1)$ symmetries were entirely separate from the $SU(2)$ of isospin, in the sense that their generators commuted with the generators $T_a$ of isospin. The question naturally arose, whether some of these symmetries could be combined in a symmetry that united some of these isospin multiplets. The winning candidate was $SU(3)$, the group of all unitary $3 \times 3$ matrices with unit determinant.[28] The $SU(2)$ transformations of isospin invariance form a subgroup, with

---

[28]  M. Gell-Mann, Cal. Tech. Synchrotron Laboratory Report CTSL–20 (1961), unpublished. Y. Ne'eman, *Nucl. Phys.* **26**, 222 (1961). [These are reproduced along with other articles on $SU(3)$ symmetry in M. Gell-Mann and Y. Ne'eman, *The Eightfold Way* (Benjamin, New York, 1964).]

the isotopic spin generators $T_a$ represented by $3 \times 3$ Hermitian matrices of the form

$$\begin{pmatrix} t_a & 0 \\ 0 & 0 \end{pmatrix},$$

where $t_a$ are the $2 \times 2$ Hermitian traceless matrices that represent the $SU(2)$ generators. There is also a $U(1)$ subgroup with a generator known as the *hypercharge*

$$Y \equiv B + S,$$

which is represented by the Hermitian traceless matrix

$$y = \begin{pmatrix} 1/3 & 0 & 0 \\ 0 & 1/3 & 0 \\ 0 & 0 & -2/3 \end{pmatrix}.$$

We can find the particle multiplets by using the tensor formalism discussed in the context of ordinary rotations at the end of Section 4.3. But there is a difference here. In general, for a group of unitary matrices in $N$ dimensions, the particle multiplets form tensors $\Psi^{n_1 n_2 \cdots}_{m_1 m_2 \cdots}$ (where the $m$s and $n$s run from 1 to $N$), with the transformation property

$$\Psi^{n_1 n_2 \cdots}_{m_1 m_2 \cdots} \mapsto \sum_{m_1' m_2' \cdots n_1' n_2' \cdots} u_{m_1' m_1} u_{m_2' m_2} \cdots u^*_{n_1' n_1} u^*_{n_2' n_2} \cdots \Psi^{n_1' n_2' \cdots}_{m_1' m_2' \cdots}.$$

In two dimensions, and only in two dimensions, there is a constant tensor (4.3.37) with two indices. When this tensor is contracted with an upper index, the index is converted into a lower index, so that it is not necessary to distinguish between upper and lower indices in two dimensions. For $N = 3$ we have to distinguish between upper and lower indices, but we can still limit ourselves to irreducible tensors that are completely symmetric in both sorts of indices, because there exists a constant antisymmetric tensor $\epsilon_{m_1 m_2 m_3}$ that otherwise would allow us to convert two upper indices into a lower index, or two lower indices into an upper index. For irreducible tensors we must also impose the condition of tracelessness

$$\Psi^{r n_2 \cdots}_{r m_2 \cdots} = 0,$$

for otherwise we could separate out a tensor $\Psi^{r n_2 \cdots}_{r m_2 \cdots}$ with one fewer upper index and one fewer lower index. For example, the nucleons, $\Lambda$, $\Sigma$s, and $\Xi$s can be united in an octet with $j = 1/2$, whose states form a traceless tensor $\Psi^n_m$, which has eight independent components. Similarly, the $\pi$s, Ks, $\bar{\text{K}}$s, and an eighth spin-zero particle, the $\eta$, form another octet, but with $j = 0$. There is also a 10-member multiplet of spin-3/2 particles that contains the $\Delta$ discussed above, corresponding to the symmetric tensor $\Psi_{m_1 m_2 m_3}$.

Since particles belonging to different species are distinguishable, we can adopt various conventions for how these particles are listed in the labels on physical state vectors. For instance, in a state containing some protons and some electrons, we could agree always to list the protons first, and then the electrons. There is no need to make the state vector antisymmetric under the interchange of protons and electrons. But when the different species all belong to the same multiplet of some internal symmetry group, in the way that protons and neutrons belong to a $t = 1/2$ multiplet of the isospin symmetry, and these particles are bosons or fermions, then the state vector must be respectively symmetric or antisymmetric under interchange of all particle labels: orbital quantum numbers (which could be positions, or momenta, or the $z$-components $m$ of orbital angular momentum) and spin $z$-components *and* the quantum numbers for the internal symmetry group.

For instance, consider a proton–neutron state:

$$\Psi_\pm = \int d\xi_1 \int d\xi_2 \ \psi_\pm(\xi_1, \xi_2) \Phi_{p,\xi_1;n,\xi_2},$$

where $\xi_1$ and $\xi_2$ label both orbital and spin quantum numbers of the two nucleons; $\int d\xi$ denotes an integral over momentum (or position) together with a sum over the spin 3-component; and the wave function $\psi_\pm$ is either symmetric or antisymmetric:

$$\psi_\pm(\xi_1, \xi_2) = \pm\psi_\pm(\xi_2, \xi_1).$$

Applying the isospin raising operator to this state gives a two-proton state:

$$(T_1 + iT_2)\Psi_\pm = \int d\xi_1 \int d\xi_2 \ \psi_\pm(\xi_1, \xi_2)\Phi_{p,\xi_1;p,\xi_2}.$$

Since protons are indistinguishable fermions, the two-proton state is antisymmetric in $\xi_1$ and $\xi_2$, so $(T_1 + iT_2)\Psi_+ = 0$ but $(T_1 + iT_2)\Psi_- \neq 0$, and hence $\Psi_+$ and $\Psi_-$ respectively have isospin zero and one. According to Eq. (4.3.34), the states of isospin zero and one are respectively odd and even in isospin 3-components, so a state that is symmetric or antisymmetric in spin and orbital quantum numbers must be respectively antisymmetric or symmetric in isospin 3-components, and hence in either case is antisymmetric under exchange of all quantum numbers. For instance, an $s$ wave state of two nucleons can only have total spin one and total isospin zero (as in the deuteron), or total spin zero and total isospin one (as in low-energy scattering of two protons or two neutrons).

* * * * *

The group $SU(3)$ has another application, not as an internal symmetry, but as a dynamical symmetry of the Hamiltonian for a harmonic oscillator in three dimensions. As described in Section 2.5, this Hamiltonian is

$$H = \hbar\omega \left[ \sum_{i=1}^{3} a_i^\dagger a_i + \frac{3}{2} \right], \tag{4.6.9}$$

where $a_i$ and $a_i^\dagger$ are lowering and raising operators, satisfying the commutation relations

$$[a_i, a_j^\dagger] = \delta_{ij}, \quad [a_i, a_j] = [a_i^\dagger, a_j^\dagger] = 0. \tag{4.6.10}$$

The Hamiltonian and commutation relations are obviously invariant under the transformations

$$a_i \mapsto \sum_j u_{ij} a_j, \quad a_i^\dagger \mapsto \sum_j u_{ij}^* a_j^\dagger, \tag{4.6.11}$$

where $u_{ij}$ is a unitary matrix, with $\sum_j u_{ij} u_{kj}^* = \delta_{ik}$. This group is $U(3)$, the group of $3 \times 3$ unitary matrices. The degenerate states with energy $(N + 3/2)\hbar\omega$ are of the form

$$a_{i_1}^\dagger a_{i_2}^\dagger \dots a_{i_N}^\dagger \Psi_0,$$

where $\Psi_0$ is the ground state with energy $3\hbar\omega/2$; under the transformation (4.6.11), these states transform as a symmetric tensor:

$$a_{i_1}^\dagger a_{i_2}^\dagger \dots a_{i_N}^\dagger \Psi_0 \mapsto \sum_{j_1 j_2 \dots j_N} u_{i_1 j_1}^* u_{i_2 j_2}^* \dots u_{i_N j_N}^* a_{j_1}^\dagger a_{j_2}^\dagger \dots a_{j_N}^\dagger \Psi_0. \tag{4.6.12}$$

The number $(N + 1)(N + 2)/2$ of independent states of energy $(N + 3/2)\hbar\omega$ found in Section 2.5 is also the number of independent components of a symmetric tensor of rank $N$ in three dimensions.

In the special case where $u_{ij} = \delta_{ij} e^{-i\varphi}$ with $\varphi$ real, the transformations (4.6.11) are the same as

$$\begin{aligned} a_i &\mapsto \exp(iH\varphi/\hbar\omega)\, a_i \exp(-iH\varphi/\hbar\omega), \\ a_i^\dagger &\mapsto \exp(iH\varphi/\hbar\omega)\, a_i^\dagger \exp(-iH\varphi/\hbar\omega), \end{aligned} \tag{4.6.13}$$

so the symmetry in this case is nothing new, just time-translation invariance. The new symmetries that are special to the three-dimensional harmonic oscillator are those for which $\mathrm{Det}\, u = 1$, forming the group $SU(3)$.

For infinitesimal transformations, we have

$$u_{ij} = \delta_{ij} + \epsilon_{ij}, \tag{4.6.14}$$

where $\epsilon_{ij}$ are here infinitesimal anti-Hermitian matrices, with $\epsilon_{ij}^* = -\epsilon_{ji}$. For $SU(3)$, these matrices are also traceless. These infinitesimal transformations must induce corresponding unitary transformations on the Hilbert space of harmonic oscillator states,

$$U(1 + \epsilon) = 1 + \sum_{ij} \epsilon_{ij} X_{ij}, \tag{4.6.15}$$

where $X_{ij}^\dagger = X_{ji}$ are symmetry generators that commute with the Hamiltonian. These symmetry generators are proportional to the operators $a_i a_j^\dagger$ mentioned in Section 2.5.

## 4.7 Inversions

We saw in Section 4.1 that the space inversion transformation $\mathbf{X}_n \mapsto -\mathbf{X}_n$ of the coordinate operators of particles (labeled $n$) is not a rotation, but a separate sort of symmetry transformation. It therefore can have consequences beyond those that can be derived from rotational invariance alone.

In a quantum theory that is invariant under space inversion, we expect there to be a unitary "parity" operator $\mathsf{P}$, with the property that

$$\mathsf{P}^{-1}\mathbf{X}_n\mathsf{P} = -\mathbf{X}_n. \tag{4.7.1}$$

In a wide class of theories, the momentum operator $\mathbf{P}_n$ can be expressed as $\mathbf{P}_n = (im_n/\hbar)[H, \mathbf{X}_n]$, so if the Hamiltonian $H$ commutes with $\mathsf{P}$, then also

$$\mathsf{P}^{-1}\mathbf{P}_n\mathsf{P} = -\mathbf{P}_n. \tag{4.7.2}$$

This transformation leaves invariant the sort of Hamiltonian we have been considering, as for instance

$$H = \sum_n \frac{\mathbf{P}_n^2}{2m_n} + V,$$

where $V$ depends only on the distances $|\mathbf{X}_n - \mathbf{X}_m|$.

As a consequence of Eqs. (4.7.1) and (4.7.2), the operator $\mathsf{P}$ commutes with the orbital angular momentum $\mathbf{L} = \sum_n \mathbf{X}_n \times \mathbf{P}_n$. Consistency with the angular-momentum commutation relations also requires that it commutes with $\mathbf{J}$ and $\mathbf{S}$.

For a system like the hydrogen atom, with a single particle in a central potential, it follows from Eq. (4.7.1) that if $\Phi_\mathbf{x}$ is an eigenstate of $\mathbf{X}$ with eigenvalue $\mathbf{x}$, then $\mathsf{P}\Phi_\mathbf{x}$ is an eigenstate of $\mathbf{X}$ with eigenvalue $-\mathbf{x}$. (Since $\mathsf{P}$ commutes with $S_3$, this state is also an eigenstate of $S_3$ with the same eigenvalue as the state $\Phi_\mathbf{x}$, so for the present we will not need to display spin indices explicitly.) Hence, apart from possible phases (about which more later),

$$\mathsf{P}\Phi_\mathbf{x} = \Phi_{-\mathbf{x}}. \tag{4.7.3}$$

A state $\Psi_\ell^m$ with orbital angular momentum $\hbar\ell$ and 3-component $\hbar m$ has a scalar product with $\Phi_\mathbf{x}$ (that is, a coordinate-space wave function) proportional to a spherical harmonic:

$$\left(\Phi_\mathbf{x}, \Psi_\ell^m\right) = R(|\mathbf{x}|)Y_\ell^m(\hat{x}). \tag{4.7.4}$$

The inversion property $Y_\ell^m(-\hat{x}) = (-1)^\ell Y_\ell^m(\hat{x})$ thus gives

$$\left(\Phi_{-\mathbf{x}}, \Psi_\ell^m\right) = (-1)^\ell\left(\Phi_{\mathbf{x}}, \Psi_\ell^m\right).$$

Inserting the operator $\mathsf{P}^{-1}\mathsf{P} = 1$ in the scalar product on the left and using Eq. (4.7.3) and the unitarity of $\mathsf{P}$, we find

$$\left(\Phi_{\mathbf{x}}, \mathsf{P}\Psi_\ell^m\right) = (-1)^\ell\left(\Phi_{\mathbf{x}}, \Psi_\ell^m\right),$$

and therefore

$$\mathsf{P}\Psi_\ell^m = (-1)^\ell\Psi_\ell^m. \tag{4.7.5}$$

This allows us to understand why, even when subtle effects like the Lamb shift and spin–orbit coupling are included, the states of hydrogen with definite $j$ also have definite values of $\ell$, rather than being mixtures of states with $\ell = j \pm 1/2$. For instance, why when all these effects are taken into account, can we still talk of the $n = 2$ states of hydrogen with $j = 1/2$ as pure $2s_{1/2}$ and $2p_{1/2}$ states? The Hamiltonian of the hydrogen atom (including spin effects and relativistic corrections) is invariant under space inversion, so space inversion applied to a one-particle state vector of definite energy gives another state vector of the same energy. With enough perturbations included to break all degeneracies between states of a given $\mathbf{J}^2$, $\mathbf{J}_z$, and $n$, the space inversion of the state vector of a state of definite energy must give a result proportional to the same state vector, which would not be true if the states of definite energy were mixtures of states with both odd and even values of $\ell$, such as states with $\ell = j + 1/2$ and $\ell = j - 1/2$.

The space inversion symmetry of atomic physics has an immediate application in the selection rules for the most common radiative transitions in atoms. As noted at the end of Section 4.4, in the approximation that the wavelength of the emitted photon is much larger than the atomic size, the transition rate is proportional to the square of the matrix element of an electric-dipole operator $\mathbf{D} = \sum_n e_n\mathbf{X}_n$ between the initial and final atomic states. It follows immediately from Eq. (4.7.1) that $\mathsf{P}^{-1}\mathbf{D}\mathsf{P} = -\mathbf{D}$. If the initial state $\Psi_a$ and final state $\Psi_b$ are eigenstates of the parity operator with eigenvalues $\pi_a$ and $\pi_b$ respectively, then

$$\pi_a\pi_b\left(\Psi_b, \mathbf{D}\Psi_a\right) = -\left(\Psi_b, \mathbf{D}\Psi_a\right),$$

so the matrix element and the transition rate vanish unless

$$\pi_a\pi_b = -1. \tag{4.7.6}$$

In the case mentioned earlier, where the transition involves just a single electron, we have $\pi_a = (-1)^{\ell_a}$ and $\pi_b = (-1)^{\ell_b}$, where $\ell_a$ and $\ell_b$ are the orbital angular momenta of the electron in the initial and final states, so in this case the parity selection rule is just that $\ell$ must change from even to odd or odd to even. For instance, in the electric-dipole approximation the radiative $3p \to 2p$ transition in hydrogen is allowed by angular-momentum conservation but forbidden by the parity selection rule. Equation (4.7.6) applies also to transitions between states with any number of charged particles.

Let us now return to the question of possible extra phase factors in transformation rules like (4.7.3) and (4.7.5). If the same extra phase factor appeared in the transformation of all states, it would have no effect, for it could be eliminated by a re-definition of the phase of the unitary operator P. There is, however, a less trivial possibility, of a phase that depends on the nature of the particles in the state, which would have important consequences for transitions in which new particles are created or destroyed. We would expect the operator P to act separately on each particle when the particles are far apart, and if P commutes with the Hamiltonian, it would then continue to act separately on each particle when they come together, so the extra phase in the transformation in a multiparticle state would be the product of the phases $\eta_n$ for the individual particles

$$P\Phi_{\mathbf{x}_1,\sigma_1;\mathbf{x}_2,\sigma_2;\dots} = \eta_1\eta_2\cdots\Phi_{-\mathbf{x}_1,\sigma_1;-\mathbf{x}_2,\sigma_2;\dots}, \qquad (4.7.7)$$

where the $\sigma$s are spin 3-components, and the phase factor $\eta_n$ depends only on the species of particle $n$. These factors are known as the *intrinsic parities* of the different particle types. The operator $P^2$ commutes with all coordinates, momenta, and spins. It could be an internal symmetry of some sort, but if it were a $U(1)$ operator that like (4.6.7) is of the form $\exp(i\alpha A)$, where $A$ is some conserved Hermitian operator, then $\exp(-i\alpha A/2)$ would also be an internal symmetry, and we could define a new space inversion operator $P' \equiv P\exp(-i\alpha A/2)$ for which $P'^2 = 1$. Dropping the prime, we suppose that P is chosen so that $P^2 = 1$. In this case, all the intrinsic parities $\eta_n$ in Eq. (4.7.7) are just either $+1$ or $-1$.

A classic example of the use of such a transformation rule is provided by the disintegration of the $1s$ state of a mesonic atom consisting of a deuterium nucleus and a negatively charged spin-zero particle, the $\pi^-$, instead of an electron. The $\pi^-$ is observed to be quickly absorbed by the deuterium nucleus, giving a pair of neutrons.[29] Because neutrons are fermions, the two-neutron state must be antisymmetric under an exchange of both spin and position, so it either has total spin one (symmetric in spins) and odd orbital angular momentum, or it has total spin zero (antisymmetric in spins) and even orbital angular momentum. But the deuterium nucleus is known to have spin one, so the $1s$ state of the d–$\pi^-$ atom has total angular momentum one, while a two-neutron state with total spin zero and even orbital angular momentum cannot have total angular momentum one. We can conclude then that the two-neutron final state here must have odd orbital angular momentum, and therefore has parity $-\eta_n^2$. This tells us then that $\eta_d\eta_{\pi^-} = -\eta_n^2$. The deuterium nucleus is known to be a mixture of $s$ and $d$ states of a proton and a neutron, so $\eta_d = \eta_p\eta_n$, and hence $\eta_p\eta_\pi = -\eta_n$. We would not expect the space inversion operator P to be part of an isotopic spin multiplet of independent inversion operators, so we expect P to commute with the

---
[29] W. Chinowsky and J. Steinberger, *Phys. Rev.* **95**, 1561 (1954).

isospin symmetries discussed in the previous section,[30] in which case $\eta_p = \eta_n$, and therefore the $\pi^-$ has intrinsic parity $-1$. Isospin invariance then tells us also that its antiparticle, the $\pi^+$, and its neutral counterpart, the $\pi^0$, also have negative intrinsic parity.

It used to be taken for granted that nature is invariant under the space inversion transformation. Then in the 1950s the use of this symmetry principle led to a serious problem. Two charged particles of similar mass were found in cosmic rays, a $\theta^+$ that decays into $\pi^+ + \pi^0$, and a $\tau^+$ that decays into $\pi^+ + \pi^+ + \pi^-$ (and also into $\pi^+ + \pi^0 + \pi^0$.) By studying the angular distribution of the $\pi$s in the final state of $\tau$ decay, it was found that these $\pi$s had no orbital angular momenta, so with $\pi$s having odd parity and spin zero, the $\tau^+$ would also have to have odd parity and spin zero. On the other hand, with two pions in the final state, if the $\theta^+$ had spin zero like the $\tau^+$ it would have to have even parity, so it seemed that the $\theta^+$ and $\tau^+$ could not be the same particle. But as measurements were improved, it was found that both the masses and the mean lifetimes of the $\theta^+$ and $\tau^+$ were indistinguishable. One could imagine some sort of symmetry that would make their masses equal, but how could their lifetimes be equal, when they decay in such different ways? Then in 1956, Tsung-Dao Lee and Chen-Ning Yang[31] proposed that the $\theta^+$ and $\tau^+$ are in fact the same particle (now called $K^+$), and that although invariance under space inversion is respected by the electromagnetic and strong nuclear forces, it is not respected by the much weaker interactions that lead to these decays. (The weakness of these interactions is shown by the long lifetime of the $K^+$ particle; it is $1.238 \times 10^{-8}$ seconds, vastly longer than the characteristic time scale $\hbar/m_K c^2 = 1.3 \times 10^{-24}$ seconds.) Lee and Yang further suggested that invariance under space inversions is badly violated in all weak interactions of elementary particles, including nuclear beta decay, and suggested experiments that soon showed that they were right.[32]

There are two other inversion symmetry transformations that commute with the strong and electromagnetic interaction Hamiltonians. One is charge-conjugation: a conserved operator $\mathsf{C}$ acting on any state simply changes every particle into its antiparticle, with a possible sign factor depending on the nature of the particles.[33] Another is time-reversal: a conserved operator $\mathsf{T}$ reverses the direction of time in the time-dependent Schrödinger equation. As we saw in

---

[30] Even apart from isospin conservation, we can always define the operator $\mathsf{P}$ so that $\eta_p = \eta_n = 1$, if necessary by including in the operator $\mathsf{P}$ a factor equal to $(-1)$ to a power given by a suitable linear combination of the conserved quantities electric charge and baryon number.

[31] T.-D. Lee and C.-N. Yang, *Phys. Rev.* **104**, 254 (1956).

[32] C. S. Wu, E. Ambler, R. W. Hayward, D. D. Hoppes, and R. P. Hudson, *Phys. Rev.* **105**, 1413 (1957); R. Garwin, L. Lederman, and M. Weinrich, *Phys. Rev.* **105**, 1415 (1957); J. I. Friedman and V. L. Telegdi, *Phys. Rev.* **105**, 1681 (1957).

[33] As mentioned in footnote 14 in Section 3.6, Dirac interpreted the negative-energy solutions of the Dirac wave equation as the wave functions of negative-energy states that are normally all filled, so that the Pauli exclusion principle prevents positive-energy electrons from falling into these negative energy states. He interpreted occasional unfilled states, or *holes*, in this sea of negative-energy states as

Section 3.6, T must be antiunitary and antilinear. The same experiments that showed that P is not respected by the weak interactions showed also that these interactions do not respect invariance under PT. Subsequent experiments also revealed a violation of CP.[34] But any quantum field theory necessarily respects invariance under CPT,[35] and as far as we know CPT is exactly conserved, so the violation of invariance under PT and CP immediately implied a violation also of invariance under C and T. Thus it appears that CPT is the only inversion under which the laws of nature are strictly invariant.

## 4.8 Algebraic Derivation of the Hydrogen Spectrum

As mentioned in Section 1.4, Pauli[36] in 1926 used the matrix mechanics of Heisenberg to give the first derivation of the energy levels of hydrogen and their degeneracies. This derivation is an outstanding example of the use of a *dynamical symmetry*: The symmetry generators not only commute with the Hamiltonian, but have commutators with each other that depend on the Hamiltonian, in such a way that we can calculate energy levels by purely algebraic means.

Pauli's derivation is based on a device that is well known in celestial mechanics, the *Runge–Lenz vector*.[37] In a potential $V(r) = -Ze^2/r$, this vector (actually the original Runge–Lenz vector multiplied by the particle mass $m$) is

$$\mathbf{R} = -\frac{Ze^2\mathbf{x}}{r} + \frac{1}{2m}\left(\mathbf{p} \times \mathbf{L} - \mathbf{L} \times \mathbf{p}\right), \qquad (4.8.1)$$

where $\mathbf{L}$ is as usual the orbital angular momentum $\mathbf{L} \equiv \mathbf{x} \times \mathbf{p}$. Classically there is no difference between $\mathbf{p} \times \mathbf{L}$ and $-\mathbf{L} \times \mathbf{p}$; it is the average of these operators that appears in the quantum-mechanical derivation Eq. (4.8.1) because this average is Hermitian, and therefore so is $\mathbf{R}$:

$$\mathbf{R}^\dagger = \mathbf{R}. \qquad (4.8.2)$$

---

antielectrons, particles known as positrons with positive energy and positive charge. Dirac's interpretation of antimatter is untenable, in part because it is now known that there are charged elementary bosons like the $W^+$ with a distinct antiparticle, the $W^-$, and the exclusion principle does not apply to bosons. Today it is pretty generally understood that the solutions of the Dirac equations are not a relativistic generalization of probability amplitudes like the Schrödinger wave function, as Dirac thought. Instead, the positive-energy solutions are matrix elements $(\Psi_0, \psi(x)\Psi_1)$ of the quantized electron field $\psi(x)$ between various one-electron states $\Psi_1$ and the vacuum $\Psi_0$, while the negative-energy solutions are matrix elements $(C\Psi_1, \psi(x)\Psi_0)$ of the electron field between the vacuum and various positron states.

[34]  J. H. Christensen, J. W. Cronin, V. L. Fitch, and R. Turlay, *Phys. Rev. Lett.* **13**, 138 (1964).

[35]  G. Lüders, *Kon. Danske Vid. Selskab Mat.-Fys. Medd.* **28**, 5 (1954); *Ann. Phys.* **2**, 1 (1957); W. Pauli, *Nuovo Cimento* **6**, 204 (1957).

[36]  W. Pauli, *Z. Physik* **36**, 336 (1926).

[37]  For its application to motion in a gravitational field, see e.g. S. Weinberg, *Gravitation and Cosmology* (Wiley, New York, 1972), Section 9.5.

Classically **R** is conserved, which has the consequence (unique to Coulomb and harmonic oscillator potentials) that the classical orbits form closed curves. The quantum-mechanical counterpart of this classical result is of course that **R** commutes with the Hamiltonian:

$$[H, \mathbf{R}] = 0, \tag{4.8.3}$$

where $H$ is the Coulomb Hamiltonian

$$H = \frac{\mathbf{p}^2}{2m} - \frac{Ze^2}{r}. \tag{4.8.4}$$

It is convenient to use the commutation relation $[L_i, p_j] = i\hbar \sum_k \epsilon_{ijk} p_k$ to rewrite Eq. (4.8.1) as

$$\mathbf{R} = -\frac{Ze^2 \mathbf{x}}{r} + \frac{1}{m}\mathbf{p} \times \mathbf{L} - \frac{i\hbar}{m}\mathbf{p}. \tag{4.8.5}$$

The angular-momentum operator is orthogonal to each of the three terms in Eq. (4.8.5), so

$$\mathbf{L} \cdot \mathbf{R} = \mathbf{R} \cdot \mathbf{L} = 0. \tag{4.8.6}$$

To calculate the square of **R**, we need formulas easily derived from the commutators among **x**, **p**, and **L**:

$$\mathbf{x} \cdot (\mathbf{p} \times \mathbf{L}) = \mathbf{L}^2, \quad (\mathbf{p} \times \mathbf{L}) \cdot \mathbf{x} = \mathbf{L}^2 + 2i\hbar\mathbf{p} \cdot \mathbf{x}, \quad (\mathbf{p} \times \mathbf{L})^2 = \mathbf{p}^2 \mathbf{L}^2,$$

$$\mathbf{p} \cdot (\mathbf{p} \times \mathbf{L}) = 0, \quad (\mathbf{p} \times \mathbf{L}) \cdot \mathbf{p} = 2i\hbar\mathbf{p}^2.$$

A straightforward calculation then gives

$$\mathbf{R}^2 = Z^2 e^4 + \left(\frac{2H}{m}\right)\left(\mathbf{L}^2 + \hbar^2\right). \tag{4.8.7}$$

So we can find the energy levels if we can find the eigenvalues of $\mathbf{R}^2$.

For this purpose, we need to work out the commutators of the components of **R** with each other. Another straightforward though tedious calculation gives

$$[R_i, R_j] = -\frac{2i}{m}\hbar \sum_k \epsilon_{ijk} H L_k. \tag{4.8.8}$$

Also, the fact that **R** is a vector tells us immediately that

$$[L_i, R_j] = i\hbar \sum_k \epsilon_{ijk} R_k. \tag{4.8.9}$$

Thus the operators **L** and $\mathbf{R}/\sqrt{-H}$ form a closed algebra. We can recognize the nature of this algebra by introducing linear combinations

$$\mathbf{A}_\pm \equiv \frac{1}{2}\left[\mathbf{L} \pm \sqrt{\frac{m}{-2H}}\mathbf{R}\right]. \tag{4.8.10}$$

Then the commutators (4.8.8) and (4.8.9) and the usual commutation relations for **L** yield

$$[A_{\pm i}, A_{\pm j}] = i\hbar \sum_k \epsilon_{ijk} A_{\pm k}, \quad [A_{\pm i}, A_{\mp j}] = 0. \tag{4.8.11}$$

So we can see that the symmetry here consists of two independent three-dimensional rotation groups. This is known as the group $SO(3) \otimes SO(3)$.

Now, from our study of the ordinary rotation group, we know that (provided the operators $\mathbf{A}_\pm$ are Hermitian) the allowed values of $\mathbf{A}_\pm^2$ take the form $\hbar^2 a_\pm(a_\pm + 1)$, where $a_\pm$ in general are independent positive integers (including zero) or half-integers; that is, $0, 1/2, 1, 3/2, \ldots$. But here we have a special condition (4.8.6), which with Eq. (4.8.10) tells us that

$$\mathbf{A}_\pm^2 = \frac{1}{4}\left[\mathbf{L}^2 + \left(\frac{m}{-2H}\right)\mathbf{R}^2\right], \tag{4.8.12}$$

so in this case $a_+ = a_-$. We will let $a$ denote their common value, and take $E$ as the corresponding eigenvalue of $H$. Then, using Eq. (4.8.7), we have

$$\hbar^2 a(a+1) = \frac{1}{4}\left[\mathbf{L}^2 + \left(\frac{m}{-2E}\right)\mathbf{R}^2\right]$$

$$= \frac{1}{4}\left[\mathbf{L}^2 + \left(\frac{m}{-2E}\right)Z^2 e^4 - (\mathbf{L}^2 + \hbar^2)\right]$$

$$= \left(\frac{m}{-8E}\right)Z^2 e^4 - \frac{\hbar^2}{4},$$

and therefore

$$\left(\frac{m}{-8E}\right)Z^2 e^4 = \hbar^2\left(a(a+1) + \frac{1}{4}\right) = \frac{\hbar^2}{4}(2a+1)^2. \tag{4.8.13}$$

We can define a principal quantum number

$$n = 2a + 1 = 1, 2, 3, \ldots, \tag{4.8.14}$$

and write Eq. (4.8.13) as a formula for the energy

$$E = -\frac{Z^2 e^4 m}{2\hbar^2 n^2}, \tag{4.8.15}$$

which of course we recognize as the energy levels of hydrogen, whose 1913 calculation by Bohr is described in Section 1.2, and whose derivation using the Schrödinger equation is given in Section 2.3.

Note that we have found only negative energies – that is, bound states. There are of course also unbound states, with $E > 0$, in which an electron is scattered by a nucleus. These states have not shown up in our calculation because, acting on states for which $H$ has a positive eigenvalue, the operators $\mathbf{A}_\pm$ given by Eq. (4.8.10) are no longer Hermitian, and this invalidates the derivation in

Section 4.2 of the familiar result that the allowed values of $\mathbf{A}_\pm^2$ can only take the form $\hbar^2 a_\pm(a_\pm + 1)$, where $a_\pm$ are positive integers or half-integers. (Mathematically, one says that the algebra furnished by the commutators of the $\mathbf{L}$ and $\mathbf{R}$ is not *compact*; that is, these are the generators of a symmetry group whose parameters do not form a compact space. It is a well-known feature of such non-compact algebras that the states connected by their generators form a continuum, which is why the allowed positive values of $E$ here form a continuum.)

We can use these algebraic results to work out not only the allowed values of energy, but also the degeneracy of each energy level. Just as for ordinary angular momentum, the eigenvalues of the operators $A_{\pm 3}$ can only take the $2a+1$ values $-a, -a+1, \ldots, a$, and since their eigenvalues are independent, there are $(2a + 1)^2 = n^2$ states with a given $n$. This is the same as the degeneracy found in Section 2.3.

This degeneracy has a pretty geometric interpretation. We have noted previously that the operators $\mathbf{A}_\pm$ are the generators of two independent three-dimensional rotation groups – that is, of $SO(3) \otimes SO(3)$. They can also be regarded as the generators of the rotation group in *four* dimensions, denoted $SO(4)$, because these are the same symmetry groups. As we saw in Eq. (4.1.10), the generators of the rotation group in any number of dimensions are operators $J_{\alpha\beta} = -J_{\beta\alpha}$, with $\alpha$ and $\beta$ running over the coordinate indices, satisfying the commutation relations

$$\frac{i}{\hbar}\Big[J_{\alpha\beta}, J_{\gamma\delta}\Big] = -\delta_{\alpha\delta}J_{\gamma\beta} + \delta_{\alpha\gamma}J_{\delta\beta} + \delta_{\beta\gamma}J_{\alpha\delta} - \delta_{\beta\delta}J_{\alpha\gamma}. \tag{4.8.16}$$

In the case of four dimensions, $\alpha$, $\beta$, etc. run from 1 to 4. If as before we let $i$, $j$, etc. run only from 1 to 3, and as in Eq. (4.1.11) take $J_{ij} \equiv \sum_k \epsilon_{ijk}L_k$, then the commutation relations with $\delta = \beta = 4$ take the form

$$[J_{i4}, J_{j4}] = -i\hbar J_{ji} = i\hbar \sum_k \epsilon_{ijk}L_k. \tag{4.8.17}$$

This is the same as Eq. (4.8.8) if we take

$$R_i = \sqrt{\frac{-2H}{m}}\, J_{i4}. \tag{4.8.18}$$

The others of the commutation relations (4.8.16) then give the commutator (4.8.9) between $L_i$ and $R_j$ and the usual commutator between $L_i$ and $L_j$. In terms of the operators (4.8.10), we have

$$J_{ij} = \sum_k \epsilon_{ijk}\Big(\mathbf{A}_{+k} + \mathbf{A}_{-k}\Big), \quad J_{k4} = \mathbf{A}_{+k} - \mathbf{A}_{-k}. \tag{4.8.19}$$

The states of the hydrogen atom with a given energy can thus be classified according to their transformation under the four-dimensional rotation group.

The condition that $a_+ = a_-$ limits these states to those transforming as four-dimensional symmetric traceless tensors. The number of independent components of a symmetric tensor of rank $r$ in four dimensions is $(3+r)!/3!r!$, while the condition of tracelessness for $r \geq 2$ requires the vanishing of a symmetric tensor with $r-2$ indices and hence with $(1+r)!/3!(r-2)!$ independent components, so the number of independent components of a symmetric traceless tensor in four dimensions is

$$\frac{(3+r)!}{3!r!} - \frac{(1+r)!}{3!(r-2)!} = (r+1)^2,$$

which is the degeneracy found earlier if we identify the states with principal quantum number $n$ as transforming like a four-dimensional symmetric traceless tensor of rank $r = n - 1$. For instance, the $n = 1$ state transforms as a four-dimensional scalar; the $n = 2$ states transform as the components of a four-dimensional vector $v_\alpha$, of which $v_i$ are the three $p$ states and $v_4$ is the $s$ state; and the $n = 3$ states transform as the components of a symmetric traceless tensor $t_{\alpha\beta}$, of which the components of the traceless part of $t_{ij}$ make up the five $d$ states, the components $t_{i4} = t_{4i}$ are the three $p$ states, and $\sum_i t_{ii} = -t_{44}$ is the one $s$ state. The relations between matrix elements of operators between states of given energy but different values of $\ell$ can be found using invariance under four-dimensional rotations, if we know the transformation properties of the operators under such rotations.

## 4.9 The Rigid Rotator

We will now take up the example of a system in which the positions of all particles are fixed, except that the whole system can rotate freely around any axis. This is not literally the case for any real system, but it is a good approximation for molecules that are subject only to excitations of very low energy. The energy required to excite the electrons in a molecule to a higher state is of the same order as for atoms, roughly $e^4 m_e / \hbar^2$, and we will see in Section 5.6 that the energy required to excite vibrations of the nuclear positions in a molecule is smaller, roughly $(m_e/m_N)^{1/2} \times e^4 m_e / \hbar^2$, where $m_N$ is a typical nuclear mass. As will be found in this section, the energy required to excite the rotational modes of a molecule is smaller still, roughly $(m_e/m_N) \times e^4 m_e / \hbar^2$. Therefore we can work out the rotational spectra of molecules by treating the positions of nuclei as if they were fixed at the minima of a potential calculated from a fixed electronic wave function.

First, let us recall the treatment of rigid rotators in classical physics. We suppose that the particles of a rigid body have positions

$$x_{ni}(t) = \sum_a R_{ia}(t) x_{na}^0, \qquad (4.9.1)$$

where $n$ labels individual particles; $i$ is a coordinate index running over the values 1, 2, 3, defined by coordinate axes fixed in the laboratory; $a$ is a coordinate index running over the values $x$, $y$, $z$, defined by coordinate axes fixed in the body; $x_{na}^0$ are a set of time-independent particle coordinates in the coordinate system fixed in the body; and $R_{ia}(t)$ is the only dynamical variable, a time-dependent rotation satisfying the usual conditions (4.1.2) for a rotation:

$$(R^{\mathrm{T}}R)_{ba} = \sum_i R_{ib}(t)R_{ia}(t) = \delta_{ab}, \tag{4.9.2}$$

from which it also follows that

$$(RR^{\mathrm{T}})_{ij} = \sum_a R_{ia}(t)R_{ja}(t) = \delta_{ij}. \tag{4.9.3}$$

The energy of rotation of this system is then given by

$$H = \frac{1}{2}\sum_{ni} m_n \dot{x}_{ni}^2 = \frac{1}{2}\sum_{niab} m_n \dot{R}_{ia}\dot{R}_{ib}x_{na}^0 x_{nb}^0, \tag{4.9.4}$$

where $m_n$ is the mass of the $n$th particle. It is convenient to introduce a constant matrix

$$N_{ab} \equiv \sum_n m_n x_{na}^0 x_{nb}^0, \tag{4.9.5}$$

so that Eq. (4.9.4) can be written

$$H = \frac{1}{2}\sum_{iab} \dot{R}_{ia}\dot{R}_{ib}N_{ab} = \frac{1}{2}\mathrm{Tr}\left(\dot{R}N\dot{R}^{\mathrm{T}}\right). \tag{4.9.6}$$

Because $R$ satisfies the condition $R^{\mathrm{T}}R = 1$, its time derivative satisfies $\dot{R}^{\mathrm{T}}R + R^{\mathrm{T}}\dot{R} = 0$, so that $\dot{R}^{\mathrm{T}}R$ is antisymmetric, and can therefore be written

$$(\dot{R}^{\mathrm{T}}R)_{ab} = \sum_i \dot{R}_{ia}R_{ib} = \sum_c \epsilon_{abc}\Omega_c, \tag{4.9.7}$$

for some $\Omega_c$. (For rotation around a fixed axis, $\Omega_c$ is in the direction of that axis, and its magnitude is the rate of rotation.) Together with Eq. (4.9.3), this gives a formula for $\dot{R}$:

$$\dot{R}_{ia} = \sum_{cd} R_{ic}\epsilon_{acd}\Omega_d. \tag{4.9.8}$$

We can use this to write the rotational energy (4.9.6) as

$$\begin{aligned} H &= \frac{1}{2}\sum_{iabcdef} R_{ic}R_{ie}\epsilon_{acd}\epsilon_{bef}\Omega_d\Omega_f N_{ab} \\ &= \frac{1}{2}\sum_{abcdf} \epsilon_{acd}\epsilon_{bcf}\Omega_d\Omega_f N_{ab}. \end{aligned}$$

This can be further simplified by using the identity

$$\sum_c \epsilon_{acd}\epsilon_{bcf} = \delta_{ab}\delta_{df} - \delta_{af}\delta_{bd}, \tag{4.9.9}$$

which gives

$$H = \frac{1}{2}\left(\sum_a \Omega_a^2 \operatorname{Tr} N - \sum_{ab} \Omega_a \Omega_b N_{ab}\right). \tag{4.9.10}$$

For this reason, we introduce a *moment-of-inertia tensor*

$$I_{ab} \equiv \delta_{ab} \operatorname{Tr} N - N_{ab}, \tag{4.9.11}$$

and write the rotational energy as

$$H = \frac{1}{2}\sum_{ab} \Omega_a \Omega_b I_{ab}. \tag{4.9.12}$$

The rotational energy (4.9.12) can also be expressed in terms of an angular-momentum vector. The components of the angular momentum in a coordinate system fixed in the laboratory are defined by

$$J_i \equiv \sum_{njk} \epsilon_{ijk} x_{nj} \dot{x}_{nk} m_n. \tag{4.9.13}$$

Using Eqs. (4.9.1), (4.9.5), and (4.9.8), this is

$$J_i = \sum_{jkab} \epsilon_{ijk} R_{ja} \dot{R}_{kb} N_{ab} = \sum_{jkab} \epsilon_{ijk}\epsilon_{bcd} R_{ja} R_{kc} \Omega_d N_{ab}.$$

We get a simpler formula for the components $\mathcal{J}_e$ of angular momentum along axes fixed in the rotating system:

$$\mathcal{J}_e \equiv \sum_i R_{ie} J_i. \tag{4.9.14}$$

The sum $\sum_{ijk} \epsilon_{ijk} R_{ie} R_{ja} R_{kc}$ is totally antisymmetric in $e, a, c$, and therefore proportional to $\epsilon_{eac}$. The proportionality constant is just the determinant of $R$, which for rotations (as distinct from inversions) is unity, so

$$\sum_{ijk} \epsilon_{ijk} R_{ie} R_{ja} R_{kc} = \epsilon_{eac}. \tag{4.9.15}$$

Using the identity (4.9.9) again, this gives

$$\mathcal{J}_a = \sum_b I_{ab} \Omega_b. \tag{4.9.16}$$

In the generic case $I_{ab}$ has an inverse, and the rotational energy (4.9.12) may be written

$$H = \frac{1}{2} \sum_{ab} \mathcal{J}_a \mathcal{J}_b I_{ab}^{-1}. \tag{4.9.17}$$

Since $I_{ab}$ is a symmetric real matrix, we can find a basis in which it is diagonal, say with components $I_x$, $I_y$, $I_z$ on the main diagonal, in which case Eq. (4.9.17) takes the form

$$H = \frac{1}{2I_x} \mathcal{J}_x^2 + \frac{1}{2I_y} \mathcal{J}_y^2 + \frac{1}{2I_z} \mathcal{J}_z^2. \tag{4.9.18}$$

We will come back at the end of this section to the special case where one of the eigenvalues of $I_{ab}$ vanishes.

In making the transition to quantum mechanics, we introduce a set of Hermitian operators $\hat{R}_{ia}$, whose eigenvalues are the components $R_{ia}$ of specific rotations $R$. (This is analogous to introducing a position operator for point particles, whose eigenvalues are specific positions. In this section we will install hats over symbols to indicate that they are operators, not c-numbers.) All these components commute with one another (but not with their time derivatives), and satisfy the constraints (4.9.2) and (4.9.3). The operators $\hat{x}_{ni}$ representing the positions of individual particles are given by the quantum version of Eq. (4.9.1):

$$\hat{x}_{ni}(t) = \sum_a \hat{R}_{ia}(t) x_{na}^0, \tag{4.9.19}$$

where for a truly rigid rotator the $x_{na}^0$ are fixed c-numbers. (For a molecule the $x_{na}^0$ are operators, but the tensors $N_{ab}$ and $I_{ab}$ are still c-numbers, calculated by taking the expectation value of the sum in Eq. (4.9.5) in a given electronic and vibrational state of the molecule.) As usual, we can define an angular-momentum operator

$$\hat{J}_i \equiv \sum_{njk} \epsilon_{ijk} \hat{x}_{nj} \dot{\hat{x}}_{nk} m_n, \tag{4.9.20}$$

with the usual commutation relations

$$[\hat{J}_i, \hat{J}_i] = i\hbar \sum_k \epsilon_{ijk} \hat{J}_k. \tag{4.9.21}$$

We can again define angular-momentum components in a basis fixed in the rotator:

$$\hat{\mathcal{J}}_a = \sum_i \hat{R}_{ia} \hat{J}_i. \tag{4.9.22}$$

Following the same reasoning as in the classical case, we can write the Hamiltonian operator as the analog of Eq. (4.9.17):

$$\hat{H} = \frac{1}{2} \sum_{ab} \hat{\mathcal{J}}_a \hat{\mathcal{J}}_b I_{ab}^{-1}. \tag{4.9.23}$$

To find the energy eigenvalues, we need the commutation relations of the operators $\hat{\mathcal{J}}_a$. We note first that under rotations of the laboratory coordinate axes, the operators $\hat{R}_{ia}$ transform not as a tensor, but as three three-vectors:

$$[\hat{J}_i, \hat{R}_{ja}] = i\hbar \sum_k \epsilon_{ijk} \hat{R}_{ka}. \qquad (4.9.24)$$

(This incidentally shows why we did not have to worry about operator-ordering in the definition (4.9.22); $\hat{J}_i$ commutes with $\hat{R}_{ja}$ in the case $i = j$.) It follows from Eqs. (4.9.21) and (4.9.24) that $\hat{\mathcal{J}}_a$ is a rotational scalar, in the sense that

$$[\hat{J}_i, \hat{\mathcal{J}}_a] = 0. \qquad (4.9.25)$$

Hence

$$[\hat{\mathcal{J}}_a, \hat{\mathcal{J}}_b] = \sum_j [\hat{\mathcal{J}}_a, \hat{R}_{jb}]\hat{J}_j = \sum_{ij} \hat{R}_{ia}[\hat{J}_i, \hat{R}_{jb}]\hat{J}_j$$
$$= i\hbar \sum_{ijk} \epsilon_{ijk} \hat{R}_{ia} \hat{R}_{kb} \hat{J}_j.$$

According to the theory of determinants, for any $3 \times 3$ matrix $M$ with non-vanishing determinant we have

$$\sum_{ijk} \epsilon_{ikj} M_{ia} M_{kb} = \text{Det } M \sum_c \epsilon_{abc} M_{cj}^{-1},$$

so, for the unimodular orthogonal matrix $\hat{R}$ of commuting operators,

$$\sum_{ijk} \epsilon_{ijk} \hat{R}_{ia} \hat{R}_{kb} = -\sum_c \epsilon_{abc} \hat{R}_{jc},$$

the minus sign arising from the ratio of $\epsilon_{ijk}$ and $\epsilon_{ikj}$. It follows then that

$$[\hat{\mathcal{J}}_a, \hat{\mathcal{J}}_b] = -i\hbar \sum_c \epsilon_{abc} \hat{\mathcal{J}}_c. \qquad (4.9.26)$$

That is, the operators $-\hat{\mathcal{J}}_a$ satisfy the same commutation relations as ordinary angular-momentum operators. Also, because $\hat{R}_{ia}$ satisfies Eq. (4.9.2), the definition (4.9.22) gives

$$\sum_i \hat{J}_i^2 = \sum_a \hat{\mathcal{J}}_a^2. \qquad (4.9.27)$$

By following the reasoning of Section 4.2, we can find states $\Psi_J^{MK}$ that are eigenstates of both $\sum_i \hat{J}_i^2$ and $\sum_a \hat{\mathcal{J}}_a^2$ with equal eigenvalues $\hbar^2 J(J+1)$, where $J$ is a positive integer, and also eigenstates of both $\hat{J}_3$ and $\hat{\mathcal{J}}_z$, with eigenvalues respectively $\hbar M$ and $\hbar K$, where $M$ and $K$ both run independently by unit steps from $-J$ to $+J$. ($J$ is an integer, because in its definition (4.9.20) we are implicitly assuming that the rotator is composed of spinless particles, whose total orbital angular momentum is **J**.)

In the general case the states $\Psi_J^{MK}$ are not eigenstates of the Hamiltonian (4.9.23). Things are much simpler for the *symmetric rotator*, for which two of the eigenvalues of the moment-of-inertia tensor $I_{ab}$ are equal. In this case, by a choice of body-fixed basis vectors, we can take this tensor to have the form

$$I = \begin{pmatrix} I_x & 0 & 0 \\ 0 & I_x & 0 \\ 0 & 0 & I_z \end{pmatrix} \tag{4.9.28}$$

and the Hamiltonian (4.9.23) is

$$\hat{H} = \frac{1}{2I_x}\left(\hat{\mathcal{J}}_x^2 + \hat{\mathcal{J}}_y^2\right) + \frac{1}{2I_z}\hat{\mathcal{J}}_z^2 = \frac{1}{2I_x}\sum_a \hat{\mathcal{J}}_a^2 + \left(\frac{1}{2I_z} - \frac{1}{2I_x}\right)\hat{\mathcal{J}}_z^2. \tag{4.9.29}$$

Thus the states $\Psi_J^{MK}$ are eigenstates of the Hamiltonian for a symmetric rotator, with energy eigenvalues

$$E(JMK) = \frac{\hbar^2 J(J+1)}{2I_x} + \left(\frac{1}{2I_z} - \frac{1}{2I_x}\right)\hbar^2 K^2. \tag{4.9.30}$$

It is a consequence of rotational invariance that these energies are independent of $M$, so that each energy level has a $(2J+1)$-fold degeneracy.

There is no similar formula for the energy eigenvalues in the general case, where all eigenvalues of $I_{ab}$ are unequal, but it is always possible to calculate the energy eigenvalues for any given $J$ by purely algebraic means. Using a basis for which $I_{ab}$ is diagonal, the Hamiltonian operator is

$$\begin{aligned} \hat{H} &= \frac{1}{2I_x}\hat{\mathcal{J}}_x^2 + \frac{1}{2I_y}\hat{\mathcal{J}}_y^2 + \frac{1}{2I_z}\hat{\mathcal{J}}_z^2 \\ &= A(\hat{\mathcal{J}}_x^2 + \hat{\mathcal{J}}_y^2 + \hat{\mathcal{J}}_z^2) + B\hat{\mathcal{J}}_z^2 + C(\hat{\mathcal{J}}_x^2 - \hat{\mathcal{J}}_y^2), \end{aligned} \tag{4.9.31}$$

where

$$A = \frac{1}{4I_x} + \frac{1}{4I_y}, \quad B = \frac{1}{2I_z} - \frac{1}{4I_x} - \frac{1}{4I_y}, \quad C = \frac{1}{4I_x} - \frac{1}{4I_y}. \tag{4.9.32}$$

We also note that

$$\hat{\mathcal{J}}_x^2 - \hat{\mathcal{J}}_y^2 = \frac{1}{2}\left(\hat{\mathcal{J}}_x + i\hat{\mathcal{J}}_y\right)^2 + \frac{1}{2}\left(\hat{\mathcal{J}}_x - i\hat{\mathcal{J}}_y\right)^2.$$

Thus in general the energy eigenstates are mixtures of $\Psi_J^{MK}$ with fixed $J$ and $M$ but with various values of $K$ differing from each other by multiples of $\pm 2$. For instance, for the case $J = 1$, in a basis with rows and columns corresponding to $K = +1$, $K = 0$, and $K = -1$, the Hamiltonian (4.9.31) is

$$\hat{H} = \hbar^2 \begin{pmatrix} 2A + B & 0 & C \\ 0 & 2A & 0 \\ C & 0 & 2A + B \end{pmatrix}.$$

The $J = 1$ energy eigenvalues $E$ and corresponding eigenstates $\Psi$ are therefore

$$E = \begin{cases} 2A + B + C, & \Psi \propto \Psi_1^{M,+1} + \Psi_1^{M,-1}, \\ 2A, & \Psi \propto \Psi_1^{M,0}, \\ 2A + B - C, & \Psi \propto \Psi_1^{M,+1} - \Psi_1^{M,-1}. \end{cases}$$

We don't need to know wave functions to calculate energy eigenvalues for the rigid rotator, but wave functions are needed for other purposes, such as the calculations of electromagnetic transition amplitudes. We will calculate the wave functions for the states $\Psi_J^{M,K}$ (whether or not these states are energy eigenstates) in a basis of states $\Phi_R^K$, defined as eigenstates of both the rotation operator $\hat{R}$ and the rotational invariant $\hat{\mathcal{J}}_z$:

$$\hat{R}_{ia}\Phi_R^K = R_{ia}\Phi_R^K, \qquad \hat{\mathcal{J}}_z\Phi_R^K = K\Phi_R^K. \tag{4.9.33}$$

It is convenient at this point to return to the formalism of Section 4.1, and for each c-number rotation $R'$ introduce a unitary operator $U(R')$ satisfying the composition law (4.1.3), which acts on any three-vector operator as in Eq. (4.1.4). In particular,

$$U^{-1}(R')\hat{R}_{ia}U(R') = \sum_j R'_{ij}\hat{R}_{ja}, \tag{4.9.34}$$

so $U(R')\Phi_R^K$ is an eigenstate of $\hat{R}_{ia}$ with eigenvalue $(R'R)_{ia}$. In particular, if we define $\Phi_1^K$ to be an eigenstate of $\hat{R}_{ia}$ with eigenvalue $\delta_{ia}$, then we can take the general eigenstate as

$$\Phi_R^K = U(R)\Phi_1^K. \tag{4.9.35}$$

Thus in this basis the wave function of the state $\Psi_J^{M,K}$ is

$$\left(\Phi_R^K, \Psi_J^{M,K}\right) = \left(\Phi_1^K, U(R^{-1})\Psi_J^{M,K}\right) = \sum_{M'} D_{M'M}^J(R^{-1})\left(\Phi_1^K, \Psi_J^{M',K}\right), \tag{4.9.36}$$

where $D_{M'M}^J(R)$ are unitary matrices[38] representing the three-dimensional rotation group, in the sense that $D(R_1)D(R_2) = D(R_1R_2)$, defined here by

---

[38] The form of these matrices of course depends on the variables chosen to parameterize rotations. For the usual case, where rotations are parameterized by Euler angles, the matrices $D_{M'M}^J(R)$ are given by numerous authors, including A. R. Edmonds, *Angular Momentum in Quantum Mechanics* (Princeton University Press, Princeton, 1957), Chapter 4; M. E. Rose, *Elementary Theory of Angular Momentum* (John Wiley & Sons, New York, 1957), Chapter IV; L. D. Landau and E. M. Lifshitz, *Quantum Mechanics – Non-Relativistic Theory*, 3rd edn. (Pergamon Press, Oxford, 1977), Section 58; Wu-Ki Tung, *Group Theory in Physics* (World Scientific, Singapore, 1985), Sections 7.3 and 8.1. We will not need explicit formulas for these matrices in what follows.

$$U(R)\Psi_J^{M,K} = \sum_{M'} D_{M'M}^J(R)\Psi_J^{M',K}. \tag{4.9.37}$$

We still need to say something about the $R$-independent coefficients $\left(\Phi_1^K, \Psi_J^{M',K}\right)$ in Eq. (4.9.36). For this purpose, we note that

$$K\left(\Phi_1^K, \Psi_J^{M',K}\right) = \left(\Phi_1^K, \hat{\mathcal{J}}_3\Psi_J^{M',K}\right) = \left(\Phi_1^K, \sum_i \hat{R}_{i3}\hat{J}_i \Psi_J^{M',K}\right).$$

Acting to the left on the state $\Phi_1^K$, the Hermitian operator $\hat{R}_{i3}$ gives a factor $\delta_{i3}$, so

$$K\left(\Phi_1^K, \Psi_J^{M',K}\right) = \left(\Phi_1^K, \hat{J}_3\Psi_J^{M',K}\right) = M'\left(\Phi_1^K, \Psi_J^{M',K}\right),$$

and therefore this matrix element vanishes unless $M' = K$:

$$\left(\Phi_1^K, \Psi_J^{M',K}\right) = c_K^J \delta_{M'K}. \tag{4.9.38}$$

Using this in Eq. (4.9.36), we find the wave function[39]

$$\left(\Phi_R^K, \Psi_J^{M,K}\right) = c_K^J D_{KM}^J(R^{-1}). \tag{4.9.39}$$

The constant factor $c_K^J$ can be found (up to an arbitrary phase) from the requirement that the wave function should be properly normalized.

We can now take up the special case in which one of the eigenvalues of $I_{ab}$ vanishes. If the eigenvalues of the matrix $N_{ab}$ defined in Eq. (4.9.5) are $N_x$, $N_y$, and $N_z$, then the eigenvalues of the moment of inertia tensor $I_{ab}$ are $N_y + N_z$, $N_z + N_x$, and $N_x + N_y$. All the $N_a$ are positive, so unless $I_{ab}$ vanishes altogether, at most one of its eigenvalues can vanish, and then only in the case where two of the $N_a$ vanish. If we choose our coordinate axes so that $N_x = N_y = 0$, then the eigenvalues of $I_{ab}$ are $I_x = I_y = N_z$ and $I_z = 0$. This is necessarily the case for a linear rotator, such as a diatomic molecule, lying along the $z$-axis, with no extension in the $x$ and $y$ directions. We have here a special case of the symmetric rotator treated earlier, whose energies are given by Eq. (4.9.30). In order to avoid infinite energies for $I_z = 0$ (or very large energies for very small $I_z$) it is necessary to consider only states with $K = 0$, for which the energies (4.9.30) are

$$E(JM0) = \frac{\hbar^2 J(J+1)}{2I_x}, \tag{4.9.40}$$

---

[39] This is the answer obtained in typical textbook treatments, such as that of L. D. Landau and E. M. Lifshitz, *Quantum Mechanics – Non-Relativistic Theory*, 3rd edn. (Pergamon Press, Oxford, 1977), Section 103, except that usually the argument of $D_{KM}^J$ is given as $R$, instead of $R^{-1}$, indicating that (perhaps to take account of the difference between rotating the system and rotating the coordinate axes) their wave functions are calculated in the basis $\Phi_{R^{-1}}^K$ rather than $\Phi_R^K$. Like most authors, Landau and Lifshitz do not specify the basis for their wave functions. Of course, wave functions can be defined in any basis we like.

and the corresponding wave functions (4.9.39) are

$$\left(\Phi_R^0, \Psi_\ell^{m,0}\right) = c_0^\ell D_{0m}^\ell (R^{-1}). \tag{4.9.41}$$

(Since $K = 0$ is an integer, $J$ and $M$ must also be integers, which are now accordingly denoted $\ell$ and $m$.) In this case the function $D_{0m}^\ell(R^{-1})$ is just proportional to an ordinary spherical harmonic:

$$D_{0m}^\ell (R^{-1}) = i^{-\ell} \sqrt{\frac{4\pi}{2\ell + 1}} Y_\ell^m(\hat{n}), \tag{4.9.42}$$

where $\hat{n}$ is the direction into which the rotation $R^{-1}$ takes the 3-axis. Since $Y_\ell^m$ is a properly normalized wave function, here we have $c_0^\ell = \sqrt{(2\ell + 1)/4\pi}$ and the rotator wave function is simply $i^{-\ell} Y_\ell^m(\hat{n})$, where here $\hat{n}$ is the direction in the laboratory frame of the $z$-axis of the rotator.

There are important limitations on the values of $\ell$ in diatomic molecules in which the two nuclei are identical. If the spins of the individual nuclei are $s'$, and these spins add up to a total spin $s$, then according to Eq. (4.3.34) (with $s'$ in place of both $j'$ and $j''$ and $s$ in place of $j$), the interchange of the two nuclear spins changes the spin wave function by a sign $(-1)^{s-2s'}$. Also, Eqs. (4.9.42) and (2.2.17) show that this interchange multiplies the orbital part of the wave function by a factor $(-1)^\ell$. But the nuclei are bosons or fermions depending on whether $2s'$ is even or odd, so the interchange of the two nuclei must change the complete wave function by a factor $(-1)^{2s'}$. Therefore we must have

$$(-1)^{s-2s'} \times (-1)^\ell = (-1)^{2s'},$$

and therefore $(-1)^\ell = (-1)^s$. Thus $\ell$ is limited to even or odd values, depending on whether the total nuclear spin is even or odd. In these two cases, the molecules are distinguished by the prefix *para* or *ortho*, respectively. For instance, in parahydrogen the total nuclear spin is $s = 0$ and $\ell$ is even, while in orthohydrogen we have $s = 1$ and $\ell$ is odd. The nucleus of deuterium has spin $s' = 1$, so deuterium molecules can be either paradeuterium, with total nuclear spin either $s = 0$ or $s = 2$ and $\ell$ even, or orthodeuterium, with total nuclear spin $s = 1$ and $\ell$ odd.

The ground state is always the *para* state, but at room temperature the energy difference between rotational levels is generally less than $k_B T$, and all of the $2s + 1$ individual *ortho* and *para* spin states are equally abundant. For instance, in hydrogen gas at room temperature there are about three orthohydrogen molecules for every parahydrogen molecule.

Finally, let's consider the order of magnitude of molecular rotational energies $E_{\mathrm{rot}}$. It is clear from Eq. (4.9.18) that in general these are of order $\hbar^2/m_N a^2$, where $m_N$ is a typical nuclear mass, and $a$ is a typical molecular dimension. At least for simple molecules, $a$ is of the same order as atomic sizes, $a \approx \hbar^2/m_e e^2$, so

$$E_{\text{rot}} \approx \frac{\hbar^2}{m_N} \left( \frac{m_e e^2}{\hbar^2} \right)^2 = \frac{m_e^2 e^4}{m_N \hbar^2},$$

which as noted earlier is less than typical electronic energies $m_e e^4 / \hbar^2$ by a factor of order $m_e / m_N$. For instance, if we take $m_N = 10 m_p$ then $E_{\text{rot}}$ is of order $10^{-3}$ eV. As a check, note that the rotational energies of the cyanogen molecule CN (whose excitation in interstellar space gave the first hint of a 3 K cosmic radiation background) are accurately given by Eq. (4.9.40), with $\hbar^2 / 2I_x = 2.35 \times 10^{-4}$ eV, in fair agreement with our crude estimate.

## Problems

1. Suppose that an electron is in a state of orbital angular momentum $\ell = 2$. Show how to construct the state vectors with total angular momentum $j = 5/2$ and corresponding 3-components $m = 5/2$ and $m = 3/2$ as linear combinations of state vectors with definite values of $S_3$ and $L_3$. Then find the state vector with $j = 3/2$ and $m = 3/2$. (All state vectors here should be properly normalized.) Summarize your results by giving values for the Clebsch–Gordan coefficients $C_{\frac{1}{2} 2}(jm; m_s m_\ell)$ in the cases $(j, m) = (5/2, 5/2)$, $(5/2, 3/2)$, and $(3/2, 3/2)$.

2. Suppose that $\mathbf{A}$ and $\mathbf{B}$ are vector operators, in the sense that

$$[J_i, A_j] = i\hbar \sum_k \epsilon_{ijk} A_k, \qquad [J_i, B_j] = i\hbar \sum_k \epsilon_{ijk} B_k.$$

Show that the cross-product $\mathbf{A} \times \mathbf{B}$ is a vector in the same sense.

3. What is the minimum value of the total angular momentum $\mathbf{J}^2$ that a state must have in order to have a non-zero expectation value for an operator $\mathcal{O}_j^m$ of spin $j$?

4. The Hamiltonian for a free particle of mass $M$ and spin $\mathbf{S}$ placed in a magnetic field $\mathbf{B}$ in the 3-direction is

$$H = \frac{\mathbf{p}^2}{2M} - g|\mathbf{B}|S_3,$$

where $g$ is a constant (proportional to the particle's magnetic moment). Give the equations that govern the time-dependence of the expectation values of all three components of $\mathbf{S}$.

5. A particle of spin $3/2$ decays into a nucleon and pion. Assume that parity is conserved in this decay. Show how the angular distribution in the final state (with spins not measured) can be used to determine the parity of the decaying particle.

6. A particle X of isospin 1 and charge zero decays into a K and a $\overline{\text{K}}$. Assume that isospin is conserved in this decay. What is the ratio of the rates of the processes $X^0 \to K^+ + \overline{K}^-$ and $X^0 \to K^0 + \overline{K}^0$?

7. Imagine that the electron has spin 3/2 instead of 1/2, but assume that the one-particle states with definite values of $n$ and $\ell$ in atoms are filled, as the atomic number increases, in the same order as in the real world. What elements with atomic numbers in the range from 1 to 21 would have chemical properties similar to those of noble gases, alkali metals, halogens, and alkali earths in the real world?

8. What is the commutator of the angular-momentum operator $\mathbf{J}$ with the generator $\mathbf{K}$ of Galilean transformations?

9. Consider an electron in a state of zero orbital angular momentum in an atom whose nucleus has spin (that is, internal angular momentum) 3/2. Express the states of the atom with total angular-momentum $z$-component $m = 1$ (of electron plus nucleus) and each possible definite value of the total angular momentum as linear combinations of states with definite values of the $z$-components of the nuclear and electron spins.

# 5

# Approximations for Energy Eigenvalues

Courses on quantum mechanics generally begin with the same time-honored examples: the free particle, the Coulomb potential and the harmonic oscillator potential, covered here in Chapter 2. This is because these are almost the only cases for which the Schrödinger equation for states of definite energy has a known exact solution. In the real world, problems are more complicated, and we have to rely on approximation schemes. Indeed, even if we could find exact solutions for complicated problems the solutions themselves would necessarily be complicated, and we would need to make approximations to understand the physical consequences of the solutions.

## 5.1 First-Order Perturbation Theory

The most widely useful approach to finding approximate solutions to complicated problems is perturbation theory. In this method one starts with a simpler problem, which can be exactly solved, and then treats the corrections to the Hamiltonian as small perturbations.

Consider an unperturbed Hamiltonian $H_0$, like that of the hydrogen atom treated in Section 2.3, which is simple enough that we can find its energy values $E_a$ and corresponding orthonormal state vectors $\Psi_a$:

$$H_0 \Psi_a = E_a \Psi_a, \tag{5.1.1}$$

$$\left( \Psi_a, \Psi_b \right) = \delta_{ab}. \tag{5.1.2}$$

Suppose we add a small term $\delta H$ to the Hamiltonian, proportional to some tiny parameter $\epsilon$. (For instance, in the case of the hydrogen atom $H_0$ was the kinetic energy operator plus a potential proportional to $1/r$, and we might take $\delta H = \epsilon U(\mathbf{x})$, where $U(\mathbf{x})$ is an arbitrary $\epsilon$-independent function of the position operator $\mathbf{x}$, representing perhaps a departure from the $1/r$ Coulomb potential due to the finite size of the proton.) The energy values then become $E_a + \delta E_a$, with corresponding state vectors $\Psi_a + \delta \Psi_a$, where $\delta E_a$ and $\delta \Psi_a$ are presumably given by power series in $\epsilon$:

$$\delta E_a = \delta_1 E_a + \delta_2 E_a + \cdots, \qquad \delta \Psi_a = \delta_1 \Psi_a + \delta_2 \Psi_a + \cdots, \qquad (5.1.3)$$

with $\delta_N E_a$ and $\delta_N \Psi_a$ proportional to $\epsilon^N$. The Schrödinger equation takes the form

$$\left( H_0 + \delta H \right) \left( \Psi_a + \delta \Psi_a \right) = \left( E_a + \delta E_a \right) \left( \Psi_a + \delta \Psi_a \right). \qquad (5.1.4)$$

To collect the terms of first order in $\epsilon$, we can drop the terms $\delta H \, \delta \Psi_a$ and $\delta E_a \, \delta \Psi_a$ in Eq. (5.1.4), whose power series start with terms of order $\epsilon^2$. We then have

$$\delta H \, \Psi_a + H_0 \, \delta_1 \Psi_a = \delta_1 E_a \, \Psi_a + E_a \, \delta_1 \Psi_a. \qquad (5.1.5)$$

To find $\delta_1 E_a$, we take the scalar product of Eq. (5.1.5) with $\Psi_a$. Because $H_0$ is Hermitian, we have

$$\left( \Psi_a, \, H_0 \, \delta_1 \Psi_a \right) = E_a \left( \Psi_a, \, \delta_1 \Psi_a \right)$$

so these terms in the scalar product cancel, and we are left with

$$\delta_1 E_a = \left( \Psi_a, \, \delta H \, \Psi_a \right). \qquad (5.1.6)$$

This is the first major result of perturbation theory: *to first order, the shift in the energy of a bound state is the expectation value in the unperturbed state of the perturbation $\delta H$.*

But this argument does not always work, even when $\delta H$ is very small. To see what may go wrong, let us calculate the change in the state vector produced by the perturbation. This time, we take the scalar product of Eq. (5.1.5) with a general unperturbed energy eigenvector $\Psi_b$. Again using the fact that $H_0$ is Hermitian, this gives

$$\left( \Psi_b, \, \delta H \, \Psi_a \right) = \delta_1 E_a \, \delta_{ab} + \left( E_a - E_b \right) \left( \Psi_b, \, \delta_1 \Psi_a \right). \qquad (5.1.7)$$

For $a = b$, this is the same as Eq. (5.1.6), so the new information is that

$$\left( \Psi_b, \, \delta H \, \Psi_a \right) = (E_a - E_b) \left( \Psi_b, \, \delta_1 \Psi_a \right) \qquad \text{for} \quad a \neq b. \qquad (5.1.8)$$

A problem arises in the case of degeneracy. Suppose there are two states $\Psi_b \neq \Psi_a$ for which $E_b = E_a$. Then Eq. (5.1.8) is inconsistent unless $\left( \Psi_b, \, \delta H \, \Psi_a \right)$ vanishes, which need not be the case. But we can always avoid this problem by a judicious choice of the degenerate unperturbed states. Suppose there are a number of states $\Psi_{a1}$, $\Psi_{a2}$, etc., all with the same energy $E_a$. The quantities $\left( \Psi_{ar}, \, \delta H \, \Psi_{as} \right)$ form an Hermitian matrix, so according to a general theorem of matrix algebra the vector space on which this matrix acts is spanned by a set of orthonormal eigenvectors $u_{rn}$ of this matrix, such that

$$\sum_r \left( \Psi_{as}, \, \delta H \, \Psi_{ar} \right) u_{rn} = \Delta_n u_{sn}. \qquad (5.1.9)$$

(See footnote 7 in Section 3.3.) We can define eigenstates of $H_0$ with the same energy $E_a$:

$$\Phi_{an} \equiv \sum_r u_{rn} \Psi_{ar}, \qquad (5.1.10)$$

for which

$$\left( \Phi_{am}, \delta H \, \Phi_{an} \right) = \sum_{rs} u_{sm}^* u_{rn} \left( \Psi_{as}, \delta H \Psi_{ar} \right) = \sum_s u_{sm}^* u_{sn} \Delta_n$$

$$= \delta_{nm} \Delta_n, \qquad (5.1.11)$$

in which we have used the orthonormality relation $\sum_s u_{sm}^* u_{sn} = \delta_{nm}$. For these states the off-diagonal matrix elements of the perturbation all vanish, so we avoid the problem of inconsistency with Eq. (5.1.8) if we start with the $\Phi$s instead of the $\Psi$s.

If we stubbornly insist on taking one of the $\Psi_{ar}$ as our unperturbed state, where some $\left( \Psi_{as}, \delta H \, \Psi_{ar} \right)$ for $s \neq r$ do not vanish, then perturbation theory doesn't work; even a tiny perturbation causes a very large change in the state vector. For instance, suppose that $H_0$ is rotationally invariant, and we add a perturbation $\delta H = \epsilon \cdot \mathbf{v}$, where $\mathbf{v}$ is some vector operator. As we saw in the previous chapter, because $H_0$ is rotationally invariant, there are $2j+1$ states with the same unperturbed energy and the same eigenvalue $\hbar^2 j(j+1)$ of $\mathbf{J}^2$. If our unperturbed state is an eigenstate of $J_3$, but $\epsilon$ is not in the 3-direction, then no matter how small $\epsilon$ is, there will be a large correction to the state vector. The perturbation forces the state into an eigenstate of $\mathbf{J} \cdot \epsilon$. But if we take the unperturbed states to be eigenstates of $\mathbf{J} \cdot \epsilon$ to begin with, then since $\delta H$ commutes with $\mathbf{J} \cdot \epsilon$ the change in the state vector will be of order $\epsilon$.

The condition that $(\Psi_a, \delta H \, \Psi_b)$ vanishes for all states with $E_a = E_b$ and $a \neq b$ determines the unperturbed states $\Psi_a$ uniquely in the case that all of the corresponding first-order energy perturbations $\delta_1 E_a = (\Psi_a, \delta H \, \Psi_a)$ are unequal. But if there are a number of different unperturbed states that all have the same zeroth-order energies *and* the same first-order energies, then any orthonormal linear combinations of these states will have the same properties and so can be taken as the unperturbed states. (This case typically arises when some symmetry requires that all matrix elements of $\delta H$ between states with a given unperturbed energy vanish.) We will see in Section 5.4 that this remaining freedom in the unperturbed state vectors is typically removed by imposing the condition that *second*-order perturbations do not produce a large change in the energy eigenvectors.

Next, let's calculate the perturbations to the state vectors. We will first consider the case of no degeneracy; that is, where the states whose energies and wave functions we want to calculate do not have the same unperturbed energies as each other or any other states. Here Eq. (5.1.8) gives immediately

$$\left(\Psi_b, \delta_1\Psi_a\right) = \frac{\left(\Psi_b, \delta H\, \Psi_a\right)}{E_a - E_b} \qquad \text{for} \quad a \neq b. \qquad (5.1.12)$$

To find the component of $\delta_1\Psi_a$ along $\Psi_a$, we need to impose the condition that $\Psi_a + \delta\Psi_a$ is properly normalized. This gives

$$1 = \left(\Psi_a + \delta\Psi_a, \Psi_a + \delta\Psi_a\right) = 1 + \left(\Psi_a, \delta_1\Psi_a\right) + \left(\delta_1\Psi_a, \Psi_a\right) + O(\epsilon^2),$$

so, to order $\epsilon$,

$$0 = \text{Re}\left(\Psi_a, \delta_1\Psi_a\right). \qquad (5.1.13)$$

We are free to choose the imaginary part of $\left(\Psi_a, \delta_1\Psi_a\right)$ to be anything we like, as this just represents a choice of phase of the whole state vector. That is, multiplying the state vector $\Psi_a$ by a phase factor $\exp(i\delta\varphi_a)$, with $\delta\varphi_a$ an arbitrary real constant of order $\epsilon$, produces a change in $\delta_1\Psi_a$ equal to $i\,\delta\varphi_a\,\Psi_a$, which changes $\left(\Psi_a, \delta_1\Psi_a\right)$ by an amount $i\,\delta\varphi_a$. So in particular, we can choose $\left(\Psi_a, \delta_1\Psi_a\right)$ to be real, in which case the normalization condition (5.1.13) becomes

$$0 = \left(\Psi_a, \delta_1\Psi_a\right). \qquad (5.1.14)$$

With Eq. (5.1.12), the completeness of the state vectors with all definite values of $H_0$ tells us that

$$\delta_1\Psi_a = \sum_b \left(\Psi_b, \delta_1\Psi_a\right)\Psi_b = \sum_{b\neq a} \Psi_b \frac{\left(\Psi_b, \delta H\, \Psi_a\right)}{E_a - E_b}. \qquad (5.1.15)$$

Next, let us consider the more complicated degenerate case, in which the states we are interested in have the same unperturbed energies as some other states. Equation (5.1.8) now tells us nothing whatever about the components of $\delta_1\Psi_a$ along unperturbed state vectors $\Psi_b$ for which $E_b = E_a$, and Eq. (5.1.12) only applies for $E_a \neq E_b$. Hence, in place of Eq. (5.1.15), we only know that

$$\delta_1\Psi_a = \sum_{c:\,E_c\neq E_a} \Psi_c \frac{\left(\Psi_c, \delta H\, \Psi_a\right)}{E_a - E_c} + \sum_{b:\,E_b = E_a} \Psi_b\left(\Psi_b, \delta_1\Psi_a\right). \qquad (5.1.16)$$

What about normalization? We can impose on the perturbed degenerate states the condition that they are orthonormal,

$$\left(\Psi_b + \delta_1\Psi_b + O(\epsilon^2),\ \Psi_a + \delta_1\Psi_a + O(\epsilon^2)\right) = \delta_{ab} \qquad \text{for} \quad E_a = E_b.$$

The terms of zeroth order in $\epsilon$ on both sides of the equation are equal, so the terms on the left of first order in $\epsilon$ must then vanish:

$$\left(\Psi_b, \delta_1\Psi_a\right) + \left(\delta_1\Psi_b, \Psi_a\right) = 0 \qquad \text{for} \quad E_a = E_b. \qquad (5.1.17)$$

That is, the Hermitian part of the matrix $\left(\Psi_b, \delta_1\Psi_a\right)$ must vanish, so that for $E_a = E_b$ we have

$$\left(\Psi_b, \delta_1\Psi_a\right) = A_{ba}, \qquad (5.1.18)$$

where $A_{ba}$ is anti-Hermitian: that is, $A_{ba} = -A_{ab}^*$. Neither the first-order Schrödinger equation (5.1.5) nor the orthonormalization condition (5.1.16) tells us anything further about the matrix $A_{ab}$.

The undetermined anti-Hermitian matrix $A_{ab}$ found in the degenerate case is a little like the undetermined phase factor $\exp(i\varphi_a)$ in the state $\Psi_a + \delta_1\Psi_a$ in the non-degenerate case. But there is a large difference. The phase factor in the non-degenerate case can be chosen to be anything we like, and in particular can be chosen to give the convenient result (5.1.14). In contrast, as we will see in Section 5.4, in the degenerate case we need to hold on to our freedom to choose $A_{ab}$ to prevent *second*-order perturbations from introducing large shifts in the *first*-order state vectors. That is, just as we had to choose the degenerate unperturbed state vectors $\Psi_a$ to make $(\Psi_b, \delta_1 H \Psi_a)$ vanish for $E_b = E_a$ and $b \neq a$ in order to allow a smooth transition to the perturbed state vectors in first order, so in Section 5.4 we will have to make a specific choice of $A_{ab}$ and hence of the first-order perturbed state vectors in order to allow a smooth transition to the perturbed state vectors in second order.

It may be somewhat surprising that a tiny perturbation to the Hamiltonian can tell us what we must take as the unperturbed energy eigenstates, but there is a similar phenomenon in classical physics. Consider a particle moving in two or more dimensions under the influence of a potential $V(\mathbf{x})$, with enough friction to bring the particle to rest at a local minimum of the potential. Suppose that the potential consists of an unperturbed term $V_0(\mathbf{x})$ plus a perturbation $\epsilon U(\mathbf{x})$. If the local minima of $V_0(\mathbf{x})$ are at isolated points $\mathbf{x}_n$, then we would expect the local minima of the complete potential to be at points $\mathbf{x}_n + \delta\mathbf{x}_n$, with $\delta\mathbf{x}_n$ of order $\epsilon$. The condition that these are local minima of the perturbed potential reads

$$0 = \left.\frac{\partial[V_0(\mathbf{x}) + \epsilon U(\mathbf{x})]}{\partial x_i}\right|_{\mathbf{x}=\mathbf{x}_n+\delta\mathbf{x}_n},$$

or, to first order in $\epsilon$,

$$0 = \left.\frac{\partial V_0(\mathbf{x})}{\partial x_i}\right|_{\mathbf{x}=\mathbf{x}_n} + \epsilon\left.\frac{\partial U(\mathbf{x})}{\partial x_i}\right|_{\mathbf{x}=\mathbf{x}_n} + \sum_j\left.\frac{\partial^2 V_0(\mathbf{x})}{\partial x_i\,\partial x_j}\right|_{\mathbf{x}=\mathbf{x}_n}(\delta\mathbf{x}_n)_j.$$

The first term vanishes because the $\mathbf{x}_n$ are local minima of the unperturbed potential, so this gives the condition on $\delta\mathbf{x}$ as

$$\sum_j\left.\frac{\partial^2 V_0(\mathbf{x})}{\partial x_i\,\partial x_j}\right|_{\mathbf{x}=\mathbf{x}_n}(\delta\mathbf{x}_n)_j = -\epsilon\left.\frac{\partial U(\mathbf{x})}{\partial x_i}\right|_{\mathbf{x}=\mathbf{x}_n}.$$

This solves the problem if $\mathcal{M}_{ij} \equiv [\partial^2 V_0/\partial x_i\, \partial x_j]_{\mathbf{x}=\mathbf{x}_n}$ is a non-singular matrix, in which case

$$(\delta\mathbf{x}_n)_i = -\epsilon \sum_j \mathcal{M}^{-1}{}_{ij} \left. \frac{\partial U(\mathbf{x})}{\partial x_j} \right|_{\mathbf{x}=\mathbf{x}_n}.$$

But if there is a vector $v_i$ for which $\sum_i v_i \mathcal{M}_{ij} = 0$, then the expansion around $\mathbf{x}_n$ breaks down unless $\sum_i v_i [\partial U/\partial x_i]_{\mathbf{x}=\mathbf{x}_n} = 0$. This problem typically arises when the local minima of the unperturbed potential are not at isolated points, and instead lie on a curve $\mathbf{x} = \mathbf{x}(s)$, so that for all $s$

$$0 = \left. \frac{\partial V_0(\mathbf{x})}{\partial x_i} \right|_{\mathbf{x}=\mathbf{x}(s)}.$$

Differentiating this with respect to $s$ gives

$$0 = \sum_j \left. \frac{\partial^2 V_0(\mathbf{x})}{\partial x_i\, \partial x_j} \right|_{\mathbf{x}=\mathbf{x}(s)} \frac{dx_j(s)}{ds}.$$

Following the same reasoning as before, the shift $\delta\mathbf{x}(s)$ in the position of the local minimum is now governed by the equation

$$\sum_j \left. \frac{\partial^2 V_0(\mathbf{x})}{\partial x_i\, \partial x_j} \right|_{\mathbf{x}=\mathbf{x}(s)} \delta x_j(s) = -\epsilon \left. \frac{\partial U(\mathbf{x})}{\partial x_i} \right|_{\mathbf{x}=\mathbf{x}(s)}.$$

Because $\partial^2 V_0(\mathbf{x})/\partial x_i\, \partial x_j$ is symmetric in $i$ and $j$, the left-hand side of this equation vanishes when multiplied with $dx_i(s)/ds$ and summed over $i$, so this equation cannot be solved unless

$$0 = \sum_i \left. \frac{dx_i(s)}{ds} \frac{\partial U(\mathbf{x})}{\partial x_i} \right|_{\mathbf{x}=\mathbf{x}(s)} = \frac{dU(\mathbf{x}(s))}{ds}.$$

That is, in order for the perturbation $\epsilon U(\mathbf{x})$ to make only a small shift in the particle's equilibrium position, the particle must not only initially be on the curve $\mathbf{x} = \mathbf{x}(s)$ where the unperturbed potential is a local minimum, but must also be at the point on this curve where the value of the *perturbation* on the curve is a local minimum.

## 5.2 The Zeeman Effect

The shift of atomic energies in the presence of an external magnetic field provides an important example of first-order perturbation theory. This is known as the *Zeeman effect*. The effect was first observed in the 1890s by the spectroscopist Pieter Zeeman[1] (1865–1943), as a splitting of the D lines of sodium mentioned at the beginning of Chapter 4 (the same spectral lines that give the

---

[1]  P. Zeeman, *Nature* **55**, 347 (1897).

light from sodium vapor lamps their orange color) in a magnetic field, but it could not be correctly calculated until the advent of quantum mechanics.

We will consider the effect of a magnetic field on the spectrum of an atom of the alkali metal type, such as sodium. In such atoms we can concentrate on the single electron outside closed shells, which feels an effective central potential due to the other electrons and the nucleus. According to classical electrodynamics, the interaction of an external magnetic field $\mathbf{B}$ with an electron moving in an orbit with orbital angular momentum $\mathbf{L}$ gives the electron an extra energy equal to $(e/2m_ec)\mathbf{B} \cdot \mathbf{L}$, so in quantum mechanics we include a term in the Hamiltonian of the form $(e/2m_ec)\mathbf{B} \cdot \mathbf{L}$, where $\mathbf{L}$ is here the angular-momentum operator. We can guess that the interaction of the magnetic field with the spin angular momentum $\mathbf{S}$ will produce an additional term in the Hamiltonian of the form $(eg_e/2m_ec)\mathbf{B} \cdot \mathbf{S}$, with a constant factor $g_e$ known as the *gyromagnetic ratio* of the electron, but there is no reason to expect that $g_e = 1$. In fact, to lowest order in the fine structure constant $e^2/\hbar c \simeq 1/137$ quantum electrodynamics gives $g_e = 2$ (a result first obtained by Dirac using his relativistic wave equation), while corrections due to processes like the emission and absorption of photons shift the predicted value to $g_e = 2.002322\ldots$, in good agreement with experiment. We therefore take the perturbation to the Hamiltonian as

$$\delta H = \frac{e}{2m_ec}\mathbf{B} \cdot \left[\mathbf{L} + g_e\mathbf{S}\right]. \tag{5.2.1}$$

To calculate the shift in the energies of the states of the atom, we need the matrix elements $\left(\Psi_{n\ell j}^{m'}, \delta H\, \Psi_{n\ell j}^{m}\right)$ of the perturbation $\delta H$ between state vectors of the same unperturbed energy $E_{n\ell j}$, where

$$H_0\Psi_{n\ell j}^{m} = E_{n\ell j}\Psi_{n\ell j}^{m}. \tag{5.2.2}$$

Here $H_0$ is the effective one-particle Hamiltonian of the electron in the absence of the magnetic field. But what must be included in this Hamiltonian? The general rule is that we can only ignore terms that produce energy shifts that are small compared with the shift produced by the perturbation in question. For typical magnetic field strengths, this means that we must include in $H_0$ not only the effective electrostatic potential produced by the nucleus and the other electrons, but also the interaction between the electron's spin and orbital angular momentum that produces the fine structure, the dependence of energy levels on $j$ for a given $n$ and $\ell$. But we can usually neglect the smaller interaction between the spins of the electron and nucleus that produces a splitting of spectral lines known as the hyperfine effect.

In calculating these expectation values, we recall that Eq. (4.4.14) tells us that for any three-vector operator $\mathbf{V}$, the matrix element $\left(\Psi_{n\ell j}^{m'}, \mathbf{V}\Psi_{n\ell j}^{m}\right)$ is in the same direction as the matrix element with $\mathbf{V}$ replaced with $\mathbf{J}$, and has the same dependence on $m$ and $m'$. In particular, this is true for the vector $\mathbf{L} + g_e\mathbf{S}$, so

$$\left(\Psi_{n\ell j}^{m'}, [\mathbf{L} + g_{\mathrm{e}}\mathbf{S}]\Psi_{n\ell j}^{m}\right) = g_{nj\ell}\left(\Psi_{n\ell j}^{m'}, \mathbf{J}\Psi_{n\ell j}^{m}\right), \tag{5.2.3}$$

where $g_{nj\ell}$ is a constant independent of $m$ and $m'$, known as the *Landé g-factor*. As mentioned in Section 4.4, this result is often explained in quantum mechanics textbooks as due to the rapid precession of the vectors $\mathbf{S}$ and $\mathbf{L}$ around the total angular momentum $\mathbf{J}$, but this odd blend of classical and quantum-mechanical reasoning is quite unnecessary; Eq. (5.2.3) is a simple consequence of the commutation relations of angular-momentum operators with vector operators.

To calculate the Landé $g$-factor, note that because $\mathbf{J}$ commutes with $\mathbf{J}^2$, the state vector $\mathbf{J}\Psi_{n\ell j}^{m}$ is itself just a linear combination of the same state vectors $\Psi_{n\ell j}^{m''}$ with various values of $m''$, so we also have

$$\sum_i \left(\Psi_{n\ell j}^{m'}, [L_i + g_{\mathrm{e}}S_i]J_i\Psi_{n\ell j}^{m}\right) = g_{nj\ell}\sum_i \left(\Psi_{n\ell j}^{m'}, J_i J_i\Psi_{n\ell j}^{m}\right). \tag{5.2.4}$$

The matrix elements on both sides are easily calculated. On the right, we use

$$\sum_i J_i J_i \Psi_{n\ell j}^{m} = \hbar^2 j(j+1)\Psi_{n\ell j}^{m},$$

while on the left, using $\mathbf{S} = \mathbf{J} - \mathbf{L}$,

$$\sum_i L_i J_i \Psi_{n\ell j}^{m} = \frac{1}{2}\Big[-\mathbf{S}^2 + \mathbf{L}^2 + \mathbf{J}^2\Big]\Psi_{n\ell j}^{m}$$

$$= \frac{\hbar^2}{2}\left[-\frac{3}{4} + \ell(\ell+1) + j(j+1)\right]\Psi_{n\ell j}^{m},$$

and, using $\mathbf{L} = \mathbf{J} - \mathbf{S}$,

$$\sum_i S_i J_i \Psi_{n\ell j}^{m} = \frac{1}{2}\Big[-\mathbf{L}^2 + \mathbf{S}^2 + \mathbf{J}^2\Big]\Psi_{n\ell j}^{m}$$

$$= \frac{\hbar^2}{2}\left[-\ell(\ell+1) + \frac{3}{4} + j(j+1)\right]\Psi_{n\ell j}^{m}.$$

(Note that, for any three-vector operator $\mathbf{V}$, we have $\mathbf{V}\cdot\mathbf{J} = \mathbf{J}\cdot\mathbf{V}$, because $[J_i, V_j] = i\hbar\sum_k \epsilon_{ijk}V_k$ vanishes for $i = j$.) Therefore Eq. (5.2.4) gives

$$\frac{1}{2}\left[-\frac{3}{4} + \ell(\ell+1) + j(j+1)\right] + g_{\mathrm{e}}\frac{1}{2}\left[-\ell(\ell+1) + \frac{3}{4} + j(j+1)\right]$$
$$= j(j+1)g_{nj\ell},$$

so that $g_{nj\ell}$ is independent of $n$, and given by

$$g_{j\ell} = 1 + (g_{\mathrm{e}} - 1)\left(\frac{j(j+1) - \ell(\ell+1) + 3/4}{2j(j+1)}\right). \tag{5.2.5}$$

Now let's return to the problem of finding the perturbed energies. According to Eqs. (5.2.1) and (5.2.3), the matrix elements we need are

$$\left(\Psi_{n\ell j}^{m'}, \delta H \, \Psi_{n\ell j}^{m}\right) = \frac{eg_{j\ell}}{2m_e c} \left(\Psi_{n\ell j}^{m'}, \mathbf{B} \cdot \mathbf{J} \Psi_{n\ell j}^{m}\right). \tag{5.2.6}$$

For $\mathbf{B}$ in a general direction, this does not satisfy the condition for the use of first-order perturbation theory found in the previous section, that the matrix element of the perturbation between different state vectors of the same unperturbed energy must vanish. We can avoid this problem by taking the unperturbed state vectors to be eigenstates of $\mathbf{B} \cdot \mathbf{J}$ instead of $J_3$, but we can also avoid the problem without introducing new state vectors in place of $\Psi_{n\ell j}^{m}$ by simply using a coordinate system in which the 3-axis is in the direction of $\mathbf{B}$. In such a coordinate system, the matrix elements (5.2.6) become

$$\left(\Psi_{n\ell j}^{m'}, \delta H \, \Psi_{n\ell j}^{m}\right) = \left(\frac{e\hbar g_{j\ell}B}{2m_e c}\right) m\delta_{m'm}. \tag{5.2.7}$$

We can therefore calculate the energy shifts using first-order perturbation theory, which gives

$$\delta E_{nj\ell m} = \left(\frac{e\hbar g_{j\ell}B}{2m_e c}\right) m. \tag{5.2.8}$$

For instance, in the D lines of sodium studied by Zeeman, there are really two spectral lines in the absence of a magnetic field, a $D_1$ line caused by a $3p_{1/2} \to 3s_{1/2}$ transition of the outer "valence" electron, and a $D_2$ line caused by the transition $3p_{3/2} \to 3s_{1/2}$. (Recall that because the potential felt by the outer electron is not simply proportional to $1/r$, there is no degeneracy between states with different values of $\ell$. Also, spin–orbit coupling gives energies a dependence on $j = \ell \pm 1/2$, indicated by a subscript, as well as on $\ell$ and on a principal quantum number $n$, which in this case has the value $n = 3$.) For the states involved, Eq. (5.2.5) gives the Landé $g$-factors (in the approximation $g_e = 2$):

$$g_{\frac{3}{2}1} = \frac{4}{3}, \quad g_{\frac{1}{2}1} = \frac{2}{3}, \quad g_{\frac{1}{2}0} = 2. \tag{5.2.9}$$

The $D_1$ and $D_2$ lines are then split into components with photon energies shifted by

$$\Delta E_1(m \to m') = E_B \left(\frac{2m}{3} - 2m'\right), \tag{5.2.10}$$

$$\Delta E_2(m \to m') = E_B \left(\frac{4m}{3} - 2m'\right), \tag{5.2.11}$$

where $E_B \equiv e\hbar B/2m_e c$. Since both the $D_1$ transition and the $D_2$ transition are between states of opposite parity and $j$ differing by 0 or 1, these are electric-dipole transitions, which as shown in Section 4.4 only allow a change in $m$

equal to zero or $\pm 1$. The $D_1$ line is then split into four components with photon energies shifted by the amounts

$$\Delta E_1(\pm 1/2 \rightarrow \pm 1/2) = \mp 2E_B/3, \tag{5.2.12}$$

$$\Delta E_1(\pm 1/2 \rightarrow \mp 1/2) = \pm 4E_B/3, \tag{5.2.13}$$

while the $D_2$ line is split into six components with photon energies shifted by the amounts

$$\Delta E_2(\pm 3/2 \rightarrow \pm 1/2) = \pm E_B, \tag{5.2.14}$$

$$\Delta E_2(\pm 1/2 \rightarrow \pm 1/2) = \mp E_B/3, \tag{5.2.15}$$

$$\Delta E_2(\pm 1/2 \rightarrow \mp 1/2) = \pm 5E_B/3. \tag{5.2.16}$$

Note that if $g_e$ were equal to unity, as would be expected classically, then Eq. (5.2.5) would give a Landé $g$-factor $g_{j\ell} = 1$ for all energy levels, so Eq. (5.2.8) would give a formula for the energy shift that depends on no properties of the energy level but the magnetic quantum number $m$:

$$\delta E_{nj\ell m} = \left(\frac{e\hbar B}{2m_e c}\right) m.$$

Both the $D_1$ line and the $D_2$ line would be split into three components, with photon energies shifted by amounts depending only on the change of the magnetic quantum number:

$$\Delta E_1(\Delta m = \pm 1) = \Delta E_2(\Delta m = \pm 1) = \pm E_B,$$
$$\Delta E_1(\Delta m = 0) = \Delta E_2(\Delta m = 0) = 0.$$

The frequency shift $E_B/h = eB/4\pi m_e c$ was derived on classical grounds by Hendrik Antoon Lorentz[2] (1853–1928), and is known as the *normal Zeeman effect*. Comparison of Lorentz's formula with the early data of Zeeman indicated that whatever charged particle inside the atom is involved in the emission of radiation has a charge/mass ratio $e/m$ about a thousand times greater than the charge/mass ratio of the hydrogen ions involved in electrolysis. This was before Thomson's discovery of the electron, and was the first indication that charges in atoms are carried by particles much lighter than atoms. But the correct splittings are those given by Eqs. (5.2.12)–(5.2.16). This is known as the *anomalous Zeeman effect*, because it is not what would be expected for $g_e = 1$.

The results derived here for the anomalous Zeeman effect are valid only for magnetic fields that are sufficiently small that the energy shift (5.2.8) is much less than the fine-structure splitting between states of the same $n$ and $\ell$ but different $j$. In the opposite limit, where the energy shift (5.2.8) is much greater than

---

[2] H. A. Lorentz, *Phil. Mag.* **43**, 232 (1897); *Ann. Physik* **43**, 278 (1897).

the fine-structure splitting (though still much less than the splittings between states with different $n$ or $\ell$), we have a larger set of essentially degenerate unperturbed states: all those with state vectors $\Psi_{n\ell m_\ell m_s}$ with eigenvalues $\hbar m_\ell$ for $L_3$ and $\hbar m_s$ for $S_3$. With the magnetic field again taken in the 3-direction, the matrix elements of the perturbation are

$$\left( \Psi_{n\ell m'_\ell m'_s}, \delta H \, \Psi_{n\ell m_\ell m_s} \right) = \left( \frac{e\hbar B}{2m_e c} \right) \Big[ m_\ell + g_e m_s \Big] \delta_{m'_\ell m_\ell} \delta_{m'_s m_s}. \qquad (5.2.17)$$

For different state vectors of the same unperturbed energy (i.e., the same values of $n$ and $\ell$) these matrix elements vanish, so we can use first-order perturbation theory for the energy shift, and find

$$\delta E_{n\ell m_\ell m_s} = \left( \frac{e\hbar B}{2m_e c} \right) \Big[ m_\ell + g_e m_s \Big]. \qquad (5.2.18)$$

The transition from energies given by Eq. (5.2.8) to energies given by Eq. (5.2.18) is known as the *Paschen–Back effect*.

## 5.3 The First-Order Stark Effect

We now turn to the shift of atomic energy levels in the presence of an external electric field, an effect discovered in 1914, and known as the *Stark effect*.[3] We will concentrate here on the Stark effect in hydrogen, where the $\ell$-independence of energies for states of a given $n$ and $j$ plays a crucial role. As we will see, the Stark effect in hydrogen provides an example in which the problem of degeneracy in first-order perturbation theory must be solved in a somewhat less trivial way than for the Zeeman effect. The Stark effect in atoms other than hydrogen (and in some hydrogen states) must be calculated using second-order perturbation theory, the subject of the next section.

The interaction of an electron with an external electrostatic potential $\varphi(\mathbf{x})$ gives it an extra energy $-e\varphi(\mathbf{x})$. Since atoms are very small compared with the scales over which $\varphi(\mathbf{x})$ varies, we can replace $\varphi(\mathbf{x})$ with the first two terms in its Taylor series. Setting the (arbitrary) value of $\varphi(\mathbf{x})$ at the position $\mathbf{x} = 0$ of the atomic nucleus equal to zero, this gives $\varphi(\mathbf{x}) = -\mathbf{E} \cdot \mathbf{x}$, where $\mathbf{E} \equiv -\nabla \varphi(0)$ is the electric field at the nucleus, so the change in the Hamiltonian may be taken as

$$\delta H = e\mathbf{E} \cdot \mathbf{X}, \qquad (5.3.1)$$

where to avoid confusion later we return here to denoting the position operator as $\mathbf{X}$.

---

[3] J. Stark, *Verh. deutsch. phys. Ges.* **16**, 327 (1914).

Once again, we take the unperturbed Hamiltonian $H_0$ to be the Hamiltonian of the hydrogen atom in the absence of the electric field, including the fine-structure splitting but neglecting the Lamb shift and the hyperfine splitting. The degenerate unperturbed state vectors are then all the state vectors $\Psi_{n\ell j}^m$ for a fixed $n$ and $j$. We need to calculate the matrix elements of the perturbation between these state vectors:

$$\left(\Psi_{n\ell' j}^{m'}, \delta H\, \Psi_{n\ell j}^m\right) = e\mathbf{E} \cdot \left(\Psi_{n\ell' j}^{m'}, \mathbf{X}\Psi_{n\ell j}^m\right). \tag{5.3.2}$$

As in the case of the Zeeman effect, to avoid non-vanishing matrix elements for $m' \neq m$, we choose the 3-axis to lie in the direction of the electric field, in which case this becomes

$$\left(\Psi_{n\ell' j}^{m'}, \delta H\, \Psi_{n\ell j}^m\right) = eE\delta_{m'm}\left(\Psi_{n\ell' j}^m, X_3\Psi_{n\ell j}^m\right). \tag{5.3.3}$$

This is still not suitable for first-order perturbation theory, because the matrix elements (5.3.3) do not vanish for $\ell' \neq \ell$. Indeed, since $\mathbf{X}$ is odd under space inversion, and space inversion gives factors $(-1)^{\ell'}$ and $(-1)^{\ell}$ when acting on the state vectors $\Psi_{n\ell' j}^m$ and $\Psi_{n\ell j}^{m'}$, respectively, the matrix element (5.3.3) vanishes unless $(-1)^{\ell'}(-1)^{\ell} = -1$, so that the *only* non-vanishing matrix elements are those for which $\ell' \neq \ell$.

For instance, in the energy levels of hydrogen with $n = 1$ and $j = 1/2$ or $n = 2$ and $j = 3/2$, there is no first-order Stark effect, because in these energy levels we only have $\ell = 0$ or $\ell = 1$, respectively. On the other hand, in the $n = 2$, $j = 1/2$ energy level of hydrogen we have both a $2s_{1/2}$ and $2p_{1/2}$ state for each $m = \pm 1/2$. Hence for $n = 2$ and $j = 1/2$ we have the non-vanishing matrix elements $\left(\Psi_{2\,1\,1/2}^{\pm 1/2}, X_3\Psi_{2\,0\,1/2}^{\pm 1/2}\right)$ and $\left(\Psi_{2\,0\,1/2}^{\pm 1/2}, X_3\Psi_{2\,1\,1/2}^{\pm 1/2}\right)$ (where as usual the state vectors are labeled $\Psi_{n\ell j}^m$, with $s = 1/2$ understood throughout). The operator $X_3$ acts on orbital angular-momentum indices but does not act on spin indices, so to calculate its matrix elements between state vectors we need to use Clebsch–Gordan coefficients to express the state vectors here in terms of state vectors $\Psi_{n\ell}^{m_\ell m_s}$ with $S_3 = \hbar m_s$ and $L_3 = \hbar m_\ell$:

$$\Psi_{n\ell j}^m = \sum_{m_\ell m_s} C_{\ell\,\frac{1}{2}}(jm; m_\ell m_s)\Psi_{n\ell}^{m_\ell m_s}. \tag{5.3.4}$$

Because $X_3$ does not involve the spin, the matrix elements of $X_3$ between state vectors with definite eigenvalues for $L_3$ and $S_3$ are

$$\left(\Psi_{n\ell}^{m_\ell m_s}, X_3\Psi_{n'\ell'}^{m_\ell' m_s'}\right)$$

$$= \delta_{m_s m_s'}\int d^3x\, R_{n\ell}(r)Y_\ell^{m_\ell *}(\theta, \phi)r\cos\theta\, R_{n'\ell'}(r)Y_{\ell'}^{m_\ell'}(\theta, \phi). \tag{5.3.5}$$

(Recall that the radial wave functions $R_{n\ell}(r)$ are real.) The operator $X_3$ commutes with both $L_3$ and $S_3$, and since the $s$-wave state vector $\Psi_{2\,0\,1/2}^{\pm 1/2}$ can only have $m_\ell = 0$, the integrals of $x_3$ between this state vector and the $p$-wave state vector $\Psi_{2\,1\,1/2}^{\pm 1/2}$ receive contributions only from the $m_\ell = 0$ components of both wave functions. The non-vanishing matrix elements are thus

$$\left(\Psi_{2\,1\,1/2}^{\pm 1/2},\, X_3\,\Psi_{2\,0\,1/2}^{\pm 1/2}\right) = \left(\Psi_{2\,0\,1/2}^{\pm 1/2},\, X_3\,\Psi_{2\,1\,1/2}^{\pm 1/2}\right)$$

$$= C_{1\frac{1}{2}}\left(\frac{1}{2} \pm \frac{1}{2}; 0 \pm \frac{1}{2}\right) C_{0\frac{1}{2}}\left(\frac{1}{2} \pm \frac{1}{2}; 0 \pm \frac{1}{2}\right) \mathcal{I},$$

$$(5.3.6)$$

where

$$\mathcal{I} \equiv \int d^3x\; r\,\cos\theta\; R_{2\,1}(r)Y_1^0(\theta)R_{2\,0}(r)Y_0^0. \qquad (5.3.7)$$

The Clebsch–Gordan coefficients in Eq. (5.3.6) are

$$C_{1\frac{1}{2}}\left(\frac{1}{2} \pm \frac{1}{2}; 0 \pm \frac{1}{2}\right) = \mp\frac{1}{\sqrt{3}}, \qquad C_{0\frac{1}{2}}\left(\frac{1}{2} \pm \frac{1}{2}; 0 \pm \frac{1}{2}\right) = 1, \quad (5.3.8)$$

so the non-zero matrix elements (5.3.3) are[4]

$$\left(\Psi_{2\,1\,1/2}^{\pm 1/2},\, \delta H\, \Psi_{2\,0\,1/2}^{\pm 1/2}\right) = \left(\Psi_{2\,0\,1/2}^{\pm 1/2},\, \delta H\, \Psi_{2\,1\,1/2}^{\pm 1/2}\right) = \mp\frac{eE\mathcal{I}}{\sqrt{3}}. \qquad (5.3.9)$$

Because there are non-vanishing matrix elements of $\delta H$ between the degenerate state vectors $\Psi_{2\,1\,1/2}^{\pm 1/2}$ and $\Psi_{2\,0\,1/2}^{\pm 1/2}$, these are not the appropriate state vectors for which to calculate perturbed energies. Instead, we must consider the orthonormal state vectors

$$\Psi_A^m \equiv \frac{1}{\sqrt{2}}\left[\Psi_{2\,1\,1/2}^m + \Psi_{2\,0\,1/2}^m\right], \qquad \Psi_B^m \equiv \frac{1}{\sqrt{2}}\left[\Psi_{2\,1\,1/2}^m - \Psi_{2\,0\,1/2}^m\right]. \quad (5.3.10)$$

The non-vanishing matrix elements of $\delta H$ between these state vectors are

$$\left(\Psi_A^{\pm 1/2},\, \delta H\, \Psi_A^{\pm 1/2}\right) = -\left(\Psi_B^{\pm 1/2},\, \delta H\, \Psi_B^{\pm 1/2}\right) = \mp\frac{eE\mathcal{I}}{\sqrt{3}}, \qquad (5.3.11)$$

while

$$\left(\Psi_A^{\pm 1/2},\, \delta H\, \Psi_B^{\pm 1/2}\right) = \left(\Psi_B^{\pm 1/2},\, \delta H\, \Psi_A^{\pm 1/2}\right) = 0. \qquad (5.3.12)$$

---

[4] The fact that the matrix elements of $\delta H$ between $j = 1/2$ state vectors depend on the value of $m = \pm 1/2$ through a sign factor $\pm$ can be understood more directly, as a consequence of the Wigner–Eckart theorem. Here $\delta H$ is proportional to $X_3$, which is the spherical component $x^\mu$ of a vector $\mathbf{X}$ with $\mu = 0$, so according to Eq. (4.4.9),

$$\left(\Psi_{2\,1\,1/2}^m,\, \delta H\, \Psi_{2\,0\,1/2}^m\right) \propto C_{1\frac{1}{2}}\left(\frac{1}{2}\, m; 0\, m\right),$$

and according to Table 4.1, this Clebsch–Gordan coefficient has the value $-2m/\sqrt{3}$.

Therefore first-order perturbation theory gives the energy shifts in *these* states as

$$\delta E_A^{\pm 1/2} = \mp \frac{eE\mathcal{I}}{\sqrt{3}}, \qquad \delta E_B^{\pm 1/2} = \pm \frac{eE\mathcal{I}}{\sqrt{3}}. \tag{5.3.13}$$

It remains to calculate the integral $\mathcal{I}$. Equations (2.1.28) and (2.3.7) give the radial wave functions as

$$R_{n\ell}(r) \propto r^\ell \exp(-r/na) \, F_{n\ell}(r/na),$$

where $a$ is the hydrogen Bohr radius given by Eq. (2.3.19), $a = \hbar^2/m_e e^2$, and Eq. (2.3.17) gives

$$F_{21}(\rho) \propto 1, \qquad F_{20}(\rho) \propto 1 - \rho.$$

Normalizing these state vectors properly, we have

$$R_{20}(r) Y_0^0 = \frac{1}{\sqrt{4\pi}} (2a)^{-3/2} \left( 2 - \frac{r}{a} \right) \exp(-r/2a),$$

$$R_{21}(r) Y_1^0(\theta) = \frac{\cos\theta}{\sqrt{4\pi}} (2a)^{-3/2} \left( \frac{r}{a} \right) \exp(-r/2a). \tag{5.3.14}$$

Then Eq. (5.3.7) gives

$$\mathcal{I} = 2\pi \int_0^\infty r^2 \, dr \int_0^\pi \sin\theta \, d\theta \, \frac{1}{4\pi} (2a)^{-3} r \cos^2\theta \left( \frac{r}{a} \right) \left( 2 - \frac{r}{a} \right) \exp(-r/a)$$

$$= -3a. \tag{5.3.15}$$

In this calculation we have tacitly assumed that the electric field is so weak that the Stark-effect energy shift is much less than the fine-structure splitting (though larger than the Lamb shift and hyperfine splittings). In the opposite limit, where the Stark-effect energy shift is much greater than the fine-structure splitting, we have degeneracy among all the state vectors $\Psi_{n\ell}^{m_\ell m_s}$ for a given value of $n$. Since $X_3$ does not act on spin indices, the spin is irrelevant here. For $n = 2$ we have non-vanishing matrix elements

$$\left( \Psi_{21}^{0 m_s}, \delta H \, \Psi_{20}^{0 m_s} \right) = \left( \Psi_{20}^{0 m_s}, \delta H \, \Psi_{21}^{0 m_s} \right) = eE\mathcal{I}. \tag{5.3.16}$$

The appropriate state vectors to use in connection with first-order perturbation theory are then

$$\Psi_A^{m_s} = \frac{1}{\sqrt{2}} \left[ \Psi_{21}^{0 m_s} + \Psi_{20}^{0 m_s} \right], \qquad \Psi_B^{m_s} = \frac{1}{\sqrt{2}} \left[ \Psi_{21}^{0 m_s} - \Psi_{20}^{0 m_s} \right], \tag{5.3.17}$$

and the energy shifts are

$$\delta E_A^{m_s} = eE\mathcal{I}, \qquad \delta E_B^{m_s} = -eE\mathcal{I}. \tag{5.3.18}$$

This is the analog of the Paschen–Back effect, and is the result that is usually quoted in quantum mechanics textbooks.

These calculations show that even a very weak electric field will thoroughly mix the $2s$ and $2p$ states. (It is only necessary that the Stark energy shift should be large compared with the Lamb shift between the $2s_{1/2}$ and $2p_{1/2}$ states.) This has the dramatic effect that the $2s$ state, which is metastable in the absence of an electric field, can rapidly decay by single-photon emission into the $1s$ state through its mixing with the $2p$ state in even a weak electric field.

## 5.4 Second-Order Perturbation Theory

We now consider the change in energies due to a perturbation $\delta H$, to second order in whatever small parameter $\epsilon$ appears in the perturbed Hamiltonian. Of course, second-order perturbations are of special interest when the first-order perturbation vanishes, as it does for the Stark shift of atomic energy levels in an electric field for the $1s_{1/2}$, $2p_{3/2}$, etc., states of hydrogen and almost all states of other atoms. Nevertheless, here we will allow for the presence of perturbations of first as well as second order.

It will be of some interest (and very little extra trouble) now to include a possible term $\delta_2 H$ in the Hamiltonian that itself is of second order in $\epsilon$, so that $H = H_0 + \delta_1 H + \delta_2 H$, with $\delta_N H$ of order $\epsilon^N$. We return to the Schrödinger equation (5.1.4), and equate the terms of second order in $\epsilon$ on both sides:

$$H_0 \, \delta_2 \Psi_a + \delta_1 H \, \delta_1 \Psi_a + \delta_2 H \, \Psi_a = E_a \, \delta_2 \Psi_a + \delta_1 E_a \, \delta_1 \Psi_a + \delta_2 E_a \, \Psi_a. \quad (5.4.1)$$

Let us again first consider the non-degenerate case, where none of the states we are interested in have the same unperturbed energies. We found in Section 5.1 that in this case the first-order perturbations to the energies and state vectors are

$$\delta_1 E_a = \left( \Psi_a, \delta_1 H \, \Psi_a \right), \quad (5.4.2)$$

$$\delta_1 \Psi_a = \sum_{b \neq a} \frac{\left( \Psi_b, \delta_1 H \, \Psi_a \right)}{E_a - E_b} \Psi_b. \quad (5.4.3)$$

To find the second-order energy shift, we take the scalar product of Eq. (5.4.1) with $\Psi_a$. Because $H_0$ is Hermitian, the term $\left( \Psi_a, H_0 \, \delta_2 \Psi_a \right)$ in the scalar product of $\Psi_a$ with the left-hand side of Eq. (5.4.1) is equal to $E_a \left( \Psi_a, \delta_2 \Psi_a \right)$, and therefore cancels this term in the scalar product of $\Psi_a$ with the right-hand side, leaving us with

$$\left( \Psi_a, \delta_1 H \, \delta_1 \Psi_a \right) + \left( \Psi_a, \delta_2 H \, \Psi_a \right) = \delta_2 E_a + \delta_1 E_a \left( \Psi_a, \delta_1 \Psi_a \right). \quad (5.4.4)$$

We drop the term proportional to $\delta_1 E_a$, because as explained in Section 5.1, we choose the phase and normalization of the perturbed state vector so that $\left( \Psi_a, \delta_1 \Psi_a \right) = 0$. Using Eq. (5.4.3) in Eq. (5.4.4) then gives

$$\delta_2 E_a = \sum_{b \neq a} \frac{\left|\left(\Psi_b, \delta_1 H \, \Psi_a\right)\right|^2}{E_a - E_b} + \left(\Psi_a, \delta_2 H \, \Psi_a\right). \tag{5.4.5}$$

When one says that an energy shift is produced by the emission and reabsorption of some virtual particle, as for instance the Lamb shift is produced by the emission and reabsorption of a virtual photon by the electron in the hydrogen atom, what is meant is that $\delta_2 E_a$ (or a higher-order correction) receives an important contribution from a state $\Psi_b$ containing that particle.

One immediate consequence of Eq. (5.4.5) is that, if $\Psi_a$ is the state of lowest energy of a system, then (in the absence of $\delta_2 H$) the second-order energy shift of its energy is always negative, because all other states have $E_b > E_a$.

As an example of the use of Eq. (5.4.5), consider a two-state system, with unperturbed energies $E_a \neq E_b$. According to Eqs. (5.4.2) and (5.4.5), in the absence of $\delta_2 H$, to second order the perturbations to these energies are

$$\delta E_a = \left(\Psi_a, \delta H \, \Psi_a\right) + \frac{\left|\left(\Psi_b, \delta H \, \Psi_a\right)\right|^2}{E_a - E_b},$$

$$\delta E_b = \left(\Psi_b, \delta H \, \Psi_b\right) - \frac{\left|\left(\Psi_b, \delta H \, \Psi_a\right)\right|^2}{E_a - E_b},$$

so second-order corrections increase the higher energy by the same amount as that by which they lower the lower energy.

We can also calculate the second-order shift in the state vectors. Taking the scalar product of Eq. (5.4.1) with $\Psi_b$ and using Eq. (5.4.3) gives, for $b \neq a$,

$$\left(\Psi_b, \delta_2 \Psi_a\right) = \frac{1}{E_a - E_b} \left[ \sum_{c \neq a} \frac{\left(\Psi_b, \delta_1 H \, \Psi_c\right)\left(\Psi_c, \delta_1 H \, \Psi_a\right)}{E_a - E_c} + \left(\Psi_b, \delta_2 H \, \Psi_a\right) \right.$$
$$\left. - \frac{\delta_1 E_a \left(\Psi_b, \delta_1 H \, \Psi_a\right)}{E_a - E_b} \right]. \tag{5.4.6}$$

The component of $\delta_2 \Psi_a$ along $\Psi_a$ can be found by imposing the condition that $\Psi_a + \delta_1 \Psi_a + \delta_2 \Psi_a + \cdots$ has unit norm. The terms in this condition of second order in $\epsilon$ tell us that

$$2 \operatorname{Re}\left(\Psi_a, \delta_2 \Psi_a\right) = -\left(\delta_1 \Psi_a, \delta_1 \Psi_a\right) = -\sum_{b \neq a} \left|\frac{\left(\Psi_b, \delta_1 H \, \Psi_a\right)}{E_a - E_b}\right|^2. \tag{5.4.7}$$

We can choose the phase of $\Psi_a + \delta_1 \Psi_a + \delta_2 \Psi_a$ so that the matrix element $\left(\Psi_a, \delta_2 \Psi_a\right)$ is real, and Eq. (5.4.7) then gives the needed formula for this matrix element. The full second-order shift in the state vector in the non-degenerate case is then

$$\delta_2 \Psi_a = \sum_{b \neq a} \frac{\Psi_b}{E_a - E_b} \left[ \sum_{c \neq a} \frac{\left(\Psi_b, \delta_1 H \ \Psi_c\right) \left(\Psi_c, \delta_1 H \ \Psi_a\right)}{E_a - E_c} + \left(\Psi_b, \delta_2 H \ \Psi_a\right) \right.$$
$$\left. - \frac{\delta_1 E_a \left(\Psi_b, \delta_1 H \ \Psi_a\right)}{E_a - E_b} \right]$$
$$- \frac{1}{2} \Psi_a \sum_{b \neq a} \left| \frac{\left(\Psi_b, \delta_1 H \Psi_a\right)}{E_a - E_b} \right|^2. \tag{5.4.8}$$

Next let's consider the more complicated degenerate case, in which some of the states in which we are interested have the same unperturbed energies. First, we note that the calculation of the second-order energy shift goes through much as in the non-degenerate case. Taking the scalar product of Eq. (5.4.1) with $\Psi_a$ again gives Eq. (5.4.4). The orthonormality condition found in Section 5.1, that the matrix $\left(\Psi_b, \delta_1 \Psi_a\right)$ for states with $E_b = E_a$ must be chosen to be anti-Hermitian, tells us that $\left(\Psi_a, \delta_1 \Psi_a\right)$ is imaginary, so it can again be made to vanish by a suitable choice of phase of $\Psi_a + \delta_1 \Psi_a$. We can use Eq. (5.1.16) for $\delta_1 \Psi_a$ in the first term $\left(\Psi_a, \delta_1 H \delta_1 \Psi_a\right)$ on the left of Eq. (5.4.4). Since the unperturbed states have been chosen so that $\left(\Psi_a, \delta_1 H \ \Psi_b\right)$ vanishes for $E_b = E_a$ but $b \neq a$, and we have chosen the phase of $\Psi_a + \delta_1 \Psi_a$ so that $\left(\Psi_a, \delta_1 \Psi_a\right)$ also vanishes, the second term in Eq. (5.1.16), which involves unknown matrix elements, does not contribute to the first term in Eq. (5.4.4). We conclude then that

$$\delta_2 E_a = \sum_{c: E_c \neq E_a} \frac{\left|\left(\Psi_a, \delta_1 H \ \Psi_c\right)\right|^2}{E_a - E_c} + \left(\Psi_a, \delta_2 H \ \Psi_a\right). \tag{5.4.9}$$

This is the same result as in the non-degenerate case, except that here we have to specify not only that the intermediate states $\Psi_c$ have $c \neq a$, but also that they have $E_c \neq E_a$.

Next, let's return to the calculation of the first-order shifts $\delta_1 \Psi_a$ in the state vectors. In Section 5.1 we were able to calculate the component of $\delta_1 \Psi_a$ along any unperturbed state $\Psi_c$ with $E_c \neq E_a$, but about its components along unperturbed states $\Psi_b$ with $E_b = E_a$, we were only able to conclude that orthonormality requires the $\left(\Psi_b, \delta_1 \Psi_a\right)$ to form an anti-Hermitian matrix. We can now go further by imposing the condition that second-order effects make only a small change in the state vectors.

Taking the scalar product of Eq. (5.4.1) with any state $\Psi_b$ for which $E_b = E_a$ but $b \neq a$ gives

$$\left( \Psi_b, \delta_2 H \, \Psi_a \right) + \left( \Psi_b, \delta_1 H \, \delta_1 \Psi_a \right) = \delta_1 E_a \left( \Psi_b, \delta_1 \Psi_a \right).$$

In the second term on the left we can insert a sum over a complete set of intermediate states $\Psi_c$ between $\delta_1 H$ and $\delta_1 \Psi_a$. Using the results of first-order perturbation theory, that $\left( \Psi_b, \delta_1 H \, \Psi_a \right) = \delta_{ab} \delta_1 E_a$ for $E_b = E_a$, and that $(\Psi_c, \delta_1 \Psi_a)$ for $E_c \neq E_a$ is given by Eq. (5.1.16), we have for $E_b = E_a$ but $b \neq a$:

$$\left( \Psi_b, \delta_2 H \, \Psi_a \right) + \delta_1 E_b \left( \Psi_b, \delta_1 \Psi_a \right) + \sum_{c: E_c \neq E_a} \frac{\left( \Psi_b, \delta_1 H \, \Psi_c \right) \left( \Psi_c, \delta_1 H \, \Psi_a \right)}{E_a - E_c}$$

$$= \delta_1 E_a \left( \Psi_b, \delta_1 \Psi_a \right). \qquad (5.4.10)$$

This result allows a complete solution for $\delta_1 \Psi_a$ in the case in which the degeneracy in zeroth order is removed in first order – that is, that if $b \neq a$ but $E_b = E_a$ then $\delta_1 E_a \neq \delta_1 E_b$. Then Eq. (5.4.10) provides a formula for the components $\left( \Psi_b, \delta_1 \Psi_a \right)$ with $E_b = E_a$ but $b \neq a$:

$$\left( \Psi_b, \delta_1 \Psi_a \right) = \frac{1}{\delta_1 E_a - \delta_1 E_b} \left[ \left( \Psi_b, \delta_2 H \, \Psi_a \right) \right.$$

$$\left. + \sum_{c: E_c \neq E_a} \frac{\left( \Psi_b, \delta_1 H \, \Psi_c \right) \left( \Psi_c, \delta_1 H \, \Psi_a \right)}{E_a - E_c} \right].$$

$$(5.4.11)$$

Inspection shows that the right-hand side is an anti-Hermitian matrix (the matrix in square brackets is Hermitian, but the energy denominator in front is antisymmetric), so this condition is allowed by the freedom in $\left( \Psi_b, \delta_1 \Psi_a \right)$ for $E_b = E_a$ that we were left with in Section 5.1 after using the Schrödinger equation and the condition of orthonormality. This still leaves $\left( \Psi_a, \delta_1 \Psi_a \right)$ undetermined, but as already noted we can choose this matrix element to vanish by a suitable choice of phase of $\Psi_a + \delta_1 \Psi_a$. So we have a complete expression for the first-order shift in the state vector in the degenerate case:

$$\delta_1 \Psi_a = \sum_{c: E_c \neq E_a} \frac{\left( \Psi_c, \delta_1 H \, \Psi_a \right)}{E_a - E_c} \Psi_c$$

$$+ \sum_{b \neq a, E_b = E_a} \frac{\Psi_b}{\delta_1 E_a - \delta_1 E_b} \left[ \left( \Psi_b, \delta_2 H \, \Psi_a \right) \right.$$

$$\left. + \sum_{c: E_c \neq E_a} \frac{\left( \Psi_b, \delta_1 H \, \Psi_c \right) \left( \Psi_c, \delta_1 H \, \Psi_a \right)}{E_a - E_c} \right] \Psi_b. \qquad (5.4.12)$$

This only applies if the zeroth-order degeneracy is removed in first order. If any of the first-order perturbations $\delta_1 E_b$ of energies for which $E_b = E_a$ are equal to $\delta_1 E_a$ then Eq. (5.4.10) tells us nothing about $\left(\Psi_b, \delta_1 \Psi_a\right)$, but instead implies that if $E_b = E_a$ and $\delta_1 E_b = \delta_1 E_a$ but $b \neq a$ then $[\delta_2^{\text{eff}} H]_{ba} = 0$, where

$$[\delta_2^{\text{eff}} H]_{ba} \equiv \left(\Psi_b, \delta_2 H \Psi_a\right) + \sum_{c: E_c \neq E_a} \frac{\left(\Psi_b, \delta_1 H \Psi_c\right)\left(\Psi_c, \delta_1 H \Psi_a\right)}{E_a - E_c}. \tag{5.4.13}$$

We noted in Section 5.1 that when there are states with the same values of the zeroth- and first-order energies we can take the unperturbed state vectors to be any orthonormal linear combination of these states. Since $\delta_2^{\text{eff}} H$ is an Hermitian matrix, by the same reasoning as we applied in Section 5.1 to $\delta_1 H$, we can choose these linear combinations to diagonalize this matrix, so that if $E_b = E_a$ and $\delta_1 E_b = \delta_1 E_a$ but $b \neq a$ then in the new basis $[\delta_2^{\text{eff}} H]_{ba} = 0$. This completely determines the unperturbed states unless some of the second-order energies $\delta_2 E_a = [\delta_2^{\text{eff}} H]_{aa}$ are equal. In this case we must look to higher orders of perturbation theory to remove the degeneracy and fix the unperturbed states.

It is generally not easy to do the sums over states in Eqs. (5.4.5) or (5.4.9). In some cases the sum can diverge; there are *ultraviolet divergences* that occur when the matrix elements $\left|\left(\Psi_b, \delta_1 H \Psi_a\right)\right|$ do not fall off rapidly enough for high-energy states $\Psi_b$ to make the sum converge, and there are *infrared divergences* that occur when there is a continuum of states $\Psi_b$ with energies $E_b$ extending down to $E_a$. The treatment of these infinities has been a major preoccupation of theoretical physicists since the 1930s.

There are two cases that allow $\delta_2 E_a$ to be more easily calculated. In the first case, the energies $E_b$ of all the states $\Psi_b$ with $b \neq a$ for which $\left(\Psi_b, \delta_1 H \Psi_a\right)$ is appreciable for a given state $\Psi_a$ are clustered at a value $E_b \simeq E_a + \Delta_a$, with $\Delta_a \neq 0$. The completeness of the orthonormal state vectors $\Psi_b$ allows us to write

$$\sum_{b \neq a} \left|\left(\Psi_b, \delta_1 H \Psi_a\right)\right|^2 = \left(\Psi_a, \delta_1 H \sum_b \Psi_b \left(\Psi_b, \delta_1 H \Psi_a\right)\right) - \left|\left(\Psi_a, \delta_1 H \Psi_a\right)\right|^2$$

$$= \left(\Psi_a, (\delta_1 H)^2 \Psi_a\right) - \left(\delta_1 E_a\right)^2 \tag{5.4.14}$$

so in the absence of degeneracy $\delta_2 E_a$ is given by what is called the *closure approximation*:

$$\delta_2 E_a \simeq \frac{1}{-\Delta_a} \sum_{b \neq a} \left|\left(\Psi_b, \delta_1 H \Psi_a\right)\right|^2 + \left(\Psi_a, \delta_2 H \Psi_a\right)$$

$$= -\frac{\left[\left(\Psi_a, (\delta_1 H)^2 \Psi_a\right) - \left(\delta_1 E_a\right)^2\right]}{\Delta_a} + \left(\Psi_a, \delta_2 H \Psi_a\right). \tag{5.4.15}$$

The second case occurs when there is a small set of states $\Psi_b$ for which $\left(\Psi_b, \delta H \, \Psi_a\right)$ is appreciable, and $E_b$ is very close though not equal to $E_a$. In this case, the sum in Eq. (5.4.5) or Eq. (5.4.9) can often be restricted to these states. For instance, the second-order Stark shift in the $2p_{3/2}$ state of hydrogen can be estimated by keeping only the $2s_{1/2}$ state, with which it is nearly degenerate, in Eq. (5.4.5).

## 5.5 The Variational Method

Some problems cannot be solved by perturbation theory, because the Hamiltonian is not close to one with known eigenvalues and eigenstates. A classic case is encountered in chemistry: there is no small parameter in which we can expand the energies and state vectors of electrons in a molecule with several nuclei. In such cases, it is often possible to get a good estimate at least of the ground state energy, by a technique known as *the variational method*. It is based on a general theorem that *the true ground state energy is less than or equal to the expectation value of the Hamiltonian in any state.*

To prove this result, recall the expression (3.1.16) for the expansion of any state vector $\Psi$ in a series of orthonormal state vectors $\Psi_n$:

$$\Psi = \sum_n \Psi_n \left(\Psi_n, \Psi\right), \quad \text{where} \quad \left(\Psi_n, \Psi_m\right) = \delta_{nm}. \tag{5.5.1}$$

We can take the $\Psi_n$ to be exact eigenvectors of the Hamiltonian

$$H\Psi_n = E_n \Psi_n. \tag{5.5.2}$$

This gives the expectation value of the Hamiltonian in the state $\Psi$ as

$$\langle H \rangle_\Psi \equiv \frac{\left(\Psi, H\Psi\right)}{\left(\Psi, \Psi\right)} = \frac{\sum_n E_n \left|\left(\Psi_n, \Psi\right)\right|^2}{\sum_n \left|\left(\Psi_n, \Psi\right)\right|^2}. \tag{5.5.3}$$

If $E_{\text{ground}}$ is the true ground state energy, then $E_n \geq E_{\text{ground}}$ for all $n$, so

$$\langle H \rangle_\Psi \geq E_{\text{ground}}, \tag{5.5.4}$$

as was to be proved.

We can check that this result is respected by the approximations we found earlier in perturbation theory. Recall that to first order in a small perturbation $\delta H$, the energy of a physical state with unperturbed state vector $\Psi_n^{(0)}$ and unperturbed energy $E_n^{(0)}$ is given by the expectation value of the total Hamiltonian

$$E_n^{(0)} + \delta E_n = E_n^{(0)} + \left(\Psi_n^{(0)}, \delta H \, \Psi_n^{(0)}\right) = \left(\Psi_n^{(0)}, (H + \delta H)\Psi_n^{(0)}\right)$$

(provided that the unperturbed state vectors have been chosen so that $\left(\Psi_n^{(0)}, \delta H \, \Psi_m^{(0)}\right) = 0$ if $E_m^{(0)} = E_n^{(0)}$ but $m \neq n$). Further, we have seen that the energy in second-order perturbation theory is *less* than this expectation value. As we have now seen, this expectation value is not only an approximation to the true energy in first-order perturbation theory, and an upper bound to the ground state energy in second-order perturbation theory – it is an exact upper bound to the ground state energy, whatever we choose for $\Psi_n^{(0)}$.

One nice thing about the variational principle is that, although the choice of a trial state vector is a matter of judgment, there is an objective way of telling which of two trial state vectors is better. Since the true ground state energy is less than the expectation value of the Hamiltonian for any trial state vector, that trial state vector that gives the smallest expectation value is better.

For a system consisting of a single particle of mass $M$ moving in three dimensions in a general potential $V(\mathbf{X})$, the Hamiltonian is

$$H = \frac{\mathbf{P}^2}{2M} + V(\mathbf{X}). \tag{5.5.5}$$

So, since $\mathbf{P}$ is Hermitian,

$$\langle H \rangle_\Psi = \frac{\sum_i \left(P_i\Psi, P_i\Psi\right)/2M + \left(\Psi, V\Psi\right)}{\left(\Psi, \Psi\right)} \tag{5.5.6}$$

$$= \langle T \rangle_\Psi + \langle V \rangle_\Psi, \tag{5.5.7}$$

where

$$\langle T \rangle_\Psi = \frac{\int d^3x \, (\hbar^2/2M) \sum_i \left|\partial\psi(\mathbf{x})/\partial x^i\right|^2}{\int d^3x \, |\psi(\mathbf{x})|^2},$$

$$\langle V \rangle_\Psi = \frac{\int d^3x \, V(\mathbf{x})|\psi(\mathbf{x})|^2}{\int d^3x \, |\psi(\mathbf{x})|^2}, \tag{5.5.8}$$

where $\psi(\mathbf{x})$ is the coordinate-space wave function $(\Phi_\mathbf{x}, \Psi)$. The mean kinetic energy $\langle T \rangle_\Psi$ is minimized by a $\psi(\mathbf{x})$ that is as flat as possible, while for an attractive potential like the Coulomb potential, the mean potential $\langle V \rangle_\Psi$ is minimized by a $\psi(\mathbf{x})$ that is concentrated near the origin. The wave function that minimizes $\langle H \rangle_\Psi$ is therefore a compromise – somewhat concentrated near the origin, but with some spread out to larger distances.

The energies of some other states besides the ground state may be given by the minimum value of the expectation value $\langle H \rangle_\Psi$ for $\Psi$ subject to certain constraints. Suppose that there is some Hermitian operator $A$ (such as $\mathbf{L}^2$) that commutes with the Hamiltonian. Then if a trial state vector $\Psi$ is an eigenstate of $A$, the expectation value of the Hamiltonian for that state vector gives an upper bound on the energies of all eigenstates of $H$ with the same eigenvalue of $A$.

Thus, for instance, taking the trial wave function $\psi(\mathbf{x})$ in Eq. (5.5.7) to have the form $R(r)Y_\ell^m(\hat{x})$, this expectation value gives an upper bound on the energies of all states of angular momentum $\ell$.

In a certain sense, the variational principle applies to all energy eigenstates. For excited states the expectation value $\langle H \rangle_\Psi$ is clearly not a minimum, but it is *stationary* under any infinitesimal variation of the state $\Psi$. The change in the expectation value when we make an infinitesimal change $\delta\Psi$ in the state vector $\Psi$ is

$$\delta\langle H\rangle_\Psi = 2\frac{\mathrm{Re}\left(\delta\Psi, H\Psi\right)}{\left(\Psi, \Psi\right)} - 2\frac{\left(\Psi, H\Psi\right)\mathrm{Re}\left(\delta\Psi, \Psi\right)}{\left(\Psi, \Psi\right)^2}$$

$$= \frac{2\,\mathrm{Re}\left(\delta\Psi, (H - \langle H\rangle_\Psi)\Psi\right)}{\left(\Psi, \Psi\right)}, \tag{5.5.9}$$

which vanishes if $\Psi$ is an eigenstate of $H$, in which case $H\Psi = (\langle H\rangle_\Psi)\Psi$.

In using the variational principle for either ground or excited states, one generally defines a trial state vector $\Psi(\lambda)$ as a function of a number of free complex parameters $\lambda_i$, and looks for values of these parameters at which $\langle H\rangle_{\Psi(\lambda)}$ is stationary in the $\lambda_i$. The variation in the trial state vector when we make a small variation $\delta\lambda_i$ in these parameters is $\delta\Psi(\lambda) = \sum_i (\partial\Psi(\lambda)/\partial\lambda_i)\,\delta\lambda_i$, so the corresponding variation in the expectation value of $H$ is given by

$$\delta\langle H\rangle_\Psi = \frac{2\,\mathrm{Re}\sum_i \delta\lambda_i \left(\partial\Psi(\lambda)/\partial\lambda_i, (H - \langle H\rangle_\Psi)\Psi\right)}{\left(\Psi, \Psi\right)}. \tag{5.5.10}$$

Since this must vanish at a stationary point for all complex $\delta\lambda_i$, we must have

$$\left(\partial\Psi/\partial\lambda_i, (H - \langle H\rangle_\Psi)\Psi\right) = 0 \tag{5.5.11}$$

for all $i$. Since the state vector $(H - \langle H\rangle_\Psi)\Psi$ is thus orthogonal to all the state vectors $\partial\Psi/\partial\lambda_i$, we can guess that if there are enough independent parameters $\lambda_i$ then $H\Psi - \langle H\rangle_\Psi\Psi$ should be small, so that $\Psi$ will be close to an eigenvector of the complete Hamiltonian with energy $\langle H\rangle_\Psi$. The more independent parameters $\lambda_i$ we introduce, the closer to $\langle H\rangle_\Psi\Psi$ the state vector $H\Psi$ is likely to be.

For a Coulomb potential there is a simple relation between the kinetic and potential energy terms in Eq. (5.5.8) at the minimum of $\langle H\rangle_\Psi$, known as the *virial theorem*. It is derived by introducing just one free parameter, the length scale, using dimensional analysis to find the dependence of expectation values on this parameter. If we normalize the trial wave function $\psi(\mathbf{x})$, so that $\int d^3x\,|\psi(\mathbf{x})|^2 = 1$, then $\psi$ has dimensionality $[\text{length}]^{-3/2}$, so it must be of the form $\psi(\mathbf{x}) = a^{-3/2} f(\mathbf{x}/a)$, where $f(\mathbf{z})$ is a dimensionless function of a dimensionless argument, and $a$ is a length that can be varied freely when we vary the wave function. By changing the variable of integration in Eq. (5.5.8) from $\mathbf{x}$ to $\mathbf{x}/a$, it is easy to see that when we vary $a$, $\langle T \rangle_\Psi$ goes as $a^{-2}$, while for a Coulomb potential $\langle V \rangle_\Psi$ goes as $a^{-1}$. Since the derivative of the sum with respect to $a$ must vanish at the true energy eigenstate, we have

$$-2\langle T \rangle_\Psi - \langle V \rangle_\Psi = 0, \qquad (5.5.12)$$

so $\langle H \rangle_\Psi = -\langle T \rangle_\Psi$. (It should perhaps be emphasized that this relation can be applied only after a stationary point of $\langle H \rangle_\Psi$ has been found; otherwise we could minimize $\langle H \rangle_\Psi$ by maximizing $\langle T \rangle_\Psi$, which is certainly not the case.) This applies to excited states as well as to the ground state, and similar results hold for multi-electron atoms, or even for molecules, provided that the only forces are Coulomb forces.

## 5.6 The Born–Oppenheimer Approximation

There are theories in which part of the Hamiltonian is suppressed by a small parameter, and yet we cannot use a perturbation theory based on the expansion of energies and eigenvalues to first or second order in this parameter. A good example is provided by molecular physics, in which the kinetic energy of nuclei is suppressed by the reciprocal of nuclear masses. Instead of ordinary perturbation theory, here we can instead use an approximation introduced by Born and J. Robert Oppenheimer (1904–1967) in 1927.[5]

The Hamiltonian for a molecule can be written[6]

$$H = T_{\text{elec}}(p) + T_{\text{nuc}}(P) + V(x, X), \qquad (5.6.1)$$

where $T_{\text{elec}}$ and $T_{\text{nuc}}$ are the kinetic energies of the electrons (labeled $n$) and nuclei (labeled $N$):

$$T_{\text{elec}}(p) = \sum_n \frac{\mathbf{p}_n^2}{2m_e}, \qquad T_{\text{nuc}}(P) = \sum_N \frac{\mathbf{P}_N^2}{2M_N}, \qquad (5.6.2)$$

and $V$ is the potential energy

$$V(x, X) = \frac{1}{2} \sum_{n \neq m} \frac{e^2}{|\mathbf{x}_n - \mathbf{x}_m|} + \frac{1}{2} \sum_{N \neq M} \frac{Z_N Z_M e^2}{|\mathbf{X}_N - \mathbf{X}_M|} - \sum_{nN} \frac{Z_N e^2}{|\mathbf{x}_n - \mathbf{X}_N|}, \qquad (5.6.3)$$

---

[5] M. Born and J. R. Oppenheimer, *Ann. Phys.* **84**, 457 (1927).

[6] In this section we are giving up our usual practice of using upper case letters for operators and lower case letters for their eigenvalues. Instead, here upper and lower case letters for coordinates and momenta refer to nuclei and electrons, respectively. We leave it to the context to clarify whether the symbols for coordinates and momenta denote operators or their eigenvalues.

where $Z_N e$ is the charge of nucleus $N$. Of course, $[x_{ni}, p_{mj}] = i\hbar \delta_{nm}\delta_{ij}$, $[X_{Ni}, P_{Mj}] = i\hbar \delta_{NM}\delta_{ij}$, and all other commutators of coordinates and/or momenta vanish. We are using upper and lower case letters for the dynamical variables of nuclei and electrons, respectively. Boldface as usual indicates three-vectors, and when boldface (and vector indices) are omitted it should be understood that $x$, $p$ and $X$, $P$ denote the whole set of dynamical variables for electrons and nuclei, respectively. We have ignored spin variables in Eqs. (5.6.1)–(5.6.3), but if necessary one can include electron and nuclear spin 3-components among the variables denoted $x$, $p$ and $X$, $P$.

We seek solutions of the Schrödinger equation:

$$\left[ T_{\text{elec}}(p) + T_{\text{nuc}}(P) + V(x, X) \right] \Psi = E\Psi. \tag{5.6.4}$$

The Born–Oppenheimer approximation exploits the suppression of the nuclear kinetic energy term by the large nuclear masses $M_N$, so let's first consider the eigenvalue problem for the reduced Hamiltonian, with $T_{\text{nuc}}$ omitted. The nuclear coordinates $X_{Ni}$ commute with this reduced Hamiltonian, so we can find simultaneous eigenvectors of both the reduced Hamiltonian and $X$:

$$\left[ T_{\text{elec}}(p) + V(x, X) \right] \Phi_{a,X} = \mathcal{E}_a(X) \Phi_{a,X}, \tag{5.6.5}$$

where the subscript $X$ here indicates the eigenvalue of the nuclear coordinate operators (which were denoted $X$ in Eq. (5.6.4)). In Eq. (5.6.5) the nuclear coordinates $\mathbf{X}_N$ can be regarded as c-number parameters, on which the reduced Hamiltonian $T_{\text{elec}} + V$ and hence also its eigenvalues and eigenfunctions depend. The reduced Hamiltonian is Hermitian, so these states can be chosen to be orthonormal, in the sense that

$$\left( \Phi_{b,X'}, \Phi_{a,X} \right) = \delta_{ab} \prod_{Ni} \delta\left( X'_{Ni} - X_{Ni} \right). \tag{5.6.6}$$

We can write the state $\Phi_{a,X}$ as a superposition of states $\Phi_{x,X}$ with definite values of the electron as well as of the nuclear coordinates

$$\Phi_{a,X} = \int dx \, \psi_a(x; X) \Phi_{x,X}. \tag{5.6.7}$$

With the $\Phi_{x,X}$ given the usual continuum normalization

$$\left( \Phi_{x',X'}, \Phi_{x,X} \right) = \prod_{ni} \delta(x_{ni} - x'_{ni}) \prod_{Nj} \delta(X_{Nj} - X'_{Nj}), \tag{5.6.8}$$

the normalization condition (5.6.6) implies that for each $X$:

$$\int dx \, \psi_a^*(x; X) \psi_b(x; X) = \delta_{ab}. \tag{5.6.9}$$

Inserting Eq. (5.6.7) in (5.6.5) gives

$$\left[T_{\text{elec}}(-i\hbar\,\partial/\partial x) + V(x, X)\right]\psi_a(x; X) = \mathcal{E}_a(X)\psi_a(x; X). \tag{5.6.10}$$

This can be regarded as an ordinary Schrödinger equation in a reduced Hilbert space, consisting of square-integrable functions of $x$.

Unfortunately, we cannot simply use first-order perturbation theory, with $T_{\text{nuc}}$ taken as the perturbation and the state vectors $\Phi_{a,X}$ taken as unperturbed energy eigenstates. This is because we are looking for discrete eigenvalues of the full Hamiltonian, for which the eigenvectors $\Psi$ would be normalizable, in the sense that $\left(\Psi, \Psi\right)$ is finite, while Eq. (5.6.6) shows that $\left(\Phi_{a,X}, \Phi_{a,X}\right)$ is infinite. We cannot expand in powers of a perturbation that converts a state vector with continuum normalization into one that is normalizable as a discrete state.

Since the $\Phi_{a,X}$ do form a complete set, the true solution $\Psi$ of the full Schrödinger equation (5.6.4) can be written

$$\Psi = \sum_a \int dX \; f_a(X)\Phi_{a,X}. \tag{5.6.11}$$

The normalization condition $(\Psi, \Psi) = 1$ here reads

$$\sum_a \int dX \; |f_a(X)|^2 = 1. \tag{5.6.12}$$

Inserting the expansion (5.6.11) in the Schrödinger equation (5.6.4), and using the reduced Schrödinger equation (5.6.5), we have

$$0 = \sum_a \int dX \; f_a(X)\left[T_{\text{nuc}}(P) + \mathcal{E}_a(X) - E\right]\Phi_{a,X}. \tag{5.6.13}$$

So far, this is exact, but it is complicated by the fact that the operator $T_{\text{nuc}}$ does not merely act on the $X$-index on $\Phi_{a,X}$. That is, acting on the basis states $\Phi_{x,X}$, an individual component of nuclear momentum gives[7]

$$P_{Ni}\Phi_{x,X} = i\hbar\frac{\partial}{\partial X_{Ni}}\Phi_{x,X}, \tag{5.6.14}$$

so that, using Eq. (5.6.7) and integrating by parts,

$$\int dX \; f_a(X)P_{N,i}\Phi_{a,X} = -i\hbar \int dx \int dX \left[\psi_a(x; X)\frac{\partial}{\partial X_{Ni}}f_a(X)\right.$$
$$\left. + f_a(X)\frac{\partial}{\partial X_{Ni}}\psi_a(x; X)\right]\Phi_{x,X}. \tag{5.6.15}$$

---

[7] A reminder: according to Eq. (3.5.11), a momentum operator $P$ acts on basis states $\Phi_X$ as $i\hbar\,\partial/\partial X$, so that
$$P\int dX \; \psi(X)\Phi_X = \int [-i\hbar\,\partial\psi(X)/\partial X]\Phi_X.$$

The Born–Oppenheimer approximation consists of dropping the derivative of $\psi_a(x; X)$ with respect to $X$ in Eq. (5.6.15), so that, using Eq. (5.6.7) again,

$$\int dX \; f_a(X) T_{\mathrm{nuc}}(P) \Phi_{a,X} \simeq \int dX \; \Phi_{a,X} \sum_N \left( \frac{-\hbar^2}{2M_N} \right) \nabla_N^2 f_a(X). \quad (5.6.16)$$

We will make this approximation and see where it leads us, and then come back to whether the solutions we find are consistent with this approximation.

With the approximation (5.6.16), the Schrödinger equation (5.6.13) becomes

$$0 = \sum_a \int dX \; \Phi_{a,X} \left[ \sum_N \left( \frac{-\hbar^2}{2M_N} \right) \nabla_N^2 + \mathcal{E}_a(X) - E \right] f_a(X). \quad (5.6.17)$$

Since the eigenvectors $\Phi_{a,X}$ of the reduced Hamiltonian are independent, each term in the sum must vanish, so for all $a$,

$$\left[ \sum_N \left( \frac{-\hbar^2}{2M_N} \right) \nabla_N^2 + \mathcal{E}_a(X) \right] f_a(X) = E f_a(X). \quad (5.6.18)$$

That is, $f_a(X)$ satisfies a Schrödinger equation in which electron dynamical variables no longer appear, except that the energy $\mathcal{E}_a(X)$ of the electronic state with fixed nuclear coordinates $X$ acts as a potential for the nuclei. For this purpose all we need to calculate about the electrons is the energy $\mathcal{E}_a(X)$, not the eigenvector $\Phi_{a,X}$. This still isn't easy, but at least we can (and usually do) find the lowest $\mathcal{E}_a(X)$ by applying the variational principle to the reduced Hamiltonian $T_{\mathrm{elec}} + V$, with nuclear coordinates held fixed.

The different electronic configurations have decoupled from each other, so that we have solutions for each $a$ in which all of the other $f_b$ vanish. From now on we will drop the index $a$, keeping our attention on just a single electronic configuration, which often is taken as the ground state, in which the electron energy $\mathcal{E}(X)$ is the lowest of the $\mathcal{E}_a(X)$.

For multi-atom molecules the function $\mathcal{E}(X)$ is pretty complicated. It may be expected to have several local minima, corresponding to different stable or metastable molecular configurations. There will be solutions of Eq. (5.6.18) with the wave function $f(X)$ concentrated around one of these minima, corresponding to various vibrational modes of the molecule in this configuration. Taking

$\mathbf{X}_N = 0$ as the coordinates of one local minimum, for each such wave function Eq. (5.6.18) may be approximated as[8]

$$\left[ \sum_N \left( \frac{-\hbar^2}{2M_N} \right) \nabla_N^2 + \frac{1}{2} \sum_{NN'ij} K_{Ni,N'j} X_{Ni} X_{Nj} \right] f(X) = Ef(X), \quad (5.6.19)$$

where

$$K_{Ni,N'j} \equiv \left[ \frac{\partial^2 \mathcal{E}(X)}{\partial X_{Ni} \partial X_{N'j}} \right]_{X=0}. \quad (5.6.20)$$

We note in passing that this program is made easier by using a result known as the *Hellmann–Feynman theorem*,[9] which states

$$\frac{\partial \mathcal{E}(X)}{\partial X_{Ni}} = \int dx \, |\psi(x; X)|^2 \frac{\partial V(x, X)}{\partial X_{Ni}}. \quad (5.6.21)$$

In other words, to calculate the first derivatives of $\mathcal{E}(X)$, as we need to do to find its local minima, we do not need to calculate derivatives of the electronic wave function $\psi(x; X)$ with respect to the nuclear coordinates $X$. To prove this, we note from Eq. (5.6.10) (dropping the subscript $a$) that

$$\mathcal{E}(X) = \int dx \, \psi^*(x; X) \Big[ T_{\text{elec}}(-i\hbar \, \partial/\partial x) + V(x, X) \Big] \psi(x; X),$$

so

$$\frac{\partial \mathcal{E}(X)}{\partial X_{Ni}} = \int dx \left[ \frac{\partial}{\partial X_{Ni}} \psi(x; X) \right]^* \Big[ T_{\text{elec}}(-i\hbar \, \partial/\partial x) + V(x, X) \Big] \psi(x; X)$$

$$+ \int dx \, \psi^*(x; X) \Big[ T_{\text{elec}}(-i\hbar \, \partial/\partial x) + V(x, X) \Big] \left[ \frac{\partial}{\partial X_{Ni}} \psi(x; X) \right]$$

$$+ \int dx \, |\psi(x; X)|^2 \frac{\partial V(x, X)}{\partial X_{Ni}}$$

$$= \mathcal{E}(X) \left\{ \int dx \left[ \frac{\partial}{\partial X_{Ni}} \psi(x; X) \right]^* \psi(x; X) \right.$$

$$+ \left. \int dx \, \psi^*(x; X) \left[ \frac{\partial}{\partial X_{Ni}} \psi(x; X) \right] \right\}$$

$$+ \int dx \, |\psi(x; X)|^2 \frac{\partial V(x, X)}{\partial X_{Ni}}.$$

---

[8] It is not necessary for our purposes, but this can be rewritten as the Schrödinger equation for a set of independent harmonic oscillators, by introducing new coordinates defined as linear combinations of the $X_{Ni}$. The wave function $f$ is then a product of harmonic oscillator wave functions, one for each new coordinate, and the energy $E$ is the sum of the corresponding harmonic oscillator energies.

[9] F. Hellmann, *Einführung in die Quantenchemie* (Franz Deutcke, Leipzig & Vienna, 1937); R. P. Feynman, *Phys. Rev.* **56**, 540 (1939).

But the normalization condition (5.6.9) is satisfied for all $X$, so

$$\int \left[\frac{\partial}{\partial X_{Ni}}\psi(x;X)\right]^* \psi(x;X) + \int dx\, \psi^*(x;X)\left[\frac{\partial}{\partial X_{Ni}}\psi(x;X)\right] = 0,$$

which yields the desired result (5.6.21).

We can now check the validity of the Born–Oppenheimer approximation, in which we neglected the derivative of $\psi_a(x;X)$ with respect to $X$ in Eq. (5.6.15). The eigenvalue equation (5.6.5) involves only electronic variables, so the only dimensional parameters in this equation are $m_e$, $e$, and $\hbar$. The distance scale over which we must vary $X$ to make an appreciable change in $\psi_a(x;X)$ is therefore the Bohr radius

$$a \approx \hbar^2/m_e e^2,$$

because this is the only quantity with the units of length that can be formed from $m_e$, $e$, and $\hbar$. On the other hand, the Schrödinger equation (5.6.19) for the vibrational wave function $f(\mathbf{x})$ of the molecule involves only the parameters $\hbar^2/M$ (where $M$ is a typical nuclear mass in this molecule) and $K$. Equation (5.6.20) shows that the units of $K$ are [energy]/[distance]$^2$, so since $K$ arises from the electronic energy, it can only be of the order of atomic binding energies, roughly $e^4 m_e/\hbar^2$, divided by $a^2$, so

$$K \approx \frac{e^4 m_e}{\hbar^2 a^2} = \frac{e^8 m_e^3}{\hbar^6}.$$

The only quantity that can be formed from $\hbar^2/M$ and $K$ that has the dimensions of length is

$$b = \left(\frac{\hbar^2}{MK}\right)^{1/4} \approx \frac{\hbar^2}{e^2 M^{1/4} m_e^{3/4}},$$

so this is the distance over which one must vary $X$ to make an appreciable change in $f_a(X)$. The ratio of the second to the first term in the square brackets in Eq. (5.6.15) is then of order

$$\frac{\text{second term}}{\text{first term}} \approx \frac{1/a}{1/b} \approx \left(\frac{m_e}{M}\right)^{1/4}.$$

This varies from 0.15 for hydrogen to 0.04 for uranium. The corrections to the Born–Oppenheimer approximation are suppressed by one or more powers of this quantity. This shows a clear failure of first-order perturbation theory; the corrections to the leading approximation here are not proportional to $1/M_N$, but to $1/M_N^{1/4}$.

There is another, perhaps more physical, way of understanding the Born–Oppenheimer approximation. The energies of excited electronic states in molecules are similar to those in atoms, of order $e^4 m_e / \hbar^2$. In contrast, the energies of the excited molecular vibrational states are of order

$$\sqrt{K \hbar^2 / M} \approx \frac{e^4 m_e^{3/2}}{\hbar^2 M^{1/2}}.$$

Hence vibrational excitation energies are smaller than electronic excitation energies by a factor of order $\sqrt{m_e / M}$. (This is why molecular spectra are generally in the infrared, while atomic spectra are in the visible or ultraviolet.) The Born–Oppenheimer approximation works because the motion of nuclei in a molecule does not involve energies large enough to excite higher electronic states.

We can carry this further. We saw in Section 4.9 that the excitation energies of rotational states of the whole molecule are of order[10] $\hbar^2 / M a^2 = m_e^2 e^4 / M \hbar^2$, which is even smaller than the vibrational energies, by an additional factor $\sqrt{m_e / M}$. Thus we have a hierarchy of energies:

Electronic:     $e^4 m_e / \hbar^2$

Vibrational:     $(m_e / M)^{1/2} \times e^4 m_e / \hbar^2$

Rotational:     $(m_e / M) \times e^4 m_e / \hbar^2$

In the language of modern elementary particle physics, in the Born–Oppenheimer approximation the electronic states are "integrated out," resulting in an "effective Hamiltonian" for the nuclear motions. Similarly, we found in Section 4.9 that to a first approximation we do not need to consider the electronic and vibrational states of molecules in calculating rotational spectra.

In much the same way, from the beginning of atomic and molecular physics, theorists employed effective Hamiltonians in which internal excitations of atomic nuclei were implicitly ignored. Born and Oppenheimer were just the first to make this sort of analysis explicit, though for them it was electronic rather than internal nuclear excitations that were ignored. Today we usually (though not always) study the internal structure of nuclei using an effective Hamiltonian in which neutrons and protons are treated as point particles, ignoring the structure of the proton and neutron as composites of quarks, since the energies required to produce excited states of the proton and neutron are larger than those encountered in ordinary nuclear phenomena. And, similarly, we use the Standard Model of elementary particles without needing to know what happens at the very high energies where gravitation becomes a strong interaction.

---

[10] These energies are of the order of the squared angular momentum divided by the moment of inertia. The angular momentum is of order $\hbar$, and the moment of inertia is of order $M a^2$, so these rotational energies are of order $\hbar^2 / M a^2 = m_e^2 e^4 / M \hbar^2$.

## 5.7 The WKB Approximation

A particle of sufficiently high momentum will have a wave function that varies very rapidly with position, much more rapidly than the potential. The Schrödinger equation can be easily solved exactly for a constant potential, so it can be solved approximately for a potential that varies much more slowly than the wave function. This is the basis of an approximation introduced independently by Gregor Wentzel[11] (1898–1978), Hendrik Kramers[12] (1894–1952), and Leon Brillouin[13] (1889–1969), known as the WKB approximation.

Consider a Schrödinger equation of the form

$$\frac{d^2 u(x)}{dx^2} + k^2(x)\, u(x) = 0, \tag{5.7.1}$$

where

$$k(x) \equiv \sqrt{\frac{2\mu}{\hbar^2}\left(E - U(x)\right)}. \tag{5.7.2}$$

This is the form of the Schrödinger equation for a particle of mass $\mu$ in one dimension, with $u(x)$ the wave function for a state of energy $E$ and with $U(x)$ the potential, and it is also the form of the Schrödinger equation for a particle of mass $\mu$ (or for two particles with reduced mass $\mu$) in three dimensions, where $x$ is the radial coordinate, $u(x)$ is $x$ times the wave function $\psi(x)$ for energy $E$, and

$$U(x) \equiv V(x) + \frac{\hbar^2}{2\mu}\frac{\ell(\ell+1)}{x^2},$$

with $V(x)$ a central potential. For the present we are assuming that $U(x) \le E$; later we will consider the case $U(x) \ge E$.

If $k(x)$ were constant, Eq. (5.7.1) would have a solution $u(x) \propto \exp(\pm ikx)$, so when $k(x)$ is slowly varying, we expect a solution of the form

$$u(x) \propto A(x) \exp\left[\pm i \int k(x)\, dx\right], \tag{5.7.3}$$

where $A(x)$ is a slowly varying amplitude. This will satisfy Eq. (5.7.1) exactly if

$$A'' \pm 2ikA' \pm ik'A = 0. \tag{5.7.4}$$

Of course, this is no easier to solve than Eq. (5.7.1), but if $A(x)$ is sufficiently slowly varying we may be able to find an approximate solution by dropping the term $A''$. We will find such a solution, and then check under what conditions it is a good approximation.

---

[11]  G. Wentzel, *Z. Physik* **38**, 518 (1926).

[12]  H. A. Kramers, *Z. Physik* **39**, 828 (1926).

[13]  L. Brillouin, *Comptes Rendus Acad. Sci.* **183**, 24 (1926).

With $A''$ neglected, Eq. (5.7.4) becomes exactly soluble, with $A(x) \propto k^{-1/2}(x)$, so that we have a pair of approximate solutions of Eq. (5.7.1):

$$u(x) \propto \frac{1}{\sqrt{k(x)}} \exp\left[\pm i \int k(x)\, dx\right]. \qquad (5.7.5)$$

These solutions are valid if the term $A''$ in Eq. (5.7.4) is indeed much smaller than $k'A$. For $A = Ck^{-1/2}$ with $C$ constant, we have

$$A'' = C\left[-\frac{k''}{2k^{3/2}} + \frac{3k'^2}{4k^{5/2}}\right],$$

so we have $|A''| \ll |k'A|$ if $|k''/k^{3/2}| \ll |k'/\sqrt{k}|$ and $|k'^2/k^{5/2}| \ll |k'/k^{1/2}|$, or in other words if

$$\left|\frac{k''}{k'}\right| \ll k, \qquad \left|\frac{k'}{k}\right| \ll k. \qquad (5.7.6)$$

These conditions simply require that the magnitude of the fractional changes in both $k'$ and $k$ in a distance $1/k$ be much less than unity.

In the classically forbidden region where $U > E$, the Schrödinger equation takes the form

$$\frac{d^2u(x)}{dx^2} - \kappa^2(x)u(x) = 0, \qquad (5.7.7)$$

where

$$\kappa(x) \equiv \sqrt{\frac{2\mu}{\hbar^2}\left(U(x) - E\right)}. \qquad (5.7.8)$$

In exactly the same way as in the case $U < E$, we can find solutions

$$u(x) \propto \frac{1}{\sqrt{\kappa(x)}} \exp\left[\pm \int \kappa(x)\, dx\right], \qquad (5.7.9)$$

which are good approximations provided

$$\left|\frac{\kappa''}{\kappa'}\right| \ll \kappa, \qquad \left|\frac{\kappa'}{\kappa}\right| \ll \kappa. \qquad (5.7.10)$$

At this point, our discussion has to divide between problems in one dimension and problems in three dimensions.

### *One Dimension*

In a typical bound-state problem in one dimension, we have $U < E$ in a finite range $a_E < x < b_E$, and $U > E$ outside this range, where the wave function must decay exponentially for $x \to \pm\infty$. The conditions (5.7.6) and (5.7.10) clearly are not satisfied near the "turning points" $a_E$ and $b_E$, where $U = E$. If the conditions (5.7.10) become satisfied for all $x$ that are sufficiently greater than $b_E$, then in order to have a normalizable solution, in this region we must have

$$u(x) \propto \frac{1}{\sqrt{\kappa(x)}} \exp\left[-\int \kappa(x)\,dx\right]. \qquad (5.7.11)$$

On the other hand, for $x$ in the range $a_E < x < b_E$, and sufficiently far from the turning points, the solution is some linear combination of the two solutions (5.7.5). To find this solution, we must ask what linear combination for $x$ sufficiently below $b_E$ fits smoothly with the solution (5.7.11) for $x$ sufficiently above $b_E$. (We will come back later to the solution below $a_E$.)

Unless $E$ takes some special value, we expect that when $x$ is near $b_E$ we have $U(x) - E \propto x - b_E$, so that for $x$ just a little above $b_E$, we have

$$\kappa(x) \simeq \beta_E \sqrt{x - b_E}, \qquad (5.7.12)$$

where $\beta_E \equiv \sqrt{2\mu U'(b_E)}/\hbar$. To be more specific, Eq. (5.7.12) is a good approximation if $b_E \le x \ll b_E + \delta_E$, where $\delta_E \equiv 2U'(b_E)/|U''(b_E)|$. In this range of $x$, it is convenient to replace $x$ with a variable

$$\phi \equiv \int_{b_E}^{x} \kappa(x')\,dx' = \frac{2\beta_E}{3}(x - b_E)^{3/2}. \qquad (5.7.13)$$

In this case, the wave equation (5.7.7) takes the form

$$\frac{d^2u}{d\phi^2} + \frac{1}{3\phi}\frac{du}{d\phi} - u = 0. \qquad (5.7.14)$$

This has two independent solutions

$$u \propto \phi^{1/3} I_{\pm 1/3}(\phi), \qquad (5.7.15)$$

where $I_\nu(\phi)$ is the Bessel function of order $\nu$ with imaginary argument:[14]

$$I_\nu(\phi) = e^{-i\pi\nu/2} J_\nu\left(e^{i\pi/2}\phi\right),$$

where $J_\nu(z)$ is the usual Bessel function of order $\nu$.

Now, as long as Eq. (5.7.12) is a good approximation, we will have

$$\frac{\kappa'}{\kappa^2} = \frac{1}{3\phi}, \qquad \frac{\kappa''}{\kappa\kappa'} = -\frac{1}{3\phi},$$

so the conditions (5.7.10) for the WKB approximation will be satisfied if $\phi \gg 1$. There will be some overlap between the regions of $x$ in which the approximation (5.7.12) and the WKB approximation are satisfied, provided $\phi(b_E + \delta_E) \gg 1$, or in other words, if

$$\frac{2\beta_E}{3}\left(\frac{2U'(b_E)}{|U''(b_E)|}\right)^{3/2} = \kappa_E L_E \gg 1, \qquad (5.7.16)$$

---

[14] See, e.g., G. N. Watson, *A Treatise on the Theory of Bessel Functions*, 2nd edn. (Cambridge University Press, Cambridge, 1944), Section 3.7.

where $\kappa_E \equiv \sqrt{2\mu|E|}/\hbar$, and $L_E$ is a length that characterizes the scale of variation of the potential,

$$L_E \equiv \frac{2^{5/2}U'^2(b_E)}{3|U''(b_E)|^{3/2}|U(b_E)|^{1/2}}. \tag{5.7.17}$$

We will assume from now on that $\kappa_E L_E \gg 1$, so that there is a region in which the WKB approximation and the approximation (5.7.12) are *both* satisfied. As we have seen, in this region we must have $\phi \gg 1$, in which case we can use the asymptotic forms of the functions (5.7.15):

$$\phi^{1/3}I_{\pm 1/3}(\phi) \to (2\pi)^{-1/2}\phi^{-1/6}\Bigg[\exp(\phi)\,(1 + O(1/\phi))$$

$$+ \exp(-\phi - i\pi/2 \mp i\pi/3)\,(1 + O(1/\phi))\Bigg]. \tag{5.7.18}$$

Note that when Eq. (5.7.12) is satisfied, $\phi^{-1/6} \propto \kappa^{-1/2}$, so the solutions (5.7.18) do indeed match the form (5.7.9) for WKB solutions. It is now clear that in order for the solution of (5.7.14) to fit smoothly with the decaying WKB solution (5.7.11) when both are valid, we must take the solution near the turning point as the linear combination

$$u \propto \phi^{1/3}\left[I_{+1/3}(\phi) - I_{-1/3}(\phi)\right]. \tag{5.7.19}$$

Similarly, on the other side of the turning point, where $x$ is in the range $b_E - \delta_E \ll x \leq b_E$, we can write

$$k(x) \simeq \beta_E\sqrt{b_E - x} \tag{5.7.20}$$

and it is convenient to introduce a variable

$$\tilde{\phi} \equiv \int_x^{b_E} k(x')\,dx' = \frac{2\beta_E}{3}(b_E - x)^{3/2}. \tag{5.7.21}$$

The Schrödinger equation (5.7.1) then becomes

$$\frac{d^2u}{d\tilde{\phi}^2} + \frac{1}{3\tilde{\phi}}\frac{du}{d\tilde{\phi}} + u = 0. \tag{5.7.22}$$

This has two independent solutions

$$u \propto \tilde{\phi}^{1/3}J_{\pm 1/3}(\tilde{\phi}), \tag{5.7.23}$$

where, again, $J_\nu(z)$ is the usual Bessel function of order $\nu$. To see what linear combination of these solutions fits smoothly with the linear combination (5.7.19), we need to consider how both behave as $x \to b_E$.

For $\phi \to 0$, the solutions $\phi^{1/3} I_{\pm 1/3}(\phi)$ have the limiting behavior

$$\phi^{1/3} I_{+1/3}(\phi) \to \frac{\phi^{2/3}}{2^{1/3}\Gamma(4/3)} = \frac{(2\beta_E/3)^{2/3}}{2^{1/3}\Gamma(4/3)}(x - b_E), \quad (5.7.24)$$

$$\phi^{1/3} I_{-1/3}(\phi) \to \frac{2^{1/3}}{\Gamma(2/3)}. \quad (5.7.25)$$

On the other hand, for $\tilde{\phi} \to 0$ the solutions $\tilde{\phi}^{1/3} J_{\pm 1/3}(\tilde{\phi})$ behave as

$$\tilde{\phi}^{1/3} J_{+1/3}(\tilde{\phi}) \to \frac{\tilde{\phi}^{2/3}}{2^{1/3}\Gamma(4/3)} = \frac{(2\beta_E/3)^{2/3}}{2^{1/3}\Gamma(4/3)}(b_E - x), \quad (5.7.26)$$

$$\tilde{\phi}^{1/3} J_{-1/3}(\tilde{\phi}) \to \frac{2^{1/3}}{\Gamma(2/3)}. \quad (5.7.27)$$

We see that $\phi^{1/3} I_{+1/3}(\phi)$ fits smoothly with $-\tilde{\phi}^{1/3} J_{+1/3}(\tilde{\phi})$, while $\phi^{1/3} I_{-1/3}(\phi)$ fits smoothly with $+\tilde{\phi}^{1/3} J_{-1/3}(\tilde{\phi})$, so the solution (5.7.19) fits smoothly with

$$u \propto \tilde{\phi}^{1/3} \left[ J_{+1/3}(\tilde{\phi}) + J_{-1/3}(\tilde{\phi}) \right]. \quad (5.7.28)$$

As long as inequality (5.7.16) is satisfied, there will be values of $x$ for which both $\tilde{\phi} \gg 1$, so that the inequalities (5.7.6) are satisfied, and also the approximation (5.7.20) is satisfied, in which case we can use the asymptotic limit of Eq. (5.7.28) for $\tilde{\phi} \gg 1$:

$$\tilde{\phi}^{1/3} \left[ J_{+1/3}(\tilde{\phi}) + J_{-1/3}(\tilde{\phi}) \right] \to \sqrt{\frac{2}{\pi}} \tilde{\phi}^{-1/6} \left[ \cos \left( \tilde{\phi} - \frac{\pi}{6} - \frac{\pi}{4} \right) \right.$$
$$\left. + \cos \left( \tilde{\phi} + \frac{\pi}{6} - \frac{\pi}{4} \right) \right],$$

so

$$u \propto \tilde{\phi}^{-1/6} \cos \left( \tilde{\phi} - \frac{\pi}{4} \right) \propto k^{-1/2}(x) \cos \left( \int_x^{b_E} k(x')\, dx' - \frac{\pi}{4} \right).$$

Everywhere between the turning points where the conditions (5.7.6) are satisfied the wave function must be a fixed linear combination of the two independent solutions (5.7.5), and so we can conclude that for all such $x$

$$u \propto k^{-1/2}(x) \cos \left( \int_x^{b_E} k(x')\, dx' - \frac{\pi}{4} \right). \quad (5.7.29)$$

The same arguments apply to the other turning point, at $x = a_E$, except that here $U(x)$ increases with *decreasing* rather than with increasing $x$, so by the same reasoning, we can conclude that everywhere between the turning points where the conditions (5.7.6) are satisfied the wave function must have the form

$$u \propto k^{-1/2}(x) \cos \left( \int_{a_E}^{x} k(x')\, dx' - \frac{\pi}{4} \right). \quad (5.7.30)$$

In order for both Eq. (5.7.29) and Eq. (5.7.30) to be correct, we must have

$$\cos\left(\int_x^{b_E} k(x')\,dx' - \frac{\pi}{4}\right) \propto \cos\left(\int_{a_E}^x k(x')\,dx' - \frac{\pi}{4}\right),$$

for all such $x$. Further, since both cosines oscillate between $+1$ and $-1$, the coefficient of proportionality can only be $+1$ or $-1$. This leaves us with just two possibilities for the arguments of the cosines:

$$\int_x^{b_E} k(x')\,dx' - \frac{\pi}{4} = \int_{a_E}^x k(x')\,dx' - \frac{\pi}{4} + n\pi$$

or else

$$\int_x^{b_E} k(x')\,dx' - \frac{\pi}{4} = -\left[\int_{a_E}^x k(x')\,dx' - \frac{\pi}{4}\right] + n\pi,$$

where $n$ is an integer, not necessarily positive. The first of these two alternatives is ruled out because the left-hand side decreases with $x$ while the right-hand side increases with $x$, so we are left with the second possibility, which can be written as

$$\int_{a_E}^{b_E} k(x')\,dx' = \left(n + \frac{1}{2}\right)\pi. \tag{5.7.31}$$

The left-hand side is positive, so here the integer $n$ can only be zero or positive-definite.

Equation (5.7.31) is almost the same as the generalization (1.2.12) of Bohr's quantization condition introduced subsequently by Sommerfeld. In a whole cycle of oscillation a particle goes from $b_E$ to $a_E$ and then back again, so the WKB approximation gives the integral in the Sommerfeld quantization condition as

$$\oint p\,dq = 2\hbar \int_{a_E}^{b_E} k(x')\,dx' = 2\pi\hbar\left(n + \frac{1}{2}\right) = h\left(n + \frac{1}{2}\right).$$

Hence Eq. (5.7.31) differs from the Sommerfeld quantization condition only by the presence of the term $1/2$ accompanying $n$. The derivation given here suggests that Eq. (5.7.31) should work well only for large $n$, in which case the term $1/2$ is inconsequential, but in fact with this term for many potentials it works surprisingly well for all $n$. In particular, for the harmonic oscillator we have $U(x) = \mu\omega^2 x^2/2$, so $E = \mu\omega^2 b_E^2/2$ and $a_E = -b_E$. The integral in Eq. (5.7.31) is then

$$\int_{a_e}^{b_e} k\,dx = \frac{\mu\omega b_E^2}{\hbar} \int_{-1}^{+1} \sqrt{1 - y^2}\,dy = \frac{\mu\omega b_E^2}{\hbar}\frac{\pi}{2} = \frac{E\pi}{\hbar\omega}$$

and Eq. (5.7.31) therefore gives $E = \hbar\omega(n + 1/2)$, which is the correct exact result for a harmonic oscillator potential.

### Three Dimensions with Spherical Symmetry

For the three-dimensional case, the radial coordinate $r$ (now using $r$ rather than $x$ for the coordinate) is of course limited to $r > 0$, so we do not have any boundary condition for $r \to -\infty$. Instead, as we saw in Section 2.1, for any potential that does not grow as fast as $1/r^2$ for $r \to 0$, the reduced wave function $u(r) \equiv r\psi(r)$ obeys the boundary condition that $u(r) \propto r^{\ell+1}$ for $r \to 0$. We generally will have an outer turning point at $r = b_E$ where $U(b_E) = E$, and the wave function must decay exponentially for $r \gg b_E$, so that in at least a range of $r$ below $b_E$ the wave function will be of the form (5.7.29):

$$
u(r) \propto k^{-1/2}(r) \cos \left( \int_r^{b_E} k(r') \, dr' - \frac{\pi}{4} \right). \tag{5.7.32}
$$

For $\ell \neq 0$ we always also have an inner turning point at $r = a_E < b_E$ where $U(a_E) = E$. The wave function (5.7.32) is then subject to the condition that it fit smoothly with a solution for $r < a_E$ that goes as $r^{\ell+1}$ rather than $r^{-\ell}$ as $r \to 0$. This can be complicated, especially because for $\ell \neq 0$ the WKB approximation does not work for $r \to 0$, where $\kappa \propto 1/r$. Things are simpler for the case $\ell = 0$, where there is no centrifugal barrier, and there may not be any inner turning point. If there is no inner turning point, then for a reasonably smooth potential the solution (5.7.32) will continue to be valid all the way down to $r = 0$. In this case, the condition that $u(r) \propto r$ for $r \to 0$ requires that the argument of the cosine in Eq. (5.7.32) must take the value $n\pi - \pi/2$ for $r = 0$, where $n$ is an integer, so that the condition for a bound state is that

$$
\int_0^{b_E} k(r') \, dr' = \left( n - \frac{1}{4} \right) \pi, \tag{5.7.33}
$$

and hence $n \geq 1$. For instance, for the $\ell = 0$ states of the Coulomb potential, we have $U(r) = -Ze^2/r$, so

$$
k(r) = \sqrt{\frac{2m_e}{\hbar^2} \left( E + Ze^2/r \right)}.
$$

For $E < 0$ there is a turning point, at $b_E = -Ze^2/E$, and

$$
\int_0^{b_E} k(r) \, dr = \sqrt{\frac{-2m_e E}{\hbar^2}} \int_0^{b_E} dr \sqrt{\frac{b_E}{r} - 1} = \frac{\pi}{2} \sqrt{-\frac{2m_e}{\hbar^2 E}} Ze^2.
$$

The condition (5.7.33) then gives

$$
E = -\frac{Z^2 e^4 m_e}{2\hbar^2 (n - 1/4)^2}.
$$

This is the same as the Bohr formula (1.2.11) for the $n$th energy level (which as shown in Chapter 2 is the correct consequence of quantum mechanics), except that $n$ is replaced here with $n - 1/4$. Thus the WKB approximation works very well for the high energy levels, for which $n \gg 1/4$, as we would expect, since for these energy levels the wave function oscillates many times. Even for moderate $n$, the WKB quantization condition (5.7.33) works pretty well for the Coulomb potential, but not as well as the Sommerfeld quantization condition (1.2.12).

## 5.8 Broken Symmetry

It sometimes happens that a Hamiltonian has a symmetry, which is shared by its eigenstates, but that the physical states that are actually realized in nature are instead nearly exact solutions of the Schrödinger equation for which the symmetry is broken. We can find examples of this in non-relativistic quantum mechanics of great importance to chemistry and molecular physics.

For instance, consider a particle of mass $m$ moving in one dimension in a potential $V(x)$ with the symmetry $V(-x) = V(x)$. If $\psi(x)$ is a solution of the Schrödinger equation with a given energy, then so is $\psi(-x)$, so in the absence of degeneracy we must have $\psi(-x) = \alpha\psi(x)$, with $\alpha$ some constant. It follows then that $\psi(x) = \alpha\psi(-x) = \alpha^2\psi(x)$, so $\alpha$ can only be $+1$ or $-1$, and the energy eigenfunctions will be either even or odd in $x$. The states of lowest energies with even or odd wave functions will generally have quite different energies.

But suppose that the potential has two minima, symmetrically spaced around the origin, separated by a high thick barrier centered at $x = 0$. This is the case for instance for the ammonia $NH_3$ molecule, where $x$ is the position of the nitrogen nucleus along a line transverse to the plane formed by the three hydrogen nuclei, and the barrier is provided by the strong repulsion between the positive charges of the nitrogen and hydrogen nuclei. If the barrier were infinitely high and thick, there would be two degenerate energy eigenstates with energies $E_0$, one with a wave function $\psi_0(x)$ that is non-zero only for $x > 0$, and the other with a wave function $\psi_0(-x)$ that is non-zero only for $x < 0$. Each of these solutions breaks the symmetry under $x \leftrightarrow -x$. From them, we could form even and odd solutions, $[\psi_0(x) \pm \psi_0(-x)]/\sqrt{2}$, that would also be degenerate, with energy $E_0$. But if the barrier is high and thick but finite, then these even and odd solutions are not degenerate, but only nearly degenerate.

To estimate the order of magnitude of the energy splitting, we can use the WKB method described in the previous section. Within the barrier, the even and odd wave functions take the form

$$\psi_\pm(x) \propto \frac{1}{\sqrt{\kappa(x)}} \left[ \exp\left( \int_0^x \kappa(x')\, dx' \right) \pm \exp\left( \int_0^{-x} \kappa(x')\, dx' \right) \right], \quad (5.8.1)$$

where for a particle of mass $m$ and energy $E$ in a potential $V(x)$,

$$\kappa(x) = \sqrt{\frac{2m}{\hbar^2}\left(V(x) - E\right)}. \tag{5.8.2}$$

This should be a good approximation within the barrier if the barrier is high enough and smooth enough that $\kappa(x)$ is much larger than the logarithmic rates of change of $\kappa(x)$ and $\kappa'(x)$.

The logarithmic derivatives of these wave functions are

$$\frac{\psi'_\pm(x)}{\psi_\pm(x)} \simeq -\frac{\kappa'(x)}{2\kappa(x)} + \kappa(x)\left[\frac{\exp(\int_0^x \kappa(x')\,dx') \mp \exp\left(\int_0^{-x}\kappa(x')\,dx'\right)}{\exp(\int_0^x \kappa(x')\,dx') \pm \exp\left(\int_0^{-x}\kappa(x')\,dx'\right)}\right]. \tag{5.8.3}$$

(For the validity of the WKB approximation it is necessary that $|\kappa'|/\kappa \ll \kappa$, so the first term in Eq. (5.8.3) is generally much less than the second term, but we keep it here anyway, because it does not raise problems for our discussion.) For a thick barrier extending from $-a$ to $+a$ with

$$\int_0^a \kappa\,dx = \int_{-a}^0 \kappa\,dx \gg 1$$

the logarithmic derivatives at the barrier edges are

$$\frac{\psi'_\pm(a)}{\psi_\pm(a)} = -\frac{\psi'_\pm(-a)}{\psi_\pm(-a)} \simeq -\frac{\kappa'(a)}{2\kappa(a)} + \kappa(a)\left[1 \mp 2\exp\left(-\int_{-a}^a \kappa(x')\,dx'\right)\right]. \tag{5.8.4}$$

The energy is determined by the condition that these logarithmic derivatives must match the logarithmic derivative of the wave function just outside the barrier. Equation (5.8.4) shows that for a thick barrier, this condition is nearly the same for the even and odd solution, the difference being a term proportional to $\exp\left(-\int_{-a}^a \kappa(x')\,dx'\right)$. Thus the even and odd wave functions have energies $E_\pm \simeq E_1 \pm \delta E$, where $E_1$ is approximately equal to the energy of both even and odd states in the limit of an infinitely thick barrier, and $\delta E$ is suppressed by a factor $\exp\left(-\int_{-a}^a \kappa(x')\,dx'\right)$.

Because $\delta E$ is very small for a thick barrier, the broken-symmetry states, with the wave function concentrated on one side or the other of the barrier, are nearly energy eigenstates. But why should these broken-symmetry states be the ones realized in nature, rather than the true energy eigenstates, which are either even or odd under the symmetry? The answer has to do with the phenomenon of decoherence, discussed in Section 3.7. The wave function will inevitably be subject to external perturbations, which for a thick barrier produce fluctuations in the phase of the wave function, with no correlation between the phase changes on the two sides of the barrier. These fluctuations cannot change a broken-symmetry wave function that is concentrated on one side of the barrier into a solution that is wholly or partly concentrated on the other side, but they rapidly change an even or odd wave function into one that is an incoherent mixture of even and odd wave functions. The states realized in the real world are the ones that are stable up to a phase under these fluctuations, and these are the broken-symmetry states.

But the broken-symmetry states, though insensitive to external perturbations, are not really stable. It is instructive to look at the time-dependence of a wave function $\psi(x, t)$ that at $t = 0$ takes the form $\psi_0(x)$, non-zero only for $x > 0$. We can write this initial wave function as

$$\psi(x, 0) = \frac{1}{2}[\psi_0(x) + \psi_0(-x)] + \frac{1}{2}[\psi_0(x) - \psi_0(-x)],$$

so at any later time $t$, the wave function is

$$
\begin{aligned}
\psi(x, t) &\simeq \frac{1}{2}[\psi_0(x) + \psi_0(-x)] \exp\left(-i(E_1 + \delta E)t/\hbar\right) \\
&\quad + \frac{1}{2}[\psi_0(x) - \psi_0(-x)] \exp\left(-i(E_1 - \delta E)t/\hbar\right) \\
&= \exp\left(-iE_1 t/\hbar\right)\left[\psi_0(x) \cos\left(\delta E t/\hbar\right) - i\psi_0(-x) \sin\left(\delta E t/\hbar\right)\right].
\end{aligned}
$$
$$(5.8.5)$$

We see that a particle given the broken-symmetry wave function $\psi_0(x)$ will at first leak through the barrier into the region $x < 0$, with an amplitude for the other wave function $\psi_0(-x)$ increasing at a rate $\Gamma = \delta E/\hbar$. Eventually the amplitude for $x < 0$ builds up, until the particle begins to leak back into the region $x > 0$. But if the barrier is very high and thick, the broken-symmetry wave function $\psi_0(x)$ can persist for an exponentially long time. Indeed, there are molecules like sugars and proteins that can exist in "chiral" configurations, configurations with a definite left-handedness or right-handedness, that are separated by barriers much thicker than for ammonia. For such molecules, the transition from one broken-symmetry state to another takes so long as to be unobservable. This is why we can encounter left- and right-handed sugars and proteins in nature.

These considerations point up a general feature of spontaneous symmetry breaking: it is always associated with systems that in some sense are very large. It is only the very large barrier in molecules like proteins and sugars that allows these molecules to have a definite handedness. In quantum field theory, it is the infinite volume of the vacuum state that allows other symmetries to be spontaneously broken.[15]

---

[15] For a discussion of this point, see S. Weinberg, *The Quantum Theory of Fields*, Vol. II (Cambridge University Press, Cambridge, 1996), Section 19.1.

## 5.9  Van der Waals Forces

There is of course no Coulomb force between electrically neutral atoms or molecules. However, even between neutral systems, there are weaker electrical forces that are of long range, in the sense that they decrease only as inverse powers of the separation, not exponentially. The first sign of such forces was found in corrections to the ideal gas equation of state, interpreted as an effect of long-range forces between molecules by Johannes Diderik van der Waals (1837–1923), in his 1873 Ph.D. thesis at the University of Leiden. These forces can arise in first-order perturbation theory between molecules with permanent electric multipole moments, but even for atoms and molecules that are without such moments, there is always a long-range force arising in second-order perturbation theory from mutually-induced electric dipole moments. This was first calculated[16] by Fritz London (1900–1954).

Consider two systems $A$ and $B$ consisting of several point particles respectively labeled $a$ and $b$, with charges $e_a$ and $e_b$. We assume that these systems are stable in isolation, and massive enough that their centers of mass have a well-defined separation vector $\mathbf{R}$. We consider separations sufficiently large that there is essentially no overlap between the spatial wave functions of the charged particles in each system, so that each charged particle can be considered to belong either to system $A$ or to system $B$. We take $\mathbf{x}_a$ to be the distance of the $a$th particle in system $A$ from the center of mass of that system, and take $\mathbf{y}_b$ to be the distance of the $b$th particle in system $B$ from the center of mass of that system. Including only electrostatic interactions between the two systems, the Hamiltonian is

$$H = H_0 + H', \tag{5.9.1}$$

where $H_0$ is the sum $H_A + H_B$ of the Hamiltonians of systems $A$ and $B$ in isolation, and

$$H' = \sum_{a \in A} \sum_{b \in B} \frac{e_a e_b}{|\mathbf{x}_a - \mathbf{y}_b + \mathbf{R}|}. \tag{5.9.2}$$

We are assuming here that the separation $R \equiv |\mathbf{R}|$ is large enough that the wave function is negligible unless $|\mathbf{x}_a| \ll R$ and $|\mathbf{y}_b| \ll R$. We can therefore expand Eq. (5.9.2) in powers of $|\mathbf{x}_a|/R$ and $|\mathbf{y}_b|/R$. For this purpose, we use the partial-wave expansion of the denominator in the directions $\hat{x}_a = \mathbf{x}_a/|\mathbf{x}_a|$, $\hat{y}_b = \mathbf{y}_b/|\mathbf{y}_b|$,

---

[16] R. Eisenschitz and F. London, *Z. Physik* **60**, 491 (1930); F. London, *Z. Physik* **63**, 245 (1930).

and $\hat{R} = \mathbf{R}/|\mathbf{R}|$. Taking account of the invariance of $|\mathbf{x}_a - \mathbf{y}_b + \mathbf{R}|$ under rotations of $\mathbf{x}_a$, $\mathbf{y}_b$, and $\mathbf{R}$, this expansion takes the form[17]

$$
\frac{1}{|\mathbf{x}_a - \mathbf{y}_b + \mathbf{R}|} = \sum_{\ell \ell' L} f_{\ell \ell' L}(|\mathbf{x}_a|, |\mathbf{y}_b|, R) \sum_{mm'M} (-1)^{L-M} C_{\ell \ell'}(LM; mm')
$$
$$
\times Y_\ell^m(\hat{x}_a) Y_{\ell'}^{m'}(\hat{y}_b) Y_L^{-M}(\hat{R}), \tag{5.9.3}
$$

where $Y_\ell^m$, etc., are the spherical harmonics described in Section 2.2, and $C_{\ell \ell'}(LM; mm')$ are the Clebsch–Gordan coefficients discussed in Section 4.3. Because a term with any given values of $\ell$ and $\ell'$ must be a power series in the Cartesian components of $\mathbf{x}_a$ and $\mathbf{y}_b$, the function $f_{\ell \ell' L}(|\mathbf{x}_a|, |\mathbf{y}_b|, R)$ must contain at least $\ell$ factors of $|\mathbf{x}_a|$ and $\ell'$ factors of $|\mathbf{y}_b|$. In fact, these are the only powers of $|\mathbf{x}_a|$ and $|\mathbf{y}_b|$ that do appear in $f_{\ell \ell' L}(|\mathbf{x}_a|, |\mathbf{y}_b|, R)$. To see this, we need only note[18] that for any vectors $\mathbf{u}$ and $\mathbf{v}$ with $|\mathbf{u}| < |\mathbf{v}|$:

$$
|\mathbf{u} - \mathbf{v}|^{-1} = \sum_{\ell=0}^{\infty} \frac{4\pi}{2\ell+1} |\mathbf{u}|^\ell |\mathbf{v}|^{-\ell-1} \sum_{m=-\ell}^{\ell} (-1)^{\ell-m} Y_\ell^m(\hat{u}) Y_\ell^{-m}(\hat{v}). \tag{5.9.4}
$$

Using this formula with $\mathbf{u} = \mathbf{x}_a$ and $\mathbf{v} = -\mathbf{R} + \mathbf{y}_b$ shows that the whole dependence of $f_{\ell \ell' L}(|\mathbf{x}_a|, |\mathbf{y}_b|, R)$ on $|\mathbf{x}_a|$ is a factor $|\mathbf{x}_a|^\ell$, while using this formula with $\mathbf{u} = \mathbf{y}_b$ and $\mathbf{v} = \mathbf{R} + \mathbf{x}_a$ shows that the whole dependence of $f_{\ell \ell' L}(|\mathbf{x}_a|, |\mathbf{y}_b|, R)$ on $|\mathbf{y}_b|$ is a factor $|\mathbf{y}_b|^{\ell'}$. Dimensional analysis tells us then that

$$
f_{\ell \ell' L}(|\mathbf{x}_a|, |\mathbf{y}_b|, R) = N_{\ell \ell' L} R^{-1-\ell-\ell'} |\mathbf{x}_a|^\ell |\mathbf{y}_b|^{\ell'}, \tag{5.9.5}
$$

where the $N_{\ell \ell' L}$ are numerical coefficients, generally of order unity, which we will not attempt to calculate except in one case. Using Eqs. (5.9.3) and (5.9.5) in Eq. (5.9.2), we find the perturbation Hamiltonian

$$
H' = \sum_{\ell \ell' L} N_{\ell \ell' L} R^{-1-\ell-\ell'}
$$
$$
\times \sum_{mm'M} (-1)^{L-M} C_{\ell \ell'}(LM; mm') Y_L^{-M}(\hat{R}) E_\ell^{m \, (A)} E_{\ell'}^{m' \, (B)}, \tag{5.9.6}
$$

---

[17] The sum over $m$ and $m'$ yields a function of $\hat{x}_a$ and $\hat{y}_b$ that transforms with angular momentum $L$, $M$, and then the sum over $M$ gives a rotational scalar. We are here using Eq. (4.3.35), with the factor $1/\sqrt{2L+1}$ included in the coefficient $f_{\ell \ell' L}$.

[18] This is equivalent to a formula given by W. Magnus and F. Oberhettinger, *Formulas and Theorems for the Functions of Mathematical Physics*, transl. J. Wermer (Chelsea Publishing Co., New York, 1949), p. 51, together with Eq. (4.3.36) for the expansion of Legendre polynomials as sums of products of spherical harmonics.

where $E_\ell^{m\,(A)}$ and $E_{\ell'}^{m'\,(B)}$ are the electric-multipole operators of systems $A$ and $B$:

$$E_\ell^{m\,(A)} \equiv \sum_{a\in A} e_a |\mathbf{x}_a|^\ell Y_\ell^m(\hat{x}_a), \qquad E_{\ell'}^{m'\,(B)} \equiv \sum_{b\in B} e_b |\mathbf{y}_b|^{\ell'} Y_{\ell'}^{m'}(\hat{y}_b). \qquad (5.9.7)$$

These operators for $\ell = 1$, $\ell = 2$, $\ell = 3$, etc. are conventionally known as the electric-dipole, -quadrupole, -octupole, etc., moments.

There are limitations on the terms that can actually appear in Eq. (5.9.6), in addition to the limitations imposed by the presence of a Clebsch–Gordan coefficient.

(i) There are no non-zero terms with $\ell = 0$ or $\ell' = 0$. A term with $\ell = 0$ or $\ell' = 0$ is proportional to $\sum_{a\in A} e_a$ or $\sum_{b\in B} e_b$ respectively, and therefore vanishes because both systems are assumed to have zero total charge.

(ii) There are no non-zero terms with $L = 0$. Any term with $L = 0$ arises from the average of Eq. (5.9.2) over the directions of $\mathbf{R}$, but this average is

$$\frac{1}{4\pi} \sum_{a\in A} \sum_{b\in B} e_a e_b \int d^2\hat{R} \, \frac{1}{|\mathbf{x}_a - \mathbf{y}_b + \mathbf{R}|} = \sum_{a\in A} \sum_{b\in B} \frac{e_a e_b}{R} \qquad (5.9.8)$$

and this vanishes because $\sum_{a\in A} e_a = \sum_{b\in B} e_b = 0$.

(iii) The only non-zero terms are those with $\ell + \ell' + L$ even. This is because Eq. (5.9.2) is manifestly even under the joint reflection $\mathbf{x}_a \mapsto -\mathbf{x}_a$, $\mathbf{y}_b \mapsto -\mathbf{y}_b$, $\mathbf{R} \mapsto -\mathbf{R}$, but according to the space reflection property (2.2.18) of the spherical harmonics, the product of spherical harmonics in Eq. (5.9.3) changes under this joint reflection by a sign $(-1)^{\ell+\ell'+L}$. Hence $N_{\ell\ell'L}$ must vanish unless $\ell + \ell' + L$ is even.

Equation (5.9.6) shows that for $R$ large the largest terms are those with $\ell + \ell'$ smallest. Taking into account the presence of the Clebsch–Gordan coefficient in Eq. (5.9.6) and the three above remarks, the leading terms are as follows.

**Dipole–Dipole.** These are terms with $\ell = \ell' = 1$, which therefore go as $R^{-3}$. Since $L = 0$ and $L = 1$ are excluded by points (ii) and (iii) above, these terms must have $L = 2$.

**Dipole–Quadrupole.** These are terms with $\ell = 1$, $\ell' = 2$, or vice versa, and therefore go as $R^{-4}$. These terms have both $L = 1$ and $L = 3$.

**Quadrupole–Quadrupole.** These are terms with $\ell = \ell' = 2$, and therefore go as $R^{-5}$. They have both $L = 2$ and $L = 4$.

**Dipole–Octupole.** These are terms with $\ell = 1$, $\ell' = 3$, or vice versa, and therefore also go as $R^{-5}$. They too have both $L = 2$ and $L = 4$.

Let us take a closer look at the dipole–dipole term, which will turn out to be most important. Expanding the denominator in Eq. (5.9.2) to first order in $\mathbf{x}_a$

and in $\mathbf{y}_b$ (and so dropping all terms that depend only on $\mathbf{x}_a$ or $\mathbf{y}_b$, which do not contribute in Eq. (5.9.2) because $\sum_{a \in A} e_a$ and $\sum_{b \in B} e_b$ are both assumed to vanish), we find

$$[H']_{\text{dipole-dipole}} = \frac{1}{R^3}\left[3\hat{R} \cdot \mathbf{D}^{(A)} \hat{R} \cdot \mathbf{D}^{(B)} - \mathbf{D}^{(A)} \cdot \mathbf{D}^{(B)}\right], \qquad (5.9.9)$$

where

$$\mathbf{D}^{(A)} \equiv \sum_{a \in A} e_a \mathbf{x}_a, \qquad \mathbf{D}^{(B)} \equiv \sum_{b \in B} e_b \mathbf{y}_b. \qquad (5.9.10)$$

Using the list of spherical harmonics in Section 2.2 and the table of Clebsch–Gordan coefficients in Section 4.3, the reader can check that the expression (5.9.9) is the same as the $\ell = \ell' = 1, L = 2$ term in the expansion (5.9.6), with $N_{112} = (4\pi)^{3/2}/3$.

In first-order perturbation theory, when systems $A$ and $B$ are in states $\Psi_\alpha$ and $\Psi_\beta$ respectively, the perturbation Hamiltonian (5.9.6) produces a potential energy given by the expectation value

$$V_1(\mathbf{R}) = \sum_{\ell\ell'L} N_{\ell\ell'L} R^{-1-\ell-\ell'} \sum_{mm'M} (-1)^{L-M} C_{\ell\ell'}(LM; mm') Y_L^{-M}(\hat{R})$$
$$\times \langle E_\ell^{m\,(A)}\rangle_\alpha \langle E_{\ell'}^{m'\,(B)}\rangle_\beta. \qquad (5.9.11)$$

The multipole operators $E_\ell^{m\,(A)}$ and $E_{\ell'}^{m'\,(B)}$ change under space inversion by factors $(-1)^\ell$ and $(-1)^{\ell'}$ respectively, so their expectation values with $\ell$ odd or $\ell'$ odd vanish if as usual the states $\Psi_\alpha$ and $\Psi_\beta$ have definite parity. Thus in the usual case, the leading term for large $R$ in first-order perturbation theory is not the dipole–dipole term, but the quadrupole–quadrupole term with $\ell = \ell' = 2$, which goes as $R^{-5}$. But as remarked at the end of Section 4.4, the expectation value of *any* operator $O_j^m$ with $j \neq 0$ vanishes for all unpolarized systems. Thus if systems $A$ and $B$ are unpolarized, then in first-order perturbation theory neither the quadrupole–quadrupole interaction nor any term in Eq. (5.9.11) contributes to the interaction energy between these systems. To find the interaction energy, we then have to go to second-order perturbation theory.

For any given multipole operators $E_\ell^{m\,(A)}$ and $E_{\ell'}^{m'\,(B)}$ including the electric-dipole operators, there are always some excited states $\Psi_{\alpha'}$ and $\Psi_{\beta'}$ for which the matrix elements $\left(\Psi_{\alpha'}, E_\ell^{m\,(A)}\Psi_\alpha\right)$ and $\left(\Psi_{\beta'}, E_{\ell'}^{m'\,(B)}\Psi_\beta\right)$ do not vanish. For instance, the electric-dipole moment has a non-vanishing matrix element between the $1s$ ground state of the hydrogen atom and the $2p$ excited state, which can be calculated from measurements of the rate of emission of Lyman-$\alpha$ photons from this excited state. Thus in second-order perturbation theory we expect the potential to be dominated for large $R$ by the dipole–dipole term, which has the least rapid decrease for $R \to \infty$. According to Eqs. (5.4.5) and (5.9.9), in second-order perturbation theory this makes a contribution to the interaction energy when systems $A$ and $B$ are in states $\Psi_\alpha$ and $\Psi_\beta$, given by

$$
V_2(\mathbf{R}) = \frac{1}{R^6} \sum_{\alpha'\beta'} \left[ E_\alpha + E_\beta - E_{\alpha'} - E_{\beta'} \right]^{-1}
$$
$$
\times \left| 3\hat{R} \cdot \left( \Psi_{\alpha'}, \mathbf{D}^{(A)} \Psi_\alpha \right) \hat{R} \cdot \left( \Psi_{\beta'}, \mathbf{D}^{(B)} \Psi_\beta \right) \right.
$$
$$
\left. - \left( \Psi_{\alpha'}, \mathbf{D}^{(A)} \Psi_\alpha \right) \cdot \left( \Psi_{\beta'}, \mathbf{D}^{(B)} \Psi_\beta \right) \right|^2 . \tag{5.9.12}
$$

There are no cancellations here that would cause this to vanish if we have to average over the 3-components of the angular momenta of states $\Psi_\alpha$ and $\Psi_\beta$. In fact, where these are the ground states, the energy denominator in Eq. (5.9.12) is negative-definite, while the numerator is positive-definite, so $V_2(\mathbf{R})$ is negative-definite. Since $|V_2(\mathbf{R})|$ also decreases monotonically with increasing $R$, this energy represents a purely attractive force between systems $A$ and $B$.

## Problems

1. Suppose that the interaction of the electron with the proton in the hydrogen atom produces a change in the potential energy of the electron of the form

$$
\Delta V(r) = V_0 \exp(-r/R),
$$

where $R$ is much smaller than the Bohr radius $a$. Calculate the shift in the energies of the $2s$ and $2p$ states of hydrogen, to first order in $V_0$.

2. It is sometimes assumed that the electrostatic potential felt by an electron in a multi-electron atom can be approximated by a shielded Coulomb potential, of the form

$$
V(r) = -\frac{Ze^2}{r} \exp(-r/R),
$$

where $R$ is the estimated radius of the atom. Use the variational method to give an approximate formula for the energy of an electron in the state of lowest energy in this potential, taking as the trial wave function

$$
\psi(\mathbf{x}) \propto \exp\left(-r/\rho\right),
$$

with $\rho$ a free parameter.

3. Calculate the shift in energy of the $2p_{3/2}$ state of hydrogen in a very weak static electric field $E$, to second order in $E$, assuming that $E$ is small enough that this shift is much less than the fine-structure splitting between the $2p_{1/2}$ and $2p_{3/2}$ states. In using second-order perturbation theory here, you can

consider only the intermediate state for which the energy-denominator is smallest.

4. The spin–orbit coupling of the electron in hydrogen produces a term in the Hamiltonian of the form

$$\Delta H = \xi(r)\mathbf{L} \cdot \mathbf{S},$$

where $\xi(r)$ is some small function of $r$. Give a formula for the contribution of $\Delta V$ to the fine-structure splitting between the $2p_{1/2}$ and $2p_{3/2}$ states in hydrogen, to first order in $\xi(r)$.

5. Using the WKB approximation, derive a formula for the energies of the bound $s$ states of a particle of mass $m$ in a potential $V(r) = -V_0 e^{-r/R}$, with $V_0$ and $R$ both positive.

# 6

# Approximations for Time-Dependent Problems

The Hamiltonian of any isolated system is time-independent, but we often have to deal with quantum-mechanical systems that are not isolated, but affected by time-dependent external fields, in which case the part of the Hamiltonian representing the interaction with these fields depends on time. Here we are not interested in calculating perturbations to the energies of bound states, because physical states are no longer characterized by definite energies. Instead, our interest is in calculating the rates at which the quantum system undergoes changes of one sort or another. Such calculations can be done exactly only in the simplest cases, so again we find it necessary to consider approximation methods, of which the simplest and most versatile is perturbation theory.

## 6.1 First-Order Perturbation Theory

We consider a Hamiltonian

$$H(t) = H_0 + H'(t), \tag{6.1.1}$$

where $H_0$ is the time-independent Hamiltonian of the system in the absence of external fields, and $H'(t)$ is a small time-dependent perturbation. The state vector $\Psi$ of the system satisfies the time-dependent Schrödinger equation

$$i\hbar \frac{d\Psi(t)}{dt} = H(t)\Psi(t). \tag{6.1.2}$$

We can find a complete orthonormal set of time-independent unperturbed state vectors

$$H_0\Psi_n = E_n\Psi_n, \qquad \left(\Psi_n, \Psi_m\right) = \delta_{nm}, \tag{6.1.3}$$

and expand $\Psi(t)$ in the $\Psi_n$,

$$\Psi(t) = \sum_n c_n(t) \exp(-i E_n t/\hbar)\, \Psi_n, \tag{6.1.4}$$

with time-dependent coefficients $c_n(t)$ from which a factor $\exp(-iE_nt/\hbar)$ has been extracted for later convenience. The perturbation $H'(t)$ acting on $\Psi_n$ may itself be expanded in the $\Psi_m$:

$$H'(t)\Psi_n = \sum_m \Psi_m \left( \Psi_m, H'(t)\Psi_n \right),$$

so the time-dependent Schrödinger equation (6.1.2) reads

$$\sum_n \left[ i\hbar \frac{dc_n(t)}{dt} + E_n c_n(t) \right] \exp(-iE_nt/\hbar)\, \Psi_n$$

$$= \sum_n c_n(t) \left[ E_n \Psi_n + \sum_m H'_{mn}(t)\Psi_m \right] \exp(-iE_nt/\hbar),$$

where

$$H'_{mn}(t) = \left( \Psi_m, H'(t)\Psi_n \right).$$

Cancelling the terms proportional to $E_n$, then interchanging the labels $m$ and $n$ on the right-hand side, and equating the coefficients of $\Psi_n$ on both sides gives a differential equation for $c_n(t)$:

$$i\hbar \frac{dc_n(t)}{dt} = \sum_m H'_{nm}(t) c_m(t) \exp(i(E_n - E_m)t/\hbar). \qquad (6.1.5)$$

So far, this has been exact. Since the rate of change (6.1.5) of $c_n(t)$ is proportional to the perturbation, to first order in this perturbation we can replace $c_m(t)$ on the right-hand side with a constant, equal to the value of $c_m(t)$ at any fixed time, say $t = 0$, in which case the solution is

$$c_n(t) \simeq c_n(0) - \frac{i}{\hbar} \sum_m c_m(0) \int_0^t dt'\, H'_{nm}(t') \exp\left( i(E_n - E_m)t'/\hbar \right). \quad (6.1.6)$$

Higher-order approximations can be obtained by iterating this procedure.

In what follows, we will see that the way that perturbation theory is used and the results obtained depend critically on the sort of time-dependence we assume for $H'(t)$. We will consider two cases: monochromatic perturbations, in which $H'(t)$ oscillates with a single frequency, and random fluctuations, for which $H'(t)$ is a stochastic variable, whose statistical properties do not change with time.

## 6.2 Monochromatic Perturbations

Let us now specialize to the case of a weak perturbation that oscillates at a single frequency $\omega/2\pi$:

$$H'(t) = -U \exp(-i\omega t) - U^\dagger \exp(i\omega t), \qquad (6.2.1)$$

with $\omega$ here taken positive. The integral in (6.1.6) is then trivial, and gives the first-order solution for the coefficients $c_n(t)$ in Eq. (6.1.4):

$$
c_n(t) = c_n(0) + \sum_m U_{nm} c_m(0) \left[ \frac{\exp\left(i(E_n - E_m - \hbar\omega)t/\hbar\right) - 1}{E_n - E_m - \hbar\omega} \right]
$$

$$
+ \sum_m U_{mn}^* c_m(0) \left[ \frac{\exp\left(i(E_n - E_m + \hbar\omega)t/\hbar\right) - 1}{E_n - E_m + \hbar\omega} \right]. \qquad (6.2.2)
$$

In particular, if all the $c_n(t)$ vanish at $t = 0$ except for $c_1(0) = 1$, then the amplitudes $c_n(t)$ for $n \neq 1$ are given by

$$
c_n(t) = U_{n1} \left[ \frac{\exp\left(i(E_n - E_1 - \hbar\omega)t/\hbar\right) - 1}{E_n - E_1 - \hbar\omega} \right]
$$

$$
+ U_{1n}^* \left[ \frac{\exp\left(i(E_n - E_1 + \hbar\omega)t/\hbar\right) - 1}{E_n - E_1 + \hbar\omega} \right]. \qquad (6.2.3)
$$

Both terms in Eq. (6.2.3) vanish at $t = 0$, and then for a while increase proportionally to $t$. The increase of the first and second terms ends when $t$ becomes of the order of $|(E_n - E_1)/\hbar - \omega|^{-1}$ or $|(E_n - E_1)/\hbar + \omega|^{-1}$, respectively, after which that term oscillates but no longer grows. The interesting case is when the final state has an energy close either to $E_1 + \hbar\omega$ or to $E_1 - \hbar\omega$, so that one of the two terms in (6.2.3) can keep growing for a long time. In the case of absorption of energy, where $E_n \simeq E_1 + \hbar\omega$, the second term stops growing long before the first term, and will consequently become relatively negligible at late times, so that

$$
c_n(t) \to U_{n1} \left[ \frac{\exp\left(i(E_n - E_1 - \hbar\omega)t/\hbar\right) - 1}{E_n - E_1 - \hbar\omega} \right].
$$

Then the probability after a sufficiently long time $t$ of finding the system in state $n \neq 1$ is

$$
\left| \left( \Psi_n, \Psi \right) \right|^2 = |c_n(t)|^2 \simeq 4|U_{n1}|^2 \frac{\sin^2\left( (E_n - E_1 - \hbar\omega)t/2\hbar \right)}{(E_n - E_1 - \hbar\omega)^2}. \qquad (6.2.4)
$$

Now, for large times we may approximate

$$
\frac{2\hbar \sin^2(Wt/2\hbar)}{\pi t W^2} \to \delta(W), \qquad (6.2.5)
$$

because this function vanishes for $t \to \infty$ like $1/t$ if $W \neq 0$, while it is so large for $W = 0$ that

$$\int_{-\infty}^{\infty} \frac{2\hbar \sin^2(Wt/2\hbar)}{\pi t W^2} \, dW = \frac{1}{\pi} \int_{-\infty}^{\infty} \frac{\sin^2 u}{u^2} \, du = 1.$$

Therefore, for large $t$ Eq. (6.2.4) gives

$$|c_n(t)|^2 = 4|U_{n1}|^2 \left( \frac{\pi t}{2\hbar} \right) \delta(E_1 + \hbar\omega - E_n),$$

and the rate of transitions to the state $n$ is therefore

$$\Gamma(1 \to n) \equiv |c_n(t)|^2/t = \frac{2\pi}{\hbar} |U_{n1}|^2 \delta(E_1 + \hbar\omega - E_n), \qquad (6.2.6)$$

a formula often known as *Fermi's golden rule*. In the case of stimulated emission of energy, where $\hbar\omega$ is close to $E_1 - E_n$, we have instead

$$\Gamma(1 \to n) = \frac{2\pi}{\hbar} |U_{1n}|^2 \delta(E_n + \hbar\omega - E_1).$$

We have treated the final states $n$ as if they are discrete. In order to use Eq. (6.2.6) in cases where the states $n$ are part of a continuum (as for a free electron produced by ionizing an atom) we may imagine that the whole system is placed in a large box. To avoid spurious effects due to the box walls, it is convenient to adopt *periodic boundary conditions*, which require that the wave function be unaffected by a translation of any of the three Cartesian coordinates, $x_i \to x_i + L_i$, where the $L_i$ are large lengths that will eventually be taken to infinity. The normalized wave function of a free particle then takes the form

$$\frac{\exp(i\mathbf{p} \cdot \mathbf{x}/\hbar)}{\sqrt{L_1 L_2 L_3}} \qquad (6.2.7)$$

with the components of $\mathbf{p}$ constrained by

$$p_i = \frac{2\pi \hbar n_i}{L_i}, \qquad (6.2.8)$$

with $n_1$, $n_2$, and $n_3$ arbitrary positive or negative integers. When we sum the rate (6.2.6) over free-particle states $n$, we are really summing over $n_1$, $n_2$, and $n_3$. Now, according to Eq. (6.2.8) the number of $n_i$ values in a range $\Delta p_i \gg \hbar/L_i$ is $L_i \Delta p_i/2\pi\hbar$, so the total number of states in a momentum-space volume $d^3 p = \Delta p_1 \Delta p_2 \Delta p_3$ is $d^3 p \, L_1 L_2 L_3/(2\pi\hbar)^3$. Thus we can sum the rate (6.2.6) over continuum states by integrating over momenta, and supplying an extra factor $L_1 L_2 L_3/(2\pi\hbar)^3$ in the rate for each free particle in the state. Equivalently, we can supply an extra factor $\sqrt{L_1 L_2 L_3}/(2\pi\hbar)^{3/2}$ in the matrix element $U_{n1}$ for each free particle in the state $n$. But the matrix element $U_{n1}$ will also contain a factor $1/\sqrt{L_1 L_2 L_3}$ from the wave function (6.2.7) for each free particle in the state $n$, so the volume factors cancel, and we are left with a factor $(2\pi\hbar)^{-3/2}$

for each free particle. Thus the rate (6.2.6) should be integrated rather than summed over the momenta of the free particles in the final states, with their wave functions taken as

$$\frac{\exp(i\mathbf{p} \cdot \mathbf{x}/\hbar)}{(2\pi\hbar)^{3/2}}, \tag{6.2.9}$$

instead of Eq. (6.2.7). This is the free-particle wave function (3.5.12), with normalization factor chosen to give the scalar product (3.5.13). (Alternatively, we can integrate over wave numbers instead of momenta, but then we must drop the factor $\hbar$ in the $3/2$ power in Eq. (6.2.9).)

The delta function in Eq. (6.2.6) fixes the sum of the free-particle energies, leaving only a finite integral over angles and energy ratios. An example is given in the next section.

## 6.3 Ionization by an Electromagnetic Wave

As an example of the use of time-dependent perturbation theory in the case of a monochromatic perturbation, consider a hydrogen atom in its ground state placed in a light wave. Just as in Section 5.3, if the wavelength of the light is much larger than the Bohr radius $a$, then the perturbation Hamiltonian depends only on the electric field at the location of the atom, which for plane polarization takes the form

$$\mathbf{E} = \mathcal{E} \exp(-i\omega t) + \mathcal{E}^* \exp(i\omega t), \tag{6.3.1}$$

with $\mathcal{E}$ constant. (We consider only the electric field, because the magnetic forces on a non-relativistic charged particle in an electromagnetic wave are less than the electric forces by a factor of order of the ratio of the particle velocity to the speed of light.) The perturbation in the Hamiltonian is then

$$H'(t) = e\mathcal{E} \cdot \mathbf{X} \exp(-i\omega t) + e\mathcal{E}^* \cdot \mathbf{X} \exp(i\omega t), \tag{6.3.2}$$

where $\mathbf{X}$ is the operator for the electron position. If we take $\mathcal{E}$ to lie in the 3-direction, with magnitude $\mathcal{E}$, then the operator $U$ in Eq. (6.2.1) is

$$U = -e\mathcal{E}X_3. \tag{6.3.3}$$

We need to calculate the matrix element of this perturbation between the normalized wave function of the ground state

$$\psi_{1s}(\mathbf{x}) = \frac{\exp(-r/a)}{\sqrt{\pi a^3}} \tag{6.3.4}$$

(where $a$ is the Bohr radius, given by Eq. (2.3.19) as $a = \hbar^2/m_e e^2 = 0.529 \times 10^{-8}$ cm) and the wave function of a free electron of momentum $\hbar\mathbf{k}_e$, normalized as described in the previous section:

$$\psi_e(\mathbf{x}) = (2\pi\hbar)^{-3/2} \exp(i\mathbf{k}_e \cdot \mathbf{x}). \tag{6.3.5}$$

We are justified in treating the emitted electron as a free particle only if it emerges with an energy much larger than the hydrogen binding energy. Otherwise, in place of Eq. (6.3.5) we should use the wave function of an unbound electron in the Coulomb field of the proton. With the binding energy of the hydrogen atom and the recoil energy of the hydrogen nucleus neglected, for a light wave number $k_\gamma$ the energy of the emitted electron equals the photon energy $\hbar c k_\gamma$, while the hydrogen binding energy (2.3.20) is $e^2/2a$, so in using Eq. (6.3.5) we are assuming that

$$k_\gamma a \gg e^2/2\hbar c \simeq 1/274. \tag{6.3.6}$$

Note that this is not inconsistent with our assumption that the light wavelength is much larger than the atomic size, which only requires that $k_\gamma a \ll 1$.

The matrix element of the perturbation (6.3.3) between the wave functions (6.3.4) and (6.3.5) is

$$U_{e,1s} = -\frac{e\mathcal{E}}{(2\pi\hbar)^{3/2}\sqrt{\pi a^3}} \int d^3x \, e^{-i\mathbf{k_e}\cdot\mathbf{x}} x_3 \exp(-r/a). \tag{6.3.7}$$

We can do the angular integral here by recalling that in general

$$\int d^3x \, e^{-i\mathbf{k}\cdot\mathbf{x}} f(r) = \frac{1}{k} \int_0^\infty 4\pi r f(r) \sin kr \, dr.$$

Differentiating this expression with respect to $k_3$ gives

$$-i \int d^3x \, e^{-i\mathbf{k}\cdot\mathbf{x}} f(r) x_3 = \frac{k_3}{k^3} \int_0^\infty 4\pi r f(r) \Big[ -\sin kr + kr \cos kr \Big] dr.$$

Applying this in Eq. (6.3.7) gives

$$U_{e,1s} = \frac{4\pi i e\mathcal{E} k_{e3}}{k_e^3 (2\pi\hbar)^{3/2}\sqrt{\pi a^3}} \int_0^\infty \exp(-r/a) \Big[ \sin k_e r - k_e r \cos k_e r \Big] r \, dr. \tag{6.3.8}$$

The integral here is given by

$$\int_0^\infty \exp(-r/a) \Big[ \sin k_e r - k_e r \cos k_e r \Big] r \, dr = \frac{8 k_e^3 a^5}{(1 + k_e^2 a^2)^3}.$$

With the final electron energy $\hbar^2 k_e^2 / 2m_e$ equal to the photon energy $\hbar c k_\gamma$, we have

$$k_e^2 a^2 \simeq \frac{2m_e c k_\gamma a^2}{\hbar} = 2k_\gamma a \cdot \frac{\hbar c}{e^2},$$

which according to Eq. (6.3.6) is much greater than one, so Eq. (6.3.8) gives

$$U_{e,1s} = \frac{8\sqrt{2} i e\mathcal{E} \cos\theta}{\pi \hbar^{3/2} k_e^5 a^{5/2}}, \tag{6.3.9}$$

where $\theta$ is the angle between $\mathbf{k_e}$ and the direction of polarization of the electromagnetic wave, taken here to be in the 3-direction.

According to Eq. (6.2.6), the differential ionization rate is

$$d\Gamma(1s \rightarrow \mathbf{k}_e) = \frac{2\pi}{\hbar}\left|U_{e,\,1s}\right|^2 \delta\left(\hbar c k_\gamma - E_e\right) \hbar^3 k_e^2 \, dk_e \, d\Omega, \qquad (6.3.10)$$

where $E_e = \hbar^2 \mathbf{k}_e^2 / 2m_e$, and $d\Omega = \sin\theta \, d\theta \, d\phi$ is the differential element of solid angle of the final electron direction, so that $\hbar^3 k_e^2 \, dk_e \, d\Omega$ is the momentum-space volume element of the final electron. (In accordance with our assumption (6.3.6), in the delta function we are neglecting the hydrogen binding energy, as well as the very small recoil energy of the hydrogen nucleus, compared with $E_e$.) Now, $dk_e = m_e \, dE_e / \hbar^2 k_e$, and the effect of the factor $dE_e \, \delta(\hbar\omega - E_e)$ in any integral over $k_e$ is just to set $k_e$ equal to the value fixed by the conservation of energy,

$$\hbar k_e = \sqrt{2m_e \hbar c k_\gamma}, \qquad (6.3.11)$$

so the differential ionization rate is

$$\frac{d\Gamma(1s \rightarrow \mathbf{k}_e)}{d\Omega} = 2\pi m_e k_e \left|U_{e,1s}\right|^2, \qquad (6.3.12)$$

with $k_e$ given by Eq. (6.3.11). Using Eq. (6.3.9) in Eq. (6.3.12) gives our final formula for the differential ionization rate,

$$\frac{d\Gamma(1s \rightarrow \mathbf{k}_e)}{d\Omega} = \frac{256 e^2 \mathcal{E}^2 m_e \cos^2\theta}{\pi \hbar^3 k_e^9 a^5}, \qquad (6.3.13)$$

valid in the range of light wave numbers with

$$\frac{1}{274} \ll k_\gamma a \ll 1. \qquad (6.3.14)$$

## 6.4 Fluctuating Perturbations

The monochromatic perturbations discussed in Section 6.2 can produce a finite transition rate between a discrete state and a continuum, as in the ionization process discussed in Section 6.3. But monochromatic perturbations cannot produce transitions between discrete states without fine-tuning the perturbation frequency. (For a perturbation that lasts a time that is short compared with the time $t$ during which we let the system evolve, the width of the frequency distribution will be large compared with $1/t$, and no fine-tuning is needed. But of course, in this case the transition probability, called $|c_n(t)|^2$ in Section 6.1, does not increase with time once the perturbation is ended, and so one cannot speak of a transition rate.) There is, however, a kind of perturbation that can span a wide range of frequencies, so that no fine-tuning is needed to produce transitions between discrete states, and yet yields a transition probability proportional to the elapsed time, so that there is a finite transition rate. It is the case of a perturbation that fluctuates randomly, but with statistical properties that do not change with time.

To be specific, suppose that the correlation between the perturbations at two different times depends only on the differences of the times, not on the times themselves:

$$\overline{H'_{nm}(t_1)H'^*_{nm}(t_2)} = f_{nm}(t_1 - t_2), \tag{6.4.1}$$

where a line over a quantity indicates an average over fluctuations. Fluctuations of this sort are called *stationary*.

In the case where $c_n(0) = \delta_{n1}$, Eq. (6.1.6) gives the transition probability to a state $n \neq 1$,

$$|c_n(t)|^2 = \frac{1}{\hbar^2} \int_0^t dt_1 \int_0^t dt_2 \, H'_{n1}(t_1)H'^*_{n1}(t_2) \exp\left(i(E_n - E_1)(t_1 - t_2)/\hbar\right), \tag{6.4.2}$$

so the average transition probability is

$$\overline{|c_n(t)|^2} = \frac{1}{\hbar^2} \int_0^t dt_1 \int_0^t dt_2 \, f_{n1}(t_1 - t_2) \exp\left(i(E_n - E_1)(t_1 - t_2)/\hbar\right). \tag{6.4.3}$$

We can write the correlation function $f_{nm}$ as a Fourier transform

$$f_{nm}(t) = \int_{-\infty}^{\infty} d\omega \, F_{nm}(\omega) \exp(-i\omega t) \tag{6.4.4}$$

so that Eq. (6.4.3) becomes

$$\overline{|c_n(t)|^2} = \frac{1}{\hbar^2} \int_{-\infty}^{\infty} d\omega \, F_{n1}(\omega) \left| \int_0^t dt_1 \exp\left[i\left((E_n - E_1)/\hbar - \omega\right)t_1\right] \right|^2$$

$$= 4 \int_{-\infty}^{\infty} d\omega \, F_{n1}(\omega) \frac{\sin^2\left[\left(E_n - E_1 - \hbar\omega\right)t/2\hbar\right]}{\left(E_n - E_1 - \hbar\omega\right)^2}. \tag{6.4.5}$$

Just as in Eq. (6.2.5), for large times we may approximate

$$\frac{2\hbar \sin^2(Wt/2\hbar)}{\pi t W^2} \rightarrow \delta(W) = \frac{1}{\hbar}\delta(W/\hbar), \tag{6.4.6}$$

so Eq. (6.4.5) gives a transition rate

$$\Gamma(1 \rightarrow n) \equiv \frac{\overline{|c_n(t)|^2}}{t} = \frac{2\pi}{\hbar^2} F_{n1}\left((E_n - E_1)/\hbar\right). \tag{6.4.7}$$

We will apply this result in the next section.

## 6.5 Absorption and Stimulated Emission of Radiation

To illustrate the general results of the previous section, let us consider an atom in a fluctuating electric field, such as that found in a gas of photons. The frequency $\omega/2\pi$ of the fluctuations that drive a transition $1 \rightarrow n$ between atomic states equals $(E_n - E_1)/h$, so the scale over which the electric field varies in space is of the order of $c/|\omega| = hc/|E_n - E_1|$. This is typically several thousands of Angstroms, much larger than atomic sizes, which are typically a few Angstroms. So it is a good approximation here, as in Eq. (5.3.1), to take the perturbation as

$$H'_{nm}(t) = e \sum_N [\mathbf{x}_N]_{nm} \cdot \mathbf{E}(t), \tag{6.5.1}$$

where $\mathbf{E}$ is the electric field at the position of the atom, the sum runs over the electrons in the atom, and

$$[\mathbf{x}_N]_{nm} = \left( \Psi_n, \mathbf{X}_N \Psi_m \right) = \int \psi_n^*(x) \mathbf{x}_N \psi_m(x) \prod_M d^3 x_M. \tag{6.5.2}$$

We assume that the fluctuations of the electric field have a correlation function of the form

$$\overline{E_i(t_1) E_j(t_2)} = \delta_{ij} \int_{-\infty}^{\infty} d\omega \, \mathcal{P}(\omega) \exp\left(-i\omega(t_1 - t_2)\right). \tag{6.5.3}$$

(In setting this proportional to $\delta_{ij}$, we are assuming that there is no preferred direction for the electric field; $\delta_{ij}$ is the most general tensor that does not depend on the orientation of the coordinate system.) Since the left-hand side is real and symmetric under the interchange of $t_1$ and $i$ with $t_2$ and $j$, we have

$$\mathcal{P}(\omega) = \mathcal{P}(-\omega) = \mathcal{P}^*(\omega). \tag{6.5.4}$$

The correlation function of the perturbation is now given by

$$\overline{H'_{nm}(t_1) H'^{*}_{nm}(t_2)} = e^2 \left| \sum_N [\mathbf{x}_N]_{nm} \right|^2 \int_{-\infty}^{\infty} d\omega \, \mathcal{P}(\omega) \exp\left(-i\omega(t_1 - t_2)\right). \tag{6.5.5}$$

That is, the function $F_{nm}(\omega)$ introduced in Eqs. (6.4.1) and (6.4.4) is

$$F_{nm}(\omega) = e^2 \left| \sum_N [\mathbf{x}_N]_{nm} \right|^2 \mathcal{P}(\omega). \tag{6.5.6}$$

Equation (6.4.7) then gives the rate at which an atom makes the transition from an initial state $m = 1$ to a higher or lower energy state $n$:

$$\Gamma(1 \rightarrow n) = \frac{2\pi e^2}{\hbar^2} \left| \sum_N [\mathbf{x}_N]_{n1} \right|^2 \mathcal{P}(\omega_{n1}), \tag{6.5.7}$$

where $\omega_{nm} = (E_n - E_m)/\hbar$.

The function $\mathcal{P}(\omega)$ can be related to the frequency distribution of energy in the fluctuating field. In radiation the magnetic field $\mathbf{B}$ has the same magnitude as the electric field, so the energy density (in unrationalized electrostatic units) is $[\mathbf{E}^2 + \mathbf{B}^2]/8\pi = \mathbf{E}^2/4\pi$. Setting $t_1 = t_2$ and summing over $i = j$ in Eq. (6.5.3), we find the average energy density of radiation

$$\rho = \frac{1}{4\pi}\overline{\mathbf{E}^2(t)} = \frac{3}{4\pi}\int_{-\infty}^{\infty} d\omega\, \mathcal{P}(\omega) = \frac{3}{2\pi}\int_{0}^{\infty} d\omega\, \mathcal{P}(\omega), \tag{6.5.8}$$

so the energy density between circular frequencies of magnitude $|\omega|$ and $|\omega| + d|\omega|$ is $(3/2\pi)\mathcal{P}(|\omega|)\, d|\omega|$. For the purposes of comparison with the results cited in Chapter 1, we can convert this into an energy distribution in frequency $\nu = |\omega|/2\pi$. The energy density between frequencies $\nu$ and $\nu + d\nu$ is

$$\rho(\nu)\, d\nu = (3/2\pi)\mathcal{P}(|\omega|)\, d|\omega| = 3\,\mathcal{P}(2\pi\nu)\, d\nu, \tag{6.5.9}$$

so we can write Eq. (6.5.7) as

$$\Gamma(1 \to n) = \frac{2\pi e^2}{3\hbar^2}\left|\sum_{N}[\mathbf{x}_N]_{n1}\right|^2 \rho(\nu_{n1}), \tag{6.5.10}$$

where $\nu_{nm} = |\omega_{nm}|/2\pi = |E_n - E_m|/h$. As we saw in Section 1.2, Einstein introduced a constant $B_1^n$ as the coefficient of $\rho(\nu_{n1})$ in the rate of absorption (if $E_n > E_1$) or stimulated emission (if $E_1 > E_n$), so in either case

$$B_1^n = \frac{2\pi e^2}{3\hbar^2}\left|\sum_{N}[\mathbf{x}_N]_{n1}\right|^2. \tag{6.5.11}$$

For hydrogen or an alkali metal, where it is essentially a single electron that interacts with radiation, this takes the familiar form

$$B_1^n = \frac{2\pi e^2}{3\hbar^2}|[\mathbf{x}]_{n1}|^2. \tag{6.5.12}$$

This agrees with the result (1.4.6), which was derived historically from the classical formula (1.4.1) for radiation from a charged oscillator and from the relation (1.2.16), which was obtained from considerations of the equilibrium of such an oscillator with black-body radiation. The historical derivation can now be reversed; using Eqs. (6.5.11) and (1.2.16), we can infer the formula (1.4.5) for the rate of spontaneous emission in a transition $1 \to n$:

$$A_1^n = \frac{4e^2|\omega_{n1}|^3}{3c^3\hbar}|[\mathbf{x}]_{n1}|^2, \tag{6.5.13}$$

without relying on an analogy with classical electrodynamics. This derivation was originally given in 1926 by Dirac.[1] The same result will be obtained in

---

[1] P. A. M. Dirac, *Proc. Roy. Soc.* A **112**, 661 (1926).

Section 11.7 by a direct calculation, in which we consider the interaction of an atom with the quantized electromagnetic field.

## 6.6 The Adiabatic Approximation

In some cases the Hamiltonian is a function $H[s]$ of one or more parameters that we will collectively label $s$, which are slowly varying functions $s(t)$ of time.[2] For instance, one might consider a spin in a slowly varying magnetic field, in which case $s(t)$ consists of the three components of the field. In such cases, we can find the solution of the time-dependent Schrödinger equation by use of what is known as the *adiabatic approximation*.[3]

For any $s$, we can find a complete orthonormal set of eigenstates $\Phi_n[s]$ of $H[s]$ with eigenvalues $E_n(s)$:

$$H[s]\Phi_n[s] = E_n[s]\Phi_n[s], \qquad \left(\Phi_n[s], \Phi_m[s]\right) = \delta_{nm}. \tag{6.6.1}$$

Since the $\Phi_n[s]$ and $\Phi_n[s']$ for any pair of parameters $s$ and $s'$ both form complete orthonormal sets, they are related by a unitary transformation. In particular, if we label the initial value of $s(t)$ at $t = 0$ as $s(0) = s_0$, then there exists a unitary operator $U[s]$ for which

$$\Phi_n[s] = U[s]\Phi_n[s_0], \qquad U[s]^{-1} = U[s]^\dagger, \qquad U[s_0] = 1, \tag{6.6.2}$$

where $U[s]$ is a sum of dyads:

$$U[s] = \sum_n \left[\Phi_n[s]\Phi_n^\dagger[s_0]\right]. \tag{6.6.3}$$

We can transform the Hamiltonian

$$\tilde{H}[s] \equiv U[s]^\dagger H[s]U[s] \tag{6.6.4}$$

so that though its eigenvalues depend on $s$, its eigenstates do not:

$$\tilde{H}[s]\Phi_n[s_0] = E_n[s]\Phi_n[s_0]. \tag{6.6.5}$$

That is, if for any operator $O$ we define

$$O_{nm} \equiv \left(\Phi_n[s_0], O\Phi_m[s_0]\right), \tag{6.6.6}$$

then in this basis the transformed Hamiltonian is

$$\tilde{H}_{nm}[s] = E_n[s]\delta_{nm}. \tag{6.6.7}$$

---

[2] In this section we use square brackets to indicate the dependence of various quantities on $s$, and parentheses to indicate dependence on time.

[3] This approximation was introduced in modern quantum mechanics by M. Born and V. Fock, Z. *Physik* **51**, 165 (1928). For a more accessible reference, see Albert Messiah, *Quantum Mechanics*, Vol. II (North-Holland Publishing Co., 1962), Chapter XVII, Sections 10–14.

The time-dependent Schrödinger equation,

$$i\hbar\frac{d}{dt}\Psi(t) = H[s(t)]\Psi(t),$$ (6.6.8)

can now be put in the form

$$i\hbar\frac{d}{dt}\tilde{\Psi}(t) = \left\{\tilde{H}[s(t)] + \Delta(t)\right\}\tilde{\Psi}(t),$$ (6.6.9)

where

$$\tilde{\Psi}(t) \equiv U[s(t)]^{\dagger}\Psi(t)$$ (6.6.10)

and

$$\Delta(t) \equiv i\hbar\left[\frac{d}{dt}U[s(t)]\right]^{\dagger}U[s(t)].$$ (6.6.11)

We note that since $U$ is unitary, $\dot{U}^{\dagger}U + U^{\dagger}\dot{U} = 0$, and so $\Delta$ is Hermitian.

At this point, it is tempting to neglect $\Delta(t)$, which involves the rate of change of the eigenvectors of $H[s(t)]$, as compared with $\tilde{H}[s(t)]$, which does not. However, this is not justified, because no matter how slowly the parameters $s(t)$ of the Hamiltonian evolve, we want to integrate the differential equation (6.6.9) out to times sufficiently late that $s(t)$ will have changed by a non-negligible amount. The length of this time interval may compensate for the smallness of $\Delta(t)$, which therefore cannot in general be neglected.

To deal with this, we perform one more unitary transformation. Define the unitary operator $V(t)$ by the differential equation

$$i\hbar\frac{d}{dt}V(t) = \tilde{H}[s(t)]V(t)$$ (6.6.12)

and the initial condition $V(0) = 1$. The solution is trivial in the basis (6.6.6):

$$V_{nm}(t) = \delta_{nm}\exp\left(i\phi_n(t)\right),$$ (6.6.13)

where $\phi_n(t)$ is a so-called *dynamical phase*:

$$\phi_n(t) = -\frac{1}{\hbar}\int_0^t E_n[s(\tau)]\,d\tau.$$ (6.6.14)

Using Eq. (6.6.12), Eq. (6.6.9) may be written

$$i\hbar\frac{d}{dt}\tilde{\tilde{\Psi}}(t) = \tilde{\Delta}(t)\tilde{\tilde{\Psi}}(t),$$ (6.6.15)

where

$$\tilde{\tilde{\Psi}}(t) \equiv V(t)^{\dagger}\tilde{\Psi}(t) = V(t)^{\dagger}U(t)^{\dagger}\Psi(t)$$ (6.6.16)

and

$$\tilde{\Delta}(t) \equiv V(t)^{\dagger}\Delta(t)V(t).$$ (6.6.17)

In the representation (6.6.6), Eq. (6.6.13) gives

$$\tilde{\Delta}_{nm}(t) = \Delta_{nm}(t) \exp\left[i\phi_m(t) - i\phi_n(t)\right]$$

$$= \Delta_{nm}(t) \exp\left[\frac{i}{\hbar} \int_0^t [E_n[s(t)] - E_m[s(t)]] \, dt\right]. \tag{6.6.18}$$

Now, if the fractional rate of change of $s(t)$ is very small compared with $(E_n[s] - E_m[s])/\hbar$ for all $n \neq m$ (which is only possible in the absence of degeneracy), then in a time that is long enough for $s(t)$ to change by an appreciable amount the phase factor in Eq. (6.6.18) will oscillate many times for $n \neq m$, preventing the build-up of the off-diagonal components of $\tilde{\Delta}$. Thus the only components of $\tilde{\Delta}$ that contribute to the long-time evolution of the state vector despite their smallness are the diagonal components, so that effectively we may make the replacement

$$\tilde{\Delta}_{nm}(t) \to \delta_{nm} \rho_n(t), \tag{6.6.19}$$

where $\rho_n(t)$ is the real quantity

$$\rho_n(t) \equiv \tilde{\Delta}_{nn}(t) = \Delta_{nn}(t) = i\hbar \left(\left[\frac{d}{dt} U[s(t)]\right]^\dagger U[s(t)]\right)_{nn}$$

$$= i\hbar \left(\frac{d}{dt} \Phi_n[s(t)], \Phi_n[s(t)]\right). \tag{6.6.20}$$

The solution of Eq. (6.6.15) is then

$$\tilde{\tilde{\Psi}}(t) = \sum_n \Phi_n[s_0] \exp[i\gamma_n(t)] \left(\Phi_n[s_0], \tilde{\tilde{\Psi}}(0)\right)$$

$$= \sum_n \Phi_n[s_0] \exp[i\gamma_n(t)] \left(\Phi_n[s_0], \Psi(0)\right), \tag{6.6.21}$$

where $\gamma_n(t)$ is the phase

$$\gamma_n(t) = -\frac{1}{\hbar} \int_0^t \rho_n(\tau) \, d\tau. \tag{6.6.22}$$

Together with Eqs. (6.6.16), (6.6.2), and (6.6.13), this gives the solution of the time-dependent Schrödinger equation (6.6.8) as

$$\Psi(t) = U(t) V(t) \tilde{\tilde{\Psi}}(t) = \sum_n U(t) \Phi_n[s_0] \left(\Phi_n[s_0], V(t) \tilde{\tilde{\Psi}}(t)\right)$$

$$= \sum_n \exp[i\phi_n(t)] \exp[i\gamma_n(t)] \Phi_n[s(t)] \left(\Phi_n[s_0], \Psi(0)\right). \tag{6.6.23}$$

That is, aside from the phases $\phi_n(t)$ and $\gamma_n(t)$, the prescription provided by the adiabatic approximation is that we are to find the time-dependence of the state vector by decomposing it into eigenstates of $H[s(t)]$, and giving each component just whatever time-dependence is needed to keep it an eigenstate of $H[s(t)]$.

As already mentioned, this only applies in the absence of degeneracy. To deal with the case of degeneracy, we can replace $n$ with a compound index $N\nu$: the energy is labeled by $N$, $M$, etc., so that $E_N \neq E_M$ if $N \neq M$, while $\nu$, $\mu$, etc. label states with a given energy. In this case, $\tilde{\Delta}$ in Eq. (6.6.15) is replaced with

$$\tilde{\Delta}_{N\nu,M\mu}(t) \to \delta_{NM} R^{(N)}_{\nu\mu}(t), \qquad (6.6.24)$$

where $R^{(N)}$ is an Hermitian operator in the space of states with energy $E_N$:

$$R^{(N)}_{\nu\mu}(t) \equiv \tilde{\Delta}_{N\mu,N\nu}(t) = \Delta_{N\mu,N\nu}(t) = i\hbar\left(\left[\frac{d}{dt}U[s(t)]\right]^\dagger U[s(t)]\right)_{N\mu,N\nu}$$

$$= i\hbar\left(\frac{d}{dt}\Phi_{N\mu}[s(t)], \Phi_{N\nu}[s(t)]\right). \qquad (6.6.25)$$

By the same reasoning that led to Eq. (6.6.23), the solution of the time-dependent Schrödinger equation (6.6.8) is here

$$\Psi(t) = \sum_N \exp[i\phi_N(t)] \sum_{\mu\nu} \Gamma^{(N)}_{\mu\nu}(t)\Phi_{N\mu}[s(t)]\Big(\Phi_{N\nu}[s_0], \Psi(0)\Big), \qquad (6.6.26)$$

where the dynamical phase $\phi_N(t)$ is given by Eq. (6.6.14), with $N$ in place of $n$, and $\Gamma^{(N)}(t)$ is a unitary matrix, defined as the solution of the equation

$$i\hbar\frac{d}{dt}\Gamma^{(N)}(t) = R^{(N)}(t)\Gamma^{(N)}(t), \qquad (6.6.27)$$

with the initial condition $\Gamma^{(N)}(0) = 1$. This unitary matrix takes the place of the phase factor $e^{i\gamma_n(t)}$ in the degenerate case.[4]

## 6.7 The Berry Phase

The non-dynamical phase $\gamma_n(t)$ appearing in the adiabatic solution (6.6.23) of the time-dependent Schrödinger equation has interesting properties and physical applications, first noted by Michael Berry.[5] First, it should be noted that $\gamma_n(t)$ is *geometric* – that is, it depends on the path through the parameter space of the Hamiltonian from $s(0)$ to $s(t)$, but not on the time-dependence of travel along this path. This can be seen by combining Eqs. (6.6.20) and (6.6.22), and writing the result as

$$\gamma_n(t) = -i\int_{C(t)} \sum_i ds_i \left(\frac{\partial}{\partial s_i}\Phi_n[s], \Phi_n[s]\right), \qquad (6.7.1)$$

[4] F. Wilczek and A. Zee, *Phys. Rev. Lett.* **52**, 2111 (1984).
[5] M. V. Berry, *Proc. Roy. Soc.* A **392**, 45 (1984).

where $C(t)$ indicates that the integral is to be taken along the path through the Hamiltonian's parameter space traced by $s(\tau)$ from $\tau = 0$ to $\tau = t$.

It is also important to note that $\gamma_n(t)$ is itself not physically significant, for we can always change the energy eigenstates $\Phi_n[s]$ by arbitrary $s$-dependent phases

$$\Phi_n[s] \rightarrow e^{i\alpha_n[s]}\Phi_n[s]. \tag{6.7.2}$$

This subjects the phase $\gamma_n(t)$ to the shift

$$\gamma_n(t) \rightarrow \gamma_n(t) + \alpha_n[s(0)] - \alpha_n[s(t)], \tag{6.7.3}$$

though of course the state vector (6.6.23) is unaffected. What is physically significant is the *classes* of phases $\gamma_n$ that are equivalent, in the sense that they can be related to one another by the transformation (6.7.3).

As Berry noted, in general these classes are non-trivial – that is, it is not generally possible to eliminate the phase $\gamma_n(t)$ by a change (6.7.2) of the basis states. To identify such cases, it is only necessary to consider the phase $\gamma_n(t)$ associated with a path $C(t)$ that begins at $t = 0$ and ends at the same point at a later time $t$. This phase is obviously independent of how we choose the phases of the energy eigenstates $\Phi_n[s]$ for $s$ at intermediate points along this curve, so if $\gamma_n(t)$ can be eliminated by a transformation like (6.7.2), then the phase $\gamma_n(t)$ associated with a closed curve must vanish, whatever phases we choose for $\Phi_n[s]$. Conversely, if the phases (6.7.1) associated with all closed curves $C(t)$ vanish, then the phase associated with a path from $s(0)$ to $s(t)$ must be the same as the phase associated with any other such path, because the difference of these phases is the phase associated with a closed curve that goes from $s(0)$ to $s(t)$ on the first path and then back to $s(0)$ along the second path. This would mean that $\gamma_n(t)$ is a function only of $s(t)$, and can therefore be eliminated by a transformation of the form (6.7.3). The phase $\gamma_n$ associated with a closed path $C$ will from now on be denoted $\gamma_n[C]$; this is often called the *Berry phase*.

The Berry phase can be put in a form that is convenient for calculation, and that makes manifest its independence of the phase convention used for the basis states $\Phi_n[s]$. According to a generalized version of Stokes' theorem, the line integral (6.7.1) may be expressed as an integral over any surface $A[C]$ bounded by the closed curve $C$:

$$\gamma_n[C] = -i \iint_{A[C]} \sum_{ij} dA_{ij} \frac{\partial}{\partial s_i} \left( \frac{\partial}{\partial s_j} \Phi_n[s], \Phi_n[s] \right), \tag{6.7.4}$$

where $dA_{ij} = -dA_{ji}$ is the tensor element of surface area.[6] For instance, in the case where the Hamiltonian depends on just three independent parameters $s_i$,

---

[6] For a flat curve $C$ in the $k$–$l$ plane in any number of dimensions, the integral $\sum_{ij} \int_{A[C]} dA_{ij}T_{ij}$ of any tensor $T_{ij}$ is equal to the ordinary integral of $T_{kl} - T_{lk}$ over the area $A[C]$ bounded by $C$. The case of a curve that is not flat can be dealt with by breaking up the area it bounds into small flat areas; the integral is the sum of the integrals over these small areas.

we have $dA_{ij} = \sum_k \epsilon_{ijk} e_k \, dA$, where $\epsilon_{ijk}$ as usual is the totally antisymmetric tensor with $\epsilon_{123} = +1$; $dA$ is the usual element of surface area; and $\mathbf{e}$ is the unit vector normal to the surface. (We use $\mathbf{e}$ rather than the conventional $\mathbf{n}$ for the unit normal to avoid confusion with the label $n$ on the state vector.) In this case, Eq. (6.7.4) is the result of the usual Stokes theorem:

$$\gamma_n[C] = -i \iint_{A[C]} dA \; \mathbf{e}[s] \cdot \left( \nabla \times (\nabla \Phi_n[s], \Phi_n[s]) \right), \qquad (6.7.5)$$

where the gradients here are taken with respect to the three $s_i$.

Returning now to the general case, we note that because $dA_{ij}$ is antisymmetric in $i$ and $j$, Eq. (6.7.4) may be written

$$\gamma_n[C] = i \iint_{A[C]} \sum_{ij} dA_{ij} \left( \frac{\partial}{\partial s_i} \Phi_n[s], \frac{\partial}{\partial s_j} \Phi_n[s] \right)$$

$$= i \iint_{A[C]} \sum_{ij} dA_{ij} \sum_m \left( \frac{\partial}{\partial s_i} \Phi_n[s], \Phi_m[s] \right) \left( \Phi_m[s], \frac{\partial}{\partial s_j} \Phi_n[s] \right).$$
$$(6.7.6)$$

By differentiating $(\Phi_n[s], \Phi_n[s]) = 1$, we see that

$$\left( \frac{\partial}{\partial s_i} \Phi_n[s], \Phi_n[s] \right) = - \left( \Phi_n[s], \frac{\partial}{\partial s_i} \Phi_n[s] \right),$$

so the contribution of the term with $m = n$ in Eq. (6.7.6) is

$$-i \iint_{A[C]} \sum_{ij} dA_{ij} \left( \frac{\partial}{\partial s_i} \Phi_n[s], \Phi_n[s] \right) \left( \frac{\partial}{\partial s_j} \Phi_n[s], \Phi_n[s] \right),$$

and this vanishes because $dA_{ij}$ is antisymmetric. On the other hand, the terms with $m \neq n$ can be put in a form not involving derivatives of the energy eigenstates. By differentiating the Schrödinger equation (6.6.1) with respect to $s_j$ and then taking the scalar product with $\Phi_m[s]$ for $m \neq n$, we find

$$\left( E_n[s] - E_m[s] \right) \left( \Phi_m[s], \frac{\partial}{\partial s_j} \Phi_n[s] \right) = \left( \Phi_m[s], \left[ \frac{\partial H[s]}{\partial s_j} \right] \Phi_n[s] \right), \quad (6.7.7)$$

so that Eq. (6.7.6) may be written

$$\gamma_n[C] = i \iint_{A[C]} \sum_{ij} dA_{ij} \sum_{m \neq n} \left( \Phi_n[s], \left[ \frac{\partial H[s]}{\partial s_i} \right] \Phi_m[s] \right)^*$$

$$\times \left( \Phi_n[s], \left[ \frac{\partial H[s]}{\partial s_j} \right] \Phi_m[s] \right)$$

$$\times (E_m[s] - E_n[s])^{-2}. \qquad (6.7.8)$$

This makes it apparent that the Berry phase is independent of the phase convention used for the energy eigenstates. Unlike the dynamical phase, the Berry phase is also independent of the scale of the Hamiltonian: multiplying $H[s]$ with a constant $\lambda$ has the effect of multiplying both $\partial H[s]/\partial s_i$ and $E_m[s]-E_n[s]$ with $\lambda$, so that the factors of $\lambda$ cancel in Eq. (6.7.8). Another advantage of Eq. (6.7.8) is that it is generally easier to calculate the derivative of the Hamiltonian with respect to the parameters $s_i$ than the derivative of the energy eigenstates. This expression for the Berry phase is real, because the area element $dA_{ij}$ is antisymmetric.

In the special case where $i$ and $j$ run over three values, Eq. (6.7.8) takes the form

$$\gamma_n[C] = \iint_{A[C]} dA\, \mathbf{e}[s] \cdot \mathbf{V}_n[s], \tag{6.7.9}$$

where $\mathbf{e}[s]$ is the unit vector normal to the surface $A[C]$ at the point $s$, and $\mathbf{V}_n[s]$ is a three-vector in parameter space:

$$\mathbf{V}_n[s] \equiv i \sum_{m \neq n} \left\{ \left( \Phi_n[s], \left[ \nabla H[s] \right] \Phi_m[s] \right)^* \times \left( \Phi_n[s], \left[ \nabla H[s] \right] \Phi_m[s] \right) \right\}$$
$$\times (E_m[s] - E_n[s])^{-2}. \tag{6.7.10}$$

This formalism has a natural application to the case of a particle or other system with non-vanishing angular momentum $\mathbf{J}$ in a slowly varying magnetic field. As mentioned earlier, the parameters $s_i$ here are the components of the magnetic field $\mathbf{B}$. We take the Hamiltonian as

$$H[\mathbf{B}] = \kappa \mathbf{B} \cdot \mathbf{J} + H_0, \tag{6.7.11}$$

where $\kappa$ is a constant, related to the magnetic moment, and $H_0$ is independent of the magnetic field or any other external field, and hence commutes with $\mathbf{J}$. The energy eigenstates are eigenstates of the component of $\mathbf{J}$ along $\mathbf{B}$ and of $\mathbf{J}^2$ and $H_0$:

$$\hat{B} \cdot \mathbf{J} \Phi_n[\mathbf{B}] = \hbar n \Phi_n[\mathbf{B}], \quad \mathbf{J}^2 \Phi_n[\mathbf{B}] = \hbar^2 j(j+1) \Phi_n[\mathbf{B}],$$
$$H_0 \Phi_n[\mathbf{B}] = E_0 \Phi_n[\mathbf{B}], \tag{6.7.12}$$

with energies

$$E_n[\mathbf{B}] = \kappa |\mathbf{B}| \hbar n + E_0, \tag{6.7.13}$$

where $n$ is an integer or half-integer, running from $-j$ to $+j$ by unit steps. In the spirit of the adiabatic approximation, we focus on one value of $n$ and one value of $E_0$ as the magnetic field changes. As promised, the factors $\kappa$ cancel in the three-vector (6.7.10), which here takes the form

$$\mathbf{V}_n[\mathbf{B}] \equiv \frac{i}{\hbar^2 |\mathbf{B}|^2} \sum_{m \neq n} \left\{ (\Phi_n[\mathbf{B}], \mathbf{J}\Phi_m[\mathbf{B}])^* \times (\Phi_n[\mathbf{B}], \mathbf{J}\Phi_m[\mathbf{B}]) \right\} (m-n)^{-2}.$$
$$\tag{6.7.14}$$

We will first calculate this three-vector at one particular value of $\mathbf{B}$ in the range $A[C]$ in field space. For this purpose, it is convenient to choose the 3-axis to lie along the direction of $\mathbf{B}$. Since $\Phi_m$ and $\Phi_n$ are then eigenstates of $J_3$, the matrix element $(\Phi_n[\mathbf{B}], \mathbf{J}\Phi_m[\mathbf{B}])$ with $n \neq m$ has components only in the 1–2 plane, and so (6.7.14) is in the 3-direction. Also, the only states $\Phi_m$ for which either $(\Phi_n[\mathbf{B}], J_1\Phi_m[\mathbf{B}])$ or $(\Phi_n[\mathbf{B}], J_2\Phi_m[\mathbf{B}])$ do not vanish have $m = n \pm 1$, and for these states $(m - n)^2 = 1$. Hence the only non-vanishing component of the vector (6.7.14) is its 3-component:

$$
\begin{aligned}
V_{n3}[\mathbf{B}] &= \frac{i}{\hbar^2 |\mathbf{B}|^2} \sum_{\pm} \left[ \left( \Phi_n[\mathbf{B}], J_1\Phi_{n\pm 1}[\mathbf{B}] \right)^* \left( \Phi_n[\mathbf{B}], J_2\Phi_{n\pm 1}[\mathbf{B}] \right) \right. \\
&\qquad\qquad \left. - \left( \Phi_n[\mathbf{B}], J_2\Phi_{n\pm 1}[\mathbf{B}] \right)^* \left( \Phi_n[\mathbf{B}], J_1\Phi_{n\pm 1}[\mathbf{B}] \right) \right] \\
&= \frac{1}{2\hbar^2 |\mathbf{B}|^2} \sum_{\pm} \left\{ \left| \left( \Phi_n[\mathbf{B}], (J_1 + iJ_2)\Phi_{n\pm 1}[\mathbf{B}] \right) \right|^2 \right. \\
&\qquad\qquad \left. - \left| \left( \Phi_n[\mathbf{B}], (J_1 - iJ_2)\Phi_{n\pm 1}[\mathbf{B}] \right) \right|^2 \right\} .
\end{aligned}
$$

According to the results of Section 4.2, the non-zero matrix elements here are

$$
\left( \Phi_n[\mathbf{B}], (J_1 + iJ_2)\Phi_{n-1}[\mathbf{B}] \right) = \hbar\sqrt{(j - n + 1)(j + n)}
$$

and

$$
\left( \Phi_n[\mathbf{B}], (J_1 - iJ_2)\Phi_{n+1}[\mathbf{B}] \right) = \hbar\sqrt{(j - n)(j + n + 1)},
$$

and so

$$
V_{n3}[\mathbf{B}] = \frac{n}{|\mathbf{B}|^2}, \quad V_{n1}[\mathbf{B}] = V_{n2}[\mathbf{B}] = 0.
$$

We can put this in a form that does not depend on our choice of the 3-axis to lie along $\mathbf{B}$:

$$
\mathbf{V}_n[\mathbf{B}] = \frac{n\mathbf{B}}{|\mathbf{B}|^3}, \tag{6.7.15}
$$

which in this form holds everywhere. The Berry phase (6.7.9) is therefore

$$
\gamma_n[C] = n \iint_{A[C]} dA \, \frac{\mathbf{B} \cdot \mathbf{e}[\mathbf{B}]}{|\mathbf{B}|^3}, \tag{6.7.16}
$$

the integral being taken over any area in the space of the magnetic field vector surrounded by the curve $C$. We can evaluate this integral using Gauss's theorem. Draw a cone (not a circular cone unless $C$ happens to be a circle) with base $A[C]$ and sides running from the origin in field space to the curve $C$. The integral (6.7.16) may be written as an integral over the whole surface of this cone, since on the sides of this cone the normal $\mathbf{e}$ is perpendicular to $\mathbf{B}$, and so these sides do not contribute to the surface integral. But then Gauss's theorem tells us that the integral over $A[C]$ of the normal component of the vector $\mathbf{B}/|\mathbf{B}|^3$ is the same as the integral of the divergence of this vector over the volume $V[C]$ of the cone:

$$\gamma_n[C] = n \int_{V[C]} d^3 B \, \boldsymbol{\nabla} \cdot \frac{\mathbf{B}}{|\mathbf{B}|^3}. \tag{6.7.17}$$

The divergence of $\mathbf{B}/|\mathbf{B}|^3$ vanishes everywhere except for a singularity $4\pi\delta^3(\mathbf{B})$ at the origin. This singularity is spherically symmetric, so the integral over $\mathbf{B}$ in Eq. (6.7.17) is just equal to $4\pi$ times the fraction of the whole sphere occupied by the cone. This fraction is the solid angle $\Omega[C]$ subtended by $C$ as seen from the origin in field space divided by $4\pi$, so the integral is just $\Omega[C]$, and the Berry phase is simply

$$\gamma_n[C] = n \, \Omega[C]. \tag{6.7.18}$$

For instance, if the magnetic field changes only in direction, keeping its 3-component fixed, then $C$ is a circle with both $B_3$ and $|\mathbf{B}|$ fixed, and

$$\gamma_n[C] = n \int_0^{\arccos(B_3/|\mathbf{B}|)} 2\pi \sin\theta \, d\theta = 2\pi n(1 - B_3/|\mathbf{B}|).$$

There are many other places in physics where a Berry phase, or a phase analogous to the Berry phase, makes an appearance.[7] We will encounter one in Section 10.4, on the Aharonov–Bohm effect.

## 6.8 Rabi Oscillations and Ramsey Interferometers

In Section 6.2 we considered a system in an initial state with energy $E_m$, exposed to a perturbation with terms proportional to $\exp(\mp i\omega t)$. We found that the probability after a time $t$ has elapsed of finding the system in a different discrete state with an energy $E_n$ increases with time, eventually becoming peaked at a frequency $\omega = \pm(E_n - E_m)/\hbar$, with the width of the peak of order $1/t$. But if we leave a system alone for a really long time, then the amplitude for the state with energy $E_n$ builds up so much that the system begins to make a transition back to energy $E_m$, and then back to energy $E_n$, and so on. This is known as a *Rabi oscillation*,[8] named for I. I. Rabi (1898–1988). As we shall see, this phenomenon gets in the way of making accurate measurements of the transition frequency $(E_n - E_m)/\hbar$, a problem solved by an interferometer[9] developed by

---

[7]  Aspects of such phases are treated in *Geometric Phases in Physics*, ed. A. Shapere and F. Wilczek (World Scientific Publishers Co., Singapore, 1989).

[8]  I. I. Rabi, *Phys. Rev.* **51**, 652 (1937).

[9]  N. F. Ramsey, *Phys. Rev.* **76**, 996 (1949). Also see N. F. Ramsey, *Molecular Beams* (Oxford University Press, London, 1956), Chapter V. For historical reviews, see D. Kleppner, *Physics Today*, January, p. 25 (2013); S. Haroche, M. Brune, and J.-M. Raimond, *Physics Today*, January, p. 27 (2013).

Norman Ramsey (1915–2011), which allows extremely accurate measurements of atomic and molecular transition frequencies.

To study Rabi oscillations we will again need to make an approximation, ignoring terms in the time-dependent Schrödinger equation whose coefficients oscillate very rapidly in time. This approximation was also used in Section 6.2, but here we will keep terms of all orders in the oscillating perturbation.

We take the perturbation to be of the form (6.2.1). The exact time-dependent Schrödinger equation (6.1.5) then takes the form

$$i\hbar \frac{d}{dt} c_n(t) = -\sum_m c_m(t) U_{nm} \exp\left(i(E_n - E_m - \hbar\omega)t/\hbar\right)$$

$$-\sum_m c_m(t) U_{mn}^* \exp\left(i(E_n - E_m + \hbar\omega)t/\hbar\right), \qquad (6.8.1)$$

where $c_n(t)$ are the components of the wave function defined by Eq. (6.1.4). We assume that the perturbation frequency $\omega$ is tuned to be close to one of the resonance frequencies, say $(E_e - E_g)/\hbar$ (where $e$ and $g$ conventionally stand for "excited state" and "ground state," though they can be any two states). As in Section 6.2, we neglect all terms in Eq. (6.8.1) with coefficients that oscillate rapidly, keeping only terms with the relatively small oscillation frequency $\pm[\omega - (E_e - E_g)/\hbar]$. Barring accidents, the only such terms in Eq. (6.8.1) are those proportional to $U_{eg}$ or $U_{eg}^*$, so with this approximation, Eq. (6.8.1) becomes

$$i\hbar \frac{d}{dt} c_e = -U_{eg} e^{-i\Delta\omega t} c_g, \quad i\hbar \frac{d}{dt} c_g = -U_{eg}^* e^{i\Delta\omega t} c_e, \qquad (6.8.2)$$

where $\Delta\omega$ is the displacement of the applied frequency from its resonance value,

$$\Delta\omega \equiv \omega - (E_e - E_g)/\hbar. \qquad (6.8.3)$$

It is easy to find an exact solution:

$$c_g(t) = Ce^{i\Delta\omega t/2}\left[-i\hbar\Omega \cos(\Omega t + \delta) - \frac{\hbar\Delta\omega}{2}\sin(\Omega t + \delta)\right], \qquad (6.8.4)$$

$$c_e(t) = CU_{eg} e^{-i\Delta\omega t/2} \sin(\Omega t + \delta), \qquad (6.8.5)$$

where $C$ and $\delta$ are arbitrary complex constants, and the frequency $\Omega$ of the Rabi oscillation is given by

$$\Omega^2 = \frac{\Delta\omega^2}{4} + \frac{|U_{eg}|^2}{\hbar^2}. \qquad (6.8.6)$$

[To find this solution, first suppose that $c_e$ takes the form (6.8.5), with unknown $\Omega$. Inserting this in the first equation of Eq. (6.8.2) then gives Eq. (6.8.4) for $c_g$. Inserting this result for $c_g$ in the second equation of Eq. (6.8.2) gives a result for $c_e$ that is consistent with Eq. (6.8.5), provided $\Omega$ satisfies Eq. (6.8.6).]

For instance, suppose that $c_g(0) = 1$ and $c_e(0) = 0$. Then $\delta = 0$ and $C = i/\hbar\Omega$, so the solution (6.8.4), (6.8.5) becomes

$$c_g(t) = e^{i\Delta\omega t/2}\left[\cos(\Omega t) - \frac{i\Delta\omega}{2\Omega}\sin(\Omega t)\right], \tag{6.8.7}$$

$$c_e(t) = \frac{iU_{eg}}{\hbar\Omega}e^{-i\Delta\omega t/2}\sin(\Omega t), \tag{6.8.8}$$

so that if the system is in state $g$ at $t = 0$, then at a later time $t$ the probability that it is in state $e$ will be

$$|c_e|^2 = \left|\frac{U_{eg}}{\hbar\Omega}\right|^2\sin^2(\Omega t). \tag{6.8.9}$$

For $|U_{eg}| \ll \hbar\,\Delta\omega/2$ we would have $\Omega \simeq \Delta\omega/2$, and Eq. (6.8.9) would be the same as the result (6.2.4) of first-order perturbation theory.

At a given time $t$ the probability (6.8.9) is peaked at $\Delta\omega = 0$, or in other words at $\omega = (E_e - E_g)/\hbar$. so we can measure the transition frequency $(E_e - E_g)/\hbar$ by finding the value of $\omega$ where the excitation probability $|c_e|^2$ reaches a maximum. But the precision of this measurement is limited to the width of the peak in the graph of $|c_e|^2$ versus $\omega$. This width is of order $1/t$ as long as the elapsed time $t$ is much less than $\hbar/|U_{eg}|$, in which case when $\Delta\omega \approx 1/t$ we can neglect the term $\hbar^2/|U_{eg}|^2$ in Eq. (6.8.6) for $\Omega^2$, so that $|\Omega| \simeq |\Delta\omega|/2$. But although we can improve the accuracy of the measurement of $(E_e - E_g)/\hbar$ up to a point by increasing the time $t$ that elapses before the excitation probability is measured, this improvement comes to an end when $t$ is of order $\hbar/|U_{eg}|$ and the precision of the measurement is of order $\hbar/|U_{eg}|$. This is not good enough to establish a really precise frequency standard.

One can do better than this by using a famous trick invented by Ramsey. In a Ramsey interferometer, a long waveguide is connected to a source of coherent microwave radiation at circular frequency $\omega$. The waveguide has two short transverse projections at its ends. An atom (or molecule) in the ground state $g$ is directed into one of these projections, so that it is exposed to a pulse of microwave radiation for a time $t_1$; it then travels outside the waveguide along its length for a much longer time $T$; it then enters the projection at the other end of the waveguide so that it is again exposed to a pulse of microwave radiation, this time for another short time $t_2$, and then passes outside the waveguide to a detector that can count atoms in the ground state $g$ or in a particular excited state $e$. As we shall now see, the probabilities of finding the atoms in these excited states are very sharply peaked at $\Delta\omega = 0$, so that by tuning $\omega$ to find this peak, one can make a very accurate measurement of the resonance frequency $(E_e - E_g)/\hbar$.

According to Eqs. (6.8.7) and (6.8.8), after the atom has been exposed to the first pulse for a time $t_1$, it will be in a coherent superposition of the ground and excited states, with amplitudes

$$c_g(t_1) = e^{i\Delta\omega t_1/2} \left[ \cos(\Omega t_1) - \frac{i\,\Delta\omega}{2\Omega} \sin(\Omega t_1) \right], \tag{6.8.10}$$

$$c_e(t_1) = \frac{i U_{eg}}{\hbar\Omega} e^{-i\Delta\omega t_1/2} \sin(\Omega t_1). \tag{6.8.11}$$

The amplitudes $c_g(t)$ and $c_e(t)$ are defined to be time-independent in the absence of perturbations, so Eqs. (6.8.10) and (6.8.11) also give the values of these amplitudes during the time from $t_1$ to $t_1 + T$ when the atom is outside the waveguide, and hence also when it re-enters the waveguide at a time $t_1 + T$. During the second pulse the amplitudes are again given by Eqs. (6.8.4) and (6.8.5), but now with the constants $C$ and $\delta$ determined by requiring that at time $t_1 + T$ the amplitudes (6.8.4) and (6.8.5) take the values (6.8.10) and (6.8.11):

$$C e^{i\Delta\omega (t_1+T)/2} \left[ -i\hbar\Omega \cos\left(\Omega(t_1 + T) + \delta\right) - \frac{\hbar\,\Delta\omega}{2} \sin\left(\Omega(t_1 + T) + \delta\right) \right]$$

$$= e^{i\Delta\omega t_1/2} \left[ \cos(\Omega t_1) - \frac{i\,\Delta\omega}{2\Omega} \sin(\Omega t_1) \right], \tag{6.8.12}$$

$$C U_{eg} e^{-i\Delta\omega (t_1+T)/2} \sin\left(\Omega(t_1 + T) + \delta\right)$$

$$= \frac{i U_{eg}}{\hbar\Omega} e^{-i\Delta\omega t_1/2} \sin(\Omega t_1). \tag{6.8.13}$$

We can derive an equation that determines the constant $\delta$ by equating the ratios of the left- and right-hand sides. After some cancellations, this gives

$$e^{i\Delta\omega T} \left[ \cot\left(\Omega(t_1 + T) + \delta\right) - i\frac{\Delta\omega}{2\Omega} \right] = \left[ \cot\left(\Omega t_1\right) - i\frac{\Delta\omega}{2\Omega} \right], \tag{6.8.14}$$

and $C$ is then given by Eq. (6.8.13):

$$C = e^{i\Delta\omega T/2} \left( \frac{i}{\hbar\Omega} \right) \frac{\sin(\Omega t_1)}{\sin\left(\Omega(t_1 + T) + \delta\right)}. \tag{6.8.15}$$

The amplitude for the excited state when the atom leaves the waveguide at the time $t_1 + t_2 + T$ is then given by Eq. (6.8.5), using the values we have found for the constants $\delta$ and $C$:

$$c_e(t_1 + t_2 + T) = C U_{eg} e^{-i\Delta\omega (t_1+t_1+T)/2} \sin\left(\Omega(t_1 + t_2 + T) + \delta\right)$$

$$= e^{-i\Delta\omega (t_1+t_1)/2} \left( \frac{i U_{eg}}{\hbar\Omega} \right)$$

$$\times \frac{\sin(\Omega t_1)\,\sin\left(\Omega(t_1 + t_2 + T) + \delta\right)}{\sin\left(\Omega(t_1 + T) + \delta\right)}$$

$$= e^{-i\Delta\omega(t_1+t_1)/2} \left( \frac{iU_{eg}}{\hbar\Omega} \right)$$

$$\times \sin(\Omega t_1) \Big[ \sin(\Omega t_2) \cot \Big( \Omega(t_1 + T) + \delta \Big)$$

$$+ \cos(\Omega t_2) \Big],$$

and therefore, using Eq. (6.8.14),

$$c_e(t_1 + t_2 + T) = e^{-i\Delta\omega(t_1+t_1)/2} \left( \frac{iU_{eg}}{\hbar\Omega} \right) \sin(\Omega t_1)$$

$$\times \left[ i\frac{\Delta\omega}{2\Omega} \sin(\Omega t_2) \left( 1 - e^{-i\Delta\omega T} \right) + e^{-i\Delta\omega T} \sin(\Omega t_2) \cot(\Omega t_1) \right.$$

$$\left. + \cos(\Omega t_2) \right]. \tag{6.8.16}$$

We will assume that $\omega$ is tuned to make $\Delta\omega$ small enough that $\hbar|\Delta\omega|$ is much less than $|U_{eg}|$, which implies that $\Omega$ is very close to $|U_{eg}|$ and $|\Delta\omega|$ is much less than $\Omega$. The probability of finding the atom in an excited state when it emerges from the waveguide is then

$$P_e \equiv |c_e(t_1 + t_2 + T)|^2 = \sin^2(\Omega t_1) \left| e^{-i\Delta\omega T} \sin(\Omega t_2) \cot(\Omega t_1) + \cos(\Omega t_2) \right|^2. \tag{6.8.17}$$

For large time intervals $T$, the phase factor $e^{-i\Delta\omega T}$ is very sensitive to changes in $\omega$, so to maximize the sensitivity of the whole expression it is usual to take the coefficient of this phase factor equal to the $T$-independent term. That is, it is best to adjust the times $t_1$ and $t_2$ so that $\sin(\Omega t_2) \cot(\Omega t_1) = \cos(\Omega t_2)$, and therefore $t_1 = t_2 \equiv \tau$, which just requires that the paths of the atom through the two projections of the waveguide should have the same length. With this assumption, Eq. (6.8.17) gives

$$P_e = \sin^2(\Omega\tau) \cos^2(\Omega\tau) \left| e^{-i\Delta\omega T} + 1 \right|^2. \tag{6.8.18}$$

We can maximize the factor $\sin^2(\Omega\tau) \cos^2(\Omega\tau)$ by taking $\Omega\tau = \pi/4$, in which case

$$P_e = \frac{1}{2} \left[ 1 + \cos\left( \Delta\omega T \right) \right]. \tag{6.8.19}$$

(In principle $\Omega$ depends on $\omega$, but because we assume that $\hbar|\Delta\omega| \ll |U_{eg}|$ this dependence is very weak, so that we can find a value of $\tau$ for which $\Omega\tau$ is very close to $\pi/4$ for all interesting values of $\omega$.)

The expression (6.8.19) has maxima equal to unity at $\Delta\omega = 2n\pi/T$, with $n$ any integer, positive or negative or zero. As $\omega$ is varied through values near $(E_e - E_g)/\hbar$, the probability $P_e$ experiences a rapid variation from one maximum to the next. Because $T$ is large, these maxima are very close together but also very narrow, so that if we could identify the maximum corresponding to $\Delta\omega = 0$ then the value of $\omega$ for which that maximum is reached would provide a very accurate measurement of the frequency $(E_e - E_g)/\hbar$. But in itself, Eq. (6.8.19) provides no clue to the identity of the maximum with $\Delta\omega = 0$.

From the beginning, it has been clear that this problem is resolved if there is some spread in the velocities of different atoms. Suppose that because of a spread in velocities, the probability that an atom spends a time between $T$ and $T + dT$ outside the waveguide between the first and the second pulses is a Gaussian:

$$P(T)\,dT = \exp\left(-(T - \overline{T})^2/\Delta T^2\right)\frac{dT}{\Delta T\,\sqrt{\pi}}, \qquad (6.8.20)$$

where $\overline{T}$ is the mean time between pulses, and $\Delta T$ is the spread in $T$. Then the fraction of atoms that leave the waveguide in the excited state is

$$\overline{P}_e = \frac{1}{2}\int_{-\infty}^{+\infty} \exp\left(-(T - \overline{T})^2/\Delta T^2\right)\frac{dT}{\Delta T\,\sqrt{\pi}}\left[1 + \cos\left(\Delta\omega\,T\right)\right],$$

$$= \frac{1}{2} + \frac{1}{2}\cos\left(\Delta\omega\,\overline{T}\right)\exp\left(-\Delta\omega^2\,\Delta T^2/4\right). \qquad (6.8.21)$$

The maximum at $\Delta\omega = 0$ still has $\overline{P}_e = 1$, but the adjacent maximum at $\Delta\omega = 2\pi/\overline{T}$ now has a smaller excitation probability,

$$\overline{P}_e = [1 + \exp(-\pi^2\,\Delta T^2/\overline{T}^2)]/2.$$

For instance if $\Delta T = 0.3\,\overline{T}$, then the maximum for $\Delta\omega = 2\pi/\overline{T}$ has $\overline{P}_e = 0.91$, which with adequate statistics should be clearly distinguishable from $\overline{P}_e = 1$. The actual distribution of $T$ will in general be different from Eq. (6.8.20) (it is actually the velocity rather than the time that has a Gaussian distribution for a thermal distribution of velocities), so the height of the maximum at $\Delta\omega = 2\pi/\overline{T}$ may be somewhat different from what we have calculated, but the measurement of $(E_e - E_g)/\hbar$ only depends on the identification of the maximum with $\Delta\omega = 0$, not on a precise knowledge of the heights of the other maxima. Some contemporary experiments have a much smaller spread in velocity, but the maximum at $\Delta\omega = 0$ can still be identified as the one that occurs at a value of $\omega$ that is fixed as the length $cT$ of the waveguide is changed.

In any case, as long as the maximum with $\Delta\omega = 0$ is identified in one way or another, Eq. (6.8.19) shows that by finding the value of $\omega$ at this maximum, we can measure the frequency $(E_e - E_g)/\hbar$ with a precision of order $1/T$, so the precision can be improved by increasing $T$, without running into any obstacle from the finite size of $|U_{eg}|$.

## 6.9 Open Systems

Closed systems are governed by time-independent Hamiltonians, so that their density matrices have a time-dependence given by the unitary transformation (3.6.24). This transformation is a special case of general linear transformations, which give the components of $\rho$ at one time as linear combinations of the components of $\rho$ at any other time. For a variety of open systems, systems that are exposed to external environments, although the time-dependence of the density matrix is more complicated than Eq. (3.6.24), it is still given by a linear relation, of the general form

$$[\rho(t)]_{MN} = \sum_{M'N'} K_{MM',NN'}(t - t')[\rho(t')]_{M'N'}, \tag{6.9.1}$$

with coefficients taken to be functions only of the elapsed time $t' - t$, under the assumption that the statistical properties of the system and its environment are time-independent. (We are here taking the physical Hilbert space to have a finite dimensionality $d$, so that the indices $M$, $N$, etc. run over $d$ values, but these considerations can often be extended to infinite-dimensional Hilbert spaces.)

As an example, suppose as in Section 6.4 that the effect of the environment is to give the Schrödinger-picture state vector $\Psi(t)$ a time-dependence governed by a rapidly and randomly fluctuating time-dependent Hamiltonian $H(t)$:

$$i\hbar \frac{d}{dt}\Psi(t) = H(t)\Psi(t).$$

The solution may be written

$$\Psi(t) = U(t, t')\Psi(t'),$$

where $U(t, t')$ is the solution of the differential equation

$$i\hbar \frac{d}{dt}U(t, t') = H(t) U(t, t')$$

with the initial condition

$$U(t', t') = 1.$$

It follows that for any given history of fluctuations, the density matrix (3.3.35) has a time-dependence given by the unitary transformation

$$\rho(t) = U(t, t')\rho(t')U^{\dagger}(t, t').$$

(We can easily see that $U$ is unitary, because with $H(t)$ Hermitian, Eq. (6.9.3) tells us that $U^{\dagger}(t, t') U(t, t')$ has vanishing rate of change, and it satisfies the initial condition $U^{\dagger}(t', t') U(t', t') = 1$.) Where $H(t)$ is rapidly and randomly fluctuating, we are less interested in individual histories of the density matrix than in its average over many fluctuations. Representing the average of any quantity over many fluctuations by a bar over that quantity, we have an averaged time-dependence

$$\overline{\rho(t)} = \overline{U(t, t')\rho(t')U^{\dagger}(t, t')}.$$

If we assume that the density matrix changes little in the characteristic time of the fluctuations in the Hamiltonian, then the average density matrix has the time-dependence (6.9.1), with

$$K_{MM',NN'}(t - t') \equiv \overline{[U(t, t')]_{MM'}[U^{\dagger}(t, t')]_{N'N}}.$$

Remarkably, whether or not the kernel $K$ takes this particular form, we can use the general properties of the kernel to derive a useful differential equation for the density matrix.[10] The necessary and sufficient condition that $\rho(t)$ given by Eq. (6.9.1) should be Hermitian for any Hermitian $\rho(t')$ is that $K$ is Hermitian, in the sense that

$$K^*_{MM',NN'}(\tau) = K_{NN',MM'}(\tau). \tag{6.9.2}$$

Also, the necessary and sufficient condition that $\rho(t)$ given by Eq. (6.9.1) should have unit trace for any $\rho(t')$ with unit trace is that

$$\sum_M K_{MM',MN'}(\tau) = \delta_{M'N'}. \tag{6.9.3}$$

Because these conditions are so general, Eq. (6.9.1) with $K$ satisfying Eqs. (6.9.2) and (6.9.3) is also used to study the evolution of closed systems in modified versions of quantum mechanics that have been introduced[11] to resolve the measurement problems discussed in Section 3.7.

From the Hermiticity condition (6.9.2), it follows that we can expand $K$ as

$$K_{MM',NN'}(\tau) = \sum_i \eta_i(\tau) u^{(i)}_{MM'}(\tau) u^{(i)*}_{NN'}(\tau), \tag{6.9.4}$$

where the $u^{(i)}_{MM'}(\tau)$ are eigenmatrices of the kernel $K_{MM',NN'}(\tau)$; the $\eta_i(\tau)$ are the corresponding real eigenvalues

$$\sum_{N'N} K_{MM',NN'}(\tau) u^{(i)}_{NN'}(\tau) = \eta_i(\tau) u^{(i)}_{MM'}(\tau) ; \tag{6.9.5}$$

and the eigenmatrices satisfy the orthonormality conditions

$$\sum_{N'N} u^{(i)*}_{NN'}(\tau) \, u^{(j)}_{NN'}(\tau) = \delta_{ij}. \tag{6.9.6}$$

The sum in Eq. (6.9.4) runs over all these eigenmatrices. The mapping (6.9.1) now reads

$$\rho_{MN}(t) = \sum_i \sum_{M'N'} \eta_i(t-t') u^{(i)}_{MM'}(t-t') \rho_{M'N'}(t') u^{(i)*}_{NN'}(t-t'), \tag{6.9.7}$$

or in a matrix notation

$$\rho(t) = \sum_i \eta_i(t-t') u^{(i)}(t-t') \rho(t') u^{(i)\dagger}(t-t'). \tag{6.9.8}$$

---

[10] The derivation described here follows the treatment of P. Pearle, *Eur. J. Phys.* **33**, 805 (2012) [arXiv:1204.2016].

[11] G. C. Ghirardi, A. Rimini, and T. Weber, *Phys. Rev. D* **34**, 470 (1986); P. Pearle, *Phys. Rev. A* **39**, 2277 (1989); G. C. Ghirardi, P. Pearle, and A. Rimini, *Phys. Rev. A* **42**, 78 (1990); P. Pearle, in *Quantum Theory: A Two-Time Success Story* (Yakir Aharonov Festschrift), eds. D. C. Struppa & J. M. Tollakson (Springer, Berlin, 2013), Chapter 9. [arXiv:1209.5082]. For a review, see A. Bassi and G. C. Ghirardi, *Physics Reports* **379**, 257 (2003).

Also, the trace condition (6.9.3) now reads

$$\sum_i \eta_i(\tau) u^{(i)\dagger}(\tau) u^{(i)}(\tau) = \mathbf{1},$$  (6.9.9)

with $\mathbf{1}$ the unit matrix.

The derivation of the differential equation for $\rho(t)$ is now an exercise in first-order perturbation theory. First, note that for $t' = t$ Eq. (6.9.1) must give $\rho(t') = \rho(t)$ for any $\rho(t)$, so in this case the kernel $K$ is

$$K_{MM',NN'}(0) = \delta_{M'M}\delta_{N'N}.$$  (6.9.10)

This has one eigenmatrix with eigenvalue $d$:

$$u^{(1)}_{MM'}(0) = \frac{1}{\sqrt{d}}\delta_{MM'}, \qquad \eta_1(0) = d,$$  (6.9.11)

and $d^2 - 1$ eigenmatrices denoted $u^{(a)}(0)$ with eigenvalue zero, taking the form of traceless matrices:

$$\sum_M u^{(a)}_{MM}(0) = 0, \qquad \eta_a(0) = 0.$$  (6.9.12)

But not any traceless matrices will do. Since the eigenvalue zero is degenerate, we must apply the rules of degenerate first-order perturbation theory worked out in Section 5.1. In order for the eigenmatrices $u^{(a)}(0)$ to connect smoothly with eigenmatrices $u^{(a)}(\tau)$ of $K(\tau)$ for small $\tau$, these eigenmatrices must be chosen to be not only eigenmatrices of $K(0)$, and hence traceless, but also such that the matrix elements of the term in $K(\tau)$ of first order in $\tau$ in the limit $\tau \to 0$ between these eigenmatrices should be diagonal:

$$\sum_{M'N'MN} u^{(b)*}_{MM'}(0) \left[ \frac{dK_{MM',NN'}(\tau)}{d\tau} \right]_{\tau=0} u^{(a)}_{NN'}(0) = \Delta_a \delta_{ab},$$  (6.9.13)

where $u^{(a)}(\tau)$ is the eigenmatrix of $K(\tau)$ that connects smoothly with $u^{(a)}(0)$. Then the corresponding eigenvalue $\eta_a(\tau)$ has derivative

$$\left[ \frac{d\eta_a(\tau)}{d\tau} \right]_{\tau=0} = \Delta_a.$$  (6.9.14)

To derive a differential equation for $\rho(t)$, we consider the limit of Eq. (6.9.1) when the elapsed time $t' - t$ becomes very small. Using Eqs. (6.9.8) and (6.9.11), and the vanishing of $\eta_a(0)$, the terms of first order in $t' - t$ in Eq. (6.9.1) give

$$\dot{\rho}(t) = \sum_a \Delta_a u^{(a)}(0)\rho(t)u^{(a)\dagger}(0) + B\rho(t) + \rho(t)B^\dagger,$$  (6.9.15)

where

$$B = \frac{1}{2d}\dot{\eta}_1(0)\mathbf{1} + d^{1/2}\dot{u}^{(1)}(0).$$  (6.9.16)

To derive a more useful formula for the matrix $B$, we use the trace condition (6.9.9). This condition is automatically satisfied for $\tau = 0$ by the eigenmatrices (6.9.11) and (6.9.12), but the derivative of Eq. (6.9.9) at $\tau = 0$ gives a non-trivial sum rule:

$$\sum_a \Delta_a u^{(a)\dagger}(0) u^{(a)}(0) + \frac{1}{d} \dot{\eta}^{(1)}(0)\mathbf{1} + d^{1/2} \dot{u}^{(1)}(0) + d^{1/2} \dot{u}^{(1)\dagger}(0) = 0,$$

or in other words

$$B + B^\dagger = -\sum_a \Delta_a u^{(a)\dagger}(0) u^{(a)}(0). \tag{6.9.17}$$

We can introduce a new sort of Hamiltonian, an Hermitian matrix $\mathcal{H}$, by defining $-i\mathcal{H}$ as the anti-Hermitian part of $B$, so that Eq. (6.9.17) reads

$$B = -i\mathcal{H} - \frac{1}{2} \sum_a \Delta_a u^{(a)\dagger}(0) u^{(a)}(0). \tag{6.9.18}$$

The differential equation (6.9.15) then takes the form

$$\dot{\rho}(t) = -i[\mathcal{H}, \rho(t)] + \sum_a \Delta_a \left[ u^{(a)}(0) \rho(t) u^{(a)}(0)^\dagger - \frac{1}{2} u^{(a)}(0)^\dagger u^{(a)}(0) \rho(t) \right.$$

$$\left. - \frac{1}{2} \rho(t) u^{(a)}(0)^\dagger u^{(a)}(0) \right]. \tag{6.9.19}$$

There is an ambiguity in the definition of the Hamiltonian, that allows us to replace the traceless matrices $u^{(a)}(0)$ in Eq. (6.9.19) with matrices $N_a$ that have any trace we like. It is easy to see that if we define

$$N_a \equiv u^{(a)}(0) + \xi_a \mathbf{1},$$

$$\mathcal{H}' \equiv \mathcal{H} - \frac{1}{2i} \sum_a \Delta_a \left( \xi_a u^{(a)}(0)^\dagger - \xi_a^* u^{(a)}(0) \right), \tag{6.9.20}$$

with $\xi_a$ any set of complex numbers, then the differential equation (6.9.19) may be rewritten as

$$\dot{\rho}(t) = -i[\mathcal{H}', \rho(t)] + \sum_a \Delta_a \left[ N_a \rho(t) N_a^\dagger - \frac{1}{2} N_a^\dagger N_a \rho(t) - \frac{1}{2} \rho(t) N_a^\dagger N_a \right]. \tag{6.9.21}$$

Since the $u^{(a)}(0)$ span the space of traceless matrices, this shows that unless we specify the traces of the matrices $N_a$, the Hamiltonian in Eq. (6.9.21) is well defined only up to the Hermitian part of a general traceless matrix.

We have not made yet any assumptions here about positivity. A matrix $A$ is said to be positive if $\sum_{MN} u_M^* A_{MN} u_N$ is positive (perhaps zero) for any $u_M$. The definition (3.3.35) makes it clear that the density matrix must be positive. (This can also be seen from the requirement that the mean value $\text{Tr}(A\rho)$ of any observable represented by a positive operator $A$ should be positive.) The density matrix $\rho(t)$ will be positive for any positive $\rho(t')$, if (though not only if[12]) all eigenvalues $\eta^{(i)}(t - t')$ are positive. This is evident if we rewrite Eq. (6.9.8) for $\eta_i(\tau) \geq 0$ in what is known as the *Kraus form*:[13]

$$\rho(t') = \sum_i A^{(i)}(t - t')\rho(t')A^{(i)\dagger}(t - t'), \qquad (6.9.22)$$

where $A^{(i)}(\tau) \equiv \sqrt{\eta_i(\tau)}u^{(i)}(\tau)$.

The eigenvalue $\eta^{(1)}(\tau)$ has the value unity for $\tau = 0$, so it is plausible that $\eta^{(1)}(\tau)$ will be positive at least for $\tau$ in some neighborhood of $\tau = 0$. On the other hand, all $\eta^{(a)}(\tau)$ vanish for $\tau = 0$, so according to Eq. (6.9.14) they will be positive at least for a range of positive $\tau$ if all $\Delta_a$ are positive, but in that case all $\eta^{(a)}(\tau)$ will be negative for small negative $\tau$. It is common to assume that all $\Delta_a$ are positive, and to use Eq. (6.9.21) only to predict the future, in which case we are assured that if $\rho(t')$ is positive then $\rho(t)$ will be positive at least for a finite range of $t$ later than $t'$, giving up any intention to use Eq. (6.9.21) to recover the past. Equation (6.9.21) can then be put in the form known as the *Lindblad equation*:[14]

$$\dot{\rho}(t) = -i[\mathcal{H}', \rho(t)] + \sum_a \left[ L_a\rho(t)L_a^\dagger - \frac{1}{2}L_a^\dagger L_a\,\rho(t) - \frac{1}{2}\rho(t)L_a^\dagger L_a \right], \quad (6.9.23)$$

where $L_a \equiv \sqrt{\Delta_a}N_a$.

There is an argument that all eigenvalues of the kernel of any physically allowed transformation of form (6.9.1) must be positive, as assumed in the derivation of the Lindblad equation. This is based on the requirement of *complete positivity*.[15] A kernel is said to be completely positive if it not only preserves the positivity of the density matrix for the system in question, but also preserves the positivity of the density matrix for a system that is expanded by including an isolated subsystem of arbitrary finite dimensionality on which the kernel acts as the unit operator. A theorem of Choi[16] shows that all eigenvalues of completely positive kernels are positive. But in the real world there are no physical states on which time-translation acts trivially except the vacuum

---

[12] The standard example of a transformation (6.9.1) for which the kernel $K$ has negative as well as positive eigenvalues but that nevertheless preserves the positivity of $\rho$ is the transposition map, with $K_{MM',NN'} = \delta_{MN'}\delta_{NM'}$. With this kernel, Eq. (6.9.1) converts $\rho$ into its transpose, which is certainly positive if $\rho$ is. But the eigenmatrices (in the sense of Eq. (6.9.5)) of this kernel are all matrices that are either symmetric or antisymmetric, with eigenvalues $+1$ and $-1$, respectively.

[13] K. Kraus, *States, Effects, and Operations – Fundamental Notions of Quantum Mechanics*, Lecture Notes in Physics 190 (Springer-Verlag, Berlin, 1983), Chapter 3.

[14] G. Lindblad, *Commun. Math. Phys.* **48**, 119 (1976); V. Gorini, A. Kossakowski and E. C. G. Sudarshan, *J. Math. Phys.* **17**, 821 (1976). The Lindblad equation can be derived as a straightforward application of an earlier result of A. Kossakowski, *Reports Math. Phys.* **3**, 247 (1972), Eq. (77).

[15] W. F. Stinespring, *Proc. Am. Math. Soc.* **6**, 211 (1955); M. D. Choi, *J. Canad. Math.* **24**, 520 (1972). For a review, see F. Benatti and R. Floreanini, *Int. J. Mod. Phys.* **B19**, 3063 (2005) [arXiv:quant-ph/0507271].

[16] M. D. Choi, *Linear Algebra and its Applications* **10**, 285 (1975)

state, which forms only a one-dimensional Hilbert space, so for some time it was not clear that the Choi theorem is physically relevant. There is, however, another requirement that does seem to be inescapably necessary, and that leads to the same conclusion about positive eigenvalues. If some system $S$ is physically realizable, then the system $S \otimes S$ consisting of two isolated copies of $S$ will presumably also be realizable. Any symmetry that acts on the density matrix of $S$ with a kernel $K$ will act on the density matrix of the combined system with a kernel given by a direct product $K \otimes K$. Benatti, Floreanini, and Romano[17] have shown that in this case, in order for $K \otimes K$ to be positive (in the sense of transforming all entangled positive Hermitian density matrices for $S \otimes S$ into positive Hermitian density matrices) it is necessary not only that $K$ be positive, but also that it be completely positive, so that all eigenvalues of $K$ are indeed positive.

The differential equation (6.9.23) has some especially interesting properties in the case where the $L_a$ are Hermitian. One feature is that it yields a non-decreasing von Neumann entropy.[18] The rate of increase of the entropy (3.3.38) is[19]

$$\frac{d}{dt} S[\rho] = -k_{\mathrm{B}} \mathrm{Tr}\left[\frac{d\rho}{dt}[1 + \ln \rho]\right] = -k_{\mathrm{B}} \mathrm{Tr}\left[\frac{d\rho}{dt} \ln \rho\right].$$

The first term in Eq. (6.9.23) makes no contribution to $dS/dt$, because $\mathrm{Tr}\left[[\mathcal{H}', \rho] \ln \rho\right] = \mathrm{Tr}\left[\mathcal{H}'[\rho, \ln \rho]\right] = 0$. We are left with

$$\frac{d}{dt} S[\rho] = -k_{\mathrm{B}} \sum_a \mathrm{Tr}\left[\left(L_a \rho L_a - L_a^2 \rho\right) \ln \rho\right]$$

$$= -k_{\mathrm{B}} \sum_a \sum_{ij} |[L_a]_{ij}|^2 (p_j - p_i) \ln p_i,$$

---

[17] F. Benatti, R. Floreanini, and R. Romano, *J. Phys. A Math. Gen.* **35**, L351 (2002).

[18] The proof given here is a modified version of the proof given by T. Banks, L. Susskind, and M. H. Peskin, *Nuclear Phys.* B **244**, 125 (1984).

[19] This follows immediately from the general rule that for any differentiable function $f(\rho)$ of an arbitrary operator function $\rho(t)$, even where $d\rho/dt$ does not commute with $\rho$, we have

$$\frac{d}{dt} \mathrm{Tr}\, f(\rho) = \mathrm{Tr}\left[f'(\rho)\frac{d\rho}{dt}\right].$$

To see this, note that if $\rho$ has eigenvalues $p_i$ with normalized eigenvectors $\Psi_i$, then

$$\mathrm{Tr}\left[f'(\rho)\frac{d\rho}{dt}\right] = \sum_i f'(p_i)\left(\Psi_i, \frac{d\rho}{dt}\Psi_i\right),$$

but because the norm of $\Psi_i$ is time-independent

$$\frac{dp_i}{dt} = \frac{d}{dt}\left(\Psi_i, \rho\Psi_i\right) = \left(\Psi_i, \frac{d\rho}{dt}\Psi_i\right) + p_i\left(\Psi_i, \frac{d}{dt}\Psi_i\right) + p_i\left(\frac{d}{dt}\Psi_i, \Psi_i\right) = \left(\Psi_i, \frac{d\rho}{dt}\Psi_i\right),$$

so

$$\mathrm{Tr}\left[f'(\rho)\frac{d\rho}{dt}\right] = \sum_i f'(p_i)\frac{dp_i}{dt} = \frac{d}{dt}\sum_i f(p_i) = \frac{d}{dt}\mathrm{Tr}\, f(\rho),$$

which is the desired relation. The final expression for $\dot{S}$ follows from the constancy of $\mathrm{Tr}\, \rho$.

where $i$ and $j$ label eigenvectors of $\rho$, and $p_i$ and $p_j$ are the corresponding eigenvalues. Since we are assuming that the $L_a$ are Hermitian, the factor $|[L_a]_{ij}|^2(p_j - p_i)$ is antisymmetric in $i$ and $j$, so the sum may be written

$$\frac{d}{dt}S[\rho] = \frac{k_B}{2}\sum_a\sum_{ij}|[L_a]_{ij}|^2(p_j - p_i)\left(\ln p_j - \ln p_i\right). \tag{6.9.24}$$

But $\ln p$ is an increasing function of $p$, so that $(p_j - p_i)\left(\ln p_j - \ln p_i\right)$ is always positive, and the entropy $S$ therefore never decreases, as was to be shown. In particular, pure states for which $S = 0$ in general evolve into ensembles of states with various probabilities, for which $S > 0$.

The late-time behavior of the density matrix provides another interesting feature of the case where all $L_\alpha$ are Hermitian. Because Eq. (6.9.23) is a linear differential equation, we expect $\rho(t)$ to be given by a sum[20]

$$\rho(t) = \sum_n \rho_n \exp(\lambda_n t), \tag{6.9.25}$$

where $\rho_n$ and $\lambda_n$ are the eigenmatrices and eigenvalues of the linear operator in Eq. (6.9.23):

$$\lambda_n\rho_n = -i[\mathcal{H}', \rho_n] + \sum_a\left[L_a\rho_n L_a^\dagger - \frac{1}{2}L_a^\dagger L_a\rho_n - \frac{1}{2}\rho_n L_a^\dagger L_a\right]. \tag{6.9.26}$$

In the case where all $L_a$ are Hermitian, we have

$$\lambda_n\,\mathrm{Tr}\left(\rho_n^\dagger\rho_n\right) = -i\,\mathrm{Tr}\left(\rho_n^\dagger[\mathcal{H}', \rho_n]\right) - \frac{1}{2}\sum_\alpha\mathrm{Tr}\left([\rho_n, L_a]^\dagger[\rho_n, L_a]\right). \tag{6.9.27}$$

The first term on the right-hand side is pure imaginary, because $\mathrm{Tr}\left(\rho_n^\dagger[\mathcal{H}, \rho_n]\right)^* = \mathrm{Tr}\left([\rho_n^\dagger, \mathcal{H}]\rho_n]\right) = \mathrm{Tr}\left(\rho_n^\dagger[\mathcal{H}, \rho_n]\right)$, while the second term is real and negative, so we can conclude that the real parts of all $\lambda_n$ are negative. Most terms in Eq. (6.9.25) therefore decay exponentially, leaving only the terms with $\mathrm{Re}\,\lambda_n = 0$, which according to Eq. (6.9.27) have $\rho_n$ commuting with all $L_a$.

This discussion gives us an idea of what sort of operators $L_a$ appear in systems that are arranged to provide a measurement of some set of observables. As we saw in Eq. (3.7.2), the effect of a measurement must be to convert the initial density matrix into a linear combination of projection operators $\Lambda_\alpha = [\Psi_\alpha\Psi_\alpha^\dagger]$ on the orthonormal eigenvectors $\Psi_\alpha$ of the observables being measured. According to the above results, in order for the density matrix to have a late-time limit of this form (aside from possible oscillations due to the "Hamiltonian" $\mathcal{H}'$) all $L_a$

---

[20] This is for the generic case, where none of the eigenvalues are degenerate. If the eigenvalue $\lambda_n$ has an $\mathcal{N}$-fold degeneracy, then the exponential $\exp(\lambda_n t)$ is accompanied by a polynomial in $t$ of order $\mathcal{N} - 1$.

must commute with the $\Lambda_\alpha$. This condition requires that the $L_a$ must be linear combinations of the $\Lambda_\alpha$:[21]

$$L_a = \sum_\alpha l_{a\alpha} \Lambda_\alpha, \tag{6.9.28}$$

with coefficients $l_{a\alpha}$ that must be real in order that the $L_a$ be Hermitian. It is plausible that because measurement involves macroscopic apparatus, the rate of change of the density matrix due to the $L_a$ is much faster than the rate of change in ordinary quantum mechanics, due to $\mathcal{H}'$. Neglecting its first term, Eq. (6.9.23) now takes the form

$$\dot\rho(t) = \sum_{\alpha\beta} C_{\alpha\beta} \left[ \Lambda_\alpha \rho(t) \Lambda_\beta - \frac{1}{2} \Lambda_\alpha \Lambda_\beta \rho(t) - \frac{1}{2} \rho(t) \Lambda_\alpha \Lambda_\beta \right], \tag{6.9.29}$$

where $C_{\alpha\beta} = \sum_a l_{a\alpha} l_{a\beta}$. We can try a solution of the form

$$\rho(t) = \sum_{\alpha\beta} f_{\alpha\beta}(t) \Lambda_\alpha \rho(0) \Lambda_\beta. \tag{6.9.30}$$

The completeness of the states $\Psi_\alpha$ implies that $\sum_\alpha \Lambda_\alpha = 1$, so the initial condition that the density matrix equals $\rho(0)$ at $t = 0$ is satisfied if $f_{\alpha\beta}(0) = 1$ for all $\alpha$ and $\beta$. Inserting (6.9.30) in Eq. (6.9.29) and again using the relation $\Lambda_\alpha \Lambda_\beta = \delta_{\alpha\beta} \Lambda_\alpha$, we find that

$$\dot f_{\alpha\beta} = \lambda_{\alpha\beta} f_{\alpha\beta}, \tag{6.9.31}$$

where

$$\lambda_{\alpha\beta} = C_{\alpha\beta} - \frac{1}{2} \left( C_{\alpha\alpha} + C_{\beta\beta} \right) = -\frac{1}{2} \sum_a \left( l_{a\alpha} - l_{a\beta} \right)^2. \tag{6.9.32}$$

The solution satisfying the initial condition $f_{\alpha\beta}(0) = 1$ is of course $f_{\alpha\beta}(t) = \exp[\lambda_{\alpha\beta} t]$, so

$$\rho(t) = \sum_{\alpha\beta} \Lambda_\alpha \rho(0) \Lambda_\beta \, \exp[\lambda_{\alpha\beta} t]. \tag{6.9.33}$$

In the generic case, where there are no different $\alpha$ and $\beta$ for which $l_{a\alpha}$ and $l_{a\beta}$ are equal for all $L_a$, all $\lambda_{\alpha\beta}$ with $\alpha \neq \beta$ are negative-definite, so all terms in

---

[21] It is obvious that this condition is sufficient, since $\Lambda_\alpha \Lambda_\beta = \delta_{\alpha\beta} \Lambda_\alpha$, so all $\Lambda$s commute with each other. To see that it is necessary, note that the condition that $L_a$ commutes with $\Lambda_\alpha$ tells us that

$$L_a \Psi_\alpha = L_a \Lambda_\alpha \Psi_\alpha = \Lambda_\alpha L_a \Psi_\alpha = \Psi_\alpha (\Psi_\alpha, L_a \Psi_\alpha),$$

so every $\Psi_\alpha$ is an eigenvector of each $L_a$. The $L_a$ are therefore just functions of the observables being measured. As we saw in Section 3.3, the most general such function is a linear combination of the projection operators $\Lambda_\alpha$.

Eq. (6.9.33) vanish for $t \to \infty$, except those terms with $\alpha = \beta$. Therefore for late times

$$\rho(t) \to \sum_{\alpha} \Lambda_{\alpha} \rho(0) \Lambda_{\alpha}. \qquad (6.9.34)$$

This is just the behavior that according to Eq. (3.7.2) is expected for a measurement of quantities whose eigenstates are $\Psi_{\alpha}$. So we see that Eq. (6.9.29) is general enough to reproduce not only the ordinary unitary evolution of the density matrix in quantum mechanics, which occurs when the $L_a$ terms in Eq. (6.9.29) are much smaller than the $\mathcal{H}'$ term, but also the change in the density matrix produced by a measurement.

## Problems

1. Consider a time-dependent Hamiltonian $H = H_0 + H'(t)$, with

   $$H'(t) = U \exp(-t/T),$$

   where $H_0$ and $U$ are time-independent operators, and $T$ is a constant. What is the probability to lowest order in $U$ that the perturbation will produce a transition from one eigenstate $n$ of $H_0$ to a different eigenstate $m$ of $H_0$ during a time interval from $t = 0$ to a time $t \gg T$?

2. Calculate the rate of ionization of a hydrogen atom in the $2p$ state in a monochromatic external electric field, averaged over the component of angular momentum in the direction of the field. (Ignore spin.)

3. Consider a Hamiltonian $H[s]$ that depends on a number of slowly varying parameters collectively called $s(t)$. What is the effect on the Berry phase $\gamma_n[C]$ for a given closed curve $C$, if $H[s]$ is replaced with $f[s]H[s]$, where $f[s]$ is an arbitrary real numerical function of the $s$?

# 7

# Potential Scattering

We do not observe the trajectories of particles within molecules or atoms or atomic nuclei. Instead, information about these systems that does not come from the energies of their discrete states we mostly have to take from scattering experiments. Indeed, as we saw in Section 1.2, at the very beginning of modern atomic physics, our understanding that the positive charge of atoms is concentrated in a small heavy nucleus came in 1911 from a scattering experiment carried out in Rutherford's laboratory, in which alpha particles emitted by radium nuclei were scattered by gold atoms. Today the exploration of the properties of elementary particles is largely carried out by studying the scattering of particles coming from high-energy accelerators.

In this chapter we will study the theory of scattering in a simple but important case, the elastic scattering of a non-relativistic particle in a local potential, but using modern techniques that can easily be extended to more general problems. The general formalism of scattering theory will be described in the following chapter.

## 7.1 In-States

We consider a non-relativistic particle of mass $\mu$ in a potential $V(\mathbf{x})$. The Hamiltonian is

$$H = H_0 + V(\mathbf{x}), \qquad (7.1.1)$$

where $H_0 = \mathbf{p}^2/2\mu$ is the kinetic energy operator, and $\mathbf{x}$ is the position operator. Later we will specialize to the case of a central potential $V(r)$, that depends only on $r \equiv |\mathbf{x}|$, but for the present it is just as easy to consider this more general case. We assume that $V(\mathbf{x}) \to 0$ for $r \to \infty$. We will not be concerned here with a particle in a bound state, which would have negative energy, but with a positive-energy particle, which comes into the potential from great distances with momentum $\hbar\mathbf{k}$, and is scattered, going out again to infinity, generally along a different direction.

In the Heisenberg picture, this situation is represented by a time-independent state vector $\Psi_{\mathbf{k}}^{\text{in}}$, the superscript "in" indicating that this state looks like it

247

consists of a particle with momentum $\hbar\mathbf{k}$ far from the scattering center if measurements are made at very early times. We have to be careful regarding what is meant by this. At very early times the particle is at a location where the potential is negligible, so it has an energy $\hbar^2\mathbf{k}^2/2\mu$, and this state vector is therefore an eigenstate of the Hamiltonian, with

$$H\Psi_{\mathbf{k}}^{\text{in}} = \frac{\hbar^2\mathbf{k}^2}{2\mu}\Psi_{\mathbf{k}}^{\text{in}}. \tag{7.1.2}$$

In the Schrödinger picture, the time-dependent state $\exp(-itH/\hbar)\,\Psi_{\mathbf{k}}^{\text{in}}$ is hence just $\Psi_{\mathbf{k}}^{\text{in}}$ times a seemingly trivial phase factor $\exp(-i\hbar t\mathbf{k}^2/2\mu)$. In order to interpret the above definition of $\Psi_{\mathbf{k}}^{\text{in}}$, we must consider the time-dependence of a *superposition* of states with a spread of energies:

$$\Psi_g(t) = \int d^3k\, g(\mathbf{k})\, \exp(-i\hbar t\mathbf{k}^2/2\mu)\, \Psi_{\mathbf{k}}^{\text{in}}, \tag{7.1.3}$$

where $g(\mathbf{k})$ is a smooth function that is peaked at some wave number $\mathbf{k}_0$. The state $\Psi_{\mathbf{k}}^{\text{in}}$ may be defined as the particular solution of the eigenvalue equation (7.1.2) that satisfies the further condition that, for any sufficiently smooth function $g(\mathbf{k})$, in the limit $t \to -\infty$,

$$\Psi_g(t) \to \int d^3k\, g(\mathbf{k})\, \exp(-i\hbar t\mathbf{k}^2/2\mu)\, \Phi_{\mathbf{k}}, \tag{7.1.4}$$

where $\Phi_{\mathbf{k}}$ are orthonormal eigenvectors of the momentum operator $\mathbf{P}$ with eigenvalue $\hbar\mathbf{k}$

$$\mathbf{P}\Phi_{\mathbf{k}} = \hbar\mathbf{k}\Phi_{\mathbf{k}}, \qquad \left(\Phi_{\mathbf{k}}, \Phi_{\mathbf{k}'}\right) = \delta^3(\hbar\mathbf{k} - \hbar\mathbf{k}'), \tag{7.1.5}$$

and hence eigenvectors of $H_0$ (not $H$!), with eigenvalues $E(|\mathbf{k}|) = \hbar^2\mathbf{k}^2/2\mu$. (Even though these states are labeled with their wave number, it proves convenient to normalize them so that their scalar product is a delta function of momentum, rather than of wave number.) The normalization condition $\left(\Psi_g, \Psi_g\right) = 1$ then is equivalent to the condition

$$\hbar^{-3}\int d^3k\, |g(\mathbf{k})|^2 = 1. \tag{7.1.6}$$

The condition (7.1.4) can be expressed by rewriting the Schrödinger equation as an integral equation. We can write equation (7.1.2) as

$$(E(|\mathbf{k}|) - H_0)\Psi_{\mathbf{k}}^{\text{in}} = V\Psi_{\mathbf{k}}^{\text{in}}.$$

This has a formal solution

$$\Psi_{\mathbf{k}}^{\text{in}} = \Phi_{\mathbf{k}} + \left(E(|\mathbf{k}|) - H_0 + i\epsilon\right)^{-1}V\Psi_{\mathbf{k}}^{\text{in}}, \tag{7.1.7}$$

where $\epsilon$ is a positive infinitesimal quantity, which is inserted to give meaning to the operator $(E(|\mathbf{k}|) - H_0 + i\epsilon)^{-1}$ when we integrate over the eigenvalues of $H_0$. It is known as the *Lippmann–Schwinger equation.*[1] (This is only a "formal" solution, because $\Psi_{\mathbf{k}}^{\text{in}}$ appears on the right-hand side as well as the left-hand side.)

Of course, we could have found a similar formal solution of the Schrödinger equation with a denominator $E(|\mathbf{k}|) - H_0 - i\epsilon$ in place of $E(|\mathbf{k}|) - H_0 + i\epsilon$. We could even have taken any average of $E(|\mathbf{k}|) - H_0 - i\epsilon$ and $E(|\mathbf{k}|) - H_0 + i\epsilon$, or dropped the first term in Eq. (7.1.7). The special feature of the particular "solution" (7.1.7) is that it also satisfies the initial condition (7.1.4).

To see this, we can expand $V\Psi_{\mathbf{k}}^{\text{in}}$ in the orthonormal free-particle states $\Phi_{\mathbf{q}}$:

$$V\Psi_{\mathbf{k}}^{\text{in}} = \hbar^3 \int d^3q \; \Phi_{\mathbf{q}} \left(\Phi_{\mathbf{q}}, V\Psi_{\mathbf{k}}^{\text{in}}\right). \tag{7.1.8}$$

Then Eq. (7.1.7) becomes

$$\Psi_{\mathbf{k}}^{\text{in}} = \Phi_{\mathbf{k}} + \hbar^3 \int d^3q \left(E(|\mathbf{k}|) - E(|\mathbf{q}|) + i\epsilon\right)^{-1} \Phi_{\mathbf{q}} \left(\Phi_{\mathbf{q}}, V\Psi_{\mathbf{k}}^{\text{in}}\right). \tag{7.1.9}$$

In calculating the integral over $\mathbf{k}$ in Eq. (7.1.3), we note that

$$\int d^3k \; g(\mathbf{k}) \frac{\exp(-i\hbar t\mathbf{k}^2/2\mu)}{E(|\mathbf{k}|) - E(q) + i\epsilon} \left(\Phi_{\mathbf{q}}, V\Psi_{\mathbf{k}}^{\text{in}}\right)$$

$$= \int d\Omega \int_0^\infty k^2 \, g(\mathbf{k}) \, dk \; \frac{\exp(-i\hbar t k^2/2\mu)}{E(k) - E(q) + i\epsilon} \left(\Phi_{\mathbf{q}}, V\Psi_{\mathbf{k}}^{\text{in}}\right),$$

where $d\Omega = \sin\theta \, d\theta \, d\phi$. We can convert the integral over $k$ to an integral over energy, using $dk = \mu \, dE / k\hbar^2$. Now, when $t \to -\infty$, the exponential oscillates very rapidly, so the only values of $E$ that contribute are those very near $E(q)$, where the denominator also varies very rapidly. Thus for $t \to -\infty$ we can set $k = q$ everywhere except in the rapidly varying exponential and denominator, giving a result proportional to

$$\int_{-\infty}^\infty \frac{\exp(-iEt/\hbar)}{E - E(q) + i\epsilon} \, dE.$$

(The range of integration has been extended to the whole real axis, which is permissible since the integral receives no appreciable contributions anyway from the range $|E - E(q)| \gg \hbar/|t|$.) For $t \to -\infty$ we can close the contour of integration with a very large semi-circle in the upper half of the complex plane, on which the integrand is negligible because, for $\text{Im } E > 0$ and $t \to -\infty$, the numerator $\exp(-iEt/\hbar)$ is exponentially small. But the only singularity of the integrand is a pole at $E = E(q) - i\epsilon$, which is in the lower half plane, so

---

[1]  B. Lippmann and J. Schwinger, *Phys. Rev.* **79**, 469 (1950).

the integral vanishes for $t \to -\infty$. This leaves only the contribution of the first term in Eq. (7.1.9), which gives Eq. (7.1.4) for $t \to -\infty$.

To clarify the significance of the condition (7.1.4), consider its scalar product with a state $\Phi_{\mathbf{x}}$ of definite position, using the usual plane-wave wave function of states of definite momentum, which as we saw in Eq. (3.5.12) takes the form

$$\left(\Phi_{\mathbf{x}}, \Phi_{\mathbf{k}}\right) = (2\pi\hbar)^{-3/2} e^{i\mathbf{k}\cdot\mathbf{x}}. \tag{7.1.10}$$

This gives, for $t \to -\infty$,

$$\left(\Phi_{\mathbf{x}}, \Psi_g(t)\right) \to (2\pi\hbar)^{-3/2} \int d^3k \, g(\mathbf{k}) \, \exp\left(i\mathbf{k}\cdot\mathbf{x} - i\hbar t \mathbf{k}^2/2\mu\right). \tag{7.1.11}$$

We will assume that the particle comes in from a great distance along the negative 3-axis, so we are interested in the limit of very large negative $t$ and $x_3$, but with $x_3/t$ held finite. However, we will also assume that the particle velocity is sufficiently closely confined to the 3-direction that, where the function $g(\mathbf{k})$ is not negligible,

$$\hbar|t|\mathbf{k}_\perp^2/2\mu \ll 1, \tag{7.1.12}$$

where $\mathbf{k}_\perp$ is the two-vector $(k_1, k_2)$. Equation (7.1.11) can then be written

$$\left(\Phi_{\mathbf{x}}, \Psi_g(t)\right) \to (2\pi\hbar)^{-3/2} \int d^2k_\perp \int_{-\infty}^{\infty} dk_3 \, g(\mathbf{k}_\perp, k_3) \exp\left(i\mathbf{k}_\perp \cdot \mathbf{x}_\perp\right)$$
$$\times \exp\left(ix_3^2\mu/2\hbar t\right) \exp\left(-i\hbar t(k_3 - \mu x_3/\hbar t)^2/2\mu\right). \tag{7.1.13}$$

The rapid oscillation of the final factor as a function of $k_3$ makes this integral negligible for $t \to -\infty$ except for contributions from $k_3$ close to its stationary point at $k_3 = \mu x_3/\hbar t$, so in the limit $t \to -\infty$ with $x_3/t$ fixed, the integral becomes

$$\left(\Phi_{\mathbf{x}}, \Psi_g(t)\right) \to (2\pi\hbar)^{-3/2} \int d^2k_\perp \, g(\mathbf{k}_\perp, \mu x_3/\hbar t) \exp\left(i\mathbf{k}_\perp \cdot \mathbf{x}_\perp\right)$$
$$\times \exp\left(ix_3^2\mu/2\hbar t\right) \int_{-\infty}^{\infty} dk_3 \exp\left(-i\hbar t(k_3 - \mu x_3/\hbar t)^2/2\mu\right)$$
$$= (2\pi\hbar)^{-3/2} \exp\left(ix_3^2\mu/2\hbar t\right) \sqrt{\frac{2\mu\pi}{i\hbar t}}$$
$$\times \int d^2k_\perp \, g(\mathbf{k}_\perp, \mu x_3/\hbar t) \exp\left(i\mathbf{k}_\perp \cdot \mathbf{x}_\perp\right). \tag{7.1.14}$$

We assume that the function $g(\mathbf{k}_\perp, k_3)$, though smooth, is strongly peaked at $k_3 = k_0$ and $\mathbf{k}_\perp = 0$, so the expression (7.1.14) is peaked at $x_3 = \hbar k_0 t/\mu$, corresponding to a particle moving along the $x_3$ axis, with velocity $\hbar k_0/\mu$.

In particular, for $t \to -\infty$ the spatial probability distribution is

$$\left|\left(\Phi_{\mathbf{x}}, \Psi_g(t)\right)\right|^2 \to \frac{\mu}{4\pi^2\hbar^4 t} \left|\int d^2k_\perp \, g(\mathbf{k}_\perp, \mu x_3/\hbar t) \, \exp\left(i\mathbf{k}_\perp \cdot \mathbf{x}_\perp\right)\right|^2,$$

$$(7.1.15)$$

and respects the conservation of probability:

$$\int d^3x \, \left|\left(\Phi_{\mathbf{x}}, \Psi_g(t)\right)\right|^2 \to \frac{\mu}{\hbar^4 t} \int d^2k_\perp \int_{-\infty}^{\infty} dx_3 \, |g(\mathbf{k}_\perp, \mu x_3/\hbar t)|^2$$

$$= \hbar^{-3} \int d^2k_\perp \int_{-\infty}^{\infty} dk_3 \, |g(\mathbf{k}_\perp, k_3)|^2 = 1. \quad (7.1.16)$$

$$* \; * \; * \; * \; *$$

We can see in greater detail how this works out by taking a simple example for the function $g(\mathbf{k})$,

$$g(\mathbf{k}) \propto \exp\left(-\frac{\Delta_0^2}{2}(\mathbf{k} - \mathbf{k}_0)^2 - i\frac{\hbar\mathbf{k} \cdot \mathbf{k}_0 t_0}{\mu} + \frac{i\hbar t_0 \mathbf{k}^2}{2\mu}\right),$$

where $t_0$ is a large negative initial time, $\mathbf{k}_0$ is in the 3-direction, and $\Delta_0$ is a constant. (The terms in the exponent proportional to $t_0$ are chosen so that, as we will see, $\Delta_0$ is the spread of the coordinate-space wave function at time $t = t_0$. These terms are stationary in $\mathbf{k}$ at $\mathbf{k} = \mathbf{k}_0$, so their presence does not invalidate the argument leading to Eq. (7.1.14).) A straightforward calculation using Eq. (7.1.11) gives a spatial probability distribution for $t \to -\infty$,

$$\left|\left(\Phi_{\mathbf{x}}, \Psi_g(t)\right)\right|^2 \propto \Delta^{-3} \exp\left(-\frac{1}{\Delta^2}\left(\mathbf{x} - (\hbar\mathbf{k}_0/\mu)t\right)^2\right),$$

where

$$\Delta \equiv \left(\Delta_0^2 + \frac{\hbar^2(t - t_0)^2}{\mu^2\Delta_0^2}\right)^{1/2}.$$

The probability distribution is thus centered on a point that moves with velocity equal to the mean momentum $\hbar\mathbf{k}_0$ divided by the mass $\mu$, reaching the scattering center $\mathbf{x} = 0$ at $t = 0$.

The spread of this distribution is $\Delta_0$ at $t = t_0$, but it begins to expand for $t - t_0 > \mu\Delta_0^2/\hbar$. This can easily be understood on simple kinematic grounds. The wave function has a spread in velocity $\Delta v$ equal to $\hbar/\mu$ times the spread in wave number, and hence of order $\hbar/\mu\Delta_0$. After a time interval $t - t_0$, this contributes an amount $\Delta v(t - t_0) \approx \hbar(t - t_0)/\mu\Delta_0$ to the spread in position. This becomes greater than the initial spread $\Delta_0$ for $t - t_0 > \mu\Delta_0^2/\hbar$.

This expansion in the wave packet does not become significant in typical cases. In order for the wave packet not to expand appreciably in the time interval from $t = t_0$ to $t = 0$, we need $\Delta_0^2 > \hbar|t_0|/\mu$. But we also must have

$\Delta_0 \ll \hbar k_0 |t_0|/\mu$, in order that $t_0$ should be sufficiently early that the wave packet does not spread all the way to the scattering center at $t = t_0$. These two conditions are compatible if $\hbar k_0^2 |t_0|/\mu \gg 1$, which just requires that the oscillation of the wave function has time to go through many cycles before the particle hits the scattering center. This requirement can be taken as part of what we mean by a scattering process.

## 7.2 Scattering Amplitudes

In the previous section we defined a state that at early times has the appearance of a particle traveling toward a collision with a scattering center. Now we must consider what this state looks like after the collision.

For this purpose, we consider the coordinate-space wave function of the state $\Psi_{\mathbf{k}}^{\text{in}}$. Returning to Eq. (7.1.7), let us write

$$V\Psi_{\mathbf{k}}^{\text{in}} = \int d^3x \; \Phi_{\mathbf{x}}\left(\Phi_{\mathbf{x}}, V\Psi_{\mathbf{k}}^{\text{in}}\right) = \int d^3x \; \Phi_{\mathbf{x}} V(\mathbf{x})\psi_{\mathbf{k}}(\mathbf{x}), \tag{7.2.1}$$

where $\psi_{\mathbf{k}}(\mathbf{x})$ is the coordinate-space wave function of the in-state,

$$\psi_{\mathbf{k}}(\mathbf{x}) \equiv \left(\Phi_{\mathbf{x}}, \Psi_{\mathbf{k}}^{\text{in}}\right). \tag{7.2.2}$$

Then, by taking the scalar product of the Lippmann–Schwinger equation (7.1.7) with a state $\Phi_{\mathbf{x}}$ of definite position, and using Eq. (7.1.10), we have

$$\psi_{\mathbf{k}}(\mathbf{x}) = (2\pi\hbar)^{-3/2}e^{i\mathbf{k}\cdot\mathbf{x}} + \int d^3y \; G_k(\mathbf{x} - \mathbf{y})V(\mathbf{y})\psi_{\mathbf{k}}(\mathbf{y}), \tag{7.2.3}$$

where $G_k$ is the Green function

$$\begin{aligned} G_k(\mathbf{x} - \mathbf{y}) &= \left(\Phi_{\mathbf{x}}, [E(k) - H_0 + i\epsilon]^{-1}\Phi_{\mathbf{y}}\right) \\ &= \int \frac{\hbar^3 d^3q}{(2\pi\hbar)^3} \frac{e^{i\mathbf{q}\cdot(\mathbf{x}-\mathbf{y})}}{E(k) - E(q) + i\epsilon} \\ &= \frac{4\pi}{(2\pi)^3} \int_0^\infty q^2\, dq \; \frac{\sin(q|\mathbf{x}-\mathbf{y}|)}{q|\mathbf{x}-\mathbf{y}|} \frac{2\mu/\hbar^2}{k^2 - q^2 + i\epsilon} \\ &= -i\frac{2\mu}{\hbar^2} \frac{1}{4\pi^2|\mathbf{x}-\mathbf{y}|} \int_{-\infty}^\infty \frac{e^{iq|\mathbf{x}-\mathbf{y}|}q\, dq}{k^2 - q^2 + i\epsilon} \\ &= -\frac{2\mu}{\hbar^2} \frac{1}{4\pi|\mathbf{x}-\mathbf{y}|} e^{ik|\mathbf{x}-\mathbf{y}|}. \end{aligned} \tag{7.2.4}$$

(The last expression is obtained by completing the contour of integration with a large semi-circle in the upper half plane, and picking up the contribution of the pole at $q = k + i\epsilon$.) For a potential $V(\mathbf{y})$ that vanishes sufficiently rapidly as $|\mathbf{y}| \to \infty$, Eq. (7.2.3) gives, for $|\mathbf{x}| \to \infty$,

$$\psi_{\mathbf{k}}(\mathbf{x}) \to (2\pi\hbar)^{-3/2} \left[ e^{i\mathbf{k}\cdot\mathbf{x}} + f_{\mathbf{k}}(\hat{x})e^{ikr}/r \right], \tag{7.2.5}$$

where $r \equiv |\mathbf{x}|$ and $f_{\mathbf{k}}(\hat{x})$ is the *scattering amplitude*,

$$f_{\mathbf{k}}(\hat{x}) = -\frac{\mu}{2\pi\hbar^2}(2\pi\hbar)^{3/2} \int d^3y \, e^{-ik\hat{x}\cdot\mathbf{y}} V(\mathbf{y})\psi_{\mathbf{k}}(\mathbf{y}). \tag{7.2.6}$$

Now let's consider how the superposition (7.1.3) behaves for late times. We consider the wave function

$$\psi_g(\mathbf{x}, t) \equiv \left( \Phi_{\mathbf{x}}, \Psi_g^{\text{in}}(t) \right) = \int d^3k \, g(\mathbf{k})\psi_{\mathbf{k}}(\mathbf{x}) \exp\left( -i\hbar t\mathbf{k}^2/2\mu \right), \tag{7.2.7}$$

in the limit $t \to +\infty$, with $r/t$ held fixed, and $\mathbf{x}$ off the 3-axis. Using Eq. (7.2.5) in this limit, Eq. (7.2.7) gives

$$\psi_g(\mathbf{x}, t) \to \frac{(2\pi\hbar)^{-3/2}}{r} \int d^2k_\perp \int_{-\infty}^{\infty} dk_3 \, g(\mathbf{k}_\perp, k_3)$$
$$\times \exp\left( ik_3 r - i\hbar t k_3^2/2\mu \right) f_{\mathbf{k}_0}(\hat{x}). \tag{7.2.8}$$

We have taken the subscript on the scattering amplitude to be $\mathbf{k}_0$, because the function $g$ is sharply peaked at this value of $\mathbf{k}$, and we have approximated $k \equiv \sqrt{k_3^2 + k_\perp^2}$ as $k \simeq k_3$ in the exponents, because $g(\mathbf{k}_\perp, k_3)$ is assumed to be negligible except for $|\mathbf{k}_\perp| \ll k_3$. As in the previous section, for large $r$ and $t$ we can set $k_3$ in $g(\mathbf{k}_\perp, k_3)$ equal to the value $k_3 = \mu r/\hbar t$ where the argument of the exponential is stationary, so that

$$\psi_g(\mathbf{x}, t) \to \frac{(2\pi\hbar)^{-3/2}}{r} f_{\mathbf{k}_0}(\hat{x}) \int d^2k_\perp \, g(\mathbf{k}_\perp, \mu r/\hbar t)$$
$$\times \int_{-\infty}^{\infty} dk_3 \exp\left( ik_3 r - i\hbar t k_3^2/2\mu \right)$$
$$= \frac{(2\pi\hbar)^{-3/2}}{r} f_{\mathbf{k}_0}(\hat{x}) \int d^2k_\perp \, g(\mathbf{k}_\perp, \mu r/\hbar t) \exp\left( i\mu r^2/2\hbar t \right)\sqrt{\frac{2\mu\pi}{i\hbar t}}. \tag{7.2.9}$$

The probability $dP(\hat{x})$ that the particle at late times is somewhere within the cone of infinitesimal solid angle $d\Omega$ around the direction $\hat{x}$ is then the integral of $|\psi_g(\mathbf{x}, t)|^2$ over this cone:

$$dP(\hat{x}, \mathbf{k}_0) = d\Omega \int_0^\infty r^2 \, dr \, |\psi_g(r\hat{x}, t)|^2$$
$$\to \frac{1}{(2\pi)^2} \frac{\mu}{\hbar^4 t} |f_{\mathbf{k}_0}(\hat{x})|^2 \int_0^\infty dr \, \left| \int d^2k_\perp \, g(\mathbf{k}_\perp, \mu r/\hbar t) \right|^2, \tag{7.2.10}$$

or, changing the variable of integration $r$ to $k_3 \equiv \mu r/\hbar t$,

$$\frac{dP(\hat{x}, \mathbf{k}_0)}{d\Omega} = \frac{1}{(2\pi)^2 \hbar^3} |f_{\mathbf{k}_0}(\hat{x})|^2 \int_0^\infty dk_3 \left| \int d^2 k_\perp\, g(\mathbf{k}_\perp, k_3) \right|^2. \qquad (7.2.11)$$

Now, the coefficient of $|f_{\mathbf{k}_0}(\hat{x})|^2$ in Eq. (7.2.11) has the dimensions of an inverse area. In fact, it is precisely the probability per unit area that the particle is in a small area centered on the 3-axis and normal to that axis:

$$\rho_\perp \equiv \lim \int_{-\infty}^\infty dx_3\, |\psi_g(0, x_3, t)|^2, \qquad (7.2.12)$$

for $t \to -\infty$. To see this, note that according to Eq. (7.1.15), with $\mathbf{x}_\perp = 0$, the quantity (7.2.12) is

$$\rho_\perp = \frac{\mu}{4\pi^2 \hbar^4 t} \int_{-\infty}^\infty dx_3 \left| \int d^2 k_\perp\, g(\mathbf{k}_\perp, \mu x_3/\hbar t) \right|^2$$

$$= \frac{1}{4\pi^2 \hbar^3} \int_{-\infty}^\infty dk_3 \left| \int d^2 k_\perp\, g(\mathbf{k}_\perp, k_3) \right|^2, \qquad (7.2.13)$$

which is the coefficient appearing in Eq. (7.2.11). Hence Eq. (7.2.11) may be written

$$\frac{dP(\hat{x}, \mathbf{k}_0)}{d\Omega} = \rho_\perp |f_{\mathbf{k}_0}(\hat{x})|^2. \qquad (7.2.14)$$

We define the *differential cross section* as the ratio

$$\frac{d\sigma(\hat{x}, \mathbf{k}_0)}{d\Omega} \equiv \frac{1}{\rho_\perp} \frac{dP(\hat{x}, \mathbf{k}_0)}{d\Omega}, \qquad (7.2.15)$$

so

$$\frac{d\sigma(\hat{x}, \mathbf{k}_0)}{d\Omega} = |f_{\mathbf{k}_0}(\hat{x})|^2. \qquad (7.2.16)$$

We can think of $d\sigma(\hat{x}, \mathbf{k}_0)$ as a tiny area normal to the 3-axis, which the particle must hit in order for it to be scattered into a solid angle $d\Omega$ around the direction $\hat{x}$. Equation (7.2.15) then says that the probability of hitting this area equals the ratio of $d\sigma$ to the effective cross-sectional area $1/\rho_\perp$ of the beam.

From now on, we shall drop the subscript 0 on $k_0$. Also, instead of writing the scattering amplitude as a function of $\mathbf{k}$ and $\hat{x}$, we will generally write it as a function of $k$ and the polar angles $\theta$ and $\phi$ of $\mathbf{x}$ around the direction of $\mathbf{k}$, so that Eq. (7.2.16) reads

$$d\sigma(\theta, \phi, k) = |f_k(\theta, \phi)|^2 \sin\theta\, d\theta\, d\phi. \qquad (7.2.17)$$

This is our general formula for the differential cross section in terms of the scattering amplitude.

Of course, to measure $d\sigma/d\Omega$, experimenters do not actually send a particle or particles toward a single target. Instead, they direct a beam of particles toward a thin slab containing some large number $N_T$ of targets. (It is necessary to specify a thin slab, to avoid the possibility of particles from the beam experiencing multiple scattering involving more than one target. This is why, in the discovery of the atomic nucleus discussed in Section 1.2, the target was chosen to be a thin gold leaf.) If scattering into some particular range of angles can occur only if a particle from the beam hits a tiny area $d\sigma$ around one of the targets, then the number of particles that are scattered into this range of angles is the number $\mathcal{N}_\mathrm{B}$ of beam particles per unit transverse area, times the total area $N_\mathrm{T}\,d\sigma$ that they have to hit.

## 7.3 The Optical Theorem

It may seem odd that the plane-wave term in Eq. (7.2.5) does not appear to be depleted by the scattering of the incident wave. Actually, in the forward direction there is an interference between the two terms in Eq. (7.2.5), which does decrease the amplitude of the plane wave beyond the scattering center, as required by the conservation of probability. In order for this to be the case, there must be a relation between the forward scattering amplitude and the total cross section for scattering. This relation is known as the *optical theorem*.[2]

To derive the theorem, we use the conservation condition for probabilities in three dimensions, which has already been discussed in Section 1.5. In coordinate space, the Schrödinger equation here is

$$-\frac{\hbar^2}{2M}\nabla^2\psi_\mathbf{k} + V(\mathbf{x})\psi_\mathbf{k} = \frac{\hbar^2\mathbf{k}^2}{2M}\psi_\mathbf{k}. \tag{7.3.1}$$

We multiply this with the complex conjugate $\psi_\mathbf{k}^*$, and then subtract the complex conjugate of the product. For a real potential this gives

$$0 = \psi_\mathbf{k}^* \nabla^2\psi_\mathbf{k} - \psi_\mathbf{k} \nabla^2\psi_\mathbf{k}^* = \nabla\cdot\left(\psi_\mathbf{k}^* \nabla\psi_\mathbf{k} - \psi_\mathbf{k} \nabla\psi_\mathbf{k}^*\right). \tag{7.3.2}$$

Using Gauss's theorem, it follows that, for a sphere of any radius $r$,

$$0 = r^2 \int_0^\pi \sin\theta\,d\theta \int_0^{2\pi} d\phi \left(\psi_\mathbf{k}^*\frac{\partial\psi_\mathbf{k}}{\partial r} - \psi_\mathbf{k}\frac{\partial\psi_\mathbf{k}^*}{\partial r}\right). \tag{7.3.3}$$

In particular, we can take $r$ large enough to use the asymptotic formula (7.2.5). In this limit, with $\mathbf{k}$ in the 3-direction and recalling that $x_3 = r\cos\theta$,

$$(2\pi\hbar)^3\psi_\mathbf{k}^*\frac{\partial\psi_\mathbf{k}}{\partial r} \to ik\cos\theta + \frac{ikf_\mathbf{k}e^{ikr(1-\cos\theta)}}{r} - \frac{f_\mathbf{k}e^{ikr(1-\cos\theta)}}{r^2}$$
$$+ \frac{ikf_\mathbf{k}^*\cos\theta\,e^{-ikr(1-\cos\theta)}}{r} + \frac{ik|f_\mathbf{k}|^2}{r^2} - \frac{|f_\mathbf{k}|^2}{r^3}$$

---

[2] The theorem has been given that name because it was first encountered in classical electrodynamics, as a relation due to Lord Rayleigh between the absorption of light and the imaginary part of the index of refraction. It was first derived for the scattering amplitude in quantum mechanics by E. Feenberg, *Phys. Rev.* **40**, 40 (1932). For a historical review, see R. G. Newton, *Amer. J. Phys.* **44**, 639 (1976).

so that

$$(2\pi\hbar)^3\left[\psi_\mathbf{k}^*\frac{\partial\psi_\mathbf{k}}{\partial r} - \psi_\mathbf{k}\frac{\partial\psi_\mathbf{k}^*}{\partial r}\right]$$

$$\to 2ik\cos\theta + \frac{ik(1+\cos\theta)e^{ikr(1-\cos\theta)}f_\mathbf{k}}{r} + \frac{ik(1+\cos\theta)e^{-ikr(1-\cos\theta)}f_\mathbf{k}^*}{r}$$

$$- \frac{e^{ikr(1-\cos\theta)}f_\mathbf{k}}{r^2} + \frac{e^{-ikr(1-\cos\theta)}f_\mathbf{k}^*}{r^2} + \frac{2ik|f_\mathbf{k}|^2}{r^2}. \tag{7.3.4}$$

For $kr \gg 1$ the exponentials $e^{\pm ikr(1-\cos\theta)}$ oscillate rapidly except where $\cos\theta = 1$, so the integral over $\theta$ in Eq. (7.3.3) receives almost its whole contribution from near $\theta = 0$. For any smooth function $g(\theta,\phi)$ of $\theta$ and $\phi$, we can therefore approximate

$$\int_0^\pi \sin\theta\, d\theta \int_0^{2\pi} d\phi\, e^{ikr(1-\cos\theta)}g(\theta,\phi) \to 2\pi g(0)\int_0^\pi \sin\theta\, d\theta\, e^{ikr(1-\cos\theta)}, \tag{7.3.5}$$

where $g(0)$ is the $\phi$-independent value of $g(\theta,\phi)$ for $\theta = 0$. Introducing the variable $\nu \equiv 1 - \cos\theta$, and replacing the limit $\nu = 2$ with $\nu = \infty$ (since the oscillation of the integral makes the contribution for $\nu$ between 2 and infinity exponentially small for large $kr$) this is

$$\int_0^\pi \sin\theta\, d\theta \int_0^{2\pi} d\phi\, e^{ikr(1-\cos\theta)}g(\theta,\phi) \to 2\pi g(0)\int_0^\infty d\nu\, e^{ikr\nu} = 2\pi i g(0)/kr. \tag{7.3.6}$$

(To evaluate the integral over $\nu$, we use the usual trick of inserting a factor $e^{-\epsilon\nu}$ with $\epsilon > 0$ in the integrand, and then letting $\epsilon$ go to zero after doing the integral.) Applying this to the solid angle integral of Eq. (7.3.4) then gives

$$(2\pi\hbar)^3\int_0^\pi \sin\theta\, d\theta \int_0^{2\pi} d\phi\left(\psi_\mathbf{k}^*\frac{\partial\psi}{\partial r} - \psi_\mathbf{k}\frac{\partial\psi_\mathbf{k}^*}{\partial r}\right)$$

$$\to \left(\frac{ik}{r}\right)\left(\frac{2\pi i}{kr}\right)2f_\mathbf{k}(0) + \left(\frac{ik}{r}\right)\left(\frac{-2\pi i}{kr}\right)2f_\mathbf{k}^*(0)$$

$$+ \frac{2ik}{r^2}\int_0^\pi \sin\theta\, d\theta \int_0^{2\pi}|f_\mathbf{k}(\theta,\phi)|^2\,d\phi + O\left(\frac{1}{r^3}\right)$$

$$\to -\frac{8\pi i}{r^2}\operatorname{Im} f_\mathbf{k}(0) + \frac{2ik}{r^2}\int_0^\pi \sin\theta\, d\theta \int_0^{2\pi} d\phi\,|f_\mathbf{k}(\theta,\phi))|^2 \tag{7.3.7}$$

and so for large $r$, Eq. (7.3.3) gives

$$\sigma_{\text{scat}} \equiv \int_0^\pi \sin\theta\, d\theta \int_0^{2\pi} d\phi\,|f_\mathbf{k}(\theta,\phi))|^2 = \frac{4\pi}{k}\operatorname{Im} f_\mathbf{k}(0). \tag{7.3.8}$$

This is a special case of what is known as the optical theorem, derived here under the condition of elastic scattering by a real potential. In this case the total cross section $\sigma_{\text{tot}}$ (defined so that, if the initial particle is confined to a transverse area $A$, then the total probability of scattering or any other reaction is $\sigma_{\text{tot}}/A$) is the same as the elastic scattering cross section $\sigma_{\text{scat}}$, so we can just as well write Eq. (7.3.8) as

$$\sigma_{\text{tot}} = \frac{4\pi}{k} \, \text{Im} \, f_{\mathbf{k}}(0). \tag{7.3.9}$$

This is the optical theorem in its most general form, which will be proved for general scattering processes in Section 8.3.

To see that Eq. (7.3.9) is what is required by the conservation of probability, let us consider a plane wave traveling in the 3-direction that strikes a thin foil of scatterers (thin enough to make multiple scattering negligible) lying in the $x-y$ plane, and calculate the wave function at a distance $z \gg 1/k$ behind the foil. For this purpose we have to add up the contributions of the individual scatterers, by multiplying the scattering amplitude with the number $\mathcal{N}$ of scatterers per unit area of the foil and integrating over the foil area. This gives a downstream wave function for $x = y = 0$:

$$\psi_{\mathbf{k}} = (2\pi\hbar)^{-3/2} \left[ e^{ikz} + \mathcal{N} \int_0^\infty \frac{b \, db}{(z^2 + b^2)^{1/2}} \right.$$
$$\left. \times \int_0^{2\pi} d\phi \, f_{\mathbf{k}}(\arctan(b/z), \phi) e^{ik(z^2 + b^2)^{1/2}} \right]$$
$$= (2\pi\hbar)^{-3/2} e^{ikz} \left[ 1 + \mathcal{N} \int_0^\infty \frac{b \, db}{(z^2 + b^2)^{1/2}} \right.$$
$$\left. \times \int_0^{2\pi} d\phi \, f_{\mathbf{k}}(\arctan(b/z), \phi) e^{ik[(z^2 + b^2)^{1/2} - z]} \right].$$

Expanding the square root in the exponent, we see that the integrand oscillates rapidly for $kb^2/z \gg 1$, so the values of $b$ that contribute appreciably to the integral are limited to an upper bound of order $\sqrt{z/k}$. Since we are assuming that $kz \gg 1$, this means that most of the integral comes from values of $b$ much less than $z$, so that it simplifies to

$$\psi_{\mathbf{k}} = (2\pi\hbar)^{-3/2} e^{ikz} \left[ 1 + \pi f_{\mathbf{k}}(0) \mathcal{N} z^{-1} \int_0^\infty db^2 \, e^{ik b^2/2z} \right]. \tag{7.3.10}$$

As usual, we interpret $\int_0^\infty e^{iax} \, dx$ by inserting a convergence factor $e^{-\epsilon x}$, calculating the integral as $1/(\epsilon - ia)$, and then setting $\epsilon = 0$, so that Eq. (7.3.10) gives

$$\psi_{\mathbf{k}} = (2\pi\hbar)^{-3/2} e^{ikz} \left[ 1 + 2i\pi f_{\mathbf{k}}(0) \mathcal{N} k^{-1} \right]. \tag{7.3.11}$$

To first order[3] in $\mathcal{N}$, the probability density in the plane wave is therefore reduced by a factor

$$(2\pi\hbar)^3 |\psi_{\mathbf{k}}|^2 = 1 - \frac{4\pi \, \text{Im} \, f_{\mathbf{k}}(0)\mathcal{N}}{k}. \tag{7.3.12}$$

This should equal $1 - P$, where $P$ is the probability that the particle is scattered or in any other way removed from the beam. This probability is given by $\sigma_{\text{tot}}/A$ times the number $\mathcal{N}A$ of scatterers in the effective area $A \equiv 1/\rho_{\text{T}}$ of the initial wave packet, so that $P = \sigma_{\text{tot}}\mathcal{N}$. Equating the quantity (7.3.12) to $1 - P$ then gives the optical theorem in its general form (7.3.9). In this form, it applies to every reaction initiated by an initial particle, relativistic or non-relativistic.

There is an immediate consequence of the optical theorem that provides important information about scattering at high energies. If the scattering amplitude $f_{\mathbf{k}}(\theta, \phi)$ is a smooth function of angles, then there must be some solid angle $\Delta\Omega$ within which the differential scattering cross section $|f_{\mathbf{k}}(\theta, \phi)|^2$ is not much less than in the forward direction – to be definite, let's say not less than $|f_{\mathbf{k}}(0)|^2/2$. Then

$$\sigma_{\text{tot}}(k) \geq \frac{1}{2}|f_{\mathbf{k}}(0)|^2 \, \Delta\Omega \geq \frac{1}{2}|\text{Im} \, f_{\mathbf{k}}(0)|^2 \, \Delta\Omega = \frac{k^2 \sigma_{\text{tot}}^2(k) \, \Delta\Omega}{32\pi^2}$$

and so

$$\Delta\Omega \leq \frac{32\pi^2}{k^2 \sigma_{\text{tot}}(k)}. \tag{7.3.13}$$

As discussed in Section 8.4, in collisions of strongly interacting particles such as protons, the total cross section becomes constant or grows slowly at high energy, so the solid angle $\Delta\Omega$ within which the differential cross section is no less than half the value in the forward direction must vanish more or less as $1/k^2$. This sharp peak of the scattering probability in the forward direction is known as the *diffraction peak*.

## 7.4 The Born Approximation

One of the advantages of the approach we have followed is that it leads immediately to a widely useful approximation, known as the *Born approximation*.[4] This approximation is generally valid for weak potentials, or more precisely, if relevant matrix elements of the potential $V$ are much less than typical matrix elements of the kinetic energy $H_0$. In this case, since Eq. (7.2.6) for the scattering amplitude already includes an explicit factor of the potential, it can be

---

[3] Terms of higher order in $\mathcal{N}$ are of the same order as terms produced by multiple scattering in the foil, which we are neglecting here.

[4] M. Born, Z. *Physik* **38**, 803 (1926).

evaluated to first order in the potential by taking the "in" wave function $\psi_{\mathbf{k}}$ as the free-particle wave function $(2\pi\hbar)^{-3/2}\exp(i\mathbf{k}\cdot\mathbf{x})$, so

$$f_{\mathbf{k}}(\hat{x}) \simeq -\frac{\mu}{2\pi\hbar^2}\int d^3y\, V(\mathbf{y})\exp\left(i(\mathbf{k}-k\hat{x})\cdot\mathbf{y}\right). \tag{7.4.1}$$

In particular, for a central potential, this gives

$$f_k(\theta,\phi) \simeq -\frac{2\mu}{\hbar^2}\int_0^\infty r^2\,dr\,V(r)\frac{\sin(qr)}{qr}, \tag{7.4.2}$$

where $\hbar q$ is the momentum transfer;

$$q \equiv |\mathbf{k}-k\hat{x}| = 2k\sin(\theta/2), \tag{7.4.3}$$

with $\theta$ the angle between the incident direction $\hat{k}$ and the direction $\hat{x}$ of scattering. The result that the amplitude is independent of the azimuthal angle $\phi$ is an obvious consequence of the symmetry of the problem under rotations about the 3-axis for central potentials, and does not depend on the Born approximation. On the other hand, the result that the scattering amplitude depends on $k$ and $\theta$ only in the combination $q$ depends not only on the potential being only a function of $r$, but also on the use of the Born approximation.

For example, consider scattering in a shielded Coulomb potential:

$$V(r) = \frac{Z_1 Z_2 e^2}{r}e^{-\kappa r}. \tag{7.4.4}$$

This is a crude approximation to the potential felt by a nucleus of charge $Z_1 e$ being scattered by an atom of atomic number $Z_2$; at small $r$ the incoming nucleus feels the full Coulomb field of the atom's nucleus, while for large $r$ that charge is screened by the atomic electrons. (A potential of this form is also known as a Yukawa potential, because Hideki Yukawa (1907–1981) showed in 1935 that a potential of this form is produced by the exchange of a spinless boson of mass $\hbar\kappa/c$ between nucleons.[5]) Using this in Eq. (7.4.2) gives

$$f_k(\theta,\phi) \simeq -\frac{2\mu Z_1 Z_2 e^2}{q\hbar^2}\int_0^\infty dr\, e^{-\kappa r}\sin(qr) = -\frac{2\mu Z_1 Z_2 e^2}{\hbar^2}\frac{1}{q^2+\kappa^2}. \tag{7.4.5}$$

In particular, the scattering amplitude for a pure Coulomb potential is given in the Born approximation by setting $\kappa = 0$ in Eq. (7.4.5). This gives a scattering cross section identical to that derived by Rutherford in his analysis of the scattering of alpha particles by gold atoms, which as discussed in Section 1.2 led in 1911 to the discovery of the atomic nucleus. Rutherford was lucky; his derivation was strictly classical, and would not have given the same result as the quantum-mechanical calculation for any potential other than the Coulomb

---

5 H. Yukawa, *Proc. Phys.-Math. Soc. (Japan)* (3) **17**, 48 (1935).

potential. We will see in Section 7.9 that the scattering amplitude receives significant corrections from effects of higher order in the potential, but for the special case of the Coulomb potential these corrections only change the phase of the scattering amplitude, and hence do not affect the Coulomb scattering cross section.

## 7.5 Phase Shifts

There is a useful representation of the scattering amplitude that is especially convenient for spherically symmetric potentials. Since the incoming wave $\exp(ikx_3)$ is invariant under rotations around the 3-axis, and the Laplacian and the potential are invariant under all rotations, the full wave function must also be invariant under rotations around the 3-axis, and hence independent of the azimuthal angle $\phi$. Expanding it in spherical harmonics, we thus encounter only terms with $m = 0$, or in other words, terms proportional to the Legendre polynomials $P_\ell(\cos\theta)$ discussed in Section 2.2. We therefore write the complete wave function as

$$\psi(r, \theta) = \sum_{\ell=0}^{\infty} R_\ell(r) P_\ell(\cos\theta). \tag{7.5.1}$$

Also, the plane-wave term in Eq. (7.2.5) has a well-known expansion:

$$\exp(ikr\cos\theta) = \sum_{\ell=0}^{\infty} i^\ell(2\ell + 1) j_\ell(kr) P_\ell(\cos\theta), \tag{7.5.2}$$

where $j_\ell(kr)$ is the *spherical Bessel function*:

$$j_\ell(z) \equiv \sqrt{\frac{\pi}{2z}} J_{\ell+1/2}(z) = (-1)^\ell z^\ell \frac{d^\ell}{(z\,dz)^\ell}\left(\frac{\sin z}{z}\right). \tag{7.5.3}$$

Equation (7.5.2) can be derived by noting that $e^{ikr\cos\theta} = e^{ikx_3}$ satisfies the wave equation $(\nabla^2 + k^2)e^{ikr\cos\theta} = 0$. According to Eqs. (2.1.16) and (2.2.1), if we write the partial wave expansion of $e^{ikr\cos\theta}$ as

$$e^{ikr\cos\theta} = \sum_{\ell=0}^{\infty} f_\ell(kr) P_\ell(\cos\theta),$$

then the coefficient $f_\ell(kr)$ must satisfy the wave equation

$$\left[\frac{1}{r^2}\frac{d}{dr}r^2\frac{d}{dr} - \frac{\ell(\ell+1)}{r^2} + k^2\right] f_\ell(kr) = 0.$$

It follows then that $\sqrt{r} f_\ell(kr)$ satisfies the Bessel differential equation for order $\ell+1/2$. With the condition that $f_\ell(kr)$ is regular at $r = 0$, this tells us that $f_\ell(kr)$ is proportional to $j_\ell(kr)$, as defined by the first equation in Eq. (7.5.3). The constant of proportionality can be found by calculating $\int_{-1}^{1} \exp(ikr\mu) \, P_\ell(\mu) \, d\mu$, and using the orthonormality property $\int_{-1}^{1} P_{\ell'}(\mu) P_\ell(\mu) \, d\mu = 2\delta_{\ell'\ell}/(2\ell+1)$. Unlike the ordinary Bessel functions, the spherical Bessel functions can be written in terms of elementary functions; for instance,

$$j_0(x) = \frac{\sin x}{x}, \quad j_1(x) = \frac{\sin x}{x^2} - \frac{\cos x}{x}, \tag{7.5.4}$$

and so on. The other solutions of the same wave equation that are not regular at the origin are spherical Neumann functions

$$n_0(x) = -\frac{\cos x}{x}, \quad n_1(x) = -\frac{\cos x}{x^2} - \frac{\sin x}{x}, \tag{7.5.5}$$

and so on.

To find the scattering amplitude, we must now consider the difference of the wave function (7.5.1) and the plane wave (7.5.2) for $r \to \infty$. If the potential vanishes sufficiently rapidly for large $r$, the reduced radial wave function $r R_\ell(r)$ for large $r$ must become proportional to a linear combination of $\cos(kr)$ and $\sin(kr)$, which without loss of generality we may write as

$$R_\ell(r) \to \frac{c_\ell(k) \sin \left( kr - \ell\pi/2 + \delta_\ell(k) \right)}{kr}, \tag{7.5.6}$$

where $c_\ell$ and $\delta_\ell$ are quantities that may depend on $k$, but not on $r$. It is easy to see that the radial wave function $R_\ell(r)$ is real, up to an overall constant factor. (With a potential that does not grow as $r \to 0$ as rapidly as $1/r^2$, the Schrödinger equation (2.1.26), multiplied with $2\mu r^2/\hbar^2 R_\ell(r)$, takes the following form for $r \to 0$:

$$\frac{1}{R_\ell(r)} \frac{d}{dr} \left( r^2 \frac{d}{dr} \right) R_\ell(r) \to \ell(\ell+1),$$

so as $r \to 0$, $R_\ell(r)$ goes as a linear combination of $r^\ell$ and $r^{-\ell-1}$. The condition of normalizability requires that we choose $R_\ell(r)$ to go purely as $r^\ell$ for $r \to 0$. For a real potential, $R_\ell^*(r)$ satisfies the same homogeneous second-order differential equation and the same initial condition on its logarithmic derivative as $R_\ell(r)$, so it must equal $R_\ell(r)$ up to a constant factor, which tells us that $R_\ell(r)$ is real, up to a complex constant factor.) Hence $c_\ell$ may be complex, but $\delta_\ell$ is necessarily real.

On the other hand, for large arguments the spherical Bessel functions appearing in the plane wave have the asymptotic behavior

$$j_\ell(kr) \to \frac{\sin \left( kr - \ell\pi/2 \right)}{kr}. \tag{7.5.7}$$

In the absence of interactions we would just have the plane-wave term in the wave function, so $R_\ell(r)$ would have to be proportional to $j_\ell(kr)$. Comparison of Eqs. (7.5.6) and (7.5.7) shows that in this case all $\delta_\ell$ would vanish. For this reason, the $\delta_\ell$ are known as *phase shifts*.

To determine the coefficients $c_\ell$, we impose the condition that for $r \to \infty$, the scattered wave $\psi(r, \theta) - \exp(ikr \cos\theta)$ can contain only terms with $r$-dependence proportional to the outgoing wave $\exp(ikr)/kr$, not the incoming wave $\exp(-ikr)/kr$. Subtracting (7.5.2) from (7.5.1), and using Eqs. (7.5.6) and (7.5.7), we see that the coefficient of $P_\ell(\cos\theta) \exp(-ikr)/2ikr$ in the scattered wave is

$$c_\ell i^\ell e^{-i\delta_\ell} - i^{2\ell}(2\ell + 1),$$

and therefore

$$c_\ell = i^\ell (2\ell + 1)e^{i\delta_\ell}. \tag{7.5.8}$$

The scattered wave then has the asymptotic behavior

$$\psi(r, \theta) - \exp(ikr \cos\theta) \to \frac{e^{ikr}}{2ikr} \sum_{\ell=0}^{\infty} (2\ell + 1) P_\ell(\cos\theta) \left(e^{2i\delta_\ell} - 1\right), \tag{7.5.9}$$

and the scattering amplitude is therefore

$$f(\theta) = \frac{1}{2ik} \sum_{\ell=0}^{\infty} (2\ell + 1) P_\ell(\cos\theta) \left(e^{2i\delta_\ell} - 1\right). \tag{7.5.10}$$

We can now verify the optical theorem. From Eq. (7.5.10) we find immediately that

$$\text{Im}\, f(0) = \frac{1}{2k} \sum_{\ell=0}^{\infty} (2\ell + 1)(1 - \cos 2\delta_\ell) = \frac{1}{k} \sum_{\ell=0}^{\infty} (2\ell + 1) \sin^2 \delta_\ell. \tag{7.5.11}$$

The orthonormality condition for the spherical harmonics gives

$$\delta_{\ell\ell'} = 2\pi \int_0^\pi Y_\ell^0(\theta) Y_{\ell'}^0(\theta) \sin\theta \, d\theta = \frac{2\ell + 1}{2} \int_0^\pi P_\ell(\cos\theta) P_{\ell'}(\cos\theta) \sin\theta \, d\theta, \tag{7.5.12}$$

so the elastic scattering cross section is

$$\sigma_{\text{scat}} = \frac{4\pi}{k^2} \sum_{\ell=0}^{\infty} (2\ell + 1) \sin^2 \delta_\ell. \tag{7.5.13}$$

The comparison of Eqs. (7.5.11) and (7.5.13) gives the optical theorem (7.3.8).

One of the things that the phase-shift formalism is good for is to analyze the behavior of the scattering amplitude at low energy. To deal with this, we will first derive a formula for the phase shift that applies at any energy, and then specialize to the case of low energy.

Suppose that the potential is negligible outside a radius $a$. (We are assuming that the potential vanishes rapidly for $r \to \infty$, so even if it is not strictly zero at any finite $r$, the results we obtain will be qualitatively reliable.) For $r > a$, the radial wave function $R_\ell(r)$ for a given $\ell$ is a solution of the free-particle wave equation, which in general is a linear combination of the spherical Bessel functions $j_\ell(kr)$ that are regular as $r \to 0$ and functions $n_\ell(kr)$ that become infinite at the origin. These functions have the asymptotic behavior for large argument

$$j_\ell(\rho) \to \frac{\sin(\rho - \ell\pi/2)}{\rho}, \quad n_\ell(\rho) \to -\frac{\cos(\rho - \ell\pi/2)}{\rho}. \tag{7.5.14}$$

Hence the linear combination that has the asymptotic behavior given by Eqs. (7.5.6) and (7.5.8) is

$$R_\ell(r) = i^\ell (2\ell + 1) e^{i\delta_\ell} \Big[ j_\ell(kr) \cos \delta_\ell - n_\ell(kr) \sin \delta_\ell \Big] \quad \text{for} \quad r > a. \tag{7.5.15}$$

The value of $R'_\ell(r)/R_\ell(r)$ at $r = a$ (where the asymptotic formulas (7.5.14) do not apply) is set by the condition that the wave function must fit smoothly with the solution of the Schrödinger equation for $r < a$ that is well behaved ($R_\ell \propto r^\ell$) at $r \to 0$, which of course depends on the details of the potential. This condition may be written

$$R'_\ell(a)/R_\ell(a) = \Delta_\ell(k), \tag{7.5.16}$$

with $\Delta_\ell(k)$ depending only on the wave function for $r < a$. Equations (7.5.15) and (7.5.16) together then give

$$\tan \delta_\ell(k) = \frac{kj'_\ell(ka) - \Delta_\ell(k)j_\ell(ka)}{kn'_\ell(ka) - \Delta_\ell(k)n_\ell(ka)}. \tag{7.5.17}$$

Now, for sufficiently small $k$, the term $k^2 R_\ell$ in the Schrödinger equation for the radial wave function has little effect, so $\Delta_\ell(k)$ becomes essentially independent of $k$ for low energy. Also, the spherical Bessel functions for small argument are

$$j_\ell(\rho) \to \frac{\rho^\ell}{(2\ell + 1)!!}, \quad n_\ell(\rho) \to -(2\ell - 1)!! \rho^{-\ell-1}, \tag{7.5.18}$$

where, for any odd integer $n$,

$$n!! \equiv n(n - 2)(n - 4) \dots 1, \tag{7.5.19}$$

with $(-1)!! \equiv 1$. Hence for $ka \ll 1$, Eq. (7.5.17) gives

$$\tan \delta_\ell \to \left( \frac{\ell - a\Delta_\ell}{a\Delta_\ell + \ell + 1} \right) \frac{(ka)^{2\ell+1}}{(2\ell + 1)!!(2\ell - 1)!!}. \tag{7.5.20}$$

This shows that $\tan \delta_\ell$ vanishes as $k^{2\ell+1}$ for $k \to 0$, and hence $\delta_\ell(k)$ either vanishes or approaches an integer multiple of $\pi$. We can go further, and say something about higher terms in $k$. Note that $\Delta_\ell$ depends on $k$ only through the presence of a term $k^2 R_\ell$ in the Schrödinger equation, so $\Delta_\ell$ is a power series in $k^2$. Also, $k^{-\ell}j_\ell(ka)$, $k^{1-\ell}j'_\ell(ka)$, $k^{\ell+1}n_\ell(ka)$, and $k^{\ell+2}n'_\ell(ka)$ are all power series in $k^2$. Hence from Eq. (7.5.17), we see that also $k^{-2\ell-1}\tan\delta_\ell$ is a power series in $k^2$.

Evidently, if there is no selection rule that suppresses $s$-wave scattering, then $\delta_0$ is the dominant phase shift for $k \to 0$. It is conventional to express $k\cot\delta_0$, rather than its reciprocal $k^{-1}\tan\delta_\ell$, as a power series in $k^2$:

$$k\cot\delta_0 \to -\frac{1}{a_s} + \frac{r_{\text{eff}}}{2}k^2 + \cdots , \qquad (7.5.21)$$

where $a_s$ and $r_{\text{eff}}$ are constants with the dimensions of length, known respectively as the *scattering length* and the *effective range*. According to Eq. (7.5.13), the cross section for $k \to 0$ approaches a constant

$$\sigma_{\text{scat}} \to 4\pi a_s^2. \qquad (7.5.22)$$

We will see in Section 8.8 that in the presence of a shallow $s$-wave bound state, it is possible to derive a formula for $a_s$ in terms of the energy of the bound state, without having to know anything about the details of the potential.

I should mention that there is an exception to these results, in the case where an $s$-wave bound state sits precisely at zero energy. In general at $k = 0$ the $\ell = 0$ radial wave function $R_0$ outside the range of the potential satisfies the Schrödinger equation $d/dr(r^2\,dR_0/dr) = 0$, so $R_0$ is a linear combination of terms that go as $1/r$ and a constant. With a bound state at zero energy, the constant term must be absent, so $R_0 \propto 1/r$ at $r = a$, and hence $\Delta_0(0) = -1/a$. In this case the denominator $a\Delta_0 + 1$ in Eq. (7.5.20) vanishes, invalidating the conclusion that $\tan\delta_0 \to 0$ for $k \to 0$. In fact, we shall show on very general grounds in Section 8.8 that in the presence of an $s$-wave bound state at zero energy, $\tan\delta_0$ at zero energy is infinite, not zero.

## 7.6 Resonances

There are other circumstances in which a phase shift will exhibit a characteristic dependence on energy, independent of the detailed form of the potential. Consider a potential $V(r)$ that has a high value much greater than the energy $E$ in a thick shell around the origin, surrounding an inner region where the potential is much smaller, with $V \ll E$. In these circumstances, the general solution of the Schrödinger equation within the barrier is a linear combination of two solutions, one solution $R_+(r, E, \ell)$ that grows exponentially with increasing $r$, and the other $R_-(r, E, \ell)$ that decays exponentially. To see this, note that at any energy $E$ below the barrier height the Schrödinger equation (2.1.29) for the reduced radial wave function $u(r, E, \ell) \equiv r R(r, E, \ell)$ within the barrier can be put in the form

$$\frac{d^2u}{dr^2} = \kappa^2 u, \tag{7.6.1}$$

where

$$\kappa^2(r, E, \ell) \equiv \frac{2\mu}{\hbar^2}\left[V(r) - E\right] + \frac{\ell(\ell+1)}{r^2} > 0. \tag{7.6.2}$$

In assuming that the barrier is high and thick, we will specifically suppose that $\kappa$ is so large that both $\kappa$ and $\kappa' \equiv \partial\kappa/\partial r$ change very little in a distance $1/\kappa$; that is,

$$\left|\frac{\kappa'}{\kappa}\right| \ll \kappa, \quad \left|\frac{\kappa''}{\kappa'}\right| \ll \kappa, \tag{7.6.3}$$

with $\kappa$ understood from now on as the positive square root of the quantity (7.6.2). Under these circumstances, we can use the WKB approximation discussed in Section 5.7 to find approximate solutions of Eq. (7.6.1), of the form

$$u_\pm(r, E, \ell) \equiv r R_\pm(r, E, \ell) = A_\pm(r, E, \ell)\exp\left(\pm\int^r \kappa(r', E, \ell)\,dr'\right), \tag{7.6.4}$$

where $A_\pm$ varies much more slowly than the argument of the exponential. (Equation (5.7.9) shows that to a good approximation, $A_\pm \propto 1/\sqrt{\kappa}$.)

These solutions are to be continued outside the barrier and into the inner region. Outside the barrier $R_+$ is much larger than $R_-$:

$$\frac{R_-(r, E, \ell)}{R_+(r, E, \ell)} = O\left(\exp\left[-2\int_{\text{barrier}} \kappa(r', E, \ell)\,dr'\right]\right) \ll 1, \tag{7.6.5}$$

the integral being taken over the whole region in which $V(r') > E$. On the other hand, the solution of the Schrödinger equation that in the inner region goes as $r^\ell$ rather than $r^{-\ell-1}$ as $r \to 0$ must take the form

$$R(r, E, \ell) = c_+(E, \ell)R_+(r, E, \ell) + c_-(E, \ell)R_-(r, E, \ell) \tag{7.6.6}$$

with coefficients $c_\pm(E, \ell)$ that are generally of the same order of magnitude.

Now recall Eq. (7.5.17) for the phase shift:

$$\tan\delta_\ell(k) = \frac{kj'_\ell(ka) - \Delta_\ell(k)j_\ell(ka)}{kn'_\ell(ka) - \Delta_\ell(k)n_\ell(ka)}, \tag{7.6.7}$$

where $\Delta_\ell(k)$ is the logarithmic derivative $\Delta_\ell(k) \equiv R'(a, E, \ell)/R(a, E, \ell)$ at a radius $a$ just outside the barrier. For generic energies below the barrier height, the wave function will be dominated by $R_+$, and $\Delta_\ell(k)$ will be equal to $R'_+(a, E, \ell)/R_+(a, E, \ell)$. For most energies, this gives $\tan\delta_\ell(E)$ a smoothly varying value, which we will call $\tan\overline{\delta}_\ell(E)$.

But suppose that in the limit of an infinitely thick barrier there would be a bound-state solution of the Schrödinger equation at an energy $E_0$ and orbital angular momentum $\ell_0$. At this energy the solution of the Schrödinger equation that goes as $r^{\ell_0}$ for $r \to 0$ must decay inside the barrier, so $c_+(E_0, \ell_0) = 0$. As long as $E$ is close enough to $E_0$ that $c_+(E, \ell_0)/c_-(E, \ell_0)$ is less than an amount of order (7.6.6), the logarithmic derivative $\Delta_{\ell_0}(k)$ will appreciably differ from $R'_+(a, E, \ell_0)/R_+(a, E, \ell_0)$, taking a value $R'_-(a, E, \ell_0)/R_-(a, E, \ell_0)$ at $E = E_0$, where $c_+$ vanishes. We conclude then that as the energy increases past $E_0$ the quantity $\tan \delta_{\ell_0}(E)$ varies rapidly, suddenly near $E = E_0$ becoming appreciably different from $\tan \overline{\delta}_{\ell_0}(E)$, and then returns to the smoothly varying value $\tan \overline{\delta}_{\ell_0}(E)$. The range in which $\tan \delta_{\ell_0}(E)$ is appreciably different from $\tan \overline{\delta}_{\ell_0}(E)$ is proportional to (7.6.6).

We will give an argument in the next section that a rapid decrease of the phase shift would violate causality. Since $\tan \delta_{\ell_0}(E)$ varies rapidly but returns to about this same value as $E$ passes $E_0$, the phase shift must increase in a narrow range of energies around $E_0$ by 180° (or possibly an integer multiple[6] of 180°), and therefore must become equal to 90° at an energy $E_R$ somewhere in that range. The phase shift can therefore be assumed to take the form

$$\delta_{\ell_0}(E) = \overline{\delta}_{\ell_0}(E) + \delta_{\ell_0}^{(R)}(E), \tag{7.6.8}$$

$$\tan \delta_{\ell_0}^{(R)}(E) = -\frac{1}{2}\frac{\Gamma}{E - E_R}, \tag{7.6.9}$$

where $\Gamma$ is a constant with the dimensions of energy, proportional to (7.6.6), and $E_R$ is an energy differing from $E_0$ by an amount at most of order $\Gamma$. (The constant of proportionality is written as $-\Gamma/2$ for later convenience. In order for Eq. (7.6.9) to give an increasing phase shift, we must have $\Gamma > 0$.) The rapid growth of the phase shift at an energy $E_R$ is like the large resonant response of a classical system to oscillatory perturbations whose frequency matches one of the natural frequencies of the system, and for this reason the divergence of $\tan \delta_{\ell_0}(E)$ at an energy $E_R$ is known as a *resonance*; $E_R$ is the resonance energy.

The non-resonant phase shift $\overline{\delta}_{\ell_0}(E)$ is typically much less than 90°. In this case, we can neglect the term $\overline{\delta}_{\ell_0}(E)$ in Eq. (7.6.8), which then gives

$$\sin^2 \delta_{\ell_0}(E) = \frac{\tan^2 \delta_{\ell_0}(E)}{1 + \tan^2 \delta_{\ell_0}(E)} = \frac{\Gamma^2/4}{(E - E_R)^2 + \Gamma^2/4},$$

so that Eq. (7.5.13) for the total cross section gives

$$\sigma_{\text{scat}} \simeq \frac{\pi(2\ell_0 + 1)}{k^2} \frac{\Gamma^2}{(E - E_R)^2 + \Gamma^2/4}. \tag{7.6.10}$$

Equation (7.6.10) is known as the *Breit–Wigner* formula.[7] We see that $\Gamma$ is the full width of the peak in the cross section at half maximum. The cross section

---

[6] In the case where $\delta_\ell(E)$ jumps up by 360°, 540°, etc., it must also pass through 270°, 540°, etc., and the scattering cross section will exhibit several peaks at nearly the same energy. This case, of several resonances that for some reason are at the same energy, will not be considered here.

[7] G. Breit and E. P. Wigner, *Phys. Rev.* **49**, 519 (1936).

at its maximum value will take the value $4\pi(2\ell_0 + 1)/k_R^2$, or roughly a square wavelength, independent of the details of the potential. A generalization of this formula to a much wider variety of problems is given in Section 8.5.

The resonance width $\Gamma$ has an important connection with the lifetime of the resonant state. Using Eqs. (7.6.8) and (7.6.9) and some elementary trigonometry, we easily see that the quantity $\exp(2i\delta_{\ell_0})$ in the scattering amplitude (7.5.10) behaves near the resonance as

$$\exp\left(2i\delta_{\ell_0}(E)\right) = \exp\left(2i\bar{\delta}_{\ell_0}(E)\right)\left[1 - \frac{i\Gamma}{E - E_R + i\Gamma/2}\right]. \qquad (7.6.11)$$

If at $t = 0$ we put the system in the nearly stable state with angular momentum $\ell_0$ and radial wave function $\int g(E)R(r, \ell_0, E)\,dE$, where $g(E)$ is a smooth function that varies slowly for $E$ near $E_R$, the resonant contribution to the time-dependent wave function $\int g(E)R(r, \ell_0, E)\exp(-iEt/\hbar)\,dE$ will have a term with a time-dependence proportional at late times to the integral

$$\int_{-\infty}^{+\infty} \frac{\exp(-iEt/\hbar)\,dE}{E - E_R + i\Gamma/2} = -2\pi i\,\exp\left(-iE_R t/\hbar - \Gamma t/2\hbar\right). \qquad (7.6.12)$$

(This integral for $t > 0$ is most easily done by completing the contour of integration with a large semi-circle in the lower half of the complex plane.) The factor $\exp(-iE_R t/\hbar)$ supports the interpretation that scattering occurs by formation of a nearly stable state with energy near $E_R$, and the factor $\exp(-\Gamma t/2\hbar)$ in the scattering amplitude, which gives a factor $\exp(-\Gamma t/\hbar)$ in the scattering probability, indicates that this state decays at a rate $\Gamma/\hbar$.

There are cases in nuclear physics of states with a barrier so thick that their decay rate $\Gamma$ is very small, small enough that nuclei in these states can be found in nature, rather than as resonances in scattering processes. The classical example is provided by nuclei that are unstable against the emission of alpha particles, first treated quantum mechanically by George Gamow[8] (1904–1968). In transitions in which the alpha particle is emitted in an $s$ wave, such as $^{238}U \to {}^{234}Th + \alpha$ and $^{226}Ra \to {}^{222}Rn + \alpha$, the barrier arises purely from the Coulomb potential, which in alpha decay is $V(r) = 2Ze^2/r$, where $Z$ is the atomic number of the final nucleus. The barrier extends from an effective nuclear radius $R$ out to a turning point where $V(r)$ equals the final kinetic energy $E_\alpha$ of the alpha particle. The barrier-penetration integral in Eq. (7.6.6) is then

$$2\int_{\text{barrier}} \kappa\,dr = 2\int_R^{2Ze^2/E_\alpha} dr\,\sqrt{\frac{2m_\alpha}{\hbar^2}\left(\frac{2Ze^2}{r} - E_\alpha\right)}. \qquad (7.6.13)$$

In many cases this exponent is quite large, giving extremely long lifetimes for alpha-emitting nuclei. The lifetime of $^{238}U$ is $4.47 \times 10^9$ years, long enough that

[8] G. Gamow, *Z. Physik* **52**, 510 (1928); also see E. U. Condon and R. W. Gurney, *Phys. Rev.* **33**, 127 (1929).

appreciable uranium has survived on earth from before the formation of the solar system. Even $^{226}$Ra has a lifetime of 1600 years, long enough for radium from a chain of radioactive decays originating with $^{238}$U to be found in association with uranium ores. (Needless to say, $\Gamma$ for $^{226}$Ra and $^{238}$U is far too small for these states ever to be seen as resonances in the scattering of alpha particles on $^{234}$Th or $^{226}$Rn.) The exponential of the quantity (7.6.13) is an extremely sensitive function of $E_\alpha$ and $Z$, which of course are known precisely, and also of $R$, which is not so well known, so this formula was historically used together with observed alpha decay rates to determine $R$.

Finally, recall that the Breit–Wigner formula (7.6.10) was derived here for the case of a negligible non-resonant phase shift $\bar{\delta}_{\ell_0}(E)$. But there are cases where $\bar{\delta}_{\ell_0}(E)$ is itself close to 90°, in which case the total phase shift rises at a resonance from 90° to 270°. Where it passes through 180°, we have a sharp dip rather than a peak in the total cross section. This effect was first observed in 1921–2 independently by Ramsauer and Townsend,[9] in the scattering of electrons by the atoms of noble gases.

## 7.7 Time Delay

The demonstration in the previous section, that a resonance of width $\Gamma$ represents a state that decays with a rate $\Gamma/\hbar$, considered the time-dependence of a superposition of scattering wave functions at a single position. To see what is going on in the scattering, we need instead to consider the time-dependence of such a superposition at late times and large distances. We did this in Section 7.2, where we derived the behavior (7.2.9) of the wave function at late times and large distances from Eqs. (7.2.5) and (7.2.7). But there we assumed that the scattering amplitude $f_{\mathbf{k}}$ depends on the wave number $k$ much more smoothly than the wave packet $g(\mathbf{k})$ or the factors $e^{ikr}$ or $\exp(-i\hbar t k^2/2\mu)$. Now we want to consider the possibility that the phase shift $\delta_\ell(E)$ for any particular angular momentum $\ell$ may vary rapidly with energy.

According to Eq. (7.5.10), the wave function (7.2.7) contains a term that for large $r$ behaves as

$$\frac{(2\pi\hbar)^{-3/2}}{2ikr} \int d^3k \, g(\mathbf{k}) \exp\left(ikr - i\hbar t k^2/2\mu + 2i\delta_\ell(E)\right)(2\ell+1)P_\ell(\cos\theta),$$

$$(7.7.1)$$

where the argument of the phase shift is $E = \hbar^2 k^2/2\mu$. At late times the integral is dominated by the value of $k$ where the argument of the exponential is stationary, at which

--------

[9] C. Ramsauer, *Ann. Physik* **4**, **64**, 513 (1921); V. A. Bailey and J. S. Townsend, *Phil. Mag.* S.6, **43**, 1127 (1922).

$$r - \hbar t k / \mu + 2\delta'_\ell(E)\hbar^2 k / \mu = 0,$$

or in other words

$$r = \frac{\hbar k}{\mu}\left(t - \Delta t\right), \qquad (7.7.2)$$

where[10]

$$\Delta t = 2\hbar \delta'_\ell(E). \qquad (7.7.3)$$

(This of course applies only if $t$ is positive as well as large; for $t$ large and negative, Eq. (7.7.2) would have no solution with $r > 0$, and this term would be absent in the asymptotic form of the wave function.) Equation (7.7.2) shows that $\Delta t$ is the time delay experienced by the incoming particle in entering and then leaving the potential.

The result (7.7.3) justifies the remark made in the previous section, that phase shifts generally can increase sharply but not decrease sharply with increasing energy. The time at which a wave packet arrives at a scattering center is uncertain by an amount of order $\mathcal{R}/v$, where $\mathcal{R}$ is the range of the potential and $v$ is the velocity of the wave packet, so it is possible to have $\Delta t$ negative if it is no greater than this in magnitude, but a negative $\Delta t$ of much larger magnitude would represent a failure of causality – the wave packet would be emerging from the potential before it entered it. With Eq. (7.7.3), this sets a crude upper limit to the rate of decrease of any phase shift with energy: $-\delta'_\ell(E) \leq \mathcal{R}/2\hbar v$.

Equation (7.7.3) has a natural application to the case of resonance. Neglecting the rate of change with energy of the non-resonant contribution $\bar{\delta}_{\ell_0}(E)$ (where $\ell_0$ is the angular momentum of the nearly stable state), Eq. (7.6.9) gives the time delay (7.7.3) near a resonance as the positive quantity

$$\Delta t = \frac{2\hbar}{1 + \tan^2 \delta^{(R)}_{\ell_0}(E)} \frac{d}{dE} \tan \delta^{(R)}_{\ell_0}(E) = \frac{\hbar\Gamma}{(E - E_R)^2 + \Gamma^2/4}. \qquad (7.7.4)$$

In particular, at the resonance peak the time delay is $4\hbar/\Gamma$. We can understand the factor 4 by noting that, according to Eq. (7.6.12), the mean time required for the leakage of a wave packet (not the probability density) out of the potential barrier is $2\hbar/\Gamma$, and it is plausible that this is also the time required for the incoming wave packet to leak into the potential barrier, giving a total time delay $4\hbar/\Gamma$.

---

[10] E. P. Wigner, *Phys. Rev.* **98**, 145 (1955).

## 7.8 Levinson's Theorem

There is a remarkable theorem[11] due to the mathematician Norman Levinson (1912–1975), which relates the behavior of the phase shift for $E > 0$ to the number of bound states with $E < 0$. It is most easily proved by supposing the system to be enclosed in a large sphere of radius $R$, on which the particle wave function must vanish. Recall that according to Eq. (7.5.6), the radial wave function for orbital angular momentum $\ell$ and positive energy $E = \hbar^2 k^2 / 2\mu$ is proportional to $\sin\left(kr - \ell\pi/2 + \delta_\ell(E)\right)$, so the boundary condition requires that these states must have $k$ equal to one of the discrete values $k_n$ for which

$$k_n R - \ell\pi/2 + \delta_\ell(E_n) = n\pi, \tag{7.8.1}$$

where $n$ is any integer for which this gives a positive value of $k_n$. The number $N_\ell(E)$ of states with orbital angular momentum $\ell$ and energies between 0 and $E$ is the number of values of $n$ for which Eq. (7.8.1) is satisfied with $0 \le E_n \le E$,

$$N_\ell(E) = \frac{1}{\pi}\left(kR + \delta_\ell(E) - \delta_\ell(0)\right). \tag{7.8.2}$$

In the absence of the interaction $V$ the phase shift vanishes, and the corresponding number of states is just $kR/\pi$, so the *change* in the number of scattering states of energy between 0 and $E$ due to the interaction is

$$\Delta N_\ell(E) = \frac{1}{\pi}\left(\delta_\ell(E) - \delta_\ell(0)\right). \tag{7.8.3}$$

Now, when we gradually turn on the interaction, physical states can neither be created nor destroyed, but states that were scattering states with energy $E > 0$ for $V = 0$ can be converted by the interaction to bound states with $E < 0$. The fact that states are neither created nor destroyed tells us that the total change $\Delta N_\ell(\infty)$ due to the interaction in the number of all positive-energy scattering states with orbital angular momentum $\ell$, plus the total number of bound states with this orbital angular momentum, must vanish, so that the number of bound states is

$$N_\ell = \frac{1}{\pi}\left(\delta_\ell(0) - \delta_\ell(\infty)\right). \tag{7.8.4}$$

This is necessarily positive, so the phase shift must either undergo no net change or suffer a net decrease as the energy rises from zero to infinity. This does not contradict the result of the previous section, which forbids only *rapid* decreases in the phase shift. Since the phase shift grows rapidly by $180°$ at each resonance,

---

[11] N. Levinson, *Kon. Danske Vid. Selskab Mat.-Fys. Medd.* **25**, 9 (1949). Levinson's proof relied on rigorous methods beyond the scope of this book. Levinson's paper shows that the result derived here does not apply if there happens to be a bound state with zero binding energy.

it must also decrease gradually away from resonances by 180° times the total number of resonances and bound states.

This is a remarkable result, but not a very useful one. It holds only for elastic scattering due to a non-relativistic central potential, but it refers to the phase shift at infinite energy, where inelastic channels are open and relativistic effects are important. There have been many attempts to generalize this theorem to models that are realistic at all energies, but so far without success.

## 7.9 Coulomb Scattering

Up to this point, in this chapter we have considered only potentials that vanish as $r \to \infty$ faster than $1/r$. But the single most important example of potential scattering is Coulomb scattering, say for a particle of charge $Z_1 e$ scattered by a scattering center of charge $Z_2 e$, for which $V(r) = Z_1 Z_2 e^2 / r$. Fortunately in this case it is possible to calculate the differential scattering cross section exactly, without needing to rely on the Born approximation or even on the partial wave expansion.

The Schrödinger equation for the Coulomb potential and a positive energy $E = \hbar^2 k^2 / 2\mu$ takes the form

$$-\frac{\hbar^2}{2\mu} \nabla^2 \psi + \frac{Z_1 Z_2 e^2}{r} \psi = \frac{\hbar^2 k^2}{2\mu} \psi. \tag{7.9.1}$$

It turns out that it is possible to find a solution of this equation that behaves well as $r \to 0$, and behaves like a plane wave plus an outgoing wave for $r \to \infty$, in the form

$$\psi(\mathbf{x}) = e^{ikz} \mathcal{F}(r - z). \tag{7.9.2}$$

A straightforward calculation shows that the Laplacian of such a wave function is

$$\nabla^2 \psi = e^{ikz} \left[ -k^2 \mathcal{F}(\rho) + \frac{2}{r} \left[ (1 - ik\rho) \mathcal{F}'(\rho) + \rho \mathcal{F}''(\rho) \right] \right], \tag{7.9.3}$$

where $\rho \equiv r - z$. The Schrödinger equation (7.9.1) thus takes the form of an ordinary differential equation

$$\rho \mathcal{F}''(\rho) + (1 - ik\rho) \mathcal{F}'(\rho) - k\xi \mathcal{F}(\rho) = 0, \tag{7.9.4}$$

where $\xi$ is the dimensionless quantity

$$\xi = \frac{Z_1 Z_2 e^2 \mu}{\hbar^2 k}. \tag{7.9.5}$$

This can be put in the form of a well-known differential equation by introducing a new independent variable

$$s \equiv ik\rho = ik(r - z). \tag{7.9.6}$$

Then Eq. (7.9.4) may be written

$$s \frac{d^2}{ds^2} \mathcal{F} + (1 - s) \frac{d}{ds} \mathcal{F} + i\xi \mathcal{F} = 0. \tag{7.9.7}$$

This is a special case of what is known as the confluent hypergeometric equation or Kummer equation:

$$s \frac{d^2}{ds^2} \mathcal{F} + (c - s) \frac{d}{ds} \mathcal{F} - a\mathcal{F} = 0, \tag{7.9.8}$$

in our case with

$$c = 1, \quad a = -i\xi. \tag{7.9.9}$$

The solution of Eq. (7.9.8) that is regular at $s = 0$ is known as the Kummer function,[12] and can be expressed as a power series

$$_1F_1(a; c; s) = 1 + \frac{a}{c} \frac{s}{1!} + \frac{a(a+1)}{c(c+1)} \frac{s^2}{2!} + \cdots . \tag{7.9.10}$$

With its normalization left to be determined, the wave function is

$$\psi(\mathbf{x}) = N e^{ikz} \, _1F_1(-i\xi; 1; ik[r - z]) \tag{7.9.11}$$

with $N$ a constant to be chosen later. The asymptotic behavior of the Kummer function for large complex argument is

$$_1F_1(a; c; s) \to \frac{\Gamma(c)}{\Gamma(c - a)} (-s)^{-a} \left[ 1 + O(1/s) \right] + \frac{\Gamma(c)}{\Gamma(a)} e^s s^{a-c} \left[ 1 + O(1/s) \right], \tag{7.9.12}$$

where $\Gamma(z)$ is the familiar gamma function, defined for Re $z > 0$ by

$$\Gamma(z) = \int_0^\infty dx \, x^{z-1} e^{-x}$$

and by analytic continuation to other values of $z$. Hence the asymptotic behavior of the wave function for large $r$ with $\cos \theta = z/r$ fixed is[13]

$$\psi \to N e^{\xi\pi/2} \left[ \frac{[k(r - z)]^{i\xi}}{\Gamma(1 + i\xi)} e^{ikz} + \frac{[k(r - z)]^{-i\xi-1}}{i\Gamma(-i\xi)} e^{ikr} \right]$$

$$= \frac{N e^{\xi\pi/2}}{\Gamma(1 + i\xi)} \left[ e^{ikz + i\xi \ln(kr(1-\cos\theta))} + f_k(\theta) \frac{e^{ikr - i\xi \ln(kr(1-\cos\theta))}}{r} \right], \tag{7.9.13}$$

---

[12] See, e.g., W. Magnus and F. Oberhettinger, *Formulas and Theorems for the Functions of Mathematical Physics*, transl. J. Webber (Chelsea Publishing Co., New York, 1949): Chapter VI, Section 1.

[13] In deriving the first line of Eq. (7.9.13), it is important to note that for $s = ik[r - z]$, the phase of $-s$ in the first term of Eq. (7.9.12) must be taken as $-\pi/2$, and the phase of $s$ in the second term of Eq. (7.9.12) must be taken as $\pi/2$.

where

$$f_k(\theta) = \frac{\Gamma(1+i\xi)}{\Gamma(-i\xi)} \frac{1}{ik(1-\cos\theta)} = -\frac{\Gamma(1+i\xi)}{\Gamma(1-i\xi)} \frac{\xi}{k(1-\cos\theta)}$$

$$= -\frac{\Gamma(1+i\xi)}{\Gamma(1-i\xi)} \frac{2Z_1 Z_2 e^2 \mu}{\hbar^2 q^2}. \tag{7.9.14}$$

Here we have used the general formula $\Gamma(1+z) = z\Gamma(z)$, and define $q^2 \equiv 2k^2(1-\cos\theta) = 4k^2\sin^2(\theta/2)$.

It is shown in the following section that the terms in the phases in Eq. (7.9.13) that go as $\ln(kr)$ are an inevitable feature of scattering by potentials that behave as $1/r$ for $r \to \infty$. The contribution of these terms becomes negligible compared with $kr$ for macroscopically large values of $r$, so Eq. (7.9.13) is effectively the same as the standard formula (7.2.5) for the asymptotic wave function, provided we take the normalization constant $N$ in Eq. (7.9.11) to have the value

$$N = \Gamma(1+i\xi)e^{-\xi\pi/2}(2\pi\hbar)^{-3/2}, \tag{7.9.15}$$

and identify $f_k(\theta)$ as the scattering amplitude.

We note that for $|\xi| \ll 1$, where the factor $\Gamma(1+i\xi)/\Gamma(1-i\xi)$ is unity, Eq. (7.9.14) gives the same scattering amplitude as the Born approximation result (7.4.5) for infinite screening radius $1/\kappa$. For all $\xi$, $\Gamma(1+i\xi)/\Gamma(1-i\xi)$ just affects the phase of the scattering amplitude, so the Born approximation here gives the correct differential cross section to all orders. The total elastic scattering cross section is infinite, meaning that every particle in the incoming beam is scattered by some amount, though in practice there always is some screening of Coulomb potentials, and the total cross section is never really infinite.

## 7.10 The Eikonal Approximation

The eikonal approximation[14] is an extension of the WKB approximation to problems in three dimensions, where no spherical symmetry is available to simplify calculations. One such problem is potential scattering, in which even for a spherically symmetric potential there is a preferred direction in space, the direction of the incoming plane wave. In its application to scattering, the eikonal approximation shows why classical mechanics can be used in some cases to calculate scattering cross sections, and also provides information about the phase of the scattering amplitude. We shall use the eikonal approximation again when we come to the Aharonov–Bohm effect in Section 10.4.

---

[14] For the eikonal approximation in optics, see M. Born and E. Wolf, *Principles of Optics* (Pergamon Press, New York, 1959).

Consider the general energy-eigenvalue problem for a single spinless[15] particle with coordinate $\mathbf{x}$:

$$H(-i\hbar\,\nabla,\mathbf{x})\psi(\mathbf{x}) = E\psi(\mathbf{x}). \qquad (7.10.1)$$

We are interested in solutions for which $\psi(\mathbf{x})$ varies much more rapidly with $\mathbf{x}$ than does the Hamiltonian $H$. Our experience with the WKB approximation suggests that we should seek a solution of the form

$$\psi(\mathbf{x}) = N(\mathbf{x})\exp\left(iS(\mathbf{x})/\hbar\right), \qquad (7.10.2)$$

where the phase $S(\mathbf{x})$ varies much more rapidly than the amplitude $N(\mathbf{x})$. If we ignore the variation of $N(\mathbf{x})$ compared with that of $S(\mathbf{x})$, then the gradient in Eq. (7.10.1) will act chiefly on the exponential in Eq. (7.10.2). In this limit, the phase should then satisfy the equation

$$H\left(\nabla S(\mathbf{x}),\mathbf{x}\right) = E. \qquad (7.10.3)$$

The problem here, which did not confront us in one dimension, is that this is just one equation for the three components of $\nabla S$. For instance, if the gradient appears in the Hamiltonian in the form of the Laplacian $\nabla^2$, then Eq. (7.10.3) tells us the magnitude of $\nabla S$ but tells us nothing about its direction. The remaining information needed to calculate $S$ is that the three-vector $\nabla S$ is a gradient. The following prescription allows us to construct a function $S(\mathbf{x})$ whose gradient satisfies Eq. (7.10.3).

First, we need an appropriate initial condition. This is provided by the condition that $S(\mathbf{x})$ should take some constant value $S_0$ on an "initial surface." This surface is not arbitrary, but is determined by the problem at hand. For instance, as we shall see, in scattering the initial surface is taken as a plane normal to the direction of the incoming beam. With $S(\mathbf{x})$ constant on the initial surface, $\nabla S(\mathbf{x})$ is normal to the initial surface at all points on the surface.

Next, we define a family of "ray paths" starting at the initial surface. These curves are defined by a pair of equations, similar to the equations of motion in classical Hamiltonian dynamics:

$$\frac{dq_i}{d\tau} = \frac{\partial H(\mathbf{p},\mathbf{q})}{\partial p_i}, \quad \frac{dp_i}{d\tau} = -\frac{\partial H(\mathbf{p},\mathbf{q})}{\partial q_i}, \qquad (7.10.4)$$

where here $\tau$ parameterizes the curves. The initial condition on these differential equations is that each trajectory starts at $\tau = 0$ with $\mathbf{q}(0)$ on the initial surface,

---

[15] For a particle with spin subject to spin-dependent forces, it is necessary to extend the treatment here to a set of coupled equations for the different spin components. The general treatment of multicomponent wave propagation in anisotropic media in the eikonal approximation is given by S. Weinberg, *Phys. Rev.* **126**, 1899 (1962).

with $\mathbf{p}(0)$ normal to the surface at that point, and with the magnitude of $\mathbf{p}(0)$ given by the condition that, at that point,

$$H(\mathbf{p}(0), \mathbf{q}(0)) = E. \tag{7.10.5}$$

Although this is a time-independent problem, we can evidently regard $\tau$ as the time required for a classical particle to travel to $\mathbf{q}(\tau)$ from the initial surface.

We assume that these ray paths without crossing fill at least a finite volume of space adjacent to the initial surface, so that for each point $\mathbf{x}$ in this volume there is a unique $\tau_{\mathbf{x}}$ such that

$$\mathbf{q}\left(\tau_{\mathbf{x}}\right) = \mathbf{x}. \tag{7.10.6}$$

The phase $S$ is then given by

$$S(\mathbf{x}) = \int_0^{\tau_{\mathbf{x}}} \mathbf{p}(\tau) \cdot \frac{d\mathbf{q}(\tau)}{d\tau} \, d\tau + S_0. \tag{7.10.7}$$

Let us check that this solves our problem. It is easy to see that for all such $\tau$,

$$H(\mathbf{p}(\tau), \mathbf{q}(\tau)) = E. \tag{7.10.8}$$

This is because the differential equations (7.10.4) imply that

$$\frac{d}{d\tau} H(\mathbf{p}(\tau), \mathbf{q}(\tau)) = \sum_i \frac{\partial H\left(\mathbf{p}(\tau), \mathbf{q}(\tau)\right)}{\partial p_i(\tau)} \frac{dp_i(\tau)}{d\tau}$$

$$+ \sum_i \frac{\partial H\left(\mathbf{p}(\tau), \mathbf{q}(\tau)\right)}{\partial q_i(\tau)} \frac{dq_i(\tau)}{d\tau}$$

$$= 0, \tag{7.10.9}$$

so since Eq. (7.10.8) is satisfied at $\tau = 0$, it is satisfied for all $\tau$, at least in a finite range.

It only remains to show that $\mathbf{p} = \nabla S$. For this purpose, we note that an infinitesimal change $\delta\mathbf{x}$ in $\mathbf{x}$ will not only change $\tau_{\mathbf{x}}$, say to $\tau_{\mathbf{x}} + \Delta\tau_{\mathbf{x}}$, but will also shift the ray path that connects the initial surface to the point $\mathbf{x}$ to a new path, having $\mathbf{q}(\tau)$ and $\mathbf{p}(\tau)$ replaced with $\mathbf{q}(\tau) + \Delta\mathbf{q}(\tau)$ and $\mathbf{p}(\tau) + \Delta\mathbf{p}(\tau)$, where $\Delta\mathbf{q}$ and $\Delta\mathbf{p}$ are infinitesimal, and

$$\delta\mathbf{x} = \left[ \frac{d\mathbf{q}(\tau)}{d\tau} \Delta\tau_{\mathbf{x}} + \Delta\mathbf{q}(\tau) \right]_{\tau=\tau_{\mathbf{x}}}. \tag{7.10.10}$$

The change in $\mathbf{x}$ produces a change in the $S(\mathbf{x})$ given by Eq. (7.10.7):

$$\delta S(\mathbf{x}) = \Delta\tau_{\mathbf{x}} \, \mathbf{p}(\tau_{\mathbf{x}}) \cdot \frac{d\mathbf{q}(\tau)}{d\tau}\bigg|_{\tau=\tau_{\mathbf{x}}}$$

$$+ \int_0^{\tau_{\mathbf{x}}} \left[ \mathbf{p}(\tau) \cdot \frac{d\Delta\mathbf{q}(\tau)}{d\tau} + \Delta\mathbf{p}(\tau) \cdot \frac{d\mathbf{q}(\tau)}{d\tau} \right] d\tau.$$

We may re-arrange this to read

$$\delta S(\mathbf{x}) = \Delta \tau_{\mathbf{x}} \, \mathbf{p}(\tau_{\mathbf{x}}) \cdot \frac{d\mathbf{q}(\tau)}{d\tau}\bigg|_{\tau = \tau_{\mathbf{x}}}$$

$$+ \int_0^{\tau_{\mathbf{x}}} \frac{d}{d\tau}\Big[\mathbf{p}(\tau) \cdot \Delta \mathbf{q}(\tau)\Big] d\tau$$

$$+ \int_0^{\tau_{\mathbf{x}}} \left[\Delta \mathbf{p}(\tau) \cdot \frac{d\mathbf{q}(\tau)}{d\tau} - \frac{d\mathbf{p}(\tau)}{d\tau} \cdot \Delta \mathbf{q}(\tau)\right] d\tau.$$

The first integral is given by the value of the integrand at the upper end-point $\tau = \tau_{\mathbf{x}}$

$$\int_0^{\tau_{\mathbf{x}}} \frac{d}{d\tau}\Big[\mathbf{p}(\tau) \cdot \Delta \mathbf{q}(\tau)\Big]\, d\tau = \mathbf{p}(\tau_{\mathbf{x}}) \cdot \Delta \mathbf{q}(\tau_{\mathbf{x}}).$$

The contribution of the lower end-point $\tau = 0$ vanishes because on the initial surface $\mathbf{p}$ is normal to the surface while $\Delta \mathbf{q}$ is tangent to the surface, so that $\mathbf{p}(0) \cdot \Delta \mathbf{q}(0) = 0$. According to the ray path equations (7.10.4), the integrand of the second integral is

$$\Delta \mathbf{p}(\tau) \cdot \frac{d\mathbf{q}(\tau)}{d\tau} - \frac{d\mathbf{p}(\tau)}{d\tau} \cdot \Delta \mathbf{q}(\tau) = \sum_i \Delta p_i(\tau) \frac{\partial H\Big(\mathbf{q}(\tau), \mathbf{p}(\tau)\Big)}{\partial p_i}$$

$$+ \sum_i \Delta q_i(\tau) \frac{\partial H\Big(\mathbf{q}(\tau), \mathbf{p}(\tau)\Big)}{\partial q_i}$$

$$= \Delta H\Big(\mathbf{q}(\tau), \mathbf{p}(\tau)\Big),$$

and this vanishes because, as we have seen, $H$ has the same value $H = E$ on all ray paths. Using Eq. (7.10.10), we are left with

$$\delta S(\mathbf{x}) = \Delta \tau_{\mathbf{x}} \, \mathbf{p}(\tau_{\mathbf{x}}) \cdot \frac{d\mathbf{q}(\tau)}{d\tau}\bigg|_{\tau = \tau_{\mathbf{x}}} + \mathbf{p}(\tau_{\mathbf{x}}) \cdot \Delta \mathbf{q}(\tau_{\mathbf{x}}) = \mathbf{p}(\tau_{\mathbf{x}}) \cdot \delta \mathbf{x} \qquad (7.10.11)$$

and so

$$\mathbf{p}(\tau_{\mathbf{x}}) = \nabla S(\mathbf{x}), \qquad (7.10.12)$$

as was to be shown.

We can learn about the amplitude $N(\mathbf{x})$ by going to the next order in gradients. Using Eq. (7.10.2), the Schrödinger equation (7.10.1) may be expressed exactly as[16]

$$H\Big(\nabla S(\mathbf{x}) - i\hbar \nabla, \, \mathbf{x}\Big) N(\mathbf{x}) = E N(\mathbf{x}). \qquad (7.10.13)$$

---

[16]  The function $H\Big(\nabla S(\mathbf{x}) - i\hbar \nabla, \, \mathbf{x}\Big)$ is defined by its power-series expansion. In this expansion, it should be understood that the operator $-i\hbar \nabla$ acts on everything to its right, including not only $N$ but also the derivatives of $S$.

With Eq. (7.10.3) satisfied, the terms of zeroth order in the gradients of $N(\mathbf{x})$ and $\nabla S(\mathbf{x})$ cancel. To first order in these gradients, the Schrödinger equation then becomes

$$\mathbf{A}(\mathbf{x}) \cdot \nabla N(\mathbf{x}) + B(\mathbf{x})N(\mathbf{x}) = 0, \qquad (7.10.14)$$

where

$$A_i(\mathbf{x}) \equiv \left[ \frac{\partial H(\mathbf{p}, \mathbf{x})}{\partial p_i} \right]_{\mathbf{p}=\nabla S(\mathbf{x})},$$

$$B(\mathbf{x}) \equiv \frac{1}{2} \sum_{ij} \left[ \frac{\partial^2 H(\mathbf{p}, \mathbf{x})}{\partial p_i \, \partial p_j} \right]_{\mathbf{p}=\nabla S(\mathbf{x})} \frac{\partial^2 S(\mathbf{x})}{\partial x_i \, \partial x_j}. \qquad (7.10.15)$$

Using Eq. (7.10.4), it follows from Eq. (7.10.14) that

$$\frac{d}{d\tau} \ln N\Big(\mathbf{q}(\tau)\Big) = -B\Big(\mathbf{q}(\tau)\Big),$$

and therefore

$$N(\mathbf{x}) = N(\mathbf{x}_0) \exp \left( - \int_0^{\tau_x} B\Big(\mathbf{q}(\tau)\Big) d\tau \right), \qquad (7.10.16)$$

where $\mathbf{x}_0$ is the point on the initial surface connected by a ray path to $\mathbf{x}$. The important thing is that $N(\mathbf{x})$ does not depend on its value at any point on the initial surface other than $\mathbf{x}_0$, so that we can speak of the wave function as being propagated from the initial surface along the ray paths.

In potential scattering we have

$$H(\mathbf{p}, \mathbf{x}) = \frac{\mathbf{p}^2}{2m} + V(\mathbf{q}),$$

so

$$\mathbf{A}(\mathbf{x}) = \frac{1}{m} \nabla S(\mathbf{x}), \quad B(\mathbf{x}) = \frac{1}{2m} \nabla^2 S(\mathbf{x}),$$

and Eq. (7.10.14) therefore gives[17]

$$0 = 2Nm\Big[\mathbf{A} \cdot \nabla N + BN\Big] = 2N\left[\nabla S \cdot \nabla N + \frac{N}{2} \nabla^2 S\right]$$

$$= \nabla \cdot \Big(N^2 \nabla S\Big). \qquad (7.10.17)$$

We can now see that the distribution of probabilities of scattering at various angles is given in the eikonal approximation by classical scattering theory. First, recall how scattering cross sections are calculated classically. Consider a beam

---

[17] The quantity $N^2 \nabla S$ is proportional to the probability current $\psi^* \nabla \psi - \psi \nabla \psi^*$ appearing in the probability conservation condition (1.5.5), so the vanishing of its divergence follows from Eq. (1.5.5) and the time-independence here of $|\psi|^2$.

of particles, coming in toward a scattering center on parallel trajectories, say along the $z$-direction. In order to be scattered into a small solid angle $\delta\Omega$ in a direction with polar and azimuthal angles $\theta$ and $\phi$, the incoming particles must initially occupy a small area $\delta A(\theta, \phi)$ transverse to the $z$-axis, proportional to $\delta\Omega$. The classical differential cross section is defined as the ratio

$$\left(\frac{d\sigma(\theta, \phi)}{d\Omega}\right)_{\text{classical}} \equiv \delta A(\theta, \phi)/\delta\Omega. \qquad (7.10.18)$$

That is, for any direction $(d\sigma/d\Omega)_{\text{classical}}\,\delta\Omega$ is the area that the particle must hit to be scattered into the solid angle $\delta\Omega$ in that direction.

For example, suppose that by solving the classical equation of motion for a spherically symmetric potential, it is found that in order for a particle that approaches the scattering center along the $z$-axis to be scattered into an angle $\theta$ it must initially travel along a line at some distance (the "impact parameter") $b(\theta)$ from the $z$-axis. Every particle that is scattered into the small solid angle $\sin\theta\,\delta\theta\,\delta\phi$ between angles $\theta$ and $\theta + \delta\theta$ and between angles $\phi$ and $\phi + \delta\phi$ will have to approach the scattering center between impact parameters $b(\theta)$ and $b(\theta) + (db(\theta)/d\theta)\,\delta\theta$ and between azimuthal angles $\phi$ and $\phi + \delta\phi$, so

$$\frac{d\sigma}{d\Omega} = |b\,db\,d\phi/\sin\theta\,d\theta\,d\phi| = \frac{b(\theta)}{\sin\theta}\left|\frac{db(\theta)}{d\theta}\right|. \qquad (7.10.19)$$

In particular, for a particle of mass $\mu$ with initial velocity $v_0$ scattered by the Coulomb potential $Z_1 Z_2 e^2/r$, the classical equations of motion give $b(\theta) = Z_1 Z_2 e^2/\mu v_0^2\tan(\theta/2)$. Using this in Eq. (7.10.19) we get a differential cross section $d\sigma/d\Omega = Z_1^2 Z_2^2 e^4/4\mu^2 v_0^4\sin^4(\theta/2)$. This is how Rutherford calculated the Coulomb scattering cross section in 1911.

Now consider how the cross section is calculated quantum mechanically in the eikonal approximation. The "initial surface" on which the phase of the wave function is constant can be taken to be a plane normal to the $z$-axis and far upstream from the scattering center. Consider the tube formed by all the classical trajectories running from a small initial area $\delta A(\theta, \phi)$ on the initial surface, past the scattering center, and then out to a great distance within a solid angle $\delta\Omega$ around the direction defined by angles $\theta$ and $\phi$. Using Gauss's theorem, it follows from Eq. (7.10.17) that the integral of the normal component of $N^2\,\nabla S$ over the surface of the tube vanishes. According to Eq. (7.10.12), this means that the integral of the normal component of $N^2\mathbf{p}$ over the surface of the tube vanishes. The sides of the tube are made up of particle trajectories, so $\mathbf{p}$ has a vanishing component normal to these sides, and therefore the only contributions to the integral come from the initial area $\delta A$, where $\mathbf{p}$ is directed along the normal but into the tube, and the final area $r^2\,d\Omega$, where $\mathbf{p}$ is directed along the normal out of the tube. Since the initial and final momentum have the same magnitude, the vanishing of the surface integral tells us simply that

$$-\delta A(\theta, \phi)\,N_{\text{initial}}^2 + r^2\,\delta\Omega\,N_{\text{final}}^2 = 0. \qquad (7.10.20)$$

To find the initial and final values of $N^2$, recall that Eq. (7.2.5) gives the wave function at large distances $r$ from the scattering center as

$$\psi_{\mathbf{k}}(\mathbf{x}) \to C\left[e^{i\mathbf{k}\cdot\mathbf{x}} + f_{\mathbf{k}}(\theta,\phi)e^{ikr}/r\right], \qquad (7.10.21)$$

where $C$ is an unimportant normalization constant, and $f_{\mathbf{k}}(\theta,\phi)$ is the scattering amplitude. Hence, comparing this with Eq. (7.10.2),

$$N_{\text{initial}} = C, \qquad N_{\text{final}} = Cf_{\mathbf{k}}(\theta,\phi)/r. \qquad (7.10.22)$$

The quantum-mechanical cross section is then given in the eikonal approximation by Eqs. (7.10.22) and (7.10.20) as

$$\left(\frac{d\sigma(\theta,\phi)}{d\Omega}\right)_{\text{eikonal}} = |f_{\mathbf{k}}(\theta,\phi)|^2 = \frac{N_{\text{final}}^2 r^2}{N_{\text{initial}}^2} = \frac{\delta A(\theta,\phi)}{\delta\Omega} = \left(\frac{d\sigma(\theta,\phi)}{d\Omega}\right)_{\text{classical}}, \qquad (7.10.23)$$

as was to be shown.

But the eikonal approximation goes beyond classical scattering theory in providing a formula for the phase of the scattering amplitude, not just its absolute value. For scattering of a particle of mass $\mu$ by a central potential $V(r)$, the Hamiltonian is

$$H = \frac{p_r^2}{2\mu} + \frac{p_\vartheta^2}{2\mu r^2} + V(r), \qquad (7.10.24)$$

from which we find that

$$\dot{r} = p_r/\mu, \qquad \dot{\vartheta} = p_\vartheta/\mu r^2, \qquad (7.10.25)$$

a dot here denoting differentiation with respect to the trajectory parameter $\tau$. There are two constants of the motion here, the energy $H$ and the angular momentum $p_\vartheta$, to which we can give the values

$$H = \hbar^2 k^2/2\mu, \qquad p_\vartheta = -\hbar k b, \qquad (7.10.26)$$

where $k$ is the wave number of the incoming wave, and $b$ is the impact parameter, the distance of closest approach to the scattering center if there were no potential. The $\vartheta$ coordinate along the trajectory can then be related to the $r$ coordinate by

$$\frac{d\vartheta}{dr} = \frac{\dot{\vartheta}}{\dot{r}} = \frac{p_\vartheta}{r^2 p_r} = -\frac{\hbar k b}{r^2 p_r}. \qquad (7.10.27)$$

Using Eqs. (7.10.26) in Eq. (7.10.24) and solving for $p_r$ gives

$$p_r = \pm\sqrt{\hbar^2 k^2 - \hbar^2 k^2 b^2/r^2 - 2\mu V(r)/r^2}. \qquad (7.10.28)$$

From Eqs. (7.10.27) and (7.10.28), we find the integrand in Eq. (7.10.7) for the phase $S/\hbar$ of the scattering amplitude:

$$[p_r\,dr + p_\vartheta\,d\vartheta]/\hbar = \pm\kappa(r)\,dr, \qquad (7.10.29)$$

where

$$\kappa(r) = \sqrt{k^2(1 - b^2/r^2) - 2\mu V(r)/\hbar^2} + \frac{k^2 b^2}{r^2\sqrt{k^2(1 - b^2/r^2) - 2\mu V(r)/\hbar^2}}.$$

$$(7.10.30)$$

It is convenient in scattering problems to take the initial surface at a large distance $R$ from the scattering center, and let the constant phase of the wave function on this surface be

$$S_0 = -\int_{r_0}^{R} \kappa(r) \, dr,$$

where $r_0$ is the point of closest approach on the classical trajectory, given by the solution of $p_r = 0$ – that is,

$$k^2(1 - b^2/r_0^2) - 2\mu V(r_0)/\hbar^2 = 0. \qquad (7.10.31)$$

The phase of the outgoing part of the wave function is then given in the eikonal approximation by Eqs. (7.10.7) and (7.10.29) as

$$S(r, \theta)/\hbar = \int_{r_0}^{r} \kappa(r) \, dr, \qquad (7.10.32)$$

it being understood that $b$ in Eq. (7.10.30) for $\kappa(r)$ is the function $b(\theta)$, the impact parameter for which the classical equations of motion give scattering at an angle $\theta$.

The integral (7.10.32) is generally quite complicated, but it gives simple results for the phase at large $r$. In scattering problems $V(r)$ must be assumed to vanish at great distances from the scattering center. Assuming that it vanishes at least as fast as $1/r$, for $r \to \infty$ Eq. (7.10.30) gives

$$\kappa(r) \to k - \frac{\mu V(r)}{\hbar^2 k} + O\left(1/r^2\right). \qquad (7.10.33)$$

We must now distinguish two cases.

• If $V(r)$ vanishes as $r \to \infty$ like $r^{-\mathcal{N}}$, with $\mathcal{N} > 1$, then the terms in Eq. (7.10.33) that go as $1/r^2$ or $V(r)$ make a contribution to the integral in Eq. (7.10.32) that becomes $r$-independent for $r \to \infty$. In this case the phase of the wave function approaches $kr + C$ for $r \to \infty$, where $C$ is $r$-independent but in general depends on $b$ as well as $k$ and hence on the scattering angle $\theta$.

• For potentials $V(r)$ that at large $r$ go as $U/r$, with $U$ constant, the integral $\int^r V(r) \, dr$ does not converge at $r \to \infty$, and for large $r$ the phase of the wave function goes as

$$S(r)/\hbar \to kr - \frac{\mu U}{\hbar^2 k} \ln r + C, \qquad (7.10.34)$$

with $C$ again in general dependent on $\theta$ as well as $k$ but not on $r$. In particular, for the Coulomb potential itself we have $U = Z_1 Z_2 e^2 = \xi \hbar^2 k/\mu$, where $\xi$

is the Coulomb scattering parameter introduced in the previous section. Thus Eq. (7.10.34) yields the $r$-dependent factor

$$e^{ikr - i\xi \ln r}$$

in the outgoing part of the wave function (7.9.13). But by using the eikonal approximation, we have seen that such $\ln r$ terms appear in the phase of the outgoing part of the wave function not just for the Coulomb potential, but also for any potential that goes as $1/r$ for $r \to \infty$.

## Problems

1. Use the Born approximation to give a formula for the $s$-wave scattering length $a_s$ for scattering of a particle of mass $\mu$ and wave number $k$ by an arbitrary central potential $V(r)$ of finite range $R$, in the limit $kR \ll 1$. Use this result and the optical theorem to calculate the imaginary part of the forward scattering amplitude to *second* order in the potential.

2. Suppose that in the scattering of a spinless non-relativistic particle of mass $\mu$ by an unknown potential, a resonance is observed at energy $E_R$, and that the elastic cross section at the peak of the resonance is found to have value $\sigma_{\max}$. Show how to use this data to give a value for the orbital angular momentum of the resonant state.

3. Give a formula for the tangent of the $\ell = 0$ phase shift for scattering by a potential

$$V(r) = \begin{cases} -V_0, & r < R, \\ 0, & r \geq R, \end{cases}$$

   for all $E > 0$, and to all orders in $V_0 > 0$.

4. Suppose that the eigenstates of an unperturbed Hamiltonian include not only continuum states of a free particle with momentum $\mathbf{p}$ and unperturbed energy $E = \mathbf{p}^2/2\mu$, but also a discrete state of angular momentum $\ell$ with a negative unperturbed energy. Suppose that when we turn on the interaction, the continuum states feel a local potential, but remain in the continuum, while also the discrete state moves to positive energy, thereby becoming unstable. What is the change in the phase shift $\delta_\ell(k)$ as the wave number $k$ increases from $k = 0$ to $k = \infty$?

5. Find an upper bound on the elastic scattering cross section in the case where the scattering amplitude $f$ is independent of angles $\theta$ and $\phi$.

# 8

# General Scattering Theory

The previous chapter described the theory of elastic scattering of a single non-relativistic particle by a local potential. There are much more general circumstances to which scattering theory is applicable. The scattering can produce additional particles; the interaction may not be a local potential; some or all of the particles involved may be moving at relativistic velocities; some may be photons; and the initial state may even contain more than two particles. This chapter will describe scattering theory at a level of generality that encompasses all these possibilities.

In this chapter we will be using the relativistic formula for energies: the energy of a particle of momentum $\mathbf{p}$ and mass $m$ is $(\mathbf{p}^2 c^2 + m^2 c^4)^{1/2}$, where $c$ is the speed of light. This is because we want to consider inelastic scattering processes, in which mass energy is converted to kinetic energy, or vice versa. It is not entirely trivial to formulate dynamical theories consistent with special relativity – the only really satisfactory approach is based on the quantum theory of fields – but as far as general principles are concerned, quantum mechanics applies equally to relativistic and non-relativistic systems.

## 8.1 The S-Matrix

We again assume that the Hamiltonian $H$ is the sum of an unperturbed Hermitian term $H_0$, describing any number of non-interacting particles, plus some sort of interaction $V$:

$$H = H_0 + V. \tag{8.1.1}$$

The only assumptions we make about $V$ are that it is Hermitian, and that its effects become negligible when the particles described by $H_0$ are all far from one another.

In Section 7.1 we defined an "in" state $\Psi_{\mathbf{k}}^{\text{in}}$ as an eigenstate of the Hamiltonian that looks like it consists of a single particle with momentum $\hbar\mathbf{k}$ far from the scattering center if measurements are made at sufficiently early times. We generalize this definition, and define "in" and "out" states $\Psi_\alpha^+$ and $\Psi_\alpha^-$ as eigenstates of the Hamiltonian

$$H\Psi_\alpha^\pm = E_\alpha \Psi_\alpha^\pm \tag{8.1.2}$$

that look like an eigenstate $\Phi_\alpha$ of the free-particle Hamiltonian

$$H_0 \Phi_\alpha = E_\alpha \Phi_\alpha \tag{8.1.3}$$

consisting of a number of particles at great distances from each other, provided measurements are made at very early times (for $\Psi_\alpha^+$) or very late times (for $\Psi_\alpha^-$). Here $\alpha$ is a compound index, standing for the types and numbers of the particles in the state, as well as all their momenta and spin 3-components (or helicities). It will be convenient to choose the states $\Phi_\alpha$ to be orthonormal

$$\left(\Phi_\beta, \Phi_\alpha\right) = \delta(\beta - \alpha). \tag{8.1.4}$$

The delta function $\delta(\alpha - \beta)$ consists of a product of Kronecker deltas for the numbers and types and spin 3-components of corresponding particles in the states $\alpha$ and $\beta$, together with three-dimensional delta functions for the momenta of the corresponding particles in these states.

The definition of $\Psi_\alpha^+$ and $\Psi_\alpha^-$ can be made more precise by specifying that if $g(\alpha)$ is a sufficiently smooth function of the momenta in the state $\alpha$, then (as a generalization of Eqs. (7.1.3) and (7.1.4))

$$\int d\alpha\, g(\alpha) \Psi_\alpha^\pm \exp(-iE_\alpha t/\hbar) \rightarrow \int d\alpha\, g(\alpha) \Phi_\alpha \exp(-iE_\alpha t/\hbar) \tag{8.1.5}$$

for $t \rightarrow \mp\infty$. (Integrals over $\alpha$ in general include sums over the numbers and types of particles along with the 3-components of their spins, as well as integrals over the momenta of all the particles in the state $\alpha$.) We can satisfy this condition by rewriting Eq. (8.1.2) as a generalization of the Lippmann–Schwinger equation (7.1.7):

$$\Psi_\alpha^\pm = \Phi_\alpha + (E_\alpha - H_0 \pm i\epsilon)^{-1} V \Psi_\alpha^\pm, \tag{8.1.6}$$

with $\epsilon$ a positive infinitesimal quantity. Equation (8.1.5) then follows by a simple extension of the argument used in Section 7.1. From Eq. (8.1.6) we have

$$\int d\alpha\, g(\alpha) \Psi_\alpha^\pm \exp(-iE_\alpha t/\hbar) = \int d\alpha\, g(\alpha) \Phi_\alpha \exp(-iE_\alpha t/\hbar)$$
$$+ \int d\alpha \int d\beta\, \frac{g(\alpha) \exp(-iE_\alpha t/\hbar)\left(\Phi_\beta, V\Psi_\alpha^\pm\right)}{E_\alpha - E_\beta \pm i\epsilon} \Phi_\beta. \tag{8.1.7}$$

The rapid oscillation of the exponential in the second term on the right-hand side kills all contributions to this integral except those from $E_\alpha$ near $E_\beta$, where the denominator varies rapidly. In particular, this allows us to extend the integral to all real $E_\alpha$, since no part of the range of integration except very near $E_\beta$ will contribute anyway for $|t| \rightarrow \infty$. This integral can be evaluated for $|t| \rightarrow \infty$ by closing the contour of integration over $E_\alpha$ with a large semi-circle in the upper half of the complex plane for $t \to -\infty$ or in the lower half of the complex plane for $t \to +\infty$, since in both cases the factor $\exp(-i E_\alpha t/\hbar)$ is exponentially damped on the semi-circle. In both cases the pole at $E_\alpha = E_\beta \mp i\epsilon$ is outside the contour of integration, so this integral vanishes, leaving us with Eq. (8.1.5). (By the way, it is the $\pm i\epsilon$ term in the denominator in Eq. (8.1.6) that has led to "in" and "out" states being conventionally denoted $\Psi_\alpha^+$ and $\Psi_\alpha^-$, respectively.)

The "in" and "out" states inhabit the same Hilbert space, and are distinguished only by how they are described, by their appearance at $t \to -\infty$ or at $t \to +\infty$. Indeed, any "in" state can be expressed as a superposition of "out" states:

$$\Psi_\alpha^+ = \int d\beta \, S_{\beta\alpha} \Psi_\beta^-. \tag{8.1.8}$$

The coefficients $S_{\beta\alpha}$ in this relation form what is known as the *S-matrix*. If we arrange a state so that it appears at $t \to -\infty$ like a free-particle state $\Phi_\alpha$, then the state is $\Psi_\alpha^+$, and Eq. (8.1.8) tells us that the state will appear at late times like the superposition $\int d\beta \, S_{\beta\alpha} \Phi_\beta$. As we will see, the S-matrix contains all information about the rates of reactions among particles of any sorts.

We can derive a useful formula for the S-matrix by considering what the "in" state looks like if measurements are made at *late* times. We again use Eq. (8.1.7) for $\Psi_\alpha^+$, but now because $t > 0$ we can only close the contour of integration of $E_\alpha$ in the second term with a large semi-circle in the *lower* half of the complex plane, so now we receive a contribution from the pole at $E_\alpha = E_\beta - i\epsilon$. Because we are integrating over a closed contour running in the clockwise direction, the contribution of this pole is $-2\pi i$ times the same integral, but with the denominator dropped, and with the integration over $E_\alpha$ replaced by setting $E_\alpha = E_\beta - i\epsilon$ in the remainder of the integrand. Since $\epsilon$ is infinitesimal, this just amounts to replacing $(E_\alpha - E_\beta + i\epsilon)^{-1}$ in Eq. (8.1.7) with $-2\pi i\delta(E_\alpha - E_\beta)$, so that for $t \to +\infty$

$$\int d\alpha \, g(\alpha) \Psi_\alpha^+ \exp(-i E_\alpha t/\hbar) \to \int d\alpha \, g(\alpha)\Phi_\alpha \exp(-i E_\alpha t/\hbar)$$

$$- 2\pi i \int d\alpha \int d\beta \, g(\alpha) \exp(-i E_\alpha t/\hbar) \left(\Phi_\beta, V\Psi_\alpha^+\right)\delta(E_\alpha - E_\beta)\Phi_\beta. \tag{8.1.9}$$

As remarked in the previous paragraph, the state $\Psi_\alpha^+$ looks at $t \to +\infty$ like the superposition $\int d\beta \, S_{\beta\alpha}\Phi_\beta$, so from Eq. (8.1.9) we have

$$S_{\beta\alpha} = \delta(\beta - \alpha) - 2\pi i\delta(E_\alpha - E_\beta)T_{\beta\alpha}, \tag{8.1.10}$$

where

$$T_{\beta\alpha} \equiv \left(\Phi_\beta, V\Psi_\alpha^+\right). \tag{8.1.11}$$

We have chosen the states $\Phi_\alpha$ to be orthonormal, and it follows then from Eq. (8.1.6) that the "in" and "out" states are also orthonormal. This is fairly obvious from the condition (8.1.5), but we can also give a more direct proof. We can evaluate the matrix element $\left(\Psi_\beta^\pm, V\Psi_\alpha^\pm\right)$ by using Eq. (8.1.6) in either the right or left side of the scalar product. The results must be equal, so (using the fact that $H_0$ and $V$ are Hermitian)

$$\left(\Psi_\beta^\pm, V\Phi_\alpha\right) + \left(\Psi_\beta^\pm, V(E_\alpha - H_0 \pm i\epsilon)^{-1}V\Psi_\alpha^\pm\right)$$
$$= \left(\Phi_\beta, V\Psi_\alpha^\pm\right) + \left(\Psi_\beta^\pm, V(E_\beta - H_0 \mp i\epsilon)^{-1}V\Psi_\alpha^\pm\right). \qquad (8.1.12)$$

We use the trivial identity

$$(E_\alpha - H_0 \pm i\epsilon)^{-1} - (E_\beta - H_0 \mp i\epsilon)^{-1} = -\frac{E_\alpha - E_\beta \pm 2i\epsilon}{(E_\alpha - H_0 \pm i\epsilon)(E_\beta - H_0 \mp i\epsilon)}$$

so that, dividing by $E_\alpha - E_\beta \pm 2i\epsilon$,

$$-\left[\frac{\left(\Phi_\alpha, V\Psi_\beta^\pm\right)}{E_\beta - E_\alpha \pm 2i\epsilon}\right]^* - \frac{\left(\Phi_\beta, V\Psi_\alpha^\pm\right)}{E_\alpha - E_\beta \pm 2i\epsilon}$$
$$= \left(\Psi_\beta^\pm, V(E_\beta - H_0 \mp i\epsilon)^{-1}(E_\alpha - H_0 \pm i\epsilon)^{-1}V\Psi_\alpha^\pm\right).$$

The only important thing about $\epsilon$ is that it is a positive infinitesimal, so we may as well replace $2\epsilon$ here with $\epsilon$. According to Eq. (8.1.6), this tells us that

$$-\left(\Phi_\alpha, [\Psi_\beta^\pm - \Phi_\beta]\right)^* - \left(\Phi_\beta, [\Psi_\alpha^\pm - \Phi_\alpha]\right) = \left([\Psi_\beta^\pm - \Phi_\beta], [\Psi_\alpha^\pm - \Phi_\alpha]\right),$$

and therefore

$$\left(\Psi_\beta^\pm, \Psi_\alpha^\pm\right) = \left(\Phi_\beta, \Phi_\alpha\right) = \delta(\alpha - \beta). \qquad (8.1.13)$$

By taking the scalar product of Eq. (8.1.8) with $\Psi_\beta^-$, we have now

$$S_{\beta\alpha} = \left(\Psi_\beta^-, \Psi_\alpha^+\right). \qquad (8.1.14)$$

Thus $S_{\beta\alpha}$ is the probability amplitude that a state that is arranged to look at $t \to -\infty$ like the free-particle state $\Phi_\alpha$ will look when measurements are made at $t \to \infty$ like the free-particle state $\Phi_\beta$.

Because $S_{\beta\alpha}$ is the matrix of scalar products of two complete orthonormal sets of state vectors, it must be unitary. We can also show this directly by multiplying Eq. (8.1.12) (for "in" states) with $\delta(E_\alpha - E_\beta)$, from which we learn that

$$\delta(E_\alpha - E_\beta)\left(T_{\alpha\beta}^* - T_{\beta\alpha}\right) = 2i\epsilon\delta(E_\alpha - E_\beta)\int d\gamma \frac{T_{\gamma\beta}^*T_{\gamma\alpha}}{(E_\alpha - E_\gamma)^2 + \epsilon^2}.$$

For infinitesimal $\epsilon$ the function $\epsilon/(x^2 + \epsilon^2)$ is negligible away from $x = 0$, while its integral over all $x$ is $\pi$, so in any integral it can be replaced with $\pi\delta(x)$. Multiplying with $-2i\pi$, replacing $\delta(E_\alpha - E_\beta)\delta(E_\alpha - E_\gamma)$ with $\delta(E_\beta - E_\gamma)\delta(E_\alpha - E_\gamma)$, and recalling Eq. (8.1.10), we have then

$$-[S_{\beta\alpha} - \delta(\alpha - \beta)] - [S_{\alpha\beta}^* - \delta(\alpha - \beta)] = \int d\gamma \, [S_{\gamma\beta} - \delta(\beta - \gamma)]^* [S_{\gamma\alpha} - \delta(\alpha - \gamma)]$$

or in other words

$$\int d\gamma \, S_{\gamma\beta}^* S_{\gamma\alpha} = \delta(\alpha - \beta). \tag{8.1.15}$$

In matrix language, $S^\dagger S = 1$, where as usual $\dagger$ denotes the transpose of the complex conjugate.

If $\alpha$ and $\beta$ were discrete states instead of members of a continuum, the unitarity of the S-matrix would yield the result that the total probability $\sum_\beta |S_{\beta\alpha}|^2$ is unity. The physical implications of unitarity in the real world, where these states form a continuum, will be discussed in Section 8.3.

$$* \ * \ * \ * \ *$$

The distinction between "in" and "out" states is contained in the sign of the $\pm i\epsilon$ term in the denominator in the Lippmann–Schwinger equation (8.1.6). To make this a bit less abstract, let's take a look at what the wave function of "out" states looks like in the case studied in Chapter 7, a non-relativistic particle of mass $\mu$ and momentum $\hbar\mathbf{k}$ being scattered by a real local potential $V(\mathbf{x})$. We saw in Section 7.2 that the coordinate-space wave scattering function $\psi_\mathbf{k}^+(\mathbf{x})$ satisfies the integral equation (7.2.3):

$$\psi_\mathbf{k}^+(\mathbf{x}) = (2\pi\hbar)^{-3/2} e^{i\mathbf{k}\cdot\mathbf{x}} + \int d^3y \, G_k^+(\mathbf{x} - \mathbf{y}) V(\mathbf{y}) \psi_\mathbf{k}^+(\mathbf{y}), \tag{8.1.16}$$

where $G_k^+(\mathbf{x} - \mathbf{y})$ is a Green function given by Eq. (7.2.4):

$$G_k^+(\mathbf{x} - \mathbf{y}) = \left( \Phi_\mathbf{x}, [E(k) - H_0 + i\epsilon]^{-1} \Phi_\mathbf{y} \right)$$

$$= -\frac{2\mu}{\hbar^2} \frac{1}{4\pi |\mathbf{x} - \mathbf{y}|} e^{ik|\mathbf{x}-\mathbf{y}|}, \tag{8.1.17}$$

and we are now including a superscript "+" to make clear that this refers only to "in" states. For "out" states, the wave function instead satisfies

$$\psi_\mathbf{k}^-(\mathbf{x}) = (2\pi\hbar)^{-3/2} e^{i\mathbf{k}\cdot\mathbf{x}} + \int d^3y \, G_k^-(\mathbf{x} - \mathbf{y}) V(\mathbf{y}) \psi_\mathbf{k}^-(\mathbf{y}), \tag{8.1.18}$$

where $G_k^-(\mathbf{x} - \mathbf{y})$ is a different Green function

$$G_k^-(\mathbf{x} - \mathbf{y}) = \left( \Phi_\mathbf{x}, [E(k) - H_0 - i\epsilon]^{-1} \Phi_\mathbf{y} \right). \tag{8.1.19}$$

Comparison of Eqs. (8.1.17) and (8.1.19) shows that

$$G_k^-(\mathbf{x} - \mathbf{y}) = G_k^{+*}(\mathbf{y} - \mathbf{x}) = -\frac{2\mu}{\hbar^2} \frac{1}{4\pi |\mathbf{x} - \mathbf{y}|} e^{-ik|\mathbf{x}-\mathbf{y}|}. \tag{8.1.20}$$

Hence the solution of Eq. (8.1.18) is simply

$$\psi_\mathbf{k}^-(\mathbf{x}) = \psi_{-\mathbf{k}}^{+*}(\mathbf{x}). \tag{8.1.21}$$

In particular, in place of Eq. (7.2.5), the asymptotic form of the "out" space wave function for large $|\mathbf{x}|$ is

$$\psi_{\mathbf{k}}^-(\mathbf{x}) \to (2\pi\hbar)^{-3/2} \left[ e^{i\mathbf{k}\cdot\mathbf{x}} + f_{-\mathbf{k}}^*(\hat{x}) e^{-ikr}/r \right], \qquad (8.1.22)$$

with $r \equiv |\mathbf{x}|$.

## 8.2 Rates

The S-matrix given by Eq. (8.1.10) evidently conserves energy. Even where the states $\alpha$ and $\beta$ are different, $S_{\beta\alpha}$ is proportional to $\delta(E_\alpha - E_\beta)$. Also, the symmetry of invariance under spatial translations tells us that the Hamiltonian $H$ commutes with the momentum operator $\mathbf{P}$, and since $H_0$ evidently commutes with $\mathbf{P}$, so does $V$; it follows then that $T_{\beta\alpha}$ and $S_{\beta\alpha}$ are proportional also to a three-dimensional delta function $\delta^3(\mathbf{P}_\alpha - \mathbf{P}_\beta)$, where $\mathbf{P}_\alpha$ and $\mathbf{P}_\beta$ are the total momenta of the states $\alpha$ and $\beta$. In the case where $\alpha$ and $\beta$ are not identical states, we can write

$$S_{\beta\alpha} = \delta(E_\alpha - E_\beta)\delta^3(\mathbf{P}_\alpha - \mathbf{P}_\beta)M_{\beta\alpha}, \qquad (8.2.1)$$

where $M_{\beta\alpha}$ is a smooth function of the momenta in the states $\alpha$ and $\beta$, containing no delta functions.[1] The presence of the delta functions in Eq. (8.2.1) poses an immediate problem: in setting the probability for the transition $\alpha \to \beta$ equal to $|S_{\beta\alpha}|^2$, what are we to make of the squares of $\delta(E_\alpha - E_\beta)$ and $\delta^3(\mathbf{P}_\alpha - \mathbf{P}_\beta)$?

The easiest way to deal with this problem is to imagine that the system is contained in a box of finite volume $V$, and that the interaction is turned on only for a finite time $T$. One consequence is that the delta functions, which as shown in Section 3.2 can be represented as

$$\delta^3(\mathbf{P}_\alpha - \mathbf{P}_\beta) \equiv \frac{1}{(2\pi\hbar)^3} \int d^3x \; e^{i(\mathbf{P}_\alpha - \mathbf{P}_\beta)\cdot\mathbf{x}/\hbar},$$

$$\delta(E_\alpha - E_\beta) \equiv \frac{1}{2\pi\hbar} \int_{-\infty}^{\infty} dt \; e^{i(E_\alpha - E_\beta)t/\hbar},$$

are instead replaced with

$$\delta_V^3(\mathbf{P}_\alpha - \mathbf{P}_\beta) \equiv \frac{1}{(2\pi\hbar)^3} \int_V d^3x \; e^{i(\mathbf{P}_\alpha - \mathbf{P}_\beta)\cdot\mathbf{x}/\hbar},$$

$$\delta_T(E_\alpha - E_\beta) \equiv \frac{1}{2\pi\hbar} \int_T dt \; e^{i(E_\alpha - E_\beta)t/\hbar}. \qquad (8.2.2)$$

---

[1] Strictly speaking, this is true only if no subsets of particles in the states $\alpha$ and $\beta$ have identical total momenta. This condition is necessary to rule out the possibility that the transition $\alpha \to \beta$ involves several distant reactions having nothing to do with each other, in which case $S_{\beta\alpha}$ would include several factors of momentum-conservation delta functions, one for each separate reaction. This possibility does not occur in the scattering of just two particles.

Then we have

$$\left[\delta_V^3(\mathbf{P}_\alpha - \mathbf{P}_\beta)\right]^2 = \frac{V}{(2\pi\hbar)^3}\delta_V^3(\mathbf{P}_\alpha - \mathbf{P}_\beta), \tag{8.2.3}$$

$$\left[\delta_T(E_\alpha - E_\beta)\right]^2 = \frac{T}{2\pi\hbar}\delta_T(E_\alpha - E_\beta). \tag{8.2.4}$$

Also, in using the square of S-matrix elements as transition probabilities, we must take the states to be suitably normalized. In coordinate space, this means that instead of giving a one-particle state $\Phi_\mathbf{p}$ of momentum $\mathbf{p}$ the wave function (6.2.9) with continuum normalization,

$$\left(\Phi_\mathbf{x}, \Phi_\mathbf{p}\right) = \frac{e^{i\mathbf{p}\cdot\mathbf{x}/\hbar}}{(2\pi\hbar)^{3/2}},$$

we take it to be normalized so that the integral of its absolute-value squared over the box is unity:

$$\left(\Phi_\mathbf{x}, \Phi_\mathbf{p}^{\text{Box}}\right) = \frac{e^{i\mathbf{p}\cdot\mathbf{x}/\hbar}}{\sqrt{V}}.$$

That is, we define the box-normalized state as

$$\Phi_\mathbf{p}^{\text{Box}} \equiv \sqrt{\frac{(2\pi\hbar)^3}{V}}\Phi_\mathbf{p}. \tag{8.2.5}$$

For multiparticle states a product of factors of $\sqrt{(2\pi\hbar)^3/V}$ appears in the relation between box-normalized and continuum-normalized states. Hence the S-matrix elements between box-normalized states are

$$S_{\beta\alpha}^{\text{Box}} = \left[\frac{(2\pi\hbar)^3}{V}\right]^{(N_\alpha + N_\beta)/2} S_{\beta\alpha}, \tag{8.2.6}$$

where $N_\alpha$ and $N_\beta$ are the numbers of particles in the initial and final states, respectively. Putting this together, we see that the probability of the transition $\alpha \to \beta$ is

$$P(\alpha \to \beta) = \left|S_{\beta\alpha}^{\text{Box}}\right|^2$$

$$= \frac{T}{2\pi\hbar}\left[\frac{(2\pi\hbar)^3}{V}\right]^{N_\alpha + N_\beta - 1} \delta_T(E_\alpha - E_\beta)\delta_V^3(\mathbf{P}_\alpha - \mathbf{P}_\beta)\left|M_{\beta\alpha}\right|^2.$$

The transition *rate* is the transition probability divided by the time $T$ during which the interaction is acting, or

$$\Gamma(\alpha \to \beta) = \frac{P(\alpha \to \beta)}{T}$$

$$= \frac{1}{2\pi\hbar}\left[\frac{(2\pi\hbar)^3}{V}\right]^{N_\alpha + N_\beta - 1} \delta_T(E_\alpha - E_\beta)\delta_V^3(\mathbf{P}_\alpha - \mathbf{P}_\beta)\left|M_{\beta\alpha}\right|^2.$$

$$\tag{8.2.7}$$

But this is still not what is generally measured. Equation (8.2.7) gives the rate of transition to a single one of the possible final states. But in a large box, these states are very close together. As we saw in Section 6.2, the number of one-particle states in a volume $d^3 p$ of momentum space is $V d^3 p/(2\pi\hbar)^3$, so the rate for transitions into a range $d\beta$ of final states is

$$d\Gamma(\alpha \to \beta) = [V/(2\pi\hbar)^3]^{N_\beta} \Gamma(\alpha \to \beta) \, d\beta$$
$$= \frac{1}{2\pi\hbar} \left[ \frac{(2\pi\hbar)^3}{V} \right]^{N_\alpha - 1} |M_{\beta\alpha}|^2 \, \delta(E_\alpha - E_\beta)\delta^3(\mathbf{P}_\alpha - \mathbf{P}_\beta) \, d\beta,$$
$$(8.2.8)$$

where $d\beta$ is here the product of the $d^3 p$ factors for each particle in the state. (We have dropped the subscripts $V$ and $T$ on the delta functions, since this formula will always be used in the limit $V \to \infty$ and $T \to \infty$, where the delta functions (8.2.2) become the ordinary delta functions.) This is our final general formula for transition rates.

The factor $(1/V)^{N_\alpha - 1}$ in Eq. (8.2.8) is just what should be expected on physical grounds. For $N_\alpha = 1$, this factor is unity, so the rate of decay of a single particle into some set $\beta$ of particles is independent of the volume in which the decay takes place

$$d\Gamma(\alpha \to \beta) = \frac{1}{2\pi\hbar} |M_{\beta\alpha}|^2 \, \delta(E_\alpha - E_\beta)\delta^3(\mathbf{P}_\alpha - \mathbf{P}_\beta) \, d\beta, \qquad (8.2.9)$$

as one would expect. For $N_\alpha = 2$ this factor is $1/V$, so the rate of producing the final state $\beta$ in the collision of two particles is proportional to the density $1/V$ of either particle at the position of the other, again as would be expected. Since this is a rate, it should actually be proportional to the rate per area $u_\alpha/V$ at which the beam of one of the particles strikes the other, where $u_\alpha$ is the relative speed of the two particles. The coefficient of $u_\alpha/V$ in the transition rate $d\Gamma(\alpha \to \beta)$ is the differential cross section

$$d\sigma(\alpha \to \beta) \equiv \frac{d\Gamma(\alpha \to \beta)}{u_\alpha/V} = \frac{(2\pi\hbar)^2}{u_\alpha} |M_{\beta\alpha}|^2 \, \delta(E_\alpha - E_\beta)\delta^3(\mathbf{P}_\alpha - \mathbf{P}_\beta) \, d\beta.$$
$$(8.2.10)$$

We will mostly work in the center-of-mass frame, in which the two particles have equal and opposite momenta – say, $\mathbf{p}$ and $-\mathbf{p}$ – in which case the relative velocity is

$$u = \frac{|\mathbf{p}|c^2}{E_1} + \frac{|\mathbf{p}|c^2}{E_2} = \frac{|\mathbf{p}|}{\mu}, \qquad \mu \equiv \frac{E_1 E_2}{c^2(E_1 + E_2)}, \qquad (8.2.11)$$

with

$$E_1 = \sqrt{\mathbf{p}^2 c^2 + m_1 c^4}, \qquad E_2 = \sqrt{\mathbf{p}^2 c^2 + m_2 c^4}.$$

In the non-relativistic case, where $E \simeq mc^2$, the quantity $\mu$ is the familiar reduced mass $m_1 m_2/(m_1 + m_2)$.

There are even physically important collision processes with *three* particles in the initial state, such as the first step $e^- + p + p \rightarrow d + \nu$ in the chain of reactions that gives heat to the Sun. The rates of such reactions are naturally proportional to the product of the densities of two of the particles at the position of the third, or $1/V^2$.

It is still necessary to explain how to deal with the factor $\delta(E_\alpha - E_\beta) \times \delta^3(\mathbf{P}_\alpha - \mathbf{P}_\beta) \, d\beta$ in Eqs. (8.2.8)–(8.2.10). For two particles in the final state, this factor is just proportional to the differential element of solid angle. Let us work in the center-of-mass frame, in which the total momentum of the initial state vanishes. Then if the final state consists of two particles of momenta $\mathbf{p}_1$ and $\mathbf{p}_1$ and energies $E_1$ and $E_2$, this factor is

$$
\delta^3(\mathbf{p}_1 + \mathbf{p}_2)\delta(E_1 + E_2 - E)\, d^3p_1\, d^3p_2 = \delta(E_1 + E_2 - E)p_1^2\, dp_1\, d\Omega_1
$$

$$
= \frac{p_1^2\, d\Omega_1}{|\partial(E_1 + E_2)/\partial p_1|}
$$

$$
= \mu p_1\, d\Omega_1, \tag{8.2.12}
$$

where $\mu$ is given by Eq. (8.2.11). In the final expression, $p_1$ is the momentum fixed by energy conservation, the solution of the equation $E_1 + E_2 = E$. (In deriving this result, we use the fact that $\delta\big(f(p)\big)dp = 1/|f'(p)|$, where $f'(p)$ is evaluated at the value of $p$ where $f(p) = 0$.)

For instance, according to Eq. (8.2.9), the rate of decay of a single particle into two particles is

$$
d\Gamma = \frac{1}{2\pi\hbar}|M_{\beta\alpha}|^2\, \mu_\beta p_\beta\, d\Omega_\beta, \tag{8.2.13}
$$

and Eq. (8.2.10) gives the differential cross section for a transition to a two-particle final state in the collision of two particles in the center-of-mass frame as

$$
d\sigma(\alpha \rightarrow \beta) = \frac{(2\pi\hbar)^2}{u_\alpha}|M_{\beta\alpha}|^2\, \mu_\beta p_\beta\, d\Omega_\beta = (2\pi\hbar)^2 \left(\frac{p_\beta}{p_\alpha}\right)\mu_\alpha\mu_\beta|M_{\beta\alpha}|^2\, d\Omega_\beta.
$$

$$
\tag{8.2.14}
$$

For the purpose of comparison with the results of the previous chapter, we note that in the case of elastic scattering of a non-relativistic particle by a fixed scattering center, there is no momentum-conservation delta function in the relation (8.2.1), which here gives

$$
S_{\mathbf{k}',\mathbf{k}} = \delta(E(k') - E(k))M_{\mathbf{k}',\mathbf{k}}, \tag{8.2.15}
$$

where $\mathbf{k}$ and $\mathbf{k}'$ are the initial and final wave numbers, and we are assuming here that $\mathbf{k}' \neq \mathbf{k}$. Comparing this with Eqs. (8.1.10) and (8.1.11) gives

$$
M_{\mathbf{k}',\mathbf{k}} = -2\pi i \left(\Phi_{\mathbf{k}'}, V\Psi_{\mathbf{k}}^+\right) = -2\pi i \int d^3x \, (2\pi\hbar)^{-3/2}e^{-i\mathbf{k}'\cdot\mathbf{x}}V(\mathbf{x})\psi_{\mathbf{k}}(\mathbf{x}).
$$

$$
\tag{8.2.16}
$$

Then Eq. (7.2.6) gives the relation between the scattering amplitude (in a slightly different notation) and the $M$-matrix element:

$$f(\mathbf{k} \to \mathbf{k}') = -2\pi \hbar i \mu M_{\mathbf{k}',\mathbf{k}}. \tag{8.2.17}$$

Here $\mu_\beta = \mu_\alpha \equiv \mu$ and $p_\alpha = p_\beta$, so in this case Eq. (8.2.14) gives the differential cross section $d\sigma = |f|^2 \, d\Omega$, as found in Section 7.2.

## 8.3 The General Optical Theorem

We now take up an important consequence of the unitarity of the S-matrix. Equation (8.2.1) applies only to the case of a reaction in which the states $\alpha$ and $\beta$ are different; more generally we have

$$S_{\beta\alpha} = \delta(\alpha - \beta) + \delta(E_\alpha - E_\beta)\delta^3(\mathbf{P}_\alpha - \mathbf{P}_\beta)M_{\beta\alpha}. \tag{8.3.1}$$

The condition of unitarity reads

$$\delta(\alpha - \beta) = \int d\gamma \, S^*_{\gamma\beta} S_{\gamma\alpha}$$

$$= \delta(\alpha - \beta) + \delta(E_\alpha - E_\beta)\delta^3(\mathbf{P}_\alpha - \mathbf{P}_\beta)\left[M_{\beta\alpha} + M^*_{\alpha\beta}\right]$$

$$+ \int d\gamma \, M^*_{\gamma\beta}M_{\gamma\alpha}\delta(E_\gamma - E_\beta)\delta^3(\mathbf{P}_\gamma - \mathbf{P}_\beta)\delta(E_\gamma - E_\alpha)\delta^3(\mathbf{P}_\gamma - \mathbf{P}_\alpha)$$

and so, for $\mathbf{P}_\beta = \mathbf{P}_\alpha$ and $E_\beta = E_\alpha$,

$$0 = M_{\beta\alpha} + M^*_{\alpha\beta} + \int d\gamma \, M^*_{\gamma\beta}M_{\gamma\alpha}\delta(E_\gamma - E_\alpha)\delta^3(\mathbf{P}_\gamma - \mathbf{P}_\alpha). \tag{8.3.2}$$

This is particularly useful in the case $\alpha = \beta$. In this case the last term of Eq. (8.3.2) is proportional to the total rate for all reactions with initial state $\alpha$, which is given by Eq. (8.2.8) as

$$\Gamma_\alpha \equiv \int d\gamma \, \Gamma(\alpha \to \gamma)$$

$$= \frac{1}{2\pi\hbar}\left[\frac{(2\pi\hbar)^3}{V}\right]^{N_\alpha - 1} \int |M_{\gamma\alpha}|^2 \, \delta(E_\alpha - E_\gamma)\delta^3(\mathbf{P}_\alpha - \mathbf{P}_\gamma)\,d\gamma. \tag{8.3.3}$$

Thus in the case $\alpha = \beta$, Eq. (8.3.2) may be written

$$\mathrm{Re}\, M_{\alpha\alpha} = -\pi\hbar\left[\frac{V}{(2\pi\hbar)^3}\right]^{N_\alpha - 1}\Gamma_\alpha. \tag{8.3.4}$$

This is the most general form of the optical theorem.

In the special case of a two-particle state $\alpha$, Eq. (8.3.4) becomes

$$\mathrm{Re}\, M_{\alpha\alpha} = -\frac{\pi\hbar}{(2\pi\hbar)^3}u_\alpha\sigma_\alpha, \tag{8.3.5}$$

where $u_\alpha$ is the relative velocity, and $\sigma_\alpha = \Gamma_\alpha/(u_\alpha/V)$ is the total cross section for all possible results of the collision of the two particles. Using Eq. (8.2.17), the imaginary part of the forward scattering amplitude is then

$$\text{Im } f(\mathbf{k}_\alpha \to \mathbf{k}_\alpha) = -2\pi\hbar\mu_\alpha \text{ Re } M_{\alpha\alpha} = \frac{\mu_\alpha u_\alpha}{4\pi\hbar}\sigma_\alpha = \frac{k_\alpha}{4\pi}\sigma_\alpha, \tag{8.3.6}$$

which is the original optical theorem, derived in Section 7.3 for the special case of potential scattering.

## 8.4 The Partial Wave Expansion

By using rotational invariance together with unitarity, we can derive a representation of the S-matrix that is much like the expression of the scattering amplitude in terms of phase shifts in the previous chapter, but now in a much more general context, including inelastic reactions and particles with spin.

We must first see how to express two-particle states $\Phi_{\mathbf{p}_1,\sigma_1;\mathbf{p}_2,\sigma_2}$ with momenta $\mathbf{p}_1$ and $\mathbf{p}_2$, spins $s_1$ and $s_2$, and spin 3-components $\sigma_1$ and $\sigma_2$, in terms of states of definite total energy $E$, total momentum $\mathbf{P}$, total angular momentum $J$, total angular-momentum 3-component $M$, orbital angular momentum $\ell$, and total spin $s$. Let us define

$$\Phi_{\mathbf{P},E,J,M,\ell,s,n} \equiv \int d^3p_1 \; \frac{1}{\sqrt{\mu|\mathbf{p}_1|}}\delta(E - E_1 - E_2)$$
$$\times \sum_{\sigma_1\sigma_2\sigma m} Y_\ell^m(\hat{p}_1)C_{s_1s_2}(s\sigma;\sigma_1\sigma_2)C_{s\ell}(JM;\sigma m)\Phi_{\mathbf{p}_1,\sigma_1;\mathbf{P}-\mathbf{p}_1,\sigma_2;n}.$$
$$\tag{8.4.1}$$

Here $n$ is a compound index, labeling the particle types, including their masses $m_1$ and $m_2$ and spins $s_1$ and $s_2$; $Y_\ell^m$ is the spherical harmonic described in Section 2.2; the $C$s are the Clebsch–Gordan coefficients described in Section 4.3; and the $E_i$ are the energies

$$E_1 \equiv \sqrt{m_1^2c^4 + \mathbf{p}_1^2c^4}, \quad E_2 \equiv \sqrt{m_2^2c^4 + (\mathbf{P} - \mathbf{p}_1)^2c^4}.$$

We will concentrate here on the center-of-mass system, for which $\mathbf{P} = 0$. In this case $\mu$ is the reduced mass defined by Eq. (8.2.11). The idea of the definition (8.4.1) is that the two spins add up to a total spin $s$ with 3-component $\sigma$, and in the center-of-mass frame with $\mathbf{P} = 0$, the total spin and the orbital angular momentum add up to a total angular momentum $J$ with 3-component $M$. As we will now see, the factor $(\mu|\mathbf{p}_1|)^{-1/2}$ is inserted to give the states (8.4.1) a simple norm.

The states $\Phi_{\mathbf{p}_1,\sigma_1;\mathbf{p}_2,\sigma_2;n}$ are taken to have the conventional continuum normalization

$$\left(\Phi_{\mathbf{p}_1',\sigma_1';\mathbf{p}_2',\sigma_2';n'},\ \Phi_{\mathbf{p}_1,\sigma_1;\mathbf{p}_2,\sigma_2;n}\right) = \delta_{n'n}\delta^3(\mathbf{p}_1'-\mathbf{p}_1)\delta^3(\mathbf{p}_2'-\mathbf{p}_2)\delta_{\sigma_1'\sigma_1}\delta_{\sigma_2'\sigma_2}. \quad (8.4.2)$$

Let us check the normalization of the states (8.4.1). In the case of interest here, where one of these states is taken to have zero total momentum, the scalar product of these states is

$$\left(\Phi_{\mathbf{P}',E',J',M',\ell',s',n'},\ \Phi_{0,E,J,M,\ell,s,n}\right) = \delta_{n'n}\delta^3(\mathbf{P}')\delta(E'-E)\int \frac{d^3p_1}{\mu|\mathbf{p}_1|}$$

$$\times\ \delta(E_1+E_2-E)\sum_{\sigma_1\sigma_2 m'm\sigma'\sigma} Y_{\ell'}^{m'}(\hat{p}_1)^* Y_{\ell}^m(\hat{p}_1)$$

$$\times\ C_{s_1s_2}(s'\sigma';\sigma_1\sigma_2)C_{s'\ell'}(J'M';\sigma'm')C_{s_1s_2}(s\sigma;\sigma_1\sigma_2)C_{s\ell}(JM;\sigma m). \quad (8.4.3)$$

Using the defining property of the delta function, we have (for $\mathbf{P}=0$)

$$\int_0^\infty p_1^2\,dp_1\,\delta(E_1+E_2-E) = \frac{p_1^2}{|(\partial/\partial p_1)(E_1+E_2)|} = p_1E_1E_2/Ec^2 = \mu p_1,$$

where here $p_1$ is the solution of the energy-conservation equation $E_1+E_2 = E$, with $E_1 \equiv \sqrt{m_1^2c^4+p_1^2c^2}$ and $E_2 \equiv \sqrt{m_2^2c^4+p_1^2c^2}$. This is canceled by the factor $1/\mu p_1$ in Eq. (8.4.3), which is why we put the square root of this factor in the definition (8.4.1). Thus Eq. (8.4.3) becomes

$$\left(\Phi_{\mathbf{P}',E',J',M',\ell',s',n'},\ \Phi_{0,E,J,M,\ell,s,n}\right) = \delta_{n'n}\delta^3(\mathbf{P}')\delta(E'-E)$$

$$\times \sum_{\sigma_1\sigma_2 m'm\sigma'\sigma}\int d^2\hat{p}_1\,Y_{\ell'}^{m'}(\hat{p}_1)^* Y_{\ell}^m(\hat{p}_1)$$

$$\times\ C_{s_1s_2}(s'\sigma';\sigma_1\sigma_2)C_{s'\ell'}(J'M';\sigma'm')C_{s_1s_2}(s\sigma;\sigma_1\sigma_2)C_{s\ell}(JM;\sigma m). \quad (8.4.4)$$

Next, we use the orthonormality properties of the spherical harmonics and Clebsch–Gordan coefficients:

$$\int d^2\hat{p}_1\,Y_{\ell'}^{m'}(\hat{p}_1)^* Y_{\ell}^m(\hat{p}_1) = \delta_{\ell'\ell}\delta_{m'm},$$

$$\sum_{\sigma_1\sigma_2} C_{s_1s_2}(s'\sigma';\sigma_1\sigma_2)C_{s_1s_2}(s\sigma;\sigma_1\sigma_2) = \delta_{s's}\delta_{\sigma'\sigma}$$

and then

$$\sum_{\sigma m} C_{s\ell}(J'M';\sigma m)C_{s\ell}(JM;\sigma m) = \delta_{J'J}\delta_{M'M},$$

so Eq. (8.4.4) becomes the desired result:

$$\left( \Phi_{\mathbf{P}',E',J',M',\ell',s',n'},\, \Phi_{0,E,J,M,\ell,s,n} \right) = \delta_{n'n}\delta^3(\mathbf{P}')\delta(E'-E)\delta_{s's}\delta_{\ell'\ell}\delta_{J'J}\delta_{M'M}.$$
(8.4.5)

The advantage of using the states (8.4.1) as a basis is that for these states the Wigner–Eckart theorem and energy and momentum conservation tell us that the S-matrix can be expressed as

$$S_{\mathbf{P}',E',J',M',\ell',s',n';0,E,J,M,\ell,s,n} = \delta^3(\mathbf{P})\delta(E'-E)\delta_{J'J}\delta_{M'M}S^J_{n'\ell's';n\ell s}(E),\quad (8.4.6)$$

where $S^J$ is a matrix with discrete indices labeling its rows and columns. It follows that in this basis, the matrix $M_{\beta\alpha}$ in Eq. (8.3.1) takes the form

$$M_{0,E,J',M',\ell',s',n';0,E,J,M,\ell,s,n} = \delta_{J'J}\delta_{M'M}\left[ S^J(E)-1 \right]_{n'\ell's';n\ell s}.$$
(8.4.7)

But to calculate cross sections, we need this matrix in the original basis of states with definite momentum for each particle. To go over to the original basis, we use Eqs. (8.4.1) and (8.4.2) to calculate the scalar product

$$\left( \Phi_{\mathbf{p}_1,\sigma_1;-\mathbf{p}_1,\sigma_2,n},\, \Phi_{\mathbf{P},E,J,M,\ell,s,n'} \right) = \frac{\delta_{nn'}}{\sqrt{\mu|\mathbf{p}_1|}}\delta^3(\mathbf{P})\delta(E-E_1-E_2)$$
$$\times \sum_{\sigma m} Y_\ell^m(\hat{p}_1)C_{s_1 s_2}(s\sigma;\sigma_1\sigma_2)C_{s\ell}(JM;\sigma m).$$
(8.4.8)

Then Eq. (8.4.5) gives

$$\Phi_{\mathbf{p}_1,\sigma_1;-\mathbf{p}_1,\sigma_2;n} = \int d^3P \int dE$$
$$\times \sum_{JM\ell sn'} \left( \Phi_{\mathbf{P},E,J,M,\ell,s,n'},\, \Phi_{\mathbf{p}_1,\sigma_1;-\mathbf{p}_1,\sigma_2;n} \right)\Phi_{\mathbf{P},E,J,M,\ell,s,n'}$$
$$= \frac{1}{\sqrt{\mu|\mathbf{p}_1|}}\sum_{JM\ell mso} Y_\ell^m(\hat{p}_1)^* C_{s_1 s_2}(s\sigma;\sigma_1\sigma_2)C_{s\ell}(JM;\sigma m)$$
$$\times \Phi_{0,E_1+E_2,J,M,\ell,s,n},$$
(8.4.9)

and from Eq. (8.4.7) we have

$$M_{\mathbf{p}_1',\sigma_1',-\mathbf{p}_1',\sigma_2',n';\mathbf{p}_1,\sigma_1,-\mathbf{p}_1,\sigma_2,n} = \frac{1}{\sqrt{\mu'|\mathbf{p}_1'|}}\frac{1}{\sqrt{\mu|\mathbf{p}_1|}}$$
$$\times \sum_{JM}\sum_{\ell'm's'\sigma'} Y_{\ell'}^{m'}(\hat{p}_1')C_{s_1's_2'}(s'\sigma';\sigma_1'\sigma_2')C_{s'\ell'}(JM;\sigma'm')$$
$$\times \sum_{\ell mso} Y_\ell^m(\hat{p}_1)^* C_{s_1 s_2}(s\sigma;\sigma_1\sigma_2)C_{s\ell}(JM;\sigma m)\left[ S^J(E)-1 \right]_{\ell',s',n';\ell,s,n}.$$
(8.4.10)

We will choose a coordinate system in which the initial momentum $\mathbf{p}_1$ is in the 3-direction, and use the property of the spherical harmonic, that in this case

$$Y_\ell^m(\hat{p}_1) = \delta_{m0}\sqrt{\frac{2\ell+1}{4\pi}}, \tag{8.4.11}$$

so that Eq. (8.4.10) simplifies slightly:

$$M_{\mathbf{p}_1',\sigma_1',-\mathbf{p}_1',\sigma_2',n';\mathbf{p}_1,\sigma_1,-\mathbf{p}_1,\sigma_2,n} = \frac{1}{\sqrt{\mu'|\mathbf{p}_1'|}}\frac{1}{\sqrt{\mu|\mathbf{p}_1|}}$$

$$\times \sum_{JM}\sum_{\ell'm's'\sigma'} Y_{\ell'}^{m'}(\hat{p}_1')C_{s_1's_2'}(s'\sigma';\sigma_1'\sigma_2')C_{s'\ell'}(JM;\sigma'm')$$

$$\times \sum_{\ell s\sigma}\sqrt{\frac{2\ell+1}{4\pi}}C_{s_1s_2}(s\sigma;\sigma_1\sigma_2)C_{s\ell}(JM;\sigma 0)\Big[S^J(E)-1\Big]_{\ell',s',n';\ell,s,n}. \tag{8.4.12}$$

This gives a complicated differential cross section, but the result becomes much simpler if we integrate over the direction of the final momentum, sum over final spin 3-components, and average over initial spin 3-components. According to Eq. (8.2.14), the total cross section for the transition $n \to n'$ when spins are not observed is

$$\sigma(n \to n'; E) = \frac{(2\pi\hbar)^2\mu\mu'}{(2s_1+1)(2s_2+1)}\left(\frac{p_1'}{p_1}\right)$$

$$\times \sum_{\sigma_1\sigma_2\sigma_1'\sigma_2'}\int d\Omega_1' \left|M_{\mathbf{p}_1',\sigma_1',-\mathbf{p}_1',\sigma_2',n';\mathbf{p}_1,\sigma_1,-\mathbf{p}_1,\sigma_2,n}\right|^2. \tag{8.4.13}$$

The sum over $J$, $M$, $\ell'$, $m'$, $s'$, $\sigma'$, $\ell$, $s$, $\sigma$ in one factor of the M-matrix in Eq. (8.4.12) is accompanied with a sum over independent variables $\overline{J}$, $\overline{M}$, $\overline{\ell'}$, $\overline{m'}$, $\overline{s'}$, $\overline{\sigma'}$, $\overline{\ell}$, $\overline{s}$, $\overline{\sigma}$ in the other factor of the M-matrix, but these double sums collapse back to single sums if in turn we use the following relations in the order listed:

$$\int Y_{\ell'}^{m'}(\hat{p}_1')Y_{\overline{\ell'}}^{\overline{m'}}(\hat{p}_1')^* d\Omega_1' = \delta_{\ell'\overline{\ell'}}\delta_{m'\overline{m'}}, \tag{8.4.14}$$

$$\sum_{\sigma_1'\sigma_2'}C_{s_1's_2'}(s'\sigma';\sigma_1'\sigma_2')C_{s_1's_2'}(\overline{s'},\overline{\sigma'};\sigma_1'\sigma_2') = \delta_{s'\overline{s'}}\delta_{\sigma'\overline{\sigma'}}, \tag{8.4.15}$$

$$\sum_{\sigma'm'}C_{s'\ell'}(JM;\sigma'm')C_{s'\ell'}(\overline{J},\overline{M};\sigma'm') = \delta_{J\overline{J}}\delta_{M\overline{M}}, \tag{8.4.16}$$

$$\sum_{\sigma_1\sigma_2}C_{s_1s_2}(s\sigma;\sigma_1\sigma_2)C_{s_1s_2}(\overline{s}\,\overline{\sigma};\sigma_1\sigma_2) = \delta_{s\overline{s}}\delta_{\sigma\overline{\sigma}}, \tag{8.4.17}$$

$$\sum_{M\sigma}C_{s\ell}(JM;\sigma 0)C_{s\overline{\ell}}(JM;\sigma 0) = \frac{2J+1}{2\ell+1}\delta_{\ell\overline{\ell}}. \tag{8.4.18}$$

After we have carried out this integral and these sums, Eq. (8.4.13) becomes

$$\sigma(n \to n'; E) = \frac{\pi}{k^2(2s_1 + 1)(2s_2 + 1)} \sum_{J\ell's'\ell s} (2J + 1) \left| \left( S^J(E) - 1 \right)_{\ell's'n',\ell sn} \right|^2,$$

(8.4.19)

where $k \equiv p_1/\hbar$ is the initial wave number. For any matrix $A$, $\sum_{N'} |A_{N'N}|^2 = (A^\dagger A)_{NN}$. so the total cross section for producing two-particle final states is

$$\sum_{n'} \sigma(n \to n'; E) = \frac{\pi}{k^2(2s_1 + 1)(2s_2 + 1)}$$

$$\times \sum_{J\ell s} (2J + 1) \left[ \left( S^{J\dagger}(E) - 1 \right) \left( S^J(E) - 1 \right) \right]_{\ell sn, \ell sn}.$$

(8.4.20)

This may be compared with the total spin-averaged cross section for all reactions, given by the general optical theorem (8.3.5):

$$\sigma_{\text{total}}(n; E) = -\frac{8\pi^2\hbar^2\mu}{p_1(2s_1 + 1)(2s_2 + 1)} \sum_{\sigma_1\sigma_2} \text{Re}\, M_{\mathbf{p}_1,\sigma_1,-\mathbf{p}_1,\sigma_2,n;\mathbf{p}_1,\sigma_1,-\mathbf{p}_1,\sigma_2,n}.$$

(8.4.21)

Using Eqs. (8.4.12) and (8.4.11) again, we then have

$$\sigma_{\text{total}}(n; E) = \frac{2\pi}{k^2(2s_1 + 1)(2s_2 + 1)} \sum_{\sigma_1\sigma_2 JM\ell's'\sigma'\ell s\sigma} \sqrt{(2\ell + 1)(2\ell' + 1)}$$

$$\times C_{s_1 s_2}(s'\sigma'; \sigma_1\sigma_2) C_{s_1 s_2}(s\sigma; \sigma_1\sigma_2) C_{s'\ell'}(JM; \sigma'0) C_{s\ell}(JM; \sigma 0)$$

$$\times \text{Re}\left[ 1 - S^J(E) \right]_{\ell's'n,\ell sn}.$$

Then Eqs. (8.4.17) and (8.4.18) (with primes instead of bars) give the total spin-averaged cross section:

$$\sigma_{\text{total}}(n; E) = \frac{2\pi}{k^2(2s_1 + 1)(2s_2 + 1)} \sum_{J\ell s} (2J + 1)\, \text{Re}\left[ 1 - S^J(E) \right]_{\ell sn, \ell sn}.$$

(8.4.22)

In general, this is *not* equal to Eq. (8.4.20), because the sum in Eq. (8.4.20) runs only over two-particle final states. The difference between (8.4.22) and (8.4.20) is the cross section for reactions in which the final state contains three or more particles:

$$\sigma_{\text{production}}(n; E) \equiv \sigma_{\text{total}}(n; E) - \sum_{n'} \sigma(n \to n'; E)$$

$$= \frac{\pi}{k^2(2s_1 + 1)(2s_2 + 1)} \sum_{J\ell s} (2J + 1) \left[ 1 - S^{J\dagger}(E) S^J(E) \right]_{\ell sn, \ell sn}.$$

(8.4.23)

It is only when the energy is too small to admit the production of extra particles that the matrix $S^J(E)$ (which was defined in the space of two-particle states) is unitary.

It sometimes happens that for a given $n$ and $E$, the only final states that can be produced from a set of initial states $\Phi_{0,E,J,M,\ell,s,n}$ are the same states as the initial ones. For instance, this is the case in the collision of two spinless particles with energy too low to allow inelastic scattering, since we necessarily have $\ell = J$, and of course $s = 0$. The same is true (ignoring weak parity violation) in the elastic scattering of particles with $s_1 = 0$ and $s_2 = 1/2$, as for instance pion–nucleon scattering below the threshold for producing extra pions,[2] since the two states with $\ell = J + 1/2$ and $\ell = J - 1/2$ have opposite parity, and therefore cannot be connected by non-zero elements of $S^J$. In any such case, the assumed vanishing of the production cross section (8.4.23) and the vanishing of $S_{\ell's'n',\ell sn}$ unless $\ell' = \ell$, $s' = s$, and $n' = n$ tells us that

$$1 = \left[S^{J\dagger}(E)S^J(E)\right]_{\ell sn,\ell sn} = \left|\left[S^J(E)\right]_{\ell sn,\ell sn}\right|^2, \qquad (8.4.24)$$

and so in these cases we can write

$$\left[S^J(E)\right]_{\ell's'n',\ell sn} = \exp(2i\delta_{J\ell sn}(E))\,\delta_{\ell'\ell}\delta_{s's}\delta_{n'n}, \qquad (8.4.25)$$

where $\delta_{J\ell sn}(E)$ is a real quantity, known (by analogy with its appearance in potential scattering) as the phase shift. Using this in Eq. (8.4.19) gives the cross section (which is here the total cross section)

$$\sigma(n \to n; E) = \frac{4\pi}{k^2(2s_1 + 1)(2s_2 + 1)} \sum_{J\ell s}(2J + 1)\sin^2\left(\delta_{J\ell sn}(E)\right). \quad (8.4.26)$$

This is a generalization of the corresponding result (7.5.13) for potential scattering, but now applicable to the case of particles with spin, or with relativistic velocities, or interactions more complicated than local potentials.

More generally, Eq. (8.4.23) tells us that $\left[S^{J\dagger}(E)S^J(E)\right]_{\ell sn,\ell sn}$ is at most unity, so in general

$$\left|\left[S^J(E)\right]_{\ell sn,\ell sn}\right|^2 \le \left[S^{J\dagger}(E)S^J(E)\right]_{\ell sn,\ell sn} \le 1. \qquad (8.4.27)$$

We can if we like write

$$\left[S^J(E)\right]_{\ell sn,\ell sn} \equiv \exp(2i\delta_{J\ell sn}(E)), \qquad (8.4.28)$$

but then in general $\text{Im}\,\delta_{J\ell sn}(E) \ge 0$.

---

[2] Strictly speaking, these remarks apply only to $\pi^+\text{p}$ or $\pi^-\text{n}$ scattering, since for the other cases we have inelastic reactions such as $\pi^-\text{p} \leftrightarrow \pi^0\text{n}$. These other cases can be treated in the same way by taking advantage of the conservation of isotopic spin as well as total angular momentum. That is, we have phase shifts for states with definite $J$, $\ell$, and total isospin $T$, with $T = 1/2$ or $T = 3/2$.

We can use this formalism to get a good insight into the behavior of the various cross sections at high energy. If the energy is so large that the wavelength $h/p$ is much smaller than the characteristic radius $R$ of the colliding particles – that is, $kR \gg 1$, where $k = p/\hbar$ – then it is plausible to invoke a classical picture of the scattering.

Suppose that two hadrons, whose cross sections are disks of radius $R_1$ and $R_2$, approach each other with momenta $\mathbf{p}_1$ and $-\mathbf{p}_1$ parallel to and at distances $b_1$ and $b_2$ from some central line. Classically, the total angular momentum is $\ell\hbar = |\mathbf{p}_1|b_1 + |\mathbf{p}_1|b_2$. The hadrons will plow into each other if $R_1 + R_2 \geq b_1 + b_2$, that is, if $\ell \leq kR$, where $k = |\mathbf{p}_1|/\hbar$ and $R = R_1 + R_2$. We suppose that in this case the particles collide destructively, with no chance of a transition $\ell s n \to \ell s n$ in which nothing happens, while for $\ell \geq kR$, there is no collision. That is, we assume that

$$S^J_{\ell s n, \ell s n} = \begin{cases} 0, & \ell < kR, \\ 1, & \ell > kR. \end{cases} \tag{8.4.29}$$

Together with Eq. (8.4.22), this gives

$$\sigma_{\text{total}}(n; E) \to \frac{2\pi}{k^2(2s_1 + 1)(2s_2 + 1)} \sum_{\ell=0}^{kR} \sum_{J,s} (2J + 1). \tag{8.4.30}$$

The values of $J$ in this sum run from $|\ell - s|$ to $\ell + s$. For $kR \gg 1$ this sum is dominated by large values of $\ell$, for which $\ell \gg s$, and hence $2J + 1 \simeq 2\ell$. The number of values of $J$ for $\ell \gg s$ is $2s + 1$. Further, the sum over $s$ runs from $s = |s_1 - s_2|$ to $s = s_1 + s_2$, so the remaining sum over $s$ is

$$\sum_{s=|s_1-s_2|}^{s_1+s_2} (2s + 1) = 2\left[\frac{(s_1 + s_2)(s_1 + s_2 + 1)}{2} - \frac{(|s_1 - s_2| - 1)|s_1 - s_2|}{2}\right]$$
$$+ s_1 + s_2 - |s_1 - s_2| + 1$$
$$= (2s_1 + 1)(2s_2 + 1).$$

Finally,

$$\sum_{\ell=0}^{kR} 2\ell = kR(kR + 1) \to (kR)^2.$$

Putting this together, Eq. (8.4.30) now gives

$$\sigma_{\text{total}}(n; E) \to 2\pi R^2. \tag{8.4.31}$$

The factor 2 in Eq. (8.4.31) may be surprising. One might have expected that high-energy particles in the center-of-mass frame experience some sort of reaction if and only if they approach each other along lines separated by no more than a distance $R$, the range of their interaction. In that case, the asymptotic value of the total cross section would be $\pi R^2$, not $2\pi R^2$. The larger cross section may be attributed to quasi-elastic scattering, with two particles in the final as well as the initial state, due to the diffraction of particles that approach each other at distances a little larger than $R$. We can estimate the relative contribution of quasi-elastic scattering and particle production if we strengthen Eq. (8.4.29), assuming that

$$
S^J_{\ell's'n',\ell s n} = \begin{cases} 0, & \ell < kR, \\ \delta_{\ell'\ell}\delta_{s's}\delta_{n'n}, & \ell > kR. \end{cases} \tag{8.4.32}
$$

In this case, Eq. (8.4.23) gives

$$
\sigma_{\text{production}}(n; E) \to \frac{\pi}{k^2(2s_1 + 1)(2s_2 + 1)} \sum_{\ell=0}^{kR} \sum_{J,s} (2J + 1) = \pi R^2. \tag{8.4.33}
$$

The result that $\sigma_{\text{production}}(n; E) \to \pi R^2$ is not surprising. Particles that collide well within the effective area $\pi R^2$ cannot merely be scattered quasi-elastically, but rather, like colliding glass spheres, must produce a shower of other particles.

The cross sections for strong-interaction scattering processes such as proton–proton scattering[3] actually do become nearly constant at very high energy. There is a slow growth of the cross sections, which may be attributed to a slow increase in $R$. We can guess that $R$ is the distance at which a potential like the Yukawa potential, $V \propto e^{-r/R_Y}/r$, falls below the kinetic energy $\hbar^2 k^2/2\mu$, which for very large $k$ gives $R \simeq R_Y \ln k$. The cross sections thus are expected to grow as $\ln^2 k$, the fastest growth allowed under very general considerations.[4] Perhaps surprisingly, this all agrees pretty well with observation.[5] Measurements of proton–proton scattering at the Large Hadron Collider at 7 TeV and in cosmic rays at 57 TeV show that the cross sections really do increase as $\ln^2 k$, while the ratio $\sigma_{\text{production}}/\sigma_{\text{total}}$ approaches $0.491 \pm 0.021$, in agreement with the ratio of Eqs. (8.4.33) and (8.4.31).

## 8.5 Resonances Revisited

In Section 7.6 we considered the scattering of a spinless non-relativistic particle by a potential with a high thick barrier surrounding an inner region in which the potential is much smaller. We found in Eq. (7.6.13) that the scattering amplitude is proportional to $(E - E_R + i\Gamma/2)^{-1}$, where $\Gamma$ is exponentially small, and $E_R$ is the energy (up to terms of order $\Gamma$) of a state that would be a stable bound state if the barrier were infinitely high or thick. By considering the time-dependence

---

[3] In proton–proton collisions there is no appreciable transition to other two-particle states, so here we do not need to distinguish between the "production" cross section (8.4.33) and the total inelastic cross section.

[4] M. Froissart, *Phys. Rev.* **135**, 1053 (1961).

[5] M. M. Block and F. Halzen, *Phys. Rev. Lett.* **107**, 212002 (2011).

of a wave packet in Eq. (7.6.12), we were able to interpret the quantity $\Gamma/\hbar$ as the decay rate of this unstable state.

This argument can be turned around and generalized. There are several possible reasons for the appearance of nearly stable states. One is the existence of a barrier, like that treated in Section 7.6, through which a particle must tunnel for the state to decay. This is the case for instance in nuclear alpha decay, such as the radioactive decay of $^{235}$U or $^{238}$U, in which the alpha particle must tunnel through a Coulomb potential due to 90 protons. A nearly stable state can also occur when the decay of the state is only possible because of an interaction that is intrinsically weak. For instance, Eq. (6.5.13) shows that the rate $\Gamma/\hbar$ at which atomic states decay by emission of a single photon is typically of order $e^2\omega^3 a^2/c^3\hbar$, where $a$ is a characteristic atomic size, and $\omega \approx e^2/a\hbar$ is the photon frequency, of the same order as the frequency with which electrons classically go around their orbits. The ratio of the decay rate to the orbital frequency is then $\Gamma/\hbar\omega \approx e^6/\hbar^3 c^3$, which is very small because $e^2/\hbar c \simeq 1/137$ is small. It is also possible for a state of a large number of particles to be nearly stable because energy conservation allows the decay only if, through some fluctuation, much of the energy of the state is concentrated on a single particle. Whatever the reason for the existence of a nearly stable state, in all such cases the existence of a state with energy $E_R$ and decay rate $\Gamma/\hbar$ implies the presence in the S-matrix of a factor $(E - E_R + i\Gamma/2)^{-1}$, so that the probability of the reaction continuing for a time $t$ will be proportional to[6]

$$\left| \int_{-\infty}^{\infty} \frac{\exp(-i\,Et/\hbar)\,dE}{E - E_R + i\Gamma/2} \right|^2 = 4\pi^2 \exp(-\Gamma t/\hbar). \qquad (8.5.1)$$

The behavior of S-matrix elements near the resonance is largely determined by the unitarity of the S-matrix, whatever the mechanism that is responsible for the nearly stable state. To analyze this, it is helpful to generalize the basis of states introduced in the previous section. For a given total energy $E$ and total momentum $\mathbf{P}$, the space occupied by the allowed individual 3-momenta has finite volume, so it is always possible to expand any multiparticle state $\Phi_{\mathbf{p}_1,\mathbf{p}_2,\mathbf{p}_3,\dots}$ in a series of states $\Phi_{E,\mathbf{P},J,M,N}$, analogous to the expansion (8.4.9) in the two-particle case. Here $E$, $\mathbf{P}$, $J$, and $M$ are again the total energy, momentum, angular momentum, and angular-momentum 3-component, and $N$ is a discrete index, a generalization of the compound index $\ell, s, n$ for two-particle states. In this basis we can write general S-matrix elements in the center-of-mass frame as

---

[6] This is calculated as usual by closing the contour of integration with a large semi-circle in the lower half plane, and picking up the contribution of the pole at $E = E_R - i\Gamma/2$. Of course, the actual integrand involves other factors, including the amplitude of the wave packet, and these may also have poles in the lower half plane, but for sufficiently narrow resonances, these poles will all be at a distance below the real axis greater than $\Gamma/2$, and therefore will not contribute at very late times.

$$S_{E'\,\mathbf{P}'\,J'\,M'\,N',\,E\,0\,J\,M\,N} = \delta(E' - E)\delta^3(\mathbf{P}')\delta_{J'J}\delta_{M'M}S^J_{N'N}(E). \qquad (8.5.2)$$

(The fact that the matrix element depends on $M$ only through the factor $\delta_{M'M}$ follows from the results of Section 4.2.) If these states are normalized so that

$$\left(\Phi_{E',\mathbf{P}',J'\,M'\,N'},\ \Phi_{E,\mathbf{P},J\,M\,N}\right) = \delta(E' - E)\delta^3(\mathbf{P}' - \mathbf{P})\delta_{J'J}\delta_{M'M}\delta_{N'N}, \qquad (8.5.3)$$

then unitarity tells us that the matrix $\mathcal{S}^J(E)$ must be unitary

$$\mathcal{S}^{J\dagger}(E)\mathcal{S}^J(E) = 1, \qquad (8.5.4)$$

where 1 is of course here the matrix with $1_{N'N} = \delta_{N'N}$.

Now, suppose that near the resonance the $\mathcal{S}^J$ matrix takes the form

$$\mathcal{S}^J(E) \simeq \mathcal{S}^{(0)} + \frac{\mathcal{R}}{E - E_{\mathrm{R}} + i\Gamma/2}, \qquad (8.5.5)$$

where $\mathcal{S}^{(0)}$ and $\mathcal{R}$ are constant matrices. We don't keep the label $J$ on $\mathcal{S}^{(0)}$ and $\mathcal{R}$, because Eq. (8.5.5) is supposed to hold only for one value of $J$, the total angular momentum of the resonant state. (The term $\mathcal{S}^{(0)}$ is analogous to $\exp(2i\bar{\delta})$, where $\bar{\delta}$ is the slowly varying non-resonant phase shift in Eq. (7.6.8).)

The matrix $\mathcal{S}^{J\dagger}(E)\mathcal{S}^J(E) - 1$ is a sum of terms proportional to $(E - E_{\mathrm{R}})/[(E - E_{\mathrm{R}})^2 + \Gamma^2/4]$, to $1/[(E - E_{\mathrm{R}})^2 + \Gamma^2/4]$, and to a constant. Since these three functions of $E$ are independent, the unitarity relation (8.5.4) requires the coefficients of each term to vanish. The constant term gives

$$\mathcal{S}^{(0)\dagger}\mathcal{S}^{(0)} = 1 \ ; \qquad (8.5.6)$$

the terms proportional to $(E - E_{\mathrm{R}})/[(E - E_{\mathrm{R}})^2 + \Gamma^2/4]$ give

$$\mathcal{S}^{(0)\dagger}\mathcal{R} + \mathcal{R}^\dagger\mathcal{S}^{(0)} = 0 \ ; \qquad (8.5.7)$$

and the terms proportional to $1/[(E - E_{\mathrm{R}})^2 + \Gamma^2/4]$ give

$$-\frac{i\Gamma}{2}\mathcal{S}^{(0)\dagger}\mathcal{R} + \frac{i\Gamma}{2}\mathcal{R}^\dagger\mathcal{S}^{(0)} + \mathcal{R}^\dagger\mathcal{R} = 0. \qquad (8.5.8)$$

These conditions can be made more perspicuous by introducing another constant matrix $\mathcal{A}$, such that

$$\mathcal{R} = -i\Gamma\mathcal{A}\mathcal{S}^{(0)}, \qquad (8.5.9)$$

which we know is possible because Eq. (8.5.6) shows that $\mathcal{S}^{(0)}$ has an inverse. Then Eqs. (8.5.7) and (8.5.8) tell us that

$$\mathcal{A}^\dagger = \mathcal{A}, \qquad \mathcal{A}^2 = \mathcal{A}. \qquad (8.5.10)$$

Because $\mathcal{A}$ is Hermitian, it can be diagonalized – that is, it can be expressed as $u\mathcal{D}u^\dagger$, where $u$ is a unitary matrix and $\mathcal{D}$ is a diagonal matrix. Further, because

$\mathcal{A}^2 = \mathcal{A}$, the elements of $\mathcal{D}$ on the diagonal are all either zero or one. That is, we can write

$$\mathcal{A}_{N'N} = \sum_r u_{N'r} u_{Nr}^*, \qquad (8.5.11)$$

the sum here running over all the eigenvalues of $\mathcal{A}$ that are one rather than zero. Because $u$ is a unitary matrix, its elements $u_{Nr}$ satisfy a normalization condition

$$\sum_N u_{Nr}^* u_{Nr'} = \left[ u^\dagger u \right]_{rr'} = \delta_{rr'}. \qquad (8.5.12)$$

Equations (8.5.5), (8.5.9), and (8.5.11) then give the matrix $S(E)$ near a resonance as

$$\mathcal{S}^J(E)_{N'N} \simeq \sum_{N''} \left[ \delta_{N'N''} - \frac{i\Gamma}{E - E_R + i\Gamma/2} \sum_r u_{N'r} u_{N''r}^* \right] S_{N''N}^{(0)}. \qquad (8.5.13)$$

So far, this has been quite general. To go further, we will now make the simplifying assumption that the scattering near the resonance is entirely dominated by the resonance, so that $\mathcal{S}^{(0)} \simeq 1$, and Eq. (8.5.13) therefore gives

$$\mathcal{S}^J(E)_{N'N} \simeq \delta_{N'N} - \frac{i\Gamma}{E - E_R + i\Gamma/2} \sum_r u_{N'r} u_{Nr}^*. \qquad (8.5.14)$$

We will further assume that the only degeneracy of the resonant state is that associated with the $2J + 1$ values of the 3-component $M$ of the total angular momentum. The index $r$ therefore takes only one value, and can henceforth be dropped. Then Eq. (8.5.14) becomes

$$\mathcal{S}^J(E)_{N'N} \simeq \delta_{N'N} - \frac{i\Gamma}{E - E_R + i\Gamma/2} u_{N'} u_N^*, \qquad (8.5.15)$$

and the normalization condition (8.5.12) is here

$$\sum_N |u_N|^2 = 1. \qquad (8.5.16)$$

Equation (8.5.15) shows that the probability of the resonant state decaying into channel $N$ is proportional to $|u_N|^2$, while Eq. (8.5.16) then tells us that the constant of proportionality is unity – that is, $|u_N|^2$ *is the probability of this decay*, known as the *branching ratio*.

In particular, for basis states containing just two particles, we can take $N$ to be the compound index $\ell$, $s$, $n$, where $\ell$ is the orbital angular momentum, $s$ is the total spin, and $n$ labels the species of the two particles, including their masses and spins. In the notation of Section 8.4, Eq. (8.5.14) gives for two-particle states

$$\mathcal{S}^J(E)_{\ell's'n',\ell s n} \simeq \delta_{\ell'\ell} \delta_{s's} \delta_{n'n} - \frac{i\Gamma}{E - E_R + i\Gamma/2} u_{\ell's'n'} u_{\ell s n}^*, \qquad (8.5.17)$$

and Eq. (8.5.16) gives

$$\sum_{\ell s n} |u_{\ell s n}|^2 + \sum_{\geq 3 \text{ particles}} |u_N|^2 = 1. \tag{8.5.18}$$

Then Eq. (8.4.19) gives the cross section for the transition $n \to n'$ (summed over final spins, and averaged over initial spins) at energies near the resonance,

$$\sigma(n \to n'; E) = \frac{\pi(2J+1)}{k^2(2s_1+1)(2s_2+1)} \frac{\Gamma_n \Gamma_{n'}}{(E - E_R)^2 + \Gamma^2/4}, \tag{8.5.19}$$

where $\Gamma_n$ is the *partial width*

$$\Gamma_n \equiv \Gamma \sum_{\ell s} |u_{\ell s n}|^2. \tag{8.5.20}$$

This is a generalization of the Breit–Wigner formula (7.6.10) derived earlier for the special case of potential scattering. Also, Eq. (8.4.22) gives the total cross section (averaged over initial spins) for *all* reactions with an initial state $n$:

$$\sigma_{\text{total}}(n; E) = \frac{\pi(2J+1)}{k^2(2s_1+1)(2s_2+1)} \frac{\Gamma_n \Gamma}{(E - E_R)^2 + \Gamma^2/4}. \tag{8.5.21}$$

Note that the ratio of the specific cross section (8.5.19) and the total cross section (8.5.21) is simply

$$\frac{\sigma(n \to n'; E)}{\sigma_{\text{total}}(n; E)} = \frac{\Gamma_{n'}}{\Gamma} = \sum_{\ell s} |u_{\ell s n'}|^2. \tag{8.5.22}$$

Whatever the final state, the probability of forming the resonant state in a collision process is the same, so Eq. (8.5.22) gives the branching ratio, the probability that the resonant state decays into the specific two-body final state $n'$. According to Eq. (8.5.18), the sum of these branching ratios is unity if the resonant state decays only into two-particle states; otherwise the sum is less than unity. Finally, since $\Gamma/\hbar$ is the total decay rate of the resonance, it follows that $\Gamma_{n'}/\hbar$ is the rate at which the resonant state decays into the specific final state $n'$.

## 8.6 Old-Fashioned Perturbation Theory

The Lippmann–Schwinger equation (8.1.6) allows an easy formal solution by iteration:

$$\Psi_\alpha^\pm = \Phi_\alpha + (E_\alpha - H_0 \pm i\epsilon)^{-1} V \Phi_\alpha$$
$$+ (E_\alpha - H_0 \pm i\epsilon)^{-1} V (E_\alpha - H_0 \pm i\epsilon)^{-1} V \Phi_\alpha + \cdots. \tag{8.6.1}$$

This in turn yields a series for the S-matrix (8.1.10) in powers of the interaction, which we shall write as

$$S_{\beta\alpha} = \delta(\alpha - \beta) - 2\pi i \delta(E_\beta - E_\alpha)\left(\Phi_\beta, \left[V + VG(E_\alpha + i\epsilon)\right]\Phi_\alpha\right), \quad (8.6.2)$$

where, for an arbitrary complex $W$,

$$G(W) = K(W) + K^2(W) + \cdots, \quad (8.6.3)$$

and

$$K(W) \equiv (W - H_0)^{-1}V. \quad (8.6.4)$$

This is called "old-fashioned perturbation theory" because it has been super-seded for most (but not all) purposes by the time-dependent perturbation theory described in the next section. The first term in square brackets in Eq. (8.6.2) provides the Born approximation discussed in Section 7.4.

A question naturally arises about the convergence of expansions such as (8.6.3). This is easy to answer if $K$ is a number; the series converges if and only if $|K| < 1$. It is also easy to answer if $K$ is a finite matrix; the series converges if and only if every eigenvalue of $K$ has an absolute value less than one. More generally, the branch of mathematics known as functional analysis tells us that operators with a property known as *complete continuity* can be approximated with arbitrary precision by finite matrices. In consequence, if $K$ is completely continuous, then the geometric series $K + K^2 + K^2 + \cdots$ will converge if all the eigenvalues of $K$ are less than one in absolute value.[7] Complete continuity has a rather abstract definition,[8] which would not be of use to us here. The important point for us is that an operator $K$ is completely continuous if (though not only if ) it has a finite value for the quantity

$$\tau_K \equiv \text{Tr}\left[K^\dagger K\right], \quad (8.6.5)$$

with the trace understood to mean the sum over all discrete indices and the integral over all continuous indices of the diagonal elements of the operator. Also, the eigenvalues $\lambda$ of $K$ all satisfy

$$|\lambda|^2 \leq \tau_K. \quad (8.6.6)$$

Hence the power series (8.6.3) converges if (but not only if) $\tau_K < 1$.

---

[7] These matters and their application to scattering theory are discussed by me in some detail, with ref-erences to the original literature, in *Lectures on Particles and Field Theory – 1964 Brandeis Summer Institute in Theoretical Physics* (Prentice-Hall, Englewood Cliffs, NJ, 1965), pp. 289–403.

[8] An operator $A$ is said to be completely continuous if for any infinite set of vectors $\Phi_\nu$, which is bounded in the sense that all norms $\left(\Phi_\nu, \Phi_\nu\right)$ are less than some number $M$, there exists a subsequence $\Phi_n$ for which $A\Phi_n$ is convergent, in the sense that for some vector $\Omega$, the norm of $A\Phi_n - \Omega$ approaches zero for $n \to \infty$.

Clearly, to have any chance of writing Eq. (8.6.3) as a series in powers of a kernel $K$ with a finite value for $\tau_K$, we must deal with the momentum-conservation delta functions in matrix elements of the operator $(W - H_0)^{-1}V$. This is no problem for theories with one particle in a fixed potential, where $K$ involves no momentum-conservation delta function. It is also no problem for two particles with no external potential. In the latter case we can define operators $\mathcal{V}$ and $\mathcal{K}$, by factoring out a delta function

$$\left(\Phi_\beta, V\Phi_\alpha\right) \equiv \delta^3(\mathbf{P}_\beta - \mathbf{P}_\alpha)\mathcal{V}_{\beta\alpha},$$

$$\left(\Phi_\beta, (W - H_0)^{-1} V \Phi_\alpha\right) \equiv \delta^3(\mathbf{P}_\beta - \mathbf{P}_\alpha)\mathcal{K}_{\beta\alpha}(W),$$

and rewrite Eqs. (8.6.2) and (8.6.3) as

$$S_{\beta\alpha} = \delta(\alpha - \beta) - 2\pi i\delta(E_\beta - E_\alpha)\delta^3(\mathbf{P}_\beta - \mathbf{P}_\alpha)\left[\mathcal{V} + \mathcal{V}\mathcal{G}(E_\alpha + i\epsilon)\right]_{\beta\alpha},$$

where, for an arbitrary complex $W$,

$$\mathcal{G}(W) = (W - H_0)^{-1}\mathcal{V} + (W - H_0)^{-1}\mathcal{V}(W - H_0)^{-1}\mathcal{V} + \cdots.$$

Since the single momentum-conservation delta function for two-body scattering has been factored out, the matrix elements of $\mathcal{K} \equiv (W - H_0)^{-1}\mathcal{V}$ will be smooth functions, at least in the sense of containing no more delta functions. It is then at least possible to have $\tau_{\mathcal{K}}$ finite, depending on the energy and the details of the potential.

It is more difficult to use the methods for problems involving three or more particles. Three-particle matrix elements of the operator $(W - H_0)^{-1}V$ contain terms in which any one of the three particles' momenta is conserved, as well as the sum of all three momenta. These terms represent the unavoidable possibility that two particles interact, leaving the third free. These delta functions can't simply be factored out of the problem, as they are not the same delta functions in each term. There are complicated ways to deal with this in any theory with a fixed number of particles, involving a rewriting of the series (8.6.3).[9] But these methods fail for theories, such as quantum field theories, with unlimited numbers of particles.

For these reasons, we will limit ourselves here to the case of a single particle in a fixed potential or the equivalent problem of two particles in the absence of an external potential. In the two-particle case we can eliminate the problem of the momentum-conservation delta functions by factoring out the delta function, as described above. For the sake of simplicity, from now on we concentrate on

---

[9] This was first worked out for the case of three particles by L. D. Faddeev, *Sov. Phys. JETP* **12**, 1014 (1961); *Sov. Phys. Doklady* **6**, 384 (1963); *Sov. Phys. Doklady* **7**, 600 (1963); and independently for arbitrary numbers of particles by S. Weinberg, *Phys. Rev.* B **133**, 232 (1964).

the case of scattering of a single non-relativistic particle by a local (though not necessarily central) potential $V(\mathbf{x})$.

Whether with one particle or two, there still is a problem with the singularity of the operator $(W - H_0)^{-1}$ when $W$ approaches real values in the spectrum of $H_0$. As noted by many authors, this can usually be dealt with by expanding in powers of a symmetrized operator, defined in the one-particle case by

$$\overline{K}(W) \equiv V^{1/2}(W - H_0)^{-1}V^{1/2}. \tag{8.6.7}$$

The S-matrix (8.6.2) can be written as

$$S_{\beta\alpha} = \delta(\alpha - \beta) - 2\pi i\, \delta(E_\beta - E_\alpha)\left(\Phi_\beta, \left[V + V^{1/2}\overline{G}(E_\alpha + i\epsilon)V^{1/2}\right]\Phi_\alpha\right), \tag{8.6.8}$$

where, for an arbitrary complex $W$,

$$\overline{G}(W) = \overline{K}(W) + \overline{K}(W)^2 + \cdots. \tag{8.6.9}$$

Using a coordinate representation, we can represent the operator $(E + i\epsilon - H_0)^{-1}$ using Eq. (7.2.4)

$$\left(\Phi_{\mathbf{x}'}, (E + i\epsilon - H_0)^{-1}\Phi_{\mathbf{x}}\right) = -\frac{2\mu}{\hbar^2}\frac{e^{ik|\mathbf{x}'-\mathbf{x}|}}{4\pi|\mathbf{x}' - \mathbf{x}|}, \tag{8.6.10}$$

where $\mu$ is the particle mass (in the two-particle case it would be the reduced mass), and $k$ is the positive root of $E = k^2/2\mu$. The trace (8.6.5) for the operator $\overline{K}$ is then

$$\tau_{\overline{K}} \equiv \mathrm{Tr}\left[\overline{K}(E + i\epsilon)^\dagger \overline{K}(E + i\epsilon)\right]$$

$$= \left(\frac{2\mu}{\hbar^2}\right)^2 \int d^3x\, d^3x'\, V(\mathbf{x}')V(\mathbf{x})\frac{1}{16\pi^2|\mathbf{x}' - \mathbf{x}|^2}. \tag{8.6.11}$$

This is convergent if $V(\mathbf{x})$ diverges no worse than $|\mathbf{x}|^{-2+\delta}$ for $|\mathbf{x}| \to 0$, and vanishes at least as fast as $|\mathbf{x}|^{-3-\delta}$ for $|\mathbf{x}| \to \infty$ (with $\delta > 0$ in both cases). For instance, for the shielded Coulomb potential $V(r) = -g\exp(-r/R)/r$, we have $\tau_{\overline{K}} = 2\mu^2 g^2 R^2/\hbar^4$. Thus the perturbation series for the S-matrix converges for $|g| < \hbar^2/\mu R\sqrt{2}$. But for the unshielded Coulomb potential $R$ is infinite, and this test for convergence does not work.

Similar techniques can be used to set limits on the binding energies of possible bound states. For this purpose, we need an expansion of the operator $[W - H]^{-1}$, known as the *resolvent*:

$$[W - H]^{-1} = [W - H_0]^{-1} + \left[K(W) + K^2(W) + \cdots\right][W - H_0]^{-1}, \tag{8.6.12}$$

where $K(W)$ is the unsymmetrized kernel (8.6.4). (We could of course write this in terms of the symmetrized kernel $V^{1/2}[W - H_0]^{-1}V^{1/2}$, but this is unnecessary here because $[W - H_0]^{-1}$ is non-singular for $W = -B < 0$.) The resolvent must become singular when $W$ equals the energy $-B$ of a bound state below the spectrum of $H_0$, because for such an energy $W - H$ annihilates the state vector of the bound state. But at an energy outside the spectrum of $H_0$, each term in Eq. (8.6.12) is finite, so the singularity in the resolvent can only come from a divergence of the series in powers of $K(-B)$. Hence a bound state with energy $-B$ is impossible if $\tau_K(-B) < 1$, where $\tau_K(-B) \equiv \text{Tr}\left[K(-B)^\dagger K(-B)\right]$. Using Eq. (8.6.10) with $k = +i\sqrt{2B\mu}/\hbar$, for a local potential we have

$$
\tau_K(-B) = \left(\frac{2\mu}{\hbar^2}\right)^2 \int d^3x \, d^3x' \, V^2(\mathbf{x}) \frac{\exp\left(-2\sqrt{2B\mu/\hbar^2}|\mathbf{x}' - \mathbf{x}|\right)}{16\pi^2|\mathbf{x}' - \mathbf{x}|^2}
$$
$$
= \left(\frac{2\mu}{\hbar^2}\right)^{3/2} \frac{1}{8\pi\sqrt{B}} \int d^3x \, V^2(\mathbf{x}). \tag{8.6.13}
$$

Hence it is only possible to have bound states with binding energies subject to the bound

$$
B \leq \left(\frac{2\mu}{\hbar^2}\right)^3 \left[\frac{1}{8\pi} \int d^3x \, V^2(\mathbf{x})\right]^2. \tag{8.6.14}
$$

It sometimes happens that $V$ itself is not small enough for transition amplitudes to be calculated using perturbation theory, but it is possible to write

$$
V = V_s + V_w, \tag{8.6.15}
$$

where $V_s$ is strong, but cannot by itself cause a given transition $\alpha \to \beta$, while $V_w$ can cause this transition, and is sufficiently weak that we can calculate the amplitude for $\alpha \to \beta$ to first order in $V_w$, though we need to include all orders in $V_s$. For instance, in nuclear beta decay, the strong nuclear interaction and even the electromagnetic interaction cannot be neglected, but they cannot themselves change neutrons into protons or vice versa, or create electrons and neutrinos. The beta decay amplitude thus would vanish if the weak nuclear interaction were absent, and since this interaction is indeed weak, the amplitude can be calculated to first order in the weak interactions. In other contexts $V_w$ might be the electromagnetic interaction, as in nuclear gamma decay. In elementary particle decay processes such as the decay of a K meson into two or three pions, $V_s$ is the strong force holding quarks and antiquarks together inside the meson, while $V_w$ is the weak force that allows quarks of one type to change into quarks of another type.

To calculate transition amplitudes to first order in $V_w$, let us first define states that would be "in" and "out" states if $V_w$ were zero:

$$
\Psi_{s\alpha}^{\pm} = \Phi_\alpha + (E_\alpha - H_0 \pm i\epsilon)^{-1} V_s \Psi_{s\alpha}^{\pm}. \tag{8.6.16}
$$

Then we can write Eq. (8.1.11) as

$$T_{\beta\alpha} = \left(\Phi_\beta, V\Psi_\alpha^+\right)$$
$$= \left([\Psi_{s\beta}^- - (E_\beta - H_0 - i\epsilon)^{-1}V_s\Psi_{s\beta}^-], V\Psi_\alpha^+\right)$$
$$= \left(\Psi_{s\beta}^-, V\Psi_\alpha^+\right) - \left(\Psi_{s\beta}^-, V_s(E_\alpha - H_0 + i\epsilon)^{-1}V\Psi_\alpha^+\right),$$

and therefore, using the Lippmann–Schwinger equation again,

$$T_{\beta\alpha} = \left(\Psi_{s\beta}^-, V\Psi_\alpha^+\right) - \left(\Psi_{s\beta}^-, V_s\Psi_\alpha^+\right) + \left(\Psi_{s\beta}^-, V_s\Phi_\alpha\right)$$
$$= \left(\Psi_{s\beta}^-, V_w\Psi_\alpha^+\right) + \left(\Psi_{s\beta}^-, V_s\Phi_\alpha\right). \tag{8.6.17}$$

This is most useful in the case mentioned earlier, where the process $\alpha \to \beta$ cannot take place in the absence of the weak interaction. In this case the last term in Eq. (8.6.17) vanishes, and we have

$$T_{\beta\alpha} = \left(\Psi_{s\beta}^-, V_w\Psi_\alpha^+\right). \tag{8.6.18}$$

So far, this is exact. Since Eq. (8.6.18) contains an explicit factor $V_w$, to first order in $V_w$ we can ignore the difference between $\Psi_\alpha^+$ and $\Psi_{s\alpha}^+$, and write Eq. (8.6.18) as

$$T_{\beta\alpha} \simeq \left(\Psi_{s\beta}^-, V_w\Psi_{s\alpha}^+\right). \tag{8.6.19}$$

This is known as *the distorted-wave Born approximation*.

For example, in nuclear beta decay, we can take $V_s$ to be the sum of the strong nuclear interaction and the electromagnetic interaction, while $V_w$ is the weak nuclear interaction. In this case $\Psi_{s\alpha}^+$ in Eq. (8.6.19) is just the state vector of the original nucleus, while $\Psi_{s\beta}^-$ is the state vector of the final nucleus and the emitted electron (or positron) and antineutrino (or neutrino). The neutrino or antineutrino does not have strong nuclear or electromagnetic interactions with the final nucleus, while the electron or positron has electromagnetic but no strong nuclear interactions with the final nucleus. In a coordinate representation, the state vector $\Psi_{s\beta}^-$ is proportional to the product of a plane wave function for the neutrino or antineutrino, which does not concern us, and the two-particle wave function of the electron or positron and final nucleus. The weak nuclear interaction acts only when the electron or positron and the nucleus are in contact, so (at least for non-relativistic electrons or positrons) the matrix element is proportional to the value of the Coulomb wave function at zero separation, given by Eqs. (7.9.11) and (7.9.10) as the quantity (7.9.15). The rate for beta decay therefore has a dependence on the quantity $\xi = \pm Z'e^2 m_e/\hbar^2 k_e$ (where $Z'e$ is the charge of the final nucleus, and the sign is plus or minus for positrons and electrons, respectively) proportional to[10]

$$\mathcal{F}(\xi) = |\Gamma(1 + i\xi)|^2 \exp(-\pi\xi) = \frac{2\pi\xi}{\exp(2\pi\xi) - 1}. \tag{8.6.20}$$

The same factor appears in the low-energy cross sections for $\nu + N \rightarrow e^- + N'$ and $\bar{\nu} + N \rightarrow e^+ + N'$.

For $|\xi| \ll 1$ the factor $\mathcal{F}$ is unity, indicating that there is neither enhancement nor suppression of the process. For $\xi \ll -1$, this factor is $2\pi|\xi|$, indicating a mild enhancement. For $\xi \gg 1$, $F \simeq 2\pi\xi \exp(-2\pi\xi)$, indicating a severe suppression. This suppression is nothing but the effect of the positive potential barrier discussed in Section 7.6.

## 8.7 Time-Dependent Perturbation Theory

The energy denominators in the old-fashioned perturbation theory discussed in the previous section give this formalism several disadvantages. Because these denominators depend on energy but not momentum, they obscure the Lorentz invariance of relativistic theories, and because the denominators depend on the energies of all the particles involved in a reaction, they obscure the independence of the rates for processes happening far from each other. Both disadvantages are avoided by describing the same perturbation series in a different formalism, known as time-dependent perturbation theory.

To derive a formula for the S-matrix in time-dependent perturbation theory, let us return to the defining condition (8.1.5) of "in" and "out" states. Using the energy eigenvalue conditions (8.1.2) and (8.1.3), we can write Eq. (8.1.5) as

$$\exp(-iHt/\hbar) \int d\alpha \, g(\alpha)\Psi_\alpha^\pm \overset{t \rightarrow \mp\infty}{\rightarrow} \exp(-iH_0 t/\hbar) \int d\alpha \, g(\alpha)\Phi_\alpha. \tag{8.7.1}$$

This can be abbreviated as

$$\Psi_\alpha^\pm = \Omega(\mp\infty)\Phi_\alpha, \tag{8.7.2}$$

where

$$\Omega(t) \equiv e^{iHt/\hbar}e^{-iH_0 t/\hbar}. \tag{8.7.3}$$

---

[10] In evaluating this, we use the reality property $\Gamma(z)^* = \Gamma(z^*)$ and the familiar recursion relation $\Gamma(1 + z) = z\Gamma(z)$ to write

$$|\Gamma(1 + i\xi)|^2 = \Gamma(1 + i\xi)\Gamma(1 - i\xi) = i\xi\Gamma(i\xi)\Gamma(1 - i\xi),$$

and then evaluate this product using the classic formula

$$\Gamma(z)\Gamma(1 - z) = \pi/\sin\pi z.$$

The limits $t \to \mp\infty$ are really only well defined when Eq. (8.7.2) is multiplied with a smooth wave-packet amplitude $g(\alpha)$ and integrated over $\alpha$, but we can understand the limit intuitively, by noting that $H$ effectively becomes equal to $H_0$ at very early or very late times, when the colliding particles are far from each other.

Using Eq. (8.1.14), we see that the S-matrix is

$$S_{\beta\alpha} = \left(\Psi_\beta^-, \Psi_\alpha^+\right) = \left(\Phi_\beta, \Omega^\dagger(+\infty)\Omega(-\infty)\Phi_\alpha\right) = \left(\Phi_\beta, U(+\infty, -\infty)\Phi_\alpha\right),$$

$$(8.7.4)$$

where

$$U(t, t') \equiv \Omega^\dagger(t)\Omega(t') = e^{iH_0 t/\hbar}e^{-iH(t-t')/\hbar}e^{-iH_0 t'/\hbar}. \qquad (8.7.5)$$

To calculate $U$, we can write Eq. (8.7.5) as a differential equation,

$$\frac{d}{dt}U(t, t') = -\frac{i}{\hbar}e^{iH_0 t/\hbar}[H - H_0]e^{-iH(t-t')/\hbar}e^{-iH_0 t'/\hbar} = -\frac{i}{\hbar}V_{\mathrm{I}}(t)U(t, t'),$$

$$(8.7.6)$$

together with the initial condition

$$U(t', t') = 1, \qquad (8.7.7)$$

where

$$V_I(t) \equiv e^{iH_0 t/\hbar}Ve^{-iH_0 t/\hbar}, \qquad (8.7.8)$$

and of course $V \equiv H - H_0$. The subscript I stands for "interaction picture," a term used to distinguish operators whose time-dependence is governed by the free-particle Hamiltonian $H_0$, in contrast to operators in the Heisenberg picture, whose time-dependence is governed by the total Hamiltonian $H$, or operators in the Schrödinger picture, which do not depend on time.

The differential equation (8.7.6) and initial condition (8.7.7) are equivalent to an integral equation

$$U(t, t') = 1 - \frac{i}{\hbar}\int_{t'}^{t} d\tau \, V_{\mathrm{I}}(\tau)U(\tau, t'), \qquad (8.7.9)$$

which can be solved (at least formally) by iteration:

$$U(t, t') = 1 - \frac{i}{\hbar}\int_{t'}^{t} d\tau \, V_I(\tau)$$

$$+ \left(-\frac{i}{\hbar}\right)^2 \int_{t'}^{t} d\tau_1 \int_{t'}^{\tau_1} d\tau_2 \, V_I(\tau_1)V_I(\tau_2) + \cdots. \qquad (8.7.10)$$

We can rewrite this by introducing a *time-ordered product*,

$$T\{V_{\mathrm{I}}(\tau)\} \equiv V_{\mathrm{I}}(\tau),$$

$$T\{V_{\mathrm{I}}(\tau_1)V_{\mathrm{I}}(\tau_2)\} \equiv \begin{cases} V_{\mathrm{I}}(\tau_1)V_{\mathrm{I}}(\tau_2), & \tau_1 > \tau_2, \\ V_{\mathrm{I}}(\tau_2)V_{\mathrm{I}}(\tau_1), & \tau_2 > \tau_1, \end{cases}$$

and in general

$$T\{V_{\mathrm{I}}(\tau_1)\dots V_{\mathrm{I}}(\tau_n)\}$$
$$\equiv \sum_P \theta(\tau_{P1}-\tau_{P2})\theta(\tau_{P2}-\tau_{P3})\dots\theta(\tau_{P[n-1]}-\tau_{Pn})V_{\mathrm{I}}(\tau_{P1})\dots V_{\mathrm{I}}(\tau_{Pn}),$$

$$(8.7.11)$$

where the sum runs over all $n!$ permutations of $1, 2, \dots, n$ into $P1, P2, \dots, Pn$, and $\theta$ is the step function

$$\theta(x) \equiv \begin{cases} 1, & x > 0, \\ 0, & x < 0. \end{cases} \qquad (8.7.12)$$

The product of step functions in Eq. (8.7.11) picks out the one term in the sum for which the $V_{\mathrm{I}}$ are time-ordered, with the $V_{\mathrm{I}}$ with the latest argument first on the left, the next-to-latest second from the left, and so on. When we integrate Eq. (8.7.11) over all $\tau_i$ from $t'$ to $t$, each of the $n!$ terms gives just the integral appearing in the $n$th-order term in Eq. (8.7.10), so

$$U(t, t') = \sum_{n=0}^{\infty} \frac{1}{n!} \left[ -\frac{i}{\hbar} \right]^n \int_{t'}^t d\tau_1 \dots \int_{t'}^t d\tau_n \, T\{V_I(\tau_1)\dots V_I(\tau_n)\}, \quad (8.7.13)$$

the $n = 0$ term being understood as the unit operator. Equation (8.7.4) then gives the Dyson perturbation series[11] for the S-matrix:

$$S_{\beta\alpha} = \sum_{n=0}^{\infty} \frac{1}{n!} \left[ -\frac{i}{\hbar} \right]^n \int_{-\infty}^{\infty} d\tau_1 \dots \int_{-\infty}^{\infty} d\tau_n$$
$$\times \left( \Phi_\beta, T\{V_I(\tau_1)\dots V_I(\tau_n)\} \Phi_\alpha \right). \qquad (8.7.14)$$

It is straightforward to calculate each term in this series – we only need to calculate the matrix element between free-particle states of the integral of a product of interaction-picture operators whose time-dependence, governed by $H_0$, is essentially trivial. Of course, when we limit the sum over $n$ to a finite number of terms, the result may or may not be a good approximation.

This formula makes Lorentz invariance transparent in at least some theories. For instance, if $V_{\mathrm{I}}(t) = \int d^3x \, \mathcal{H}(\mathbf{x}, t)$, where $\mathcal{H}$ is a scalar function of field variables, then Eq. (8.7.14) gives

$$S_{\beta\alpha} = \sum_{n=0}^{\infty} \frac{1}{n!} \left[ -\frac{i}{\hbar} \right]^n \int d^4x_1 \dots \int d^4x_n$$
$$\times \left( \Phi_\beta, T\{\mathcal{H}(x_1)\dots\mathcal{H}(x_n)\}\Phi_\alpha \right), \qquad (8.7.15)$$

---

[11] F. J. Dyson, *Phys. Rev.* **75**, 486, 1736 (1949).

the integrals now running over all space and time. This at least appears Lorentz-invariant, though we still have to worry about the time-ordering in Eq. (8.7.15). The statement that a spacetime point $\{\mathbf{x}', t'\}$ is at a later time than a point $\{\mathbf{x}, t\}$ is Lorentz-invariant if $\{\mathbf{x}', t'\}$ is inside the light cone centered at $\{\mathbf{x}, t\}$ – that is, if $(\mathbf{x}'-\mathbf{x})^2 < c^2(t'-t)^2$. Thus the time-ordering in Eq. (8.7.15) is Lorentz-invariant if $\mathcal{H}(\mathbf{x}, t)$ commutes with $\mathcal{H}(\mathbf{x}', t')$ whenever $(\mathbf{x}' - \mathbf{x})^2 \geq c^2(t' - t)^2$. (This is a sufficient, but not a necessary condition, for there are important theories in which non-vanishing terms in the commutators of $\mathcal{H}(\mathbf{x}, t)$ with $\mathcal{H}(\mathbf{x}', t')$ for $(\mathbf{x}' - \mathbf{x})^2 \geq c^2(t' - t)^2$ are canceled by terms in the Hamiltonian that cannot be written as the integrals of scalars.)

Equation (8.7.14) also makes the independence of distant processes transparent. Suppose that the transition $\alpha \rightarrow \beta$ consists of two separate transitions $a \rightarrow b$ and $A \rightarrow B$, with all the particles in the states $a$ and $b$ far from all the particles in the states $A$ and $B$. If we assume that interactions become negligible between sufficiently distant particles, then each $V_{\mathrm{I}}(t)$ in Eq. (8.7.14) acts either on the particles in the states $a$ and $b$ or on the particles in the states $A$ and $B$, but not on both. If $V_{\mathrm{I}}(\mathbf{x}, t)$ acts only on the particles in the states $a$ and $b$ while $V_{\mathrm{I}}(\mathbf{x}', t')$ acts only on the particles in the states $A$ and $B$, then these operators commute, and their time-ordered product can be replaced by an ordinary product. For a given term of $n$th order in Eq. (8.7.14), we must sum over the number $m$ of operators that act on the particles in the states $a$ and $b$ from $m = 0$ to $m = n$, with the remaining $n - m$ operators acting on the particles in the states $A$ and $B$. The number of ways of selecting the $m$ operators acting on $a$ and $b$ from the $n - m$ operators acting on $A$ and $B$ is $n!/m!(n - m)!$, so

$$
\begin{aligned}
S_{bB,aA} &= \sum_{n=0}^{\infty} \frac{1}{n!} \left[ -\frac{i}{\hbar} \right]^n \int_{-\infty}^{\infty} d\tau_1 \ldots \int_{-\infty}^{\infty} d\tau_n \sum_{m=0}^{n} \frac{n!}{m!(n-m)!} \\
&\quad \times \left( \Phi_b, T\{V_I(\tau_1) \ldots V_I(\tau_m)\}\Phi_a \right) \left( \Phi_B, T\{V_I(\tau_{m+1}) \ldots V_I(\tau_n)\} \Phi_A \right) \\
&= S_{ba} \, S_{BA}.
\end{aligned}
$$

This factorization ensures that the rates for the various final states $b$ produced from the initial state $a$ do not depend on the existence of the transition $A \rightarrow B$. It is not easy to see this essential factorization in old-fashioned perturbation theory.

In the exceptional cases where the $V_{\mathrm{I}}$ with different $\tau$-arguments all commute with one another, we can drop the time-ordering in Eq. (8.7.14), so that the sum is just the usual convergent series for the exponential function

$$
S_{\beta\alpha} = \left( \Phi_\beta, \exp\left[ \frac{-i}{\hbar} \int_{-\infty}^{\infty} d\tau \, V_{\mathrm{I}}(\tau) \right] \Phi_\alpha \right).
$$

Even where (as is usual) this simple result does not hold, it is common to abbreviate the result (8.7.14) as

$$S_{\beta\alpha} = \left( \Phi_\beta, T \left\{ \exp \left[ \frac{-i}{\hbar} \int_{-\infty}^{\infty} d\tau \; V_{\mathrm{I}}(\tau) \right] \right\} \Phi_\alpha \right), \tag{8.7.16}$$

the $T$ indicating that this quantity is to be evaluated by time-ordering each term in the power series for the expression in curly brackets.

For a very simple example, where the $V_{\mathrm{I}}(\tau_i)$ do *not* commute with one another, consider the classic example of a single non-relativistic particle being scattered by a local potential. Here $H_0$ is the kinetic energy, a function $H_0 = \mathbf{p}^2/2\mu$ of the momentum operator, and $V$ is a function $V(\mathbf{x})$ of the position operator. Since the relation Eq. (8.7.8) between the interaction in the interaction picture and in the Schrödinger picture is a similarity transformation, it gives (at least for any potential that can be expressed as a power series)

$$V_{\mathrm{I}}(\tau) = V \Big( \mathbf{x}_{\mathrm{I}}(\tau) \Big), \tag{8.7.17}$$

where $\mathbf{x}_{\mathrm{I}}(\tau)$ is the position operator in the interaction picture

$$\mathbf{x}_{\mathrm{I}}(t) \equiv e^{iH_0t/\hbar} \mathbf{x} e^{-iH_0t/\hbar}. \tag{8.7.18}$$

This operator satisfies the differential equation

$$\frac{d}{dt} \mathbf{x}_{\mathrm{I}}(t) = \frac{i}{\hbar} e^{iH_0t/\hbar}[H_0, \mathbf{x}] e^{-iH_0t/\hbar} = \frac{1}{\mu} e^{iH_0t/\hbar} \mathbf{p} e^{-iH_0t/\hbar} = \mathbf{p}/\mu, \tag{8.7.19}$$

and the obvious initial condition

$$\mathbf{x}_{\mathrm{I}}(0) = \mathbf{x}, \tag{8.7.20}$$

so

$$\mathbf{x}_{\mathrm{I}}(t) = \mathbf{x} + \mathbf{p}t/\mu, \tag{8.7.21}$$

and therefore

$$V_{\mathrm{I}}(\tau) = V \Big( \mathbf{x} + \mathbf{p}\tau/\mu \Big). \tag{8.7.22}$$

(Here $\mathbf{x}$ and $\mathbf{p}$ are the time-independent position and momentum operators in the Schrödinger picture.)

Because this involves both $\mathbf{x}$ and $\mathbf{p}$, the $\mathbf{x}_{\mathrm{I}}(\tau)$ with different $\tau$s do not commute with each other. Instead

$$[x_{\mathrm{I}i}(\tau), x_{\mathrm{I}j}(\tau')] = \frac{i\hbar}{\mu} \Big( \tau' - \tau \Big) \delta_{ij}. \tag{8.7.23}$$

Therefore the $V_{\mathrm{I}}(\tau)$ with different $\tau$s do not commute with each other, and so this is *not* an example where the Dyson series is simply the expansion of an exponential function.

Although the S-matrix is a central concern of particle physics, it is not the only thing worth calculating. We sometimes need to calculate the expectation value of a Heisenberg-picture operator $\mathcal{O}_H(t)$ (which may be given by a product of operators, all at the same time $t$), in a state $\Psi_\alpha^+$ that is defined by its appearance at very early times. (This is the problem that particularly concerns us in calculating correlation functions in cosmology, where $\alpha$ is usually taken as the vacuum state.) This entails a different version of time-dependent perturbation theory, known as the "in–in" formalism.[12] Any Heisenberg-picture operator can be expressed in terms of the corresponding interaction-picture operator by

$$\mathcal{O}_H(t) = e^{iHt/\hbar}\mathcal{O}e^{-iHt/\hbar} = e^{iHt/\hbar}e^{-iH_0t/\hbar}\mathcal{O}_I(t)e^{iH_0t/\hbar}e^{-iHt/\hbar}$$
$$= \Omega(t)\mathcal{O}_I(t)\Omega^\dagger(t). \tag{8.7.24}$$

We use this together with Eqs. (8.7.2) and (8.7.5) to write the expectation value as

$$\left(\Psi_\alpha^+, \mathcal{O}_H(t)\Psi_\alpha^+\right) = \left(\Phi_\alpha, \Omega^\dagger(-\infty)\Omega(t)\mathcal{O}_I(t)\Omega^\dagger(t)\Omega(-\infty)\Phi_\alpha\right)$$
$$= \left(\Phi_\alpha, U^\dagger(t, -\infty)\mathcal{O}_I(t)U(t, -\infty)\Phi_\alpha\right). \tag{8.7.25}$$

Then, using the perturbation series (8.7.13) for $U(t, -\infty)$, we have

$$\left(\Psi_\alpha^+, \mathcal{O}_H(t)\Psi_\alpha^+\right) = \left(\Phi_\alpha, \left[T\left\{\exp\left[\frac{-i}{\hbar}\int_{-\infty}^t d\tau\, V_I(\tau)\right]\right\}\right]^\dagger\right.$$
$$\left. \times \mathcal{O}_I(t)T\left\{\exp\left[\frac{-i}{\hbar}\int_{-\infty}^t d\tau\, V_I(\tau)\right]\right\}\Phi_\alpha\right), \tag{8.7.26}$$

where $T\{\cdot\}$ has the same meaning as in Eq. (8.7.16); that is, we must time-order the $V_I$ operators in the power-series expansion of the exponential. The adjoint of the first time-ordered product in Eq. (8.7.26) means that the interaction operators in this part of the expression are not time-ordered, but anti-time-ordered; that is, the operator first on the left is the one with the *earliest* argument, and so on. Thus the structure of the "in–in" expectation value (8.7.26) is very different from that of the Dyson expansion (8.7.16) for the S-matrix.

---

12 J. Schwinger, *Proc. Nat. Acad. Sci. USA* **46**, 1401 (1960); *J. Math. Phys.* **2**, 407 (1961); K. T. Mahanthappa, *Phys. Rev.* **126**, 329 (1962); P. M. Bakshi and K. T. Mahanthappa, *J. Math. Phys.* **4**, 1, 12 (1963); L. V. Keldysh, *Sov. Phys. JETP* **20**, 1018 (1965); D. Boyanovsky and H. J. de Vega, *Ann. Phys.* **307**, 335 (2003); B. DeWitt, *The Global Approach to Quantum Field Theory* (Clarendon Press, Oxford, 2003), Section 31. For a review, with applications to cosmological correlations, see S. Weinberg, *Phys. Rev. D* **72**, 043514 (2005) [hep-th/0506236].

## 8.8 Shallow Bound States

Sometimes when a bound state is sufficiently weakly bound, we can obtain results for scattering amplitudes just from a knowledge of the binding energy, with no detailed information about the interaction. For this purpose, we use a tool known as the *Low equation*.[13]

To derive the Low equation, we operate on the Lippmann–Schwinger equation (8.1.6) with the interaction $V$, so that

$$V\Psi_\alpha^\pm = V\Phi_\alpha + V[E_\alpha - H_0 \pm i\epsilon]^{-1}V\Psi_\alpha^\pm. \tag{8.8.1}$$

We can write the solution of this equation as

$$V\Psi_\alpha^\pm = T(E_\alpha \pm i\epsilon)\Phi_\alpha, \tag{8.8.2}$$

where $T(W)$ is the solution of the operator equation

$$T(W) = V + V(W - H_0)^{-1}T(W). \tag{8.8.3}$$

We recall that the S-matrix is given according to Eqs. (8.1.10) and (8.1.11) as

$$S_{\beta\alpha} = \delta(\beta - \alpha) - 2\pi i \delta(E_\beta - E_\alpha)T_{\beta\alpha}, \tag{8.8.4}$$

where

$$T_{\beta\alpha} \equiv \left(\Phi_\beta, V\Psi_\alpha^+\right) = \left(\Phi_\beta, T(E_\alpha + i\epsilon)\Phi_\alpha\right). \tag{8.8.5}$$

So far, there is nothing new here, except for a little formalism. Now note that with some elementary algebra, we can write the solution of the operator equation (8.8.3) as

$$T(W) = V + V(W - H)^{-1}V. \tag{8.8.6}$$

We can evaluate the resolvent operator $(W - H)^{-1}$ by inserting a sum over a complete set of independent eigenstates of $H$. These include the scattering "in" states $\Psi_\alpha^+$, and any bound states. (We do not include the "out" states $\Psi_\alpha^-$ here, because they are not independent; $\Psi_\alpha^-$ can be written as the superposition $\int d\beta\, S_{\alpha\beta}^* \Psi_\beta^+$.) Thus

$$\left(\Phi_\beta, T(W)\Phi_\alpha\right) = V_{\beta\alpha} + \int db \frac{\left(\Phi_\beta, V\Psi_b\right)\left(\Phi_\alpha, V\Psi_b\right)^*}{W - E_b} + \int d\gamma \frac{T_{\beta\gamma}T_{\alpha\gamma}^*}{W - E_\gamma}, \tag{8.8.7}$$

where $V_{\beta\alpha} \equiv \left(\Phi_\beta, V\Phi_\alpha\right)$, and $b$ labels the properties of the various bound states, including their total momentum. In particular, setting $W = E_\alpha + i\epsilon$, Eq. (8.8.7) gives

---

[13] The equation is named for Francis Low. I have not been able to find a reference to the place where it was first published.

$$T_{\beta\alpha} = V_{\beta\alpha} + \int db \, \frac{\left(\Phi_\beta, V\Psi_b\right)\left(\Phi_\alpha, V\Psi_b\right)^*}{E_\alpha - E_b} + \int d\gamma \, \frac{T_{\beta\gamma} T_{\alpha\gamma}^*}{E_\alpha - E_\gamma + i\epsilon}. \quad (8.8.8)$$

(We don't need the $i\epsilon$ in the denominator of the bound-state term, since the energy of any bound state must be outside the spectrum of $H_0$.) Equation (8.8.8) is known as the Low equation.

The Low equation is a non-linear integral equation for $T_{\beta\alpha}$, in which a non-zero value for $T_{\beta\alpha}$ is driven by the first two terms in Eq. (8.8.8). For a shallow bound state, whose energy is very near the continuum, it is plausible that the bound-state term in Eq. (8.8.8) will dominate over the potential term, and give $T_{\beta\gamma}$ and $T_{\alpha\gamma}$ particularly large values when $E_\gamma$ is nearest the bound-state energies – that is, near the minimum continuum energy – provided these two particles have $\ell = 0$, to avoid suppression of the matrix elements by factors $k^\ell$. Thus, when $\alpha$ is a two-particle state with $\ell = 0$, and $\beta$ is a state of two particles of the same two species as $\alpha$, it is plausible to limit $\gamma$ to two-particle states of the same two species. (I have in mind here the low-energy scattering of a proton and a neutron, where the shallow bound state is the deuteron, but will continue for a while to keep the analysis more general.) As in Section 8.4, these two-particle states can be labeled by their total energy, their total momentum $\mathbf{P}$, their total spin $s$, their orbital angular momentum $\ell = 0$, their total angular momentum $J = s$, the 3-component $\sigma$ of the total angular momentum (and total spin), and the species of the two particles. Dropping the labels $\ell = 0$, $s$, and the two species labels, which will be the same throughout, the free-particle states can be denoted $\Phi_{E,\mathbf{P},\sigma}$, and the scattering "in" states can be denoted $\Psi_{E,\mathbf{P},\sigma}^+$. The bound states that contribute in Eq. (8.8.8) must also have a spin $s$. If we assume that there is only one such bound state, we can drop the label $s$ and $\ell = 0$, and denote the bound state only by its total momentum and spin 3-component, as $\Psi_{\mathbf{P},\sigma}$, with the energy a fixed function of $\mathbf{P}$. The relevant matrix elements in the center-of-mass system then have the form

$$T_{E',\mathbf{P}',\sigma';E,0,\sigma} = T(E', E)\delta^3(\mathbf{P}')\delta_{\sigma',\sigma}, \quad (8.8.9)$$

and

$$\left(\Phi_{E,0,\sigma}, V\Psi_{\mathbf{P},\sigma'}\right) = \mathcal{G}(E)\delta^3(\mathbf{P})\delta_{\sigma'\sigma}. \quad (8.8.10)$$

From now on we will understand $E$ as the energy measured relative to the total rest mass in the two-particle state, so that it is integrated from zero to infinity, and the bound-state energy in the center-of-mass frame is $-B$, with $B$ the binding energy. Neglecting the potential term in Eq. (8.8.8), the Low equation now reads

$$T(E', E) = \frac{\mathcal{G}(E')\mathcal{G}^*(E)}{E + B} + \int_0^\infty dE'' \, \frac{T(E', E'')T^*(E, E'')}{E - E'' + i\epsilon}. \quad (8.8.11)$$

Now, as we have explained, we are interested in this equation in the case where $E$ and $E'$ are small, comparable in magnitude to the binding energy $B$. In this case, it is presumably a good approximation to write

$$\mathcal{G}(E) = \sqrt{p(E)}\, g, \tag{8.8.12}$$

where $g$ is a constant, and $p(E)$ is the momentum of either particle in the center-of-mass system when the total energy is $E$. With non-relativistic kinematics, $p(E) = \sqrt{2\mu E}$, where $\mu$ is the reduced mass. The factor $p(E)$ is needed, because we expect $V\Psi_{0,\sigma}$ to have matrix elements with two-particle states of individual momenta $\mathbf{p}$ and $-\mathbf{p}$ that are analytic in $\mathbf{p}$ near $\mathbf{p} = 0$, and as shown in Eq. (8.4.9), these two-particle states are given by the states $\Phi_{E,0,\sigma}$ times a factor proportional to $1/\sqrt{|\mathbf{p}|}$. The Low equation (8.8.11) now reads

$$T(E', E) = \frac{\sqrt{p(E')p(E)}\,|g|^2}{E+B} + \int_0^\infty dE'' \frac{T(E', E'')T^*(E, E'')}{E - E'' + i\epsilon}. \tag{8.8.13}$$

Inspection of this equation shows that it can be solved with an *ansatz*

$$T(E', E) = \sqrt{p(E')p(E)}\, t(E), \tag{8.8.14}$$

so that Eq. (8.8.13) is satisfied if

$$t(E) = \frac{|g|^2}{E+B} + \int_0^\infty dE' \, p(E') \frac{|t(E')|^2}{E - E' + i\epsilon}. \tag{8.8.15}$$

This can actually be solved exactly. As shown at the end of this section, the solution for an arbitrary positive function $p(E)$ is

$$t(E) = \left[ \frac{E+B}{|g|^2} + (E+B)^2 \int_0^\infty \frac{p(E')\,dE'}{(E'+B)^2(E'-E-i\epsilon)} \right]^{-1},$$

as long as $p(E)$ does not grow too fast as $E \to \infty$. For the case $p(E) = \sqrt{2\mu E}$, this gives

$$t(E) = \left[ \frac{E+B}{|g|^2} + \frac{\pi(B-E)}{2}\sqrt{\frac{2\mu}{B}} + i\pi\sqrt{2\mu E} \right]^{-1}. \tag{8.8.16}$$

We can calculate the coupling $g$ of the bound state to its constituents, by using the condition that the bound-state vector $\Psi_{\mathbf{P},\sigma}$ is normalized, in the sense that

$$\left( \Psi_{\mathbf{P}',\sigma'}, \Psi_{0,\sigma} \right) = \delta^3(\mathbf{P}')\delta_{\sigma'\sigma}. \tag{8.8.17}$$

The bare two-particle state $\Phi_{E,0,\sigma}$ is an eigenstate of $H_0$ with eigenvalue $E$, while the bound state $\Psi_{0,\sigma}$ is an eigenstate of $H$ with eigenvalue $-B$, so

$$\left( \Phi_{E,0,\sigma}, V\Psi_{\mathbf{P}',\sigma'} \right) = \left( \Phi_{E,0,\sigma}, [H - H_0]\Psi_{\mathbf{P}',\sigma'} \right) = -(E+B)\left( \Phi_{E,0,\sigma}, \Psi_{\mathbf{P}',\sigma'} \right),$$

or, using Eqs. (8.8.10) and (8.8.12),

$$\left(\Phi_{E,0,\sigma}, \Psi_{\mathbf{P}',\sigma'}\right) = -\delta^3(\mathbf{P}')\delta_{\sigma'\sigma}\frac{g\sqrt{p(E)}}{E+B}. \tag{8.8.18}$$

Thus, expanding in bare two-particle states, Eq. (8.8.17) gives

$$1 = |g|^2 \int_0^\infty \frac{p(E)\,dE}{(E+B)^2}$$

and so[14]

$$|g|^2 = \frac{1}{\pi}\sqrt{\frac{2B}{\mu}}. \tag{8.8.19}$$

Using this in the solution (8.8.16) of the Low equation, we have

$$t(E) = \frac{1}{\pi\sqrt{2\mu}}\left[\sqrt{B} + i\sqrt{E}\right]^{-1}. \tag{8.8.20}$$

We now have to convert this result into a formula for the $\ell = 0$ phase shift. Equations (8.4.7) and (8.4.25) give the center-of-mass scattering amplitude in the basis used here (suppressing the indices $\ell = 0$, $s$, $n$, and $J = s$) as

$$M_{0,E,\sigma';0,E,\sigma} = \delta_{\sigma'\sigma}\left[e^{2i\delta(E)} - 1\right].$$

Also, comparing Eqs. (8.3.1) and (8.8.4), and using Eq. (8.8.9), we have

$$\delta^3(\mathbf{P})M_{\mathbf{P},E,\sigma';0,E,\sigma} = -2\pi i\, T_{E,0,\sigma';E,\mathbf{P},\sigma'} = -2\pi i\, T(E,E)\delta^3(\mathbf{P})\delta_{\sigma',\sigma},$$

so Eqs. (8.8.9) and (8.8.14) give

$$e^{2i\delta(E)} - 1 = -2\pi i T(E,E) = -2\pi i\sqrt{2\mu E}\, t(E). \tag{8.8.21}$$

Using the solution (8.8.20), we have then

$$e^{2i\delta(E)} - 1 = -2i\sqrt{E}\left[\sqrt{B} + i\sqrt{E}\right]^{-1}. \tag{8.8.22}$$

Taking the reciprocal, we find that a term $-1/2$ appears on both sides, so after cancelling this term, we have

$$\cot\delta = -\sqrt{B/E}. \tag{8.8.23}$$

Note that this result is real, and so is consistent with the unitarity of the S-matrix, a non-trivial consistency condition that would not be satisfied in the Born

---

[14] More generally, if in addition to the continuum the eigenstates of $H_0$ include an elementary particle state with the same quantum numbers as the bound state, $|g|$ is less than the value given in Eq. (8.8.19) by a factor $1 - Z$, where $Z$ is the probability that an examination of the bound state will find it in the elementary particle state rather than the two-particle state. The case $Z \neq 0$ is studied in detail by S. Weinberg, *Phys. Rev.* B**137**, 672 (1965).

approximation. The result (8.8.23) may be compared with the effective range expansion (7.5.21). Setting $E = \hbar^2 k^2 / 2\mu$, we have $k \cot \delta = -\sqrt{2\mu B}/\hbar$, so the scattering length is

$$a_s = \hbar/\sqrt{2\mu B}, \tag{8.8.24}$$

and the effective range and all higher terms in the expansion are negligible. These are precise results in the limit of vanishing $B$ and $E$, with $E/B$ fixed.

As mentioned earlier, the classic application of this calculation is to low-energy proton–neutron scattering in the state with the same total spin $s = 1$ as the deuteron. Here $\mu = m_n m_p / (m_n + m_p) \simeq m_p/2$ and $B = 2.2246$ MeV, so Eq. (8.8.24) gives $a_s = 4.31 \times 10^{-13}$ cm. On the other hand, experiment gives $a_s = 5.41 \times 10^{-13}$ cm. The measured effective range is not zero, but considerably smaller: $r_{\text{eff}} = 1.75 \times 10^{-13}$ cm. The range of nuclear forces is of the order of $10^{-13}$ cm, so the accuracy of these predictions is as good as could be expected.

Incidentally, note that for $B \to 0$, Eq. (8.8.23) gives $\cot \delta \to 0$, so $\delta \to 90°$, perhaps plus a multiple of $180°$. This is an exception to the low-energy limits discussed in Section 7.5.

$$* * * * *$$

We return here to the solution of the non-linear integral equation (8.8.15). We define a function for general complex $z$:

$$f(z) \equiv \frac{|g|^2}{z + B} + \int_0^\infty dE' \, p(E') \frac{|t(E')|^2}{z - E'}, \tag{8.8.25}$$

so that

$$t(E) = f(E + i\epsilon). \tag{8.8.26}$$

We note that $-f(z)$ is analytic in the upper half plane, where it has positive-definite imaginary part

$$\text{Im}\left[-f(z)\right] = \text{Im}\, z \left[\frac{|g|^2}{|z + B|^2} + \int_0^\infty dE' \, p(E') \frac{|t(E')|^2}{|z - E'|^2}\right]. \tag{8.8.27}$$

The same is then also true of $1/f(z)$. A general theorem[15] tells us that any such function must have the representation

$$f^{-1}(z) = f^{-1}(z_0) + (z - z_0) f^{-1'}(z_0) + (z - z_0)^2 \int_{-\infty}^\infty dE' \frac{\sigma(E')}{(E' - z_0)^2 (E' - z)}, \tag{8.8.28}$$

---

[15] A. Herglotz, *Ver. Verhandl. Sachs. Ges. Wiss. Leipzig, Math.-Phys.* **63**, 501 (1911); J. A. Shohat and J. D. Tamarkin, *The Problem of Moments* (American Mathematical Society, New York, 1943), Chapter II.

where $\sigma(E)$ is real and positive, and $z_0$ is arbitrary. (A formula of this sort is called a "twice-subtracted dispersion relation.") It is convenient to choose $z_0 = -B$. We know that $f^{-1}(-B) = 0$ and $f^{-1'}(-B) = 1/|g|^2$, so

$$f^{-1}(z) = \frac{z+B}{|g|^2} + (z+B)^2 \int_{-\infty}^{\infty} dE \, \frac{\sigma(E)}{(E+B)^2(E-z)}. \tag{8.8.29}$$

Now, what is $\sigma(E)$? Let us first tentatively assume that $f(z)$ has no zeros on the real axis. Then Eq. (8.8.29) gives

$$\sigma(E) = \frac{1}{\pi} \operatorname{Im} f^{-1}(E+i\epsilon) = -\frac{\operatorname{Im} f(E+i\epsilon)}{\pi |f(E+i\epsilon)|^2}$$

$$= \begin{cases} p(E), & E \geq 0, \\ 0, & E \leq 0. \end{cases} \tag{8.8.30}$$

Using this in Eq. (8.8.29) gives

$$f(z) = \left[ \frac{z+B}{|g|^2} + (z+B)^2 \int_0^{\infty} \frac{p(E') \, dE'}{(E'+B)^2(E'-z)} \right]^{-1}. \tag{8.8.31}$$

Setting $z = E + i\epsilon$ gives $t(E)$, and taking $p(E) = \sqrt{2\mu E}$ then yields Eq. (8.8.16).

This solution is not unique, for we have assumed above that $f(z)$ has no zeros on the real axis. But any other solution will become indistinguishable from the one found here in the limit as $B$ is taken much smaller than the position of such zeros.

## 8.9 Time Reversal of Scattering Processes

As we saw in Sections 3.6 and 4.7, in many contexts it is a good approximation to assume a symmetry under the reversal of time, represented in quantum mechanics by an antilinear and antiunitary operator T. Where time reversal is a good symmetry, the operator T commutes with the Hamiltonian (with both terms, $H_0$ and $V$), but anticommutes with the momentum and angular-momentum operators, so it converts a free-particle state $\Phi_\alpha$ into another free-particle state:

$$\mathsf{T}\Phi_\alpha = \Phi_{\mathcal{T}\alpha}, \tag{8.9.1}$$

where $\mathcal{T}\alpha$ denotes a state of the same particles as $\alpha$, but with all momenta and spin $z$-components reversed. However, matters are more complicated when interactions are taken into account. We define the "in" and "out" states $\Psi_\alpha^+$ and $\Psi_\alpha^-$ as eigenstates of the Hamiltonian that look like the free-particle state $\Phi_\alpha$ at early and late times, respectively, so the time-reversal operator T acting on these states should give eigenstates of the Hamiltonian with the same energy that look like the free-particle state $\Phi_{T\alpha}$ at *late and early* times, respectively. That is,

$$\mathsf{T}\Psi_\alpha^\pm = \Psi_{T\alpha}^\mp. \tag{8.9.2}$$

We can verify this by applying the operator $\mathsf{T}$ to the Lippmann–Schwinger equation (8.1.6). Using Eq. (8.9.1) and keeping in mind that $\mathsf{T}$ is not linear but antilinear, we find that

$$\mathsf{T}\Psi_\alpha^\pm = \Phi_{T\alpha} + (E_\alpha - E_\beta \mp i\epsilon)^{-1}V\mathsf{T}\Psi_\alpha^\pm, \tag{8.9.3}$$

so $\mathsf{T}\Psi_\alpha^\pm$ satisfies the same Lippmann–Schwinger equation as $\Psi_{T\alpha}^\mp$.

Because $\mathsf{T}$ is antiunitary, time-reversal invariance does *not* tell us that $S_{\beta\alpha}$ equals the S-matrix $S_{T\beta\,T\alpha}$ for the same reaction with spins and momenta reversed. Instead, recalling the defining property (3.4.10) of antiunitary operators, we have

$$S_{\beta\alpha} = (\Psi_\beta^-, \Psi_\alpha^+) = (\mathsf{T}\Psi_\alpha^+, \mathsf{T}\Psi_\beta^-) = (\Psi_{T\alpha}^-, \Psi_{T\beta}^+) = S_{T\alpha,T\beta}. \tag{8.9.4}$$

This is known as the *Principle of Detailed Balance.*

By itself, this tells us nothing about any one transition with $\alpha \neq \beta$. We get useful information about individual transitions if time-reversal invariance is combined with certain approximations. For instance, to first order in the interaction $V$, for $\beta \neq \alpha$ Eq. (8.6.2) gives the Born-approximation result $S_{\beta\alpha} = -2\pi i\delta(E_\alpha - E_\beta)\left(\Phi_\beta, V\Phi_\alpha\right)$, so since $V$ is Hermitian, in this approximation we have $S_{\alpha\beta} = -S_{\beta\alpha}^*$, and therefore the time-reversal invariance result (8.9.4) gives

$$S_{\beta\alpha} = -S_{T\beta\,T\alpha}^*. \tag{8.9.5}$$

The minus sign and complex conjugation don't matter when we calculate rates, which involve absolute squares of S-matrix elements, so in the Born approximation time-reversal invariance does tell us that the rate for any process equals the rate for the *same* process when all spins and momenta are reversed.

This result can be generalized by using a much more widely applicable approximation, the distorted-wave Born approximation discussed in Section 8.6. This approximation applies when we can write the interaction $V$ as a sum

$$V = V_s + V_w, \tag{8.9.6}$$

where the term $V_s$ is much stronger than the term $V_w$, but cannot by itself produce the reaction in question. (As shown by the examples discussed in Section 8.6, $V_s$ and $V_w$ are not always the strong and weak nuclear interactions, though they often are.) According to Eq. (8.6.19) in all such cases the distorted-wave Born approximation gives the scattering amplitude for any reaction $\alpha \to \beta$ to first order in $V_w$ but to all orders in $V_s$, as

$$T_{\beta\alpha} = \left(\Psi_{s\beta}^-, V_w\Psi_{s\alpha}^+\right), \tag{8.9.7}$$

where $T_{\beta\alpha}$ is the amplitude appearing in the general formula (8.1.10) for the S-matrix

$$S_{\beta\alpha} = \delta(\alpha - \beta) - 2\pi i \delta(E_\alpha - E_\beta)T_{\beta\alpha} \tag{8.9.8}$$

and the subscript s on state vectors indicates that these "in" and "out" state vectors are solutions of the Lippmann–Schwinger equation (8.1.6) with only $V_s$ included in the interaction $V$.

If we now assume that the time-reversal operator T commutes with $V_w$ as well as $V_s$ and $H_0$, and recall that T is antiunitary, we have

$$T_{\beta\alpha} = \left( \mathsf{T}\Psi_{s\alpha}^+, V_w \, \mathsf{T}\Psi_{s\beta}^- \right) = \left( \Psi_{s\,T\alpha}^-, V_w \, \Psi_{s\,T\beta}^+ \right),$$

and using the fact that $V_w$ is Hermitian, this gives

$$T_{\beta\alpha} = \left( \Psi_{s\,T\beta}^+, V_w \, \Psi_{s\,T\alpha}^- \right)^* . \tag{8.9.9}$$

This is what we need, except that we now have an "in" state on the left and an "out'" state on the right. We can fix this, by recalling the relation (8.1.8) between "in" and "out'" states and using the detailed-balance relation (8.9.4) for strong scattering:

$$\Psi_{s\,T\beta}^+ = \int d\beta' \, S_{T\beta\,T\beta'}^{\mathrm{s}} \Psi_{s\,T\beta'}^- = \int d\beta' \, S_{\beta'\,\beta}^{\mathrm{s}} \Psi_{s\,T\beta'}^- ,$$

$$\Psi_{s\,T\alpha}^- = \int d\alpha' \, S_{T\alpha'\,T\alpha}^{\mathrm{s}*} \Psi_{s\,T\alpha'}^+ = \int d\alpha' \, S_{\alpha'\,\alpha}^{\mathrm{s}*} \Psi_{s\,T\alpha'}^+ ,$$

where $S^{\mathrm{s}}$ is the S-matrix calculated including only $V_s$ in the interaction $V$. So now, using Eq. (8.9.9) again,

$$T_{\beta\alpha} = \int d\alpha' \int d\beta' \, S_{\beta'\beta}^{\mathrm{s}} S_{\alpha\alpha'}^{\mathrm{s}} T_{T\beta',T\alpha'}^* . \tag{8.9.10}$$

This now relates the process $\alpha \to \beta$ to *the same process* $T\alpha \to T\beta$ with spins and momenta reversed, which is what we wanted.

It should be noted that the integrals over $\alpha'$ and $\beta'$ in Eq. (8.9.10) (which consist of integrals over momenta, and sums over discrete variables) run only over states that can be produced respectively from $\alpha$ and $\beta$ by the strong interaction $V_s$. In particular, in a case like beta decay, in which the initial state $\alpha$ is a discrete eigenstate of $H_0 + V_s$ that would be stable in the absence of the weak interaction $V_w$, and the same is true of the final state $\beta$ except for the presence of particles like photons, electrons, and/or neutrinos, on which $V_s$ has no effect, the S-matrix factors in Eq. (8.9.10) are delta functions, and we have

$$T_{\beta\alpha} = T_{T\beta,T\alpha}^* , \tag{8.9.11}$$

just as in the Born approximation.

More generally, we may be able to choose a basis of states like that discussed in Section 8.4 for which the "strong" S-matrix $S^s_{\beta'\beta}$ is diagonal

$$S^s_{\beta'\beta} = e^{2i\delta_\beta} \delta(\beta' - \beta), \qquad (8.9.12)$$

where $\delta_\beta$ is a real phase shift. If the initial state $\alpha$ is a discrete eigenstate of $V_s$ that would be stable in the absence of $V_w$, then Eq. (8.9.10) tells us that

$$T_{\beta\alpha} = e^{2i\delta_\beta} T^*_{\mathcal{T}\beta,\mathcal{T}\alpha}. \qquad (8.9.13)$$

This is known as the *Watson–Fermi theorem*.[16] It can be used together with data on processes such as the K-meson decay mode $K \to 2\pi + e + \nu$ to measure the phase shifts for processes such as pion–pion scattering that are not easy to measure by other means.[17]

## Problems

1. Consider a general Hamiltonian $H_0 + V$, where $H_0$ is the free-particle energy. Define a state $\Psi^0_\alpha$ by the modified Lippmann–Schwinger equation

    $$\Psi^0_\alpha = \Phi_\alpha + \frac{E_\alpha - H_0}{(E_\alpha - H_0)^2 + \epsilon^2} V \Psi^0_\alpha,$$

    where $\Phi_\alpha$ is an eigenstate of $H_0$ with eigenvalue $E_\alpha$, and $\epsilon$ is a positive infinitesimal quantity. Define

    $$A_{\beta\alpha} \equiv \left( \Phi_\beta, V \Psi^0_\alpha \right).$$

    (a) Show that $A_{\beta\alpha} = A^*_{\alpha\beta}$ for $E_\beta = E_\alpha$.

    (b) For the simple case of a non-relativistic particle with energy $\mathbf{k}^2 \hbar^2/2\mu$ in a local potential $V(\mathbf{x})$, calculate the asymptotic behavior of the coordinate-space wave function $\left( \Phi_\mathbf{x}, \Psi^0_\mathbf{k} \right)$ of the state $\Psi^0_\mathbf{k}$ for $\mathbf{x} \to \infty$. Express the result in terms of matrix elements of $A$.

2. Consider a separable interaction, whose matrix elements between free-particle states have the form

    $$\left( \Phi_\beta, V \Phi_\alpha \right) = f(\alpha) f^*(\beta),$$

    where $f(\alpha)$ is some general function of the momenta and other quantum numbers characterizing the free-particle state $\Phi_\alpha$.

    (a) Find an exact solution of the Lippmann–Schwinger equation for the "in" state in this theory.

16 K. Watson, *Phys. Rev.* **88**, 1163 (1952); E. Fermi, *Nuovo Cimento* **2**, Suppl. 1, 17 (1965).
17 N. Cabibbo and A. Maksymowicz, *Phys. Rev.* **B137**, 438 (1965); **168**, 1926 (1968).

(b) Use the result of (a) to calculate the S-matrix.

(c) Verify the unitarity of the S-matrix.

3. The scattering of $\pi^+$ on protons at energies less than a few hundred MeV is purely elastic, and receives appreciable contributions only from orbital angular momenta $\ell = 0$ and $\ell = 1$.

    (a) List all the phase shifts that enter in the amplitude for $\pi^+$–proton scattering at these low energies. (Recall that the spins of the pion and proton are zero and $1/2$, respectively.)

    (b) Give a formula for the differential scattering cross section in terms of these phase shifts.

4. By direct calculation, show that the terms of first and second order in the interaction in time-dependent perturbation theory give the same results for the S-matrix as the first- and second-order terms in old-fashioned perturbation theory.

5. Assume isospin conservation, and suppose that the only appreciable phase shift in the scattering of pions on nucleons is the one with quantum numbers $J = 3/2$, $\ell = 1$, and $T = 3/2$. Calculate the differential cross sections for the reactions $\pi^+ + p \to \pi^+ + p$, $\pi^+ + n \to \pi^+ + n$, $\pi^+ + n \to \pi^0 + p$, and $\pi^- + n \to \pi^- + n$ in terms of this phase shift.

6. The $\Lambda^0$ is a particle of spin $1/2$ and mass $1116 \text{ GeV}/c^2$. It decays only through the weak nuclear forces, into an isotopic spin-$1/2$ state of a nucleon and a pion. Find the phases of the amplitudes for decay into states with $\ell = 0$ and $\ell = 1$, in terms of the phase shifts for $s$-wave and $p$-wave pion–nucleon scattering with total angular momentum $j = 1/2$ and total isospin $t = 1/2$, at total energy $1116 \text{ GeV}$. (This process does not conserve parity, but you can assume time-reversal invariance.)

# 9

# The Canonical Formalism

To carry out calculations in quantum mechanics, we need a formula for the Hamiltonian as a function of operators whose commutation relations are known. So far, we have dealt with simple systems, for which it is easy to guess such a formula. For a system of non-relativistic spinless particles interacting through a potential $V$ that depends only on particle separations, the classical formula for the energy suggests that we should take

$$H = \sum_n \frac{\mathbf{p}_n^2}{2m_n} + V(\mathbf{x}_1 - \mathbf{x}_2, \mathbf{x}_1 - \mathbf{x}_3, \ldots),$$

where $\mathbf{x}_n$ and $\mathbf{p}_n$ are the position and momentum of the $n$th particle. We saw in Section 3.5 that the commutator of the total momentum operator $\mathbf{P} = \sum_n \mathbf{p}_n$ with the coordinate of the $n$th particle in any system is given by Eq. (3.5.3), and from this it was a short jump to guess the commutation relation (3.5.6) of the momenta and positions of individual particles:

$$[x_{ni}, p_{mj}] = i\hbar \delta_{nm} \delta_{ij}.$$

But our task can be much harder in more complicated theories, dealing with velocity-dependent interactions, or interactions of particles with fields, or interactions of fields with each other.

This problem is generally dealt with by the rules of the canonical formalism. As we will see in Section 9.1, the equations of motion in classical systems can usually be derived from a function of generalized coordinate variables and their time-derivatives, known as the Lagrangian. The great advantage of the Lagrangian formalism, described in Section 9.2, is that it allows us to derive the existence of conserved quantities from symmetry principles. One of these conserved quantities is the Hamiltonian, discussed in Section 9.3. The Hamiltonian is expressed in terms of generalized coordinates and generalized momenta. As shown in Section 9.4, these variables must satisfy certain commutation relations in order for the conserved quantities provided by the Lagrangian formalism to act as the generators of symmetry transformations with which they are associated, and in particular for the Hamiltonian to act as the generator of time translations.

I will illustrate all these points by reference to the theory of non-relativistic particles in a local potential. In this case, the application of the canonical formalism is pretty simple. It becomes more complicated for systems satisfying a constraint, such as a particle constrained to move on a surface. Constrained systems are discussed in Section 9.5. An alternative version of the canonical formalism, the path-integral formalism, is derived in Section 9.6.

## 9.1 The Lagrangian Formalism

It is common to find that the dynamical equations that govern the general coordinate variables $q_N(t)$ describing a classical physical system can be derived from a variational principle, which states that an integral

$$I[q] \equiv \int_{-\infty}^{\infty} L\Big(q(t), \dot{q}(t), t\Big) \, dt \qquad (9.1.1)$$

is stationary with respect to all infinitesimal variations $q_N(t) \mapsto q_N(t) + \delta q_N(t)$, for which all $\delta q_N(t)$ vanish at the end-points of the integral, $t \to \pm\infty$. The function or functional $L$ is known as the Lagrangian of the theory, while the functional $I[q]$ is called the action. In a theory of particles, $N$ is a compound index $ni$, with $q_N(t)$ the $i$th component $x_{ni}(t)$ of the position of the $n$th particle at time $t$. In a theory of fields, $N$ is a compound label $n\mathbf{x}$, with $q_N(t)$ the value of the $n$th field at a position $\mathbf{x}$ and time $t$. We will treat $N$ as a discrete index, but we will find it easy in Chapter 11 to adapt the formulas we derive here to the case of fields.

We are here letting $L$ have an explicit dependence on time, to take account of the possibility that the system is affected by time-dependent external fields, but in the case of an isolated system $L$ depends on time only through its dependence on $q(t)$ and $\dot{q}(t)$.

The condition that (9.1.1) should be stationary gives

$$0 = \delta I[q]$$

$$= \sum_N \int_{-\infty}^{\infty} \left[ \frac{\partial L\Big(q(t), \dot{q}(t), t\Big)}{\partial q_N(t)} \delta q_N(t) + \frac{\partial L\Big(q(t), \dot{q}(t), t\Big)}{\partial \dot{q}_N(t)} \delta \dot{q}_N(t) \right] dt.$$

The variation in the time-derivative is the time-derivative of the variation, so we can integrate the second term by parts. Since the variations vanish at the end-points of the integral, the result is

$$0 = \sum_N \int_{-\infty}^{\infty} \left[ \frac{\partial L\Big(q(t), \dot{q}(t), t\Big)}{\partial q_N(t)} - \frac{d}{dt} \frac{\partial L\Big(q(t), \dot{q}(t), t\Big)}{\partial \dot{q}_N(t)} \right] \delta q_N(t) \, dt.$$

$$(9.1.2)$$

This must hold for any infinitesimal functions $q_N(t)$ that vanish as $t \to \pm\infty$, so for each $N$ and each finite $t$ we must have

$$\frac{\partial L\big(q(t), \dot{q}(t)\big)}{\partial q_N(t)} = \frac{d}{dt}\frac{\partial L\big(q(t), \dot{q}(t), t\big)}{\partial \dot{q}_N(t)}. \tag{9.1.3}$$

For instance, for a classical system consisting of a number of non-relativistic particles with masses $m_n$, interacting through a potential that depends only on position, the Newtonian equations of motion are

$$m_n \ddot{x}_{ni}(t) = -\frac{\partial V}{\partial x_{ni}(t)}. \tag{9.1.4}$$

These are just the Lagrangian equations (9.1.3), if we take the Lagrangian as

$$L = \sum_n \frac{m_n}{2}\dot{\mathbf{x}}_n^2 - V. \tag{9.1.5}$$

One of the nice things about the Lagrangian formalism is that it makes it easy to use any coordinates we like. For instance, consider a single particle of mass $m$ moving in two dimensions in a potential $V(r)$ that depends only on the radial coordinate. Here we can take the $q_N$ to be the polar coordinates $r$ and $\theta$, and write the Lagrangian (9.1.5) as

$$L = \frac{m}{2}\Big[\dot{r}^2 + r^2\dot{\theta}^2\Big] - V(r). \tag{9.1.6}$$

The Lagrangian equations of motion (9.1.3) in these coordinates are

$$0 = \frac{d}{dt}\frac{\partial L}{\partial \dot{r}} - \frac{\partial L}{\partial r} = m\ddot{r} - mr\dot{\theta}^2 + V'(r), \tag{9.1.7}$$

$$0 = \frac{d}{dt}\frac{\partial L}{\partial \dot{\theta}} - \frac{\partial L}{\partial \theta} = \frac{d}{dt}\Big(mr^2\dot{\theta}\Big). \tag{9.1.8}$$

We see in Eq. (9.1.7) the effect of centrifugal force, and in Eq. (9.1.8) the second law of Kepler, in both cases derived without having to convert the Cartesian equations of motion (9.1.4) directly into polar coordinates.

A more challenging example of the Lagrangian formalism is provided by the theory of charged particles in an electromagnetic field, discussed in the next chapter.

## 9.2 Symmetry Principles and Conservation Laws

The great advantage of the Lagrangian formalism is that it provides a simple connection between symmetry principles and the existence of conserved quantities. Every continuous symmetry of the action implies the existence of a quantity that, according to the equations of motion, does not change with time. This general result is due to Emmy Noether (1882–1935), and is known as *Noether's theorem.*[1]

Consider any infinitesimal transformation of the variables $q_N(t)$,

$$q_N \to q_N + \epsilon \mathcal{F}_N(q, \dot{q}), \tag{9.2.1}$$

where $\epsilon$ is an infinitesimal constant, and the $\mathcal{F}_N$ are functions of the $q$s and $\dot{q}$s that depend on the nature of the symmetry in question. This is a symmetry of the Lagrangian if

$$0 = \sum_N \left[ \frac{\partial L}{\partial q_N} \mathcal{F}_N + \frac{\partial L}{\partial \dot{q}_N} \dot{\mathcal{F}}_N \right]. \tag{9.2.2}$$

Using the Lagrangian equations (9.1.3) of motion in the first term, this is

$$0 = \sum_N \left[ \left( \frac{d}{dt} \frac{\partial L}{\partial \dot{q}_N} \right) \mathcal{F}_N + \frac{\partial L}{\partial \dot{q}_N} \dot{\mathcal{F}}_N \right] = \frac{dF}{dt}, \tag{9.2.3}$$

where $F$ is the conserved quantity

$$F \equiv \sum_N \frac{\partial L}{\partial \dot{q}_N} \mathcal{F}_N(q, \dot{q}). \tag{9.2.4}$$

For instance, as long as the potential $V$ depends only on differences of particle coordinates, the Lagrangian (9.1.5) is invariant under translations

$$x_{ni} \to x_{ni} + \epsilon_i \tag{9.2.5}$$

with the same $\epsilon_i$ for each particle label $n$. Then, for each $i$, we have a conserved quantity, the $i$th component of the total momentum

$$P_i = \sum_n \frac{\partial L}{\partial \dot{x}_{ni}} = \sum_n m_n \dot{x}_{ni}. \tag{9.2.6}$$

Similarly, if $V$ is rotationally invariant, then the Lagrangian (9.1.5) is invariant under the infinitesimal rotations

$$\mathbf{x}_n \to \mathbf{x}_n + \mathbf{e} \times \mathbf{x}_n, \tag{9.2.7}$$

with the same infinitesimal 3-vector $\mathbf{e}$ for each particle label $n$. It follows that

$$\frac{d}{dt}\mathbf{L} = 0, \tag{9.2.8}$$

where

$$\mathbf{e} \cdot \mathbf{L} = \sum_{ni} \frac{\partial L}{\partial \dot{x}_{ni}} [\mathbf{e} \times \mathbf{x}_n]_i = \sum_n m_n \dot{\mathbf{x}}_n \cdot [\mathbf{e} \times \mathbf{x}_n].$$

---

[1] E. Noether, *Nachr. König. Gesell. Wiss. zu Göttingen, Math.-phys. Klasse* 235 (1918).

Recalling that the triple scalar product of any vectors $\mathbf{a}$, $\mathbf{b}$, and $\mathbf{c}$ has the symmetry property $\mathbf{a} \cdot [\mathbf{b} \times \mathbf{c}] = \mathbf{b} \cdot [\mathbf{c} \times \mathbf{a}]$, we see that

$$\mathbf{L} = \sum_n m_n \mathbf{x}_n \times \dot{\mathbf{x}}_n. \tag{9.2.9}$$

This is only the orbital angular momentum, and of course it is not necessarily conserved if the interaction involves the spin operators $\mathbf{S}_n$ of the particles, because in that case the Lagrangian is not invariant under transformations like (9.2.7) unless we also include transformations of the spin.

More generally, we can consider transformations that are *not* symmetries of the Lagrangian, but that are symmetries of the action. It is important to be clear about what is meant by this. In saying that an infinitesimal transformation is a symmetry of the action, we do not mean only that the transformation leaves the action invariant when the equations of motion are satisfied, because *all* infinitesimal transformations leave the action invariant when the equations of motion are satisfied – that is how the equations of motion are derived in the Lagrangian formalism. A symmetry of the action is a transformation that leaves the action invariant, whether or not the equations of motion are satisfied. In this case, instead of Eq. (9.2.2), we must have

$$\sum_N \left[ \frac{\partial L}{\partial q_N} \mathcal{F}_N + \frac{\partial L}{\partial \dot{q}_N} \dot{\mathcal{F}}_N \right] = \frac{dG}{dt}, \tag{9.2.10}$$

where $G(t)$ is some function of the $q_N(t)$ and $\dot{q}_N(t)$, and perhaps also of $t$, that takes equal values (such as zero) at $t = \pm\infty$, so that $\int \dot{G}\,dt = 0$. To repeat, Eq. (9.2.10) is required to be satisfied whether or not $q_N(t)$ and $\dot{q}_N(t)$ obey the equations of motion (9.1.13). Where they are satisfied, the left-hand side of Eq. (9.2.10) equals $dF/dt$, and so this invariance condition yields the conservation law

$$0 = \frac{d}{dt}[F - G], \tag{9.2.11}$$

with $F$ again given by Eq. (9.2.4). We will see an example of such a symmetry of the action in the next section.

## 9.3 The Hamiltonian Formalism

From the Lagrangian we can construct the quantity known as the Hamiltonian, whose usefulness we have seen repeatedly in the foregoing chapters. The Hamiltonian is conserved if the Lagrangian has no *explicit* dependence on time, and more generally its time-dependence arises solely from any explicit time-dependence of the Lagrangian. The Hamiltonian is defined by

$$H \equiv \sum_N \dot{q}_N \frac{\partial L}{\partial \dot{q}_N} - L. \tag{9.3.1}$$

Using the Lagrangian equations of motion (9.1.3), its rate of change is

$$\frac{dH}{dt} = \sum_N \ddot{q}_N \frac{\partial L}{\partial \dot{q}_N} + \sum_N \dot{q}_N \frac{\partial L}{\partial q_N} - \frac{dL}{dt}.$$

But the total rate of change of the Lagrangian is

$$\frac{dL}{dt} = \frac{\partial L}{\partial t} + \sum_N \ddot{q}_N \frac{\partial L}{\partial \dot{q}_N} + \sum_N \dot{q}_N \frac{\partial L}{\partial q_N},$$

where $\partial L / \partial t$ is the rate of change of the Lagrangian due to any explicit time-dependence, as in the case of time-dependent external fields. Hence

$$\frac{dH}{dt} = -\frac{\partial L}{\partial t}, \tag{9.3.2}$$

and in particular the Hamiltonian is conserved for isolated systems, where the Lagrangian has no explicit time-dependence.

The constancy of the Hamiltonian in cases where $L$ has no explicit time-dependence can be regarded as a consequence of the invariance of the action in such cases under a symmetry transformation: time translation. When we shift the time coordinate by an infinitesimal $\epsilon$, the change in any variable $q_N(t)$ is $\epsilon \dot{q}_N(t)$, so in the notation of Eq. (9.2.1), we have here $\mathcal{F}_N(t) = \dot{q}_N(t)$, and the quantity (9.2.4) is

$$F = \sum_N \frac{\partial L}{\partial q_N} \dot{q}_N.$$

This is not time-independent, because time-translation is a symmetry not of the Lagrangian, but only of the action. Here we have

$$\sum_N \left[ \frac{\partial L}{\partial q_N} \mathcal{F}_N + \frac{\partial L}{\partial \dot{q}_N} \dot{\mathcal{F}}_N \right] = \sum_N \left[ \frac{\partial L}{\partial q_N} \dot{q}_N + \frac{\partial L}{\partial \dot{q}_N} \ddot{q}_N \right] = \frac{dL}{dt},$$

so the quantity $G$ in Eq. (9.2.10) is here just $G = L$, and the conserved quantity in Eq. (9.2.1) is

$$F - G = \sum_N \frac{\partial L}{\partial q_N} \dot{q}_N - L = H.$$

Instead of the second-order differential equations of motion of the Lagrangian formalism, we can use the Hamiltonian formalism to write the equations of motion as first-order differential equations for twice as many variables: the $q_N$, and their "canonical conjugates,"

$$p_N = \frac{\partial L}{\partial \dot{q}_N}. \tag{9.3.3}$$

For this purpose, we must think of the Hamiltonian as a function $H(q, p)$ of the $q_N$ and $p_N$, with $\dot{q}_N$ in Eq. (9.3.1) regarded as a function of the $q_N$ and $p_N$ given by solving Eq. (9.3.3) for $\dot{q}_N$. That is, Eq. (9.3.1) should be interpreted as

$$H(q, p) = \sum_N \dot{q}_N(q, p) p_N - L\Big(q, \dot{q}(q, p)\Big). \qquad (9.3.4)$$

Then

$$\frac{\partial H}{\partial q_N} = \sum_M \frac{\partial \dot{q}_M}{\partial q_N} p_M - \frac{\partial L}{\partial q_N} - \sum_M \frac{\partial L}{\partial \dot{q}_M} \frac{\partial \dot{q}_M}{\partial q_N}.$$

The first and third terms cancel according to Eq. (9.3.3), and the Lagrangian equation of motion (9.1.3) then gives

$$\dot{p}_N = -\frac{\partial H}{\partial q_N}. \qquad (9.3.5)$$

Also,

$$\frac{\partial H}{\partial p_N} = \dot{q}_N + \sum_M p_M \frac{\partial \dot{q}_M}{\partial p_N} - \sum_M \frac{\partial L}{\partial \dot{q}_M} \frac{\partial \dot{q}_M}{\partial p_N}.$$

Now the second and third terms cancel, leaving us with

$$\dot{q}_N = \frac{\partial H}{\partial p_N}. \qquad (9.3.6)$$

Equations (9.3.5) and (9.3.6) are the general equations of motion in the Hamiltonian formalism.

For a very simple example, consider the Lagrangian (9.1.5):

$$L = \sum_n \frac{m_n}{2} \dot{\mathbf{x}}_n^2 - V(\mathbf{x}),$$

where here $q_{ni} \equiv [\mathbf{x}_n]_i$. Equation (9.3.3) here gives the familiar result $\mathbf{p}_n = m_n \dot{\mathbf{x}}_n$, which can be solved without much difficulty to give $\dot{\mathbf{x}}_n = \mathbf{p}_n/m_n$. The Hamiltonian (9.3.1) is then

$$H = \sum_n \frac{1}{m_n} \mathbf{p}_n^2 - L = \sum_n \frac{1}{2m_n} \mathbf{p}_n^2 + V(\mathbf{x}).$$

This is the familiar Hamiltonian on which we based our calculations in Chapter 2. The equations of motion (9.3.5) and (9.3.6) are here

$$\dot{p}_{ni} = -\frac{\partial V}{\partial x_{ni}}, \qquad \dot{x}_{ni} = p_{ni}/m_n,$$

which together yield the equations of motion (9.1.4).

The Hamiltonian formalism can be used in any coordinate system. For instance, for the two-dimensional system with Lagrangian (9.1.6), the canonical conjugates to $r$ and $\theta$ are

$$p_r = m\dot{r}, \qquad p_\theta = mr^2\dot{\theta} \tag{9.3.7}$$

and the Hamiltonian is

$$H = \frac{p_r^2}{2m} + \frac{p_\theta^2}{2mr^2} + V(r). \tag{9.3.8}$$

According to Eq. (9.3.5), the fact that the Hamiltonian does not depend on $\theta$ tells us immediately that $p_\theta$ is constant, in agreement with Kepler's second law.

## 9.4 Canonical Commutation Relations

Up to this point, our discussion in this chapter has been in classical terms, though it applies equally well to quantum-mechanical operators in the Heisenberg picture. Now we must make the transition to quantum mechanics by imposing suitable commutation relations on the $q_N$ and $p_N$.

To motivate these commutation relations, we return to the implementation of symmetry principles in quantum mechanics. For the present, we shall restrict ourselves to symmetries of the Lagrangian like space translation or rotation, for which the functions $\mathcal{F}_N$ introduced in Section 9.2 depend only on the $q$s, not the $\dot{q}$s. That is, we assume that the Lagrangian is invariant under an infinitesimal transformation

$$q_N \to q_N + \epsilon \mathcal{F}_N(q). \tag{9.4.1}$$

In order to realize this symmetry as a quantum-mechanical unitary transformation

$$[1 - i\epsilon F/\hbar]^{-1} q_N [1 - i\epsilon F/\hbar] = q_N + \epsilon \mathcal{F}_n(q), \tag{9.4.2}$$

we need an operator $F$ to serve as a generator of the symmetry, in the sense that

$$[F, q_N] = -i\hbar \mathcal{F}_N(q). \tag{9.4.3}$$

(The factor $-i/\hbar$ is extracted from $F$ in Eq. (9.4.2), to maintain an analogy with the formula (3.5.2) for the unitary operator that represents translations.) We saw in Section 9.2 that the invariance of the Lagrangian under the transformation (9.4.1) implies the existence of a conserved quantity (9.2.4), which we can now write

$$F = \sum_N p_N \mathcal{F}_N(q). \tag{9.4.4}$$

Such operators $F$ satisfy the commutation relation (9.4.3) for all symmetries of the form (9.4.1) if we impose the canonical commutation relations

$$[q_N(t), p_{N'}(t)] = i\hbar\delta_{NN'}, \tag{9.4.5}$$

$$[q_N(t), q_{N'}(t)] = [p_N(t), p_{N'}(t)] = 0. \tag{9.4.6}$$

The commutation relation of $p$s with each other in Eq. (9.4.6) is not needed to obtain Eq. (9.4.3), but with it, in simple cases, the operators (9.4.4) generate simple transformations of the $p_N$ as well as of the $q_N$. For the case of non-relativistic particles (labeled $n$) in a translation-invariant potential (where $N$ is the compound index $ni$), there is a symmetry under translations, in which Eq. (9.4.1) takes the form (9.2.5), and the generator (9.2.6) takes the form

$$\mathbf{P} = \sum_n \mathbf{p}_n. \tag{9.4.7}$$

In this case, it is obvious from Eq. (9.4.6) that the $\mathbf{p}_n$ are all translation-invariant,

$$[\mathbf{P}, \mathbf{p}_n] = 0. \tag{9.4.8}$$

Likewise, for non-relativistic spinless particles in a rotationally invariant potential, there is a symmetry under rotations, in which Eq. (9.4.1) takes the form (9.2.7), and the generator (9.2.9) takes the form

$$\mathbf{L} = \sum_n \mathbf{x}_n \times \mathbf{p}_n. \tag{9.4.9}$$

(Because this is a cross-product of vectors, it does not involve products of the same components of position and momentum, so the order of these operators is here immaterial.) In this case, $\mathbf{L}$ acts as a generator of rotations on both positions and momenta

$$[L_i, x_{nj}] = i\hbar \sum_k \epsilon_{ijk} x_{nk}, \qquad [L_i, p_{nj}] = i\hbar \sum_k \epsilon_{ijk} p_{nk}, \tag{9.4.10}$$

where as usual $\epsilon_{ijk}$ is the totally antisymmetric quantity with $\epsilon_{123} = 1$. (To prove this, write Eq. (9.4.9) as $L_i = \sum_n \epsilon_{ij'k'} x_{nj'} p_{nk'}$.)

In theories of particles with spin, an operator that involves spins in scalar combinations such as $\mathbf{s}_n \cdot \mathbf{p}_m$ or $\mathbf{s}_n \cdot \mathbf{x}_m$ will be rotationally invariant, but will not commute with the orbital angular momentum $\mathbf{L}$. The spin matrices $\mathbf{s}_n$ are defined to satisfy the usual commutation relations,

$$[s_{ni}, s_{n'j}] = i\hbar\delta_{nn'} \sum_k \epsilon_{ijk} s_{nk}, \qquad [s_{ni}, x_{n'j}] = [s_{ni}, p_{n'j}] = 0,$$

so the operator $\mathbf{J} \equiv \mathbf{L} + \sum_n \mathbf{s}_n$ generates rotations on spins as well as coordinates and momenta

$$[J_i, x_{nj}] = i\hbar \sum_k \epsilon_{ijk} x_{nk}, \qquad [J_i, p_{nj}] = i\hbar \sum_k \epsilon_{ijk} p_{nk},$$

$$[J_i, s_{nj}] = i\hbar \sum_k \epsilon_{ijk} s_{nk}. \tag{9.4.11}$$

Thus **J** commutes with any rotationally invariant operator.

The symmetry of time-translation invariance again requires special treatment, because it is a symmetry of the action but not of the Lagrangian, and because the functions $\mathcal{F}_N$ in the transformation rule (9.2.1) depend on the time-derivatives $\dot{q}_N$. We note that, as a consequence of the commutation relations (9.4.5) and (9.4.6), for any function $f(q, p)$ of the $q_N$ and $p_N$, we have

$$[f(q, p), q_N] = -i\hbar \frac{\partial f(q, p)}{\partial p_N}, \tag{9.4.12}$$

$$[f(q, p), p_N] = i\hbar \frac{\partial f(q, p)}{\partial q_N}. \tag{9.4.13}$$

(To prove Eq. (9.4.12), note that if we move $q_N$ in the product $f(q, p)q_N$ to the left past all the $p$s in $f(q, p)$, for each $p_N$ in $f(q, p)$ we get a term $-i\hbar$ times the function $f(q, p)$ with that $p_N$ omitted. The sum of these terms is the same as $-i\hbar \, \partial f(q, p)/\partial p_N$. The proof of Eq. (9.4.13) is similar. The derivatives must be calculated by removing factors of $p_N$ or $q_N$, leaving the order of all other operators unchanged. For instance $\partial q_2 p_1 p_2 / \partial p_1 = q_2 p_2$.) The Hamiltonian equations of motion (9.3.5) and (9.3.6) thus can be written

$$\dot{p}_N = \frac{i}{\hbar}[H(q, p), p_N], \qquad \dot{q}_N = \frac{i}{\hbar}[H(q, p), q_N], \tag{9.4.14}$$

so the Hamiltonian is the generator of time-translations. It follows also that for any function $f(q, p)$ that does not depend explicitly on time,

$$\dot{f}(q, p) = \frac{i}{\hbar}[H(q, p), f(q, p)]. \tag{9.4.15}$$

In particular, since **P** commutes with any translationally invariant Hamiltonian, it is conserved in the absence of external fields. The spin matrices in the Heisenberg picture are *defined* to have a time-dependence matching Eq. (9.4.14):

$$\dot{\mathbf{s}}_n = \frac{i}{\hbar}[H, \mathbf{s}_n]. \tag{9.4.16}$$

From Eqs. (9.4.15) and (9.4.16) we have the same for the total angular momentum $\mathbf{J} = \mathbf{L} + \sum_n \mathbf{s}_n$,

$$\dot{\mathbf{J}} = \frac{i}{\hbar}[H, \mathbf{J}], \tag{9.4.17}$$

so **J** is conserved if the Hamiltonian is rotationally invariant, as it will be for isolated systems.

We can generalize Eqs. (9.4.12) and (9.4.13) to give a formula for the commutator of two functions of both $q$s and $p$s:

$$[f(q, p), g(q, p)] = i\hbar[f(q, p), g(q, p)]_{\mathrm{P}}, \tag{9.4.18}$$

where $[f(q, p), g(q, p)]_P$ denotes the quantity known in classical dynamics as the *Poisson bracket*

$$[f(q, p), g(q, p)]_P \equiv \sum_N \left[ \frac{\partial f(q, p)}{\partial q_N} \frac{\partial g(q, p)}{\partial p_N} - \frac{\partial g(q, p)}{\partial q_N} \frac{\partial f(q, p)}{\partial p_N} \right].$$

(9.4.19)

(When we move $f(q, p)$ to the right past $g(q, p)$ we get a sum of terms: according to Eq. (9.4.12) for each $q_N$ in $g(q, p)$ we get a factor $-i\hbar \, \partial f(q, p)/\partial p_N$ times $g(q, p)$ with that $q_N$ omitted, which gives the second term in Eq. (9.4.19), and according to Eq. (9.4.13) for each $p_N$ in $g(q, p)$ we get a factor $+i\hbar \, \partial f(q, p)/\partial q_N$ times $g(q, p)$ with that $p_N$ omitted, which gives the first term in Eq. (9.4.19). Again, in quantum mechanics one must specify the order of the $q$s and $p$s in the Poisson bracket, which is best done on a case-by-case basis.)

Commutators have certain algebraic properties:

$$[f, g] = -[g, f], \tag{9.4.20}$$

$$[f, gh] = [f, g]h + g[f, h], \tag{9.4.21}$$

and the *Jacobi identity*

$$[f, [g, h]] + [g, [h, f]] + [h, [f, g]] = 0. \tag{9.4.22}$$

It is easy to check directly that the Poisson bracket (9.4.19) satisfies the same algebraic conditions.

As we saw in Section 1.4, on the basis of an analogy with the Poisson brackets of quantum mechanics, Dirac in 1926 generalized the commutation relations guessed at by Heisenberg to the full set (9.4.5), (9.4.6). But it would be difficult to argue that this analogy or the canonical formalism itself has the status of a fundamental principle of physics, especially since there are physical quantities like spin to which the canonical formalism does not apply. On the other hand, in the present state of physics symmetry principles seem as fundamental as anything we know. That is why in this section the canonical commutation relations have been motivated by the necessity of constructing quantum-mechanical operators that generate symmetry transformation, rather than by an analogy with Poisson brackets.

## 9.5 Constrained Hamiltonian Systems

So far we have considered systems with equal numbers of independent $q$s and $p$s, but in general these canonical variables may be subject to constraints. We will see an important physical example of such a constrained system in Chapter 11, but for the present we will illustrate the problem with a somewhat artificial but revealing example: a non-relativistic particle that is constrained to remain on a surface described by a constraint

$$f(\mathbf{x}) = 0, \tag{9.5.1}$$

where $f(\mathbf{x})$ is some smooth function of position. For instance, for a particle constrained to move on a sphere of radius $R$, we could take $f(\mathbf{x}) = \mathbf{x}^2 - R^2$.

We can take the Lagrangian as

$$L(\mathbf{x}, \dot{\mathbf{x}}) = \frac{m}{2}\dot{\mathbf{x}}^2 - V(\mathbf{x}) + \lambda f(\mathbf{x}), \tag{9.5.2}$$

where $V(\mathbf{x})$ is a local potential and $\lambda$ is an additional coordinate. The Lagrangian equations of motion for $\mathbf{x}$ are

$$m\ddot{\mathbf{x}} = -\nabla V + \lambda \nabla f = 0. \tag{9.5.3}$$

Also, since no time derivative of $\lambda$ appears in the Lagrangian, the equation of motion for $\lambda$ just says that $\partial L/\partial\lambda = 0$, which yields the constraint (9.5.1). (Note that $\nabla f(\mathbf{x})$ is in the direction of the normal to the surface (9.5.1) at $\mathbf{x}$, because for any infinitesimal vector $\mathbf{u}$ that is tangent to this surface at $\mathbf{x}$, both $f(\mathbf{x} + \mathbf{u})$ and $f(\mathbf{x})$ must vanish, so $f(\mathbf{x} + \mathbf{u}) - f(\mathbf{x}) = \mathbf{u}\cdot\nabla f(\mathbf{x}) = 0$. Hence Eq. (9.5.3) embodies the physical requirement that constraining the particle to the surface (9.5.1) can only produce forces normal to this surface.)

Equation (9.5.1) is what is known as a primary constraint, imposed directly by the nature of the system. There is also a secondary constraint, imposed by the condition that the primary constraint remains satisfied as the particle moves: for all $\mathbf{x}$ on the surface,

$$\frac{df}{dt} = \dot{\mathbf{x}}\cdot\nabla f(\mathbf{x}) = 0. \tag{9.5.4}$$

Then there is also the condition that this secondary constraint remains satisfied:

$$\ddot{\mathbf{x}}\cdot\nabla f + (\dot{\mathbf{x}}\cdot\nabla)^2 f = 0. \tag{9.5.5}$$

(The quantity $(\dot{\mathbf{x}}\cdot\nabla)^2 f$ does not generally vanish, because Eq. (9.5.4) only requires that $\dot{\mathbf{x}}\cdot\nabla f$ must vanish when $\mathbf{x}$ is on the surface, so that its gradient in directions off the surface need not vanish.) Equation (9.5.5) is not counted as a new constraint, because it just serves to determine $\lambda$. Using the equation of motion (9.5.3) in Eq. (9.5.5) gives

$$\lambda = \frac{1}{(\nabla f)^2}\left[\nabla f\cdot\nabla V - m(\dot{\mathbf{x}}\cdot\nabla)^2 f\right], \tag{9.5.6}$$

so the equation of motion becomes

$$m\ddot{\mathbf{x}} = -\nabla V + \nabla f\,\frac{\nabla f\cdot\nabla V}{(\nabla f)^2} - \frac{m}{(\nabla f)^2}\nabla f\,(\dot{\mathbf{x}}\cdot\nabla)^2 f. \tag{9.5.7}$$

The reader can check that this equation depends only on the surface to which the particle is constrained, not on the particular function $f(\mathbf{x})$ whose vanishing is used to describe this constraint. That is, if we introduce a new function $g(\mathbf{x}) = G\big(f(\mathbf{x})\big)$, where $G$ is any smooth function of $f$ with a unique zero at $f = 0$, then from the equation of motion with $g(\mathbf{x})$ in place of $f(\mathbf{x})$, we can derive the equation of motion in the form (9.5.7) involving $f$.

Since $\partial L / \partial \dot{\lambda} = 0$, the Hamiltonian for this system is simply

$$H(\mathbf{x}, \mathbf{p}) = \mathbf{p} \cdot \dot{\mathbf{x}} - L,$$

where

$$\mathbf{p} = m\dot{\mathbf{x}}.$$

Using the constraint (9.5.1), this is simply

$$H(\mathbf{x}, \mathbf{p}) = \frac{\mathbf{p}^2}{2m} + V(\mathbf{x}). \tag{9.5.8}$$

But we cannot here impose the usual canonical commutation relations $[x_i, p_j] = i\hbar\delta_{ij}$, because this would be inconsistent with both the primary constraint (9.5.1) and the secondary constraint (9.5.4), which now reads

$$\mathbf{p} \cdot \nabla f = 0. \tag{9.5.9}$$

So what commutation rules *should* we use?

A general answer was suggested by Dirac[2] for a large class of constrained Hamiltonian systems. Suppose there are a number of primary and secondary constraints, which can be expressed in the form

$$\chi_r(q, p) = 0. \tag{9.5.10}$$

For instance, in the problem discussed above, there are two $\chi$s, with

$$\chi_1 = f(\mathbf{x}), \qquad \chi_2 = \mathbf{p} \cdot \nabla f(\mathbf{x}). \tag{9.5.11}$$

Dirac distinguished two cases, distinguished by the properties of the matrix

$$C_{rs}(q, p) \equiv [\chi_r(q, p), \chi_s(q, p)]_{\mathrm{P}}, \tag{9.5.12}$$

where $[f, g]_{\mathrm{P}}$ denotes the Poisson bracket, defined by Eq. (9.4.19):

$$[f(q, p), g(q, p)]_{\mathrm{P}} \equiv \sum_N \left[ \frac{\partial f(q, p)}{\partial q_N} \frac{\partial g(q, p)}{\partial p_N} - \frac{\partial g(q, p)}{\partial q_N} \frac{\partial f(q, p)}{\partial p_N} \right],$$

$$\tag{9.5.13}$$

with the constraints applied only *after* the partial derivatives are calculated. Constraints for which there exists some $u_s$ for which $\sum_s C_{rs} u_s = 0$ for all $r$ are called *first-class constraints*, and must be dealt with by imposing conditions that reduce the number of independent variables. (For instance, in the example of a particle constrained to a surface, if we kept $\lambda$ as an independent variable instead of imposing the condition (9.5.6), then the constraints in this example would be first class. We will see another example of a first-class constraint in Chapter 11, eliminated by a choice of gauge for the electromagnetic potentials.) When this has been done, the constraints are of the *second class*, defined by the condition that

$$\mathrm{Det}\, C \neq 0, \tag{9.5.14}$$

---

[2] P. A. M. Dirac, *Lectures on Quantum Mechanics* (Yeshiva University, New York, 1964).

so that the matrix $C$ has an inverse $C^{-1}$. Dirac proposed that in a theory with only second-class constraints, instead of commutators being given by $i\hbar$ times the Poisson bracket, as in Eq. (9.4.18), they are given by

$$[f(q,p), g(q,p)] = i\hbar[f(q,p), g(q,p)]_D, \qquad (9.5.15)$$

where $[f(q,p), g(q,p)]_D$ is the *Dirac bracket*[3]

$$[f(q,p), g(q,p)]_D \equiv [f(q,p), g(q,p)]_P - \sum_{rs}[f(q,p), \chi_r(q,p)]_P$$

$$\times\, C_{rs}^{-1}(q,p)[\chi_s(q,p), g(q,p)]_P. \qquad (9.5.16)$$

In particular, in place of the usual canonical commutation relations, Dirac's proposal requires that

$$[q_N, p_M] = i\hbar\left[\delta_{NM} - \sum_{rs}\frac{\partial\chi_r}{\partial p_N}C_{rs}^{-1}\frac{\partial\chi_s}{\partial q_M}\right], \qquad (9.5.17)$$

and

$$[q_N, q_M] = i\hbar\sum_{rs}\frac{\partial\chi_r}{\partial p_N}C_{rs}^{-1}\frac{\partial\chi_s}{\partial p_M}, \qquad (9.5.18)$$

$$[p_N, p_M] = i\hbar\sum_{rs}\frac{\partial\chi_r}{\partial q_N}C_{rs}^{-1}\frac{\partial\chi_s}{\partial q_M}. \qquad (9.5.19)$$

(Where the Dirac bracket involves non-commuting operators, it is necessary to be careful with their ordering. Once again, this has to be dealt with on a case-by-case basis.) Conversely, the general commutation relation (9.5.15) follows from Eqs. (9.5.17)–(9.5.19).

This proposal satisfies a number of necessary conditions on commutators. First, the Dirac bracket has the same algebraic properties (9.4.20)–(9.4.22) as commutators:

$$[f, g]_D = -[g, f]_D, \qquad (9.5.20)$$
$$[f, gh]_D = [f, g]_D h + g[f, h]_D, \qquad (9.5.21)$$
$$[f, [g, h]_D]_D + [g, [h, f]_D]_D + [h, [f, g]_D]_D = 0. \qquad (9.5.22)$$

Further, the assumption (9.5.15) is consistent with the constraints. Note that the Dirac bracket of any constraint function, say $\chi_r(q,p)$, with any other function $g(q,p)$ is given by Eqs. (9.5.12) and (9.5.16) as

$$[\chi_r, g]_D = [\chi_r, g]_P - \sum_{r's}C_{rr'}C_{r's}^{-1}[\chi_s, g]_P = 0, \qquad (9.5.23)$$

---

[3] There are various circumstances in which Eq. (9.5.15) can be derived from the usual canonical commutation relations for a reduced set of canonical variables; see T. Maskawa and H. Nakajima, *Prog. Theor. Phys.* **56**, 1295 (1976); S. Weinberg, *The Quantum Theory of Fields*, Vol. I (Cambridge University Press, Cambridge, 1995), Appendix to Chapter 7.

so that Eq. (9.5.15) is consistent with the condition that the operator $\chi_r$ vanishes.

Let's see how this works for the above example of a particle constrained to a surface. The Poisson bracket of the constraint functions (9.5.11) is

$$C_{12} = -C_{21} = [\chi_1, \chi_2]_D = (\nabla f)^2, \tag{9.5.24}$$

and of course $C_{11} = C_{22} = 0$, so the inverse $C$-matrix has elements

$$C_{12}^{-1} = -C_{21}^{-1} = -(\nabla f)^{-2}, \quad C_{11}^{-1} = C_{22}^{-1} = 0. \tag{9.5.25}$$

Thus (9.5.17) gives

$$[x_i, p_j] = i\hbar \left[ \delta_{ij} - \frac{\partial f}{\partial x_i} (\nabla f)^{-2} \frac{\partial f}{\partial x_j} \right]. \tag{9.5.26}$$

Also, since $\chi_1$ does not depend on $\mathbf{p}$, Eq. (9.5.18) here gives

$$[x_i, x_j] = 0. \tag{9.5.27}$$

It takes a little more effort to calculate the commutator of the $p$s. According to Eq. (9.5.19), we have

$$[p_i, p_j] = -i\hbar \left[ \frac{\partial f}{\partial x_i} (\nabla f)^{-2} \frac{\partial}{\partial x_j} (\mathbf{p} \cdot \nabla f) - i \leftrightarrow j \right]. \tag{9.5.28}$$

In general, this does not vanish. For instance, if we constrain the particle to remain on a sphere of radius $R$, so that $f(\mathbf{x}) = \mathbf{x}^2 - R^2$, then Eq. (9.5.28) gives

$$[p_i, p_j] = -i \frac{\hbar}{R^2} \left( x_i p_j - x_j p_i \right).$$

The difference between these commutation relations and the usual ones is the non-vanishing of the commutator (9.5.28), and the presence of the second term in Eq. (9.5.26), which is needed for the commutator of $\mathbf{p} \cdot \nabla f$ with $x_i$ to be consistent with the vanishing of $\mathbf{p} \cdot \nabla f$.

We can now work out the equations of motion in this example. Because the Hamiltonian $H$ is the generator of time-translations, we must as usual have $\dot{\mathcal{O}} = (i/\hbar)[H, \mathcal{O}]$ for any operator $\mathcal{O}$. Using the commutation relations (9.5.26)–(9.5.28) and Eq. (9.5.8) for $H$, we have

$$\dot{x}_i = \frac{i}{2m\hbar} [\mathbf{p}^2, x_i] = \frac{1}{m} p_j \left[ \delta_{ij} - \frac{\partial f}{\partial x_i} (\nabla f)^{-2} \frac{\partial f}{\partial x_j} \right],$$

and since $\mathbf{p} \cdot \nabla f = 0$, this gives the familiar result

$$\dot{\mathbf{x}} = \mathbf{p}/m. \tag{9.5.29}$$

On the other hand,

$$
\dot{p}_j = \frac{i}{\hbar} \left[ \left( \frac{\mathbf{p}^2}{2m} + V(\mathbf{x}) \right), \, p_j \right]
$$

$$
= \frac{1}{m(\nabla f)^2} \frac{\partial f}{\partial x_j} (\mathbf{p} \cdot \nabla)^2 f - \sum_i \frac{\partial V}{\partial x_i} \left[ \delta_{ij} - \frac{\partial f}{\partial x_i} (\nabla f)^{-2} \frac{\partial f}{\partial x_j} \right]
$$

or in other words,

$$
\dot{\mathbf{p}} = -\frac{1}{m(\nabla f)^2} \, \nabla f \, (\mathbf{p} \cdot \nabla)^2 f - \nabla V + \nabla f \, \frac{\nabla f \cdot \nabla V}{(\nabla f)^2}. \tag{9.5.30}
$$

Thus Dirac's assumption (9.5.15) yields the same equations of motion (9.5.7) as provided by the classical Lagrangian for this model.

## 9.6 The Path-Integral Formalism

In his Ph.D. thesis,[4] Richard Feynman (1918–1988) proposed a formalism, according to which the amplitude for a transition from one configuration of a set of particles at an initial time to another configuration at a final time is given by an integral over all the paths that particles can take in going from the initial to the final configuration. Feynman seems to have intended this path-integral formalism as an alternative to the usual formulation of quantum mechanics, but as later realized, it can be derived from the usual canonical formalism.

Let us consider a set of Heisenberg-picture operators $Q_N(t)$ and their canonical conjugates $P_N(t)$, satisfying the usual commutation relations (9.4.5) and (9.4.6):

$$
[Q_N(t), P_M(t)] = i\hbar \delta_{NM}, \tag{9.6.1}
$$

$$
[Q_N(t), Q_M(t)] = [P_N(t), P_M(t)] = 0. \tag{9.6.2}
$$

(We are now using upper case letters to distinguish the operators from their eigenvalues, which are denoted with lower case letters.) We can introduce a complete orthonormal set of eigenvectors of all the $Q_N(t)$:

$$
Q_N(t)\Psi_{q,t} = q_N \Psi_{q,t}, \tag{9.6.3}
$$

$$
\left( \Psi_{q',t}, \Psi_{q,t} \right) = \delta(q - q') \equiv \prod_N \delta(q_N - q'_N). \tag{9.6.4}
$$

Suppose we want to calculate the probability amplitude $\left( \Psi_{q',t'}, \Psi_{q,t} \right)$ for the system to go from a state in which the $Q_N(t)$ have eigenvalues $q_N$ to a state

---

[4] R. P. Feynman, *The Principle of Least Action in Quantum Mechanics* (Princeton University, 1942; University Microfilms Publication No. 2948, Ann Arbor, MI). Also see R. P. Feynman and A. R. Hibbs, *Quantum Mechanics and Path Integrals* (McGraw-Hill, New York, 1965).

in which the $Q_N(t')$ have eigenvalues $q'_N$, where $t' > t$. For this purpose, we introduce into the time interval from $t$ to $t'$ a large number $\mathcal{N}$ of times $\tau_n$, with $t' > \tau_1 > \tau_2 > \cdots > \tau_{\mathcal{N}} > t$, and use the completeness of the states $\Psi_{q,\tau}$ to write

$$\left(\Psi_{q',t'}, \Psi_{q,t}\right)$$
$$= \int dq_1\, dq_2 \dots dq_{\mathcal{N}} \left(\Psi_{q',t'}, \Psi_{q_1,\tau_1}\right)\left(\Psi_{q_1,\tau_1}, \Psi_{q_2,\tau_2}\right) \cdots \left(\Psi_{q_{\mathcal{N}},\tau_{\mathcal{N}}}, \Psi_{q,t}\right),$$
$$(9.6.5)$$

where $\int dq_n$ is an abbreviation for $\prod_N \int dq_{N,n}$. (The subscripts on the $q$s in Eq. (9.6.5) are values of the index $n$, labeling different times, rather than values of the index $N$, which labels different canonical variables.) So now we need to calculate the scalar product $\left(\Psi_{q',\tau'}, \Psi_{q,\tau}\right)$ for a general $q'$ and $q$ (not necessarily related to the $q$ and $q'$ in Eq. (9.6.5)) when $\tau'$ is very slightly larger than $\tau$.

For this purpose, we recall that the Heisenberg-picture operators have a time-dependence given by

$$Q_N(\tau') = e^{iH(\tau'-\tau)/\hbar}Q_N(\tau)e^{-iH(\tau'-\tau)/\hbar}, \tag{9.6.6}$$

so

$$\Psi_{q',\tau'} = e^{iH(\tau'-\tau)/\hbar}\Psi_{q',\tau}, \tag{9.6.7}$$

and therefore

$$\left(\Psi_{q',\tau'}, \Psi_{q,\tau}\right) = \left(\Psi_{q',\tau}, e^{-iH(\tau'-\tau)/\hbar}\Psi_{q,\tau}\right). \tag{9.6.8}$$

(Note that the argument of the exponential in Eq. (9.6.7) is $iH(\tau'-\tau)/\hbar$ rather than $-iH(\tau'-\tau)/\hbar$ because $\Psi_{q',\tau'}$ is not the Schrödinger-picture state vector at time $\tau'$, but is rather defined as an eigenstate of a Heisenberg-picture operator at this time.) Now, the Hamiltonian $H$ may be written as a function of the Schrödinger-picture operators $Q_N$ and $P_N$, or since the Hamiltonian commutes with itself, it can just as well be written as the same function of $Q_N(\tau)$ and $P_N(\tau)$ for any $\tau$. To evaluate the matrix element (9.6.8) we need to insert a complete orthonormal set of eigenstates of the $P_N(t)$ to the right of the exponential,

$$\left(\Psi_{q',\tau'}, \Psi_{q,\tau}\right) = \int dp \left(\Psi_{q',\tau}, \exp\left[-iH\left(Q(\tau), P(\tau)\right)(\tau'-\tau)/\hbar\right]\Phi_{p,\tau}\right)$$
$$\times \left(\Phi_{p,\tau}, \Psi_{q,\tau}\right),$$

where $\int dp \equiv \prod_N \int dp_N$, and

$$P_N(\tau)\Phi_{p,\tau} = p_N\Phi_{p,\tau}, \tag{9.6.9}$$

$$\left(\Phi_{p',\tau}, \Phi_{p,\tau}\right) = \delta(p - p') \equiv \prod_N \delta(p_N - p'_N). \tag{9.6.10}$$

We can always use the commutation relations (9.6.1) and (9.6.2) to write the Hamiltonian in a form with all $Q$s to the left of all $P$s, in which case the operators $Q(\tau)$ and $P(\tau)$ in the Hamiltonian can be replaced with their eigenvalues:[5]

$$\left(\Psi_{q',\tau'},\,\Psi_{q,\tau}\right) = \int dp \, \exp\left[-iH(q',p)(\tau'-\tau)/\hbar\right]\left(\Psi_{q',\tau},\,\Phi_{p,\tau}\right)\left(\Phi_{p,\tau},\,\Psi_{q,\tau}\right). \tag{9.6.11}$$

Just as for ordinary plane waves, the scalar products remaining in Eq. (9.6.11) take the simple form

$$\left(\Psi_{q',\tau},\,\Phi_{p,\tau}\right) = \prod_N \frac{e^{ip_N q'_N/\hbar}}{\sqrt{2\pi\hbar}}, \quad \left(\Phi_{p,\tau},\,\Psi_{q,\tau}\right) = \prod_N \frac{e^{-ip_N q_N/\hbar}}{\sqrt{2\pi\hbar}},$$

so Eq. (9.6.11) now reads

$$\left(\Psi_{q',\tau'},\,\Psi_{q,\tau}\right) = \int \prod_N \frac{dp_N}{2\pi\hbar} \exp\left[\,-iH(q',p)(\tau'-\tau)/\hbar \right.$$
$$\left. +i\sum_N p_N(q'_N - q_N)/\hbar\right],$$

or in the form in which we need it in Eq. (9.6.5),

$$\left(\Psi_{q_n,\tau_n},\,\Psi_{q_{n+1},\tau_{n+1}}\right) = \int \prod_N \frac{dp_{N,n}}{2\pi\hbar}$$
$$\times \exp\left[\,-\frac{i}{\hbar}H(q_n,p_n)(\tau_n - \tau_{n+1})\right.$$
$$\left. +\frac{i}{\hbar}\sum_N p_{N,n}(q_{N,n} - q_{N,n+1})\right], \tag{9.6.12}$$

with the understanding that

$$q_0 = q', \quad \tau_0 = t', \quad q_{n+1} = q, \quad \tau_{n+1} = \tau.$$

We can now use Eq. (9.6.12) for the matrix elements in Eq. (9.6.5), which gives

$$\left(\Psi_{q',t'},\,\Psi_{q,t}\right) = \int \left[\prod_N \prod_{n=1}^{\mathcal{N}} dq_{N,n}\right]\left[\int \prod_N \prod_{n=0}^{\mathcal{N}} \frac{dp_{N,n}}{2\pi\hbar}\right]$$
$$\times \exp\left[\,-\frac{i}{\hbar}\sum_{n=0}^{\mathcal{N}} H(q_n,p_n)(\tau_n - \tau_{n+1})\right.$$
$$\left. +\frac{i}{\hbar}\sum_N \sum_{n=0}^{\mathcal{N}} p_{N,n}(q_{N,n} - q_{N,n+1})\right]. \tag{9.6.13}$$

---

[5] Because $H$ appears in the exponential, this is only valid for infinitesimal $\tau' - \tau$, in which case the exponential is a linear function of $H$.

We can introduce c-number functions $q_N(\tau)$ and $p_N(\tau)$ that interpolate between the $\tau_n$, in such a way that

$$q_N(\tau_n) = q_{N,n}, \qquad p_N(\tau_n) = p_{N,n}. \tag{9.6.14}$$

Further, we can take the difference of successive $\tau$s to be an infinitesimal $d\tau$:

$$\tau_{n-1} - \tau_n = d\tau, \tag{9.6.15}$$

so that, to first order in $d\tau$,

$$q_{N,n} - q_{N,n+1} = \dot{q}_N(\tau_n)\, d\tau,$$
$$H(q_n, p_n)(\tau_n - \tau_{n+1}) = H(q(\tau_n), p(\tau_n))\, d\tau,$$

and therefore Eq. (9.6.13) may be written

$$\left(\Psi_{q',t'}, \Psi_{q,t}\right) = \int_{q(t)=q;\; q(t')=q'} \prod_\tau dq(\tau) \int \prod_\tau \frac{dp(\tau)}{2\pi\hbar}$$
$$\times \exp\left[\frac{i}{\hbar}\int_t^{t'} d\tau \left(\sum_N p_N(\tau)\dot{q}_N(\tau) - H\big(q(\tau), p(\tau)\big)\right)\right], \tag{9.6.16}$$

where

$$\int \prod_\tau dq(\tau) \int \prod_\tau \frac{dp(\tau)}{2\pi\hbar} \equiv \int \prod_N \prod_{n=1}^{\mathcal{N}} dq_{N,n} \int \prod_N \prod_{n=0}^{\mathcal{N}} \frac{dp_{N,n}}{2\pi\hbar}.$$

That is, this is a *path integral*, an integral over all functions $q_N(\tau)$ and $p_N(\tau)$, with $q_N(\tau)$ constrained by the conditions that $q_N(t) = q_N$ and $q_N(t') = q'_N$.

One of the nice things about the path-integral formalism is that it allows an easy passage from quantum mechanics to the classical limit. In macroscopic systems, we generally have

$$\int_t^{t'} d\tau \left(\sum_N p_N(\tau)\dot{q}_N(\tau) - H\big(q(\tau), p(\tau)\big)\right) \gg \hbar.$$

The phase of the exponential in Eq. (9.6.16) is then very large, so that the exponential oscillates very rapidly, killing all contributions to the path integral except from paths where the phase is stationary with respect to small variations in the path. The condition that the phase is stationary with respect to variations of the $q_N(\tau)$ that leave the values at the initial and final times unchanged is that

$$0 = \int_t^{t'} \left[\sum_N p_N(\tau)\, \delta\dot{q}_N(\tau) - \frac{\partial H}{\partial q_N(\tau)}\, \delta q_N(\tau)\right]$$
$$= \int_t^{t'} \left[-\sum_N \dot{p}_N(\tau) - \frac{\partial H}{\partial q_N(\tau)}\right] \delta q_N(\tau),$$

so

$$\dot{p}_N = -\frac{\partial H}{\partial q_N}.$$

Also, the condition that the phase is stationary with respect to arbitrary variations of the $p_N(\tau)$ is that

$$\dot{q}_N = \frac{\partial H}{\partial p_N}.$$

Of course, we recognize these as the classical equations of motion.

Feynman was motivated in part by the aim of expressing transition probabilities in quantum mechanics in terms of the Lagrangian rather than the Hamiltonian. (As discussed in Section 8.7, in Lorentz-invariant theories the Lagrangian unlike the Hamiltonian is typically the integral of a scalar density.) But the integrand of the integral in the exponential in Eq. (9.6.16) is *not* the Lagrangian, because $p_N(t)$ here is an independent integration variable, not the quantity $\partial L/\partial \dot{q}_N$. There is one commonly encountered case in which the integral over $p(\tau)$ can be evaluated by simply setting $p_N = \partial L/\partial \dot{q}_N$, so that the integrand really is the Lagrangian. This is the case in which the Hamiltonian is the sum of a term of second order in the $p$s, with constant coefficients, plus possible terms of first and zeroth order in the $p$s, so that the exponential is a Gaussian function of the $p$s. The integral of a Gaussian function is given in general by the formula

$$\int_{-\infty}^{\infty} \prod_r d\xi_r \; \exp\left\{ i\left[ \frac{1}{2}\sum_{rs} K_{rs}\xi_r\xi_s + \sum_r L_r\xi_r + M \right] \right\}$$
$$= \left[ \mathrm{Det}(K/2i\pi) \right]^{-1/2} \exp\left\{ i\left[ \frac{1}{2}\sum_{rs} K_{rs}\xi_{0r}\xi_{0s} + \sum_r L_r\xi_{0r} + M \right] \right\},$$
$$\tag{9.6.17}$$

where $\xi_{0r}$ is the value of $\xi_r$ at which the argument of the exponential is stationary:

$$\sum_s K_{rs}\xi_{0s} + L_r = 0. \tag{9.6.18}$$

The value of $p_N(\tau)$ at which the integrand in Eq. (9.6.16) is stationary satisfies the condition that

$$\dot{q}_N(\tau) = \frac{\partial H\big(q(\tau), p(\tau)\big)}{\partial p_N(\tau)}, \tag{9.6.19}$$

whose solution makes $\sum_N p_N(\tau)\dot{q}_N(\tau) - H\Big(q(\tau), p(\tau)\Big)$ equal to the Lagrangian. So the integral over the $p$s in Eq. (9.6.16) gives

$$\Big(\Psi_{q',t'}, \Psi_{q,t}\Big) = C \int_{q(t)=q;\ q(t')=q'} \prod_\tau dq(\tau) \exp\left[\frac{i}{\hbar} \int_t^{t'} d\tau\, L\Big(q(\tau), \dot{q}(\tau)\Big)\right],$$
(9.6.20)

with $C$ a constant of proportionality that is independent of $q$ and $q'$, and independent of the terms in the Hamiltonian that are linear in or independent of the $p$s. It does, however, depend on the time interval $t' - t$, and on its splitting into $\mathcal{N}+1$ segments of length $d\tau$. For instance, for a non-relativistic particle moving in a potential in $D$ dimensions, the term in the Hamiltonian that is quadratic in $p$ is $\mathbf{p}^2/2m$, which according to Eq. (9.6.17) is all we need in order to calculate $C$. In this case[6]

$$C = \left[\frac{1}{2\pi\hbar} \int_{-\infty}^{\infty} dp\, \exp\left(-\frac{ip^2\, d\tau}{2m\hbar}\right)\right]^{(\mathcal{N}+1)D}$$
$$= \left[\frac{m}{2i\pi\hbar\, d\tau}\right]^{(\mathcal{N}+1)D/2}.$$
(9.6.21)

The remaining path integration in Eq. (9.6.20) is generally not easy. The cases where it can be done easily are that of a free particle (or free field), or a particle in a harmonic oscillator potential, for which the Lagrangian is quadratic in $\dot{q}_N$ and $q_N$. Here again, with a quadratic Lagrangian, the integral can be done up to a constant factor by setting $q(t)$ equal to the function for which the integral of the Lagrangian is stationary with respect to small variations in the functions $q_N(\tau)$ for which $q_N(t') = q'_N$ and $q_N(t) = q_N$ are fixed – that is, for which $q_N(\tau)$ satisfies the classical equations of motion

$$\frac{d}{d\tau} \frac{\partial L(\tau)}{\partial \dot{q}_N(\tau)} = \frac{\partial L(\tau)}{\partial q_N(\tau)},$$

with $q_N(t') = q'_N$ and $q_N(t) = q_N$. For instance, for a free particle in $D$ dimensions, we have $L = m\dot{\mathbf{x}}^2/2$, and the solution of the classical equations of motion has constant velocity

$$\dot{\mathbf{x}}(\tau) = \left(\frac{\mathbf{x}' - \mathbf{x}}{t' - t}\right).$$

Hence Eq. (9.6.20) gives

$$\Big(\Psi_{\mathbf{x}',t'}, \Psi_{\mathbf{x},t}\Big) = BC \exp\left(\frac{im(\mathbf{x}' - \mathbf{x})^2}{2(t' - t)\hbar}\right),$$
(9.6.22)

---

[6] Feynman and Hibbs, *op. cit.*, give an indirect argument for this result, rather than obtaining it from the integral over $p$s, which does not appear in their book.

where $B$ is, like $C$, a constant independent of $\mathbf{x}'$ and $\mathbf{x}$. A rather tedious calculation along the lines of our calculation of $C$ gives[7]

$$B = \mathcal{N}^{-D/2} \left( \frac{m}{2i\pi\hbar\, d\tau} \right)^{-D\mathcal{N}/2},$$

so, since $\mathcal{N}\, d\tau = t' - t$,

$$BC = \left( \frac{m}{2i\pi\hbar(t' - t)} \right)^{D/2}. \tag{9.6.23}$$

We can check this, by noting that (9.6.22) must approach the delta function $\delta^D(\mathbf{x}' - \mathbf{x})$ in the limit as $t' \to t$. That is, for any smooth function $f(\mathbf{x})$, in this limit we must have

$$\int d^D x \left( \frac{m}{2i\pi\hbar(t' - t)} \right)^{D/2} \exp\left( \frac{im(\mathbf{x}' - \mathbf{x})^2}{2(t' - t)\hbar} \right) f(\mathbf{x}) \to f(\mathbf{x}').$$

For $t' \to t$ the exponential varies very rapidly with $\mathbf{x}$ except at $\mathbf{x} = \mathbf{x}'$, so the integral can be done by setting the argument of $f$ equal to $\mathbf{x}'$, and all we need to show is that

$$\int d^D x \left( \frac{m}{2i\pi\hbar(t' - t)} \right)^{D/2} \exp\left( \frac{im(\mathbf{x}' - \mathbf{x})^2}{2(t' - t)\hbar} \right) = 1,$$

which follows from the standard formula for the integrals of Gaussian functions. The $\mathbf{x}'$-dependence of the matrix element (9.6.22) can be understood by noting that this matrix element is nothing but the wave function of the state $\Psi_{\mathbf{x},\tau}$, defined as an eigenstate of the $\mathbf{x}(\tau)$, in a basis in which the $\mathbf{x}(t')$ are diagonal. Thus this matrix element must satisfy the Schrödinger equation

$$-\left( \frac{\hbar^2 \nabla'^2}{2m} \right) \left( \Psi_{\mathbf{x}',t'}, \Psi_{\mathbf{x},t} \right) = i\hbar \frac{\partial}{\partial t'} \left( \Psi_{\mathbf{x}',t'}, \Psi_{\mathbf{x},t} \right),$$

and it does. Thus the path-integral formalism allows us to find the solution of the Schrödinger equation, without ever writing down the Schrödinger equation.

In an experiment in which a particle is made to pass from a point $\mathbf{x}$ on one side of a screen in which there are several holes to a point $\mathbf{x}'$ on the other side, there is not just one trajectory $\mathbf{x}(\tau)$ for which the action $\int L(\tau)\, d\tau$ is stationary, but a trajectory for each hole. The path-integral formalism thus allows us to understand the interference pattern produced in such an experiment without wave mechanics, but instead as a consequence of the superposition of contributions of several possible classical paths.

More generally, for non-quadratic Lagrangians, the path integral (9.6.20) cannot be calculated analytically. One way of dealing with this problem is to expand

---

[7] Feynman and Hibbs, *op. cit.* pp. 43–44.

in powers of the non-quadratic part of the Lagrangian, which yields a Lagrangian version of time-dependent perturbation theory. The other approach is to divide the range of integration from $t$ to $t'$ into a finite number of segments of duration $\Delta\tau$, and calculate the integral of $\exp(iL(\tau)\,\Delta\tau/\hbar)$ over particle coordinates at each segment end numerically. In quantum field theories one would also have to represent space as a lattice of points, and integrate over fields numerically at each point in the spacetime lattice. This approach can reveal features of a problem that are not accessible through perturbation theory.[8]

## Problems

1. Consider the theory of a single particle with Lagrangian

$$L = \frac{m}{2}\dot{\mathbf{x}}^2 + \dot{\mathbf{x}}\cdot\mathbf{f}(\mathbf{x}) - V(\mathbf{x}),$$

where $\mathbf{f}(\mathbf{x})$ and $V(\mathbf{x})$ are arbitrary vector and scalar functions of position.

- Find the equation of motion satisfied by $\mathbf{x}$.
- Find the Hamiltonian, as a function of $\mathbf{x}$ and its canonical conjugate $\mathbf{p}$.
- What is the Schrödinger equation satisfied by the coordinate-space wave function $\psi(\mathbf{x}, t)$?

2. Show that Poisson brackets and Dirac brackets both satisfy the Jacobi identity.

3. Consider a one-dimensional harmonic oscillator, with Hamiltonian

$$H = \frac{p^2}{2m} + \frac{m\omega^2 x^2}{2}.$$

Use the path-integral formalism to calculate the probability amplitude for a transition from a position $x$ at time $t$ to a position $x'$ at time $t' > t$.

---

[8] For applications of lattice methods to field theory, see M. Creutz, *Quarks, Gluons, and Lattices* (Cambridge University Press, Cambridge, 1985); T. DeGrand and C. DeTar, *Lattice Methods for Quantum Chromodynamics* (World Scientific Press, Singapore, 2006).

# 10

# Charged Particles in Electromagnetic Fields

In this chapter we take up the problem of charged non-relativistic particles in an external electromagnetic field – that is, a field produced by some macroscopic system whose quantum fluctuations are negligible. This problem is of great physical importance in itself, and it also provides an example in which the canonical commutation relations are somewhat surprising.

## 10.1 Canonical Formalism for Charged Particles

Consider a set of non-relativistic spinless particles with masses $m_n$ and charges $e_n$, in a classical external electric field $\mathbf{E}(\mathbf{x}, t)$ and magnetic field $\mathbf{B}(\mathbf{x}, t)$. (Effects of spin are considered in Section 10.3.) Because it is easy, we will also include in the theory a local potential $\mathcal{V}$ depending on some or all of the various particle coordinates. The equations of motion of the particles are

$$m_n \ddot{\mathbf{x}}_n(t) = e_n \left[ \mathbf{E}\Big(\mathbf{x}_n(t), t\Big) + \frac{1}{c}\dot{\mathbf{x}}_n(t) \times \mathbf{B}\Big(\mathbf{x}_n(t), t\Big) \right] - \nabla_n \mathcal{V}\Big(\mathbf{x}(t)\Big) . \quad (10.1.1)$$

It is not possible to write a simple Lagrangian for this system directly in terms of $\mathbf{E}$ and $\mathbf{B}$; instead we must introduce a vector potential $\mathbf{A}(\mathbf{x}, t)$ and scalar potential $\phi(\mathbf{x}, t)$, for which

$$\mathbf{E} = -\frac{1}{c}\dot{\mathbf{A}} - \nabla\phi , \quad \mathbf{B} = \nabla \times \mathbf{A} . \quad (10.1.2)$$

(This is always possible, because $\mathbf{E}$ and $\mathbf{B}$ satisfy the homogeneous Maxwell equations $\nabla \times \mathbf{E} + \dot{\mathbf{B}}/c = 0$ and $\nabla \cdot \mathbf{B} = 0$.)

Let us tentatively take the Lagrangian as

$$L(t) = \sum_n \left[ \frac{m_n}{2}\dot{\mathbf{x}}_n^2(t) - e_n\phi\Big(\mathbf{x}_n(t), t\Big) + \frac{e_n}{c}\dot{\mathbf{x}}_n(t) \cdot \mathbf{A}\Big(\mathbf{x}_n(t), t\Big) \right] - \mathcal{V}(\mathbf{x}) ,$$

$$(10.1.3)$$

and check whether it gives the right equations of motion (10.1.1). Here $\phi$ and $\mathbf{A}$ are external fields, not dynamical variables. (They will become dynamical variables when we quantize the electromagnetic field in the next chapter.) Therefore we are concerned here with the differential equations (9.1.3) only where the $q_N(t)$ are the coordinates $x_{ni}(t)$. For the Lagrangian (10.1.3), we have (leaving the time argument of $\mathbf{x}_n$ to be understood)

$$\frac{\partial L(t)}{\partial x_{ni}} = -e_n \frac{\partial \phi(\mathbf{x}_n, t)}{\partial x_{ni}} + \frac{e_n}{c} \sum_j \dot{x}_{nj} \frac{\partial A_j(\mathbf{x}_n, t)}{\partial x_{ni}} - \frac{\partial V(\mathbf{x})}{\partial x_{ni}} , \qquad (10.1.4)$$

$$\frac{\partial L(t)}{\partial \dot{x}_{ni}} = m_n \dot{x}_{ni} + \frac{e_n}{c} A_i(\mathbf{x}_n, t) , \qquad (10.1.5)$$

and so

$$\frac{d}{dt} \frac{\partial L(t)}{\partial \dot{x}_{ni}} = m_n \ddot{x}_{ni} + \frac{e_n}{c} \frac{\partial A_i(\mathbf{x}_n, t)}{\partial t} + \frac{e_n}{c} \sum_j \frac{\partial A_i(\mathbf{x}_n, t)}{\partial x_{nj}} \dot{x}_{nj} . \qquad (10.1.6)$$

The equations of motion (9.1.3) are then

$$m_n \ddot{x}_{ni} = -e_n \frac{\partial \phi(\mathbf{x}_n, t)}{\partial x_{ni}} - \frac{e_n}{c} \frac{\partial A_i(\mathbf{x}_n, t)}{\partial t}$$
$$+ \frac{e_n}{c} \sum_j \dot{x}_{nj} \left[ \frac{\partial A_j(\mathbf{x}_n, t)}{\partial x_{ni}} - \frac{\partial A_i(\mathbf{x}_n, t)}{\partial x_{nj}} \right] - \frac{\partial V(\mathbf{x})}{\partial x_{ni}} . \qquad (10.1.7)$$

We recognize that, according to Eq. (10.1.2), the coefficients of $e_n$ in the first two terms on the right add up to give the electric field. Also, the sum in the third term on the right is

$$\sum_j \dot{x}_{nj} \left[ \frac{\partial A_j(\mathbf{x}_n, t)}{\partial x_{ni}} - \frac{\partial A_i(\mathbf{x}_n, t)}{\partial x_{nj}} \right] = \sum_{jk} \dot{x}_{nj} \epsilon_{ijk} [\nabla \times \mathbf{A}(\mathbf{x}_n, t)]_k$$
$$= [\dot{\mathbf{x}}_n \times \mathbf{B}(\mathbf{x}_n, t)]_i ,$$

where as usual $\epsilon_{ijk}$ is the totally antisymmetric tensor with $\epsilon_{123} = 1$. Hence the equation of motion (10.1.7) derived from this Lagrangian is indeed the same as Eq. (10.1.1).

To calculate energy levels, we need to construct a Hamiltonian. According to Eq. (10.1.5), here the time derivative of the coordinate is a function of both the coordinate and its canonical conjugate:

$$\dot{\mathbf{x}}_n = \frac{1}{m_n} \left[ \mathbf{p}_n - \frac{e_n}{c} \mathbf{A}(\mathbf{x}_n, t) \right] . \qquad (10.1.8)$$

Equation (9.3.1) then gives the Hamiltonian as

$$H(\mathbf{x}, \mathbf{p}, t) = \sum_n \frac{1}{m_n} \mathbf{p}_n \cdot \left[ \mathbf{p}_n - \frac{e_n}{c} \mathbf{A}(\mathbf{x}_n, t) \right]$$

$$- \sum_n \left\{ \frac{1}{2m_n} \left[ \mathbf{p}_n - \frac{e_n}{c} \mathbf{A}(\mathbf{x}_n, t) \right]^2 - e_n \phi \left( \mathbf{x}_n, t \right) \right.$$

$$\left. + \frac{e_n}{m_n c} \left[ \mathbf{p}_n - \frac{e_n}{c} \mathbf{A}(\mathbf{x}_n, t) \right] \cdot \mathbf{A} \left( \mathbf{x}_n, t \right) \right\}$$

$$+ \mathcal{V}(\mathbf{x}) ,$$

or more simply

$$H(\mathbf{x}, \mathbf{p}, t) = \sum_n \frac{1}{2m_n} \left[ \mathbf{p}_n - \frac{e_n}{c} \mathbf{A}(\mathbf{x}_n, t) \right]^2 + \sum_n e_n \phi \left( \mathbf{x}_n, t \right) + \mathcal{V}(\mathbf{x}) . \quad (10.1.9)$$

If we now used Eq. (10.1.8) to write the first term as $\sum_n m_n \dot{\mathbf{x}}_n^2/2$, then it would appear as if the dynamics of these particles was unaffected by the vector potential, but this is wrong; in using the Hamiltonian to derive dynamical equations, we must consider it as in Eq. (9.3.4), as a function of the $\mathbf{x}_n$ and $\mathbf{p}_n$, and not as a function of the $\mathbf{x}_n$ and $\dot{\mathbf{x}}_n$. In particular, it is $\mathbf{p}_n$ and not $m_n \dot{\mathbf{x}}_n$ that appears in the canonical commutation relations

$$[x_{ni}, p_{mj}] = i\hbar \delta_{nm} \delta_{ij} , \quad (10.1.10)$$

$$[x_{ni}, x_{mj}] = [p_{ni}, p_{mj}] = 0 . \quad (10.1.11)$$

We will use this Hamiltonian and these commutation relations in Section 10.3 to find the energy levels of a charged particle in a uniform magnetic field.

The presence of the vector potential in the Hamiltonian (10.1.9) does not invalidate the conservation of probability, but it does require a change in the probability current introduced in Eq. (1.5.5). For simplicity, consider a system containing just a single particle with mass $m$ and charge $-e$. (For atomic nuclei, replace $-e$ with $Ze$.) In the Schrödinger equation for the coordinate-space wave function $\psi$ we replace $\mathbf{p}$ with $-i\hbar \nabla$, as required by the commutation relations, so that

$$- i\hbar \frac{\partial \psi(\mathbf{x}, t)}{\partial t} = H(\mathbf{x}, -i\hbar \nabla, t) \psi(\mathbf{x}, t), \quad (10.1.12)$$

where

$$H(\mathbf{x}, -i\hbar \nabla, t) = \frac{1}{2m} \left[ -i\hbar \nabla + \frac{e}{c} \mathbf{A}(\mathbf{x}, t) \right]^2 - e\phi \left( \mathbf{x}, t \right) + \mathcal{V}(\mathbf{x}) . \quad (10.1.13)$$

Thus the rate of change of the probability density is

$$\frac{|\partial \psi(\mathbf{x}, t)|^2}{\partial t} = \frac{i}{\hbar} \left( \psi^*(\mathbf{x}, t) H(\mathbf{x}, -i\hbar \nabla, t) \psi(\mathbf{x}, t) \right.$$

$$\left. - \psi(\mathbf{x}, t) H(\mathbf{x}, +i\hbar \nabla, t) \psi^*(\mathbf{x}, t) \right) . \quad (10.1.14)$$

The terms $\mathcal{V}$, $-e\phi$, and $(e^2/2mc^2)\mathbf{A}^2$ in $H$ all cancel on the right-hand side, just leaving us with the terms of first and second order in gradients. A straightforward calculation then allows us to put Eq. (10.1.14) in the form of a conservation law analogous to Eq. (1.5.5):

$$\frac{|\partial\psi(\mathbf{x}, t)|^2}{\partial t} + \boldsymbol{\nabla} \cdot \boldsymbol{\mathcal{J}}(\mathbf{x}, t) = 0 , \qquad (10.1.15)$$

where $\boldsymbol{\mathcal{J}}(\mathbf{x}, t)$ is the probability current

$$\boldsymbol{\mathcal{J}} = \frac{-i\hbar}{2m} \left[ \psi^* \left[ \boldsymbol{\nabla} + \frac{ie}{\hbar c}\mathbf{A} \right] \psi - \psi \left( \left[ \boldsymbol{\nabla} + \frac{ie}{\hbar c}\mathbf{A} \right] \psi \right)^* \right] . \qquad (10.1.16)$$

## 10.2 Gauge Invariance

Different vector and scalar potentials can yield the same electric and magnetic fields. Specifically, inspection of Eqs. (10.1.2) shows that we can change the potentials by a *gauge transformation*

$$\mathbf{A}(\mathbf{x}, t) \mapsto \mathbf{A}'(\mathbf{x}, t) = \mathbf{A}(\mathbf{x}, t) + \boldsymbol{\nabla}\alpha(\mathbf{x}, t) , \qquad (10.2.1)$$

$$\phi(\mathbf{x}, t) \mapsto \phi'(\mathbf{x}, t) = \phi(\mathbf{x}, t) - \frac{1}{c}\frac{\partial}{\partial t}\alpha(\mathbf{x}, t) \qquad (10.2.2)$$

(where $\alpha(\mathbf{x}, t)$ is an arbitrary real function), with no change in the electric and magnetic fields. It is therefore striking that, although the Lagrangian (10.1.3) depends on the specific choice of vector and scalar potentials, the equations of motion derived from this Lagrangian depend only on the electric and magnetic fields. We can understand this by noting that, under the transformation (10.2.1), (10.2.2), the Lagrangian is transformed to

$$L(t) \mapsto L'(t) = L(t) + \sum_n \frac{e_n}{c} \left[ \frac{\partial\alpha(\mathbf{x}_n, t)}{\partial t} + \dot{\mathbf{x}}_n \cdot \boldsymbol{\nabla}_n\alpha(\mathbf{x}_n, t) \right]$$

$$= L(t) + \frac{d}{dt} \sum_n \frac{e_n}{c}\alpha(\mathbf{x}_n, t) . \qquad (10.2.3)$$

The Lagrangian is thus not gauge-invariant, but the action $\int dt\, L(t)$ *is* gauge-invariant (provided we take $\alpha(\mathbf{x}, t)$ to vanish for $t \to \pm\infty$), and since the field equations are the statement that the action is stationary with respect to small variations of the dynamical parameters that vanish as $t \to \pm\infty$, they too are gauge-invariant.

The Hamiltonian, though, is not gauge-invariant. If we make the change of gauge (10.2.1), (10.2.2) in the Hamiltonian (10.1.9), we obtain a new Hamiltonian:

$$H'(\mathbf{x}, \mathbf{p}, t) = \sum_n \frac{1}{2m_n} \left[ \mathbf{p}_n - \frac{e_n}{c} \mathbf{A}(\mathbf{x}_n, t) - \frac{e_n}{c} \nabla \alpha(\mathbf{x}_n, t) \right]^2$$

$$+ \sum_n e_n \phi \left( \mathbf{x}_n, t \right) - \sum_n \frac{e_n}{c} \frac{\alpha(\mathbf{x_n}, t)}{dt} + V(\mathbf{x}) . \tag{10.2.4}$$

Now, according to the commutation relations (10.1.10), (10.1.11), we can define a unitary operator

$$U(t) \equiv \exp \left[ i \sum_n \frac{e_n}{\hbar c} \alpha(\mathbf{x_n}, t) \right] , \tag{10.2.5}$$

for which

$$U(t) \mathbf{p}_n(t) U^{-1}(t) = \mathbf{p}_n(t) - \frac{e_n}{c} \nabla \alpha(\mathbf{x}_n, t) . \tag{10.2.6}$$

The Hamiltonian (10.2.4) in the new gauge may therefore be expressed as

$$H'(\mathbf{x}, \mathbf{p}, t) = U(t) H(\mathbf{x}, \mathbf{p}, t) U^{-1}(t) + i\hbar \left[ \frac{d}{dt} U(t) \right] U^{-1}(t) , \tag{10.2.7}$$

with the second term on the right providing the next-to-last term in Eq. (10.2.4). (We are taking the $\mathbf{x}_n$ and $\mathbf{p}_n$ here as time-independent operators in the Schrödinger picture, which allows us to write the time-derivative in the second term in Eq. (10.2.7) as $d/dt$ instead of $\partial/\partial t$.) It is then easy to see that, if $\Psi(t)$ satisfies the time-dependent Schrödinger equation in the original gauge

$$i\hbar \frac{d}{dt} \Psi(t) = H(t) \Psi(t) , \tag{10.2.8}$$

then the unitarily transformed state vector

$$\Psi'(t) \equiv U(t) \Psi(t) \tag{10.2.9}$$

satisfies the time-dependent Schrödinger equation in the new gauge:

$$i\hbar \frac{d}{dt} \Psi'(t) = U(t) H(t) \Psi(t) + i\hbar \left[ \frac{d}{dt} U(t) \right] \Psi(t) = H'(t) \Psi'(t) . \tag{10.2.10}$$

Recall that $\mathbf{x}_n$ is the operator that multiplies the coordinate-space wave function with the $n$th coordinate vector, so the transformation (10.2.9) is a position-dependent change of phase of the coordinate-space wave functions, with no change in the probability density in coordinate space. There is also no change in the probability current (10.1.16) for a single particle of charge $-e$ and mass $m$. The gauge transformation (10.2.1), (10.2.2) induces on the wave function of this particle a change of phase by a factor $\exp(-ie\alpha/\hbar c)$, so the effect in Eq. (10.1.6) of the change in the vector potential is canceled by the change of the gradient of $\psi$.

It is of special interest to consider the effect of a gauge transformation on the energy eigenvalues of the Hamiltonian in the case of time-independent electric and magnetic fields, for which the Hamiltonian is time-independent. To keep the fields time-independent, we will take the gauge transformation to be also time-independent.[1] In this case, Eq. (10.2.7) is just a unitary transformation, $H' = UHU^{-1}$, so if $\Psi$ is an eigenstate of $H$ with eigenvalue $E$, then $\Psi' = U\Psi$ is an eigenstate of $H'$ with the same eigenvalue $E$. In cases where energies are well defined, they are gauge-invariant.

## 10.3 Landau Energy Levels

As an example of the use of the theory of charged particles in an electromagnetic field described in previous sections, we will now take up a classic problem first treated in 1930 by Lev Landau (1908–1968): the quantum theory of motion in two dimensions of an electron in a uniform magnetic field.[2] Since electrons have spin, we must add a term $-\mu_e \mathbf{s} \cdot \mathbf{B}/(\hbar/2)$ to the Hamiltonian, where $\mu_e$ is a parameter known as the magnetic moment of the electron. The Hamiltonian for an electron (with charge $-e$) in a general electromagnetic field is then

$$H = \frac{1}{2m_e}\left(\mathbf{p} + \frac{e}{c}\mathbf{A}(\mathbf{x}, t)\right)^2 - e\phi(\mathbf{x}, t) - \frac{2\mu_e}{\hbar}\mathbf{s} \cdot \mathbf{B}(\mathbf{x}, t) . \tag{10.3.1}$$

We are here neglecting any interaction between electrons, so that it is adequate to consider one electron at a time. We assume that the magnetic field is in the $+z$-direction, and has a constant value $B_z$. We also include an electric field along the $z$-direction, which depends only on $z$, and has the function of confining the electron in this direction, whether to a thin sheet or to the whole thickness of a slab of material. We can then take the vector and scalar potentials to have the form

$$A_y = xB_z , \quad A_x = A_z = 0 , \quad \phi = \phi(z) . \tag{10.3.2}$$

(This choice is of course not unique, but as shown in Section 10.2, the eigenvalues of the Hamiltonian are independent of the choice of potentials giving the assumed electric and magnetic fields.) With these potentials, the Hamiltonian (10.3.1) takes the form

$$H = \frac{1}{2m_e}\left(p_x^2 + (p_y + eB_zx/c)^2 + p_z^2\right) - e\phi(z) - 2\mu_e s_z B_z/\hbar . \tag{10.3.3}$$

This Hamiltonian commutes with the operators $p_y$ and $s_z$, and with

$$\mathcal{H} \equiv \frac{p_z^2}{2m_e} - e\phi(z) , \tag{10.3.4}$$

---

[1] The transformed fields will also be time-independent if we let $\alpha(\mathbf{x}, t) = \lambda t$, with $\lambda$ independent of $\mathbf{x}$ and $t$. This amounts to a change of an arbitrary additive constant in the electrostatic potential, and shifts all energies in a system of total charge $Q$ by the same amount, $-\lambda Q/c$.

[2] L. Landau, *Z. Physik* **64**, 629 (1930).

so we can look for states $\Psi$ that are eigenstates of all these operators,

$$\mathcal{H}\Psi = \mathcal{E}\Psi , \quad s_z\Psi = \pm\frac{\hbar}{2}\Psi , \quad p_y\Psi = \hbar k_y\Psi , \tag{10.3.5}$$

as well as

$$H\Psi = E\Psi . \tag{10.3.6}$$

The Schrödinger equation (10.3.6) then reads

$$\frac{1}{2m_{\rm e}}\left(p_x^2 + (\hbar k_y + e B_z x/c)^2\right)\Psi = (E - \mathcal{E} \pm \mu_{\rm e} B_z)\Psi . \tag{10.3.7}$$

We can put this in a more familiar form, by writing it as

$$\left[\frac{1}{2m_{\rm e}}p_x^2 + \frac{m_{\rm e}\omega^2}{2}(x - x_0)^2\right]\Psi = (E - \mathcal{E} \pm \mu_{\rm e} B_z)\Psi , \tag{10.3.8}$$

where

$$\omega = \frac{e B_z}{m_{\rm e} c} , \quad x_0 = -\frac{\hbar k_y c}{e B_z} . \tag{10.3.9}$$

(The parameter $\omega$ is the circular frequency of classical electron orbits in a magnetic field $B_z$, and is therefore known as the cyclotron frequency.) Of course, we recognize Eq, (10.3.8) as the Schrödinger equation for a harmonic oscillator, discussed in Section 2.5. (Even though $p_x$ in Eq. (10.3.7) is not simply equal to $m_{\rm e}\dot{x}$, it does satisfy the commutation relation $[x, p_x] = i\hbar$, and therefore acts as the differential operator $-i\hbar\,\partial/\partial x$ on the coordinate-space wave function, just as for the ordinary harmonic oscillator.) The presence of $x_0$ in Eq. (10.3.8) has no effect on the energy eigenvalues, as it can be absorbed into a re-definition of the coordinate, $x \mapsto x' = x - x_0$. So the energies are given by

$$E = \mathcal{E} \mp \mu_{\rm e} B_z + \hbar\omega\left(n + \frac{1}{2}\right) , \tag{10.3.10}$$

where $n = 0, 1, 2, \ldots$.

This takes an interesting form if we use the actual value of the electron magnetic moment

$$\mu_{\rm e} = -\frac{e\hbar(1 + \delta)}{2m_{\rm e} c} , \tag{10.3.11}$$

where $\delta = 0.001165923(8)$ is a small radiative correction. Equation (10.3.10) then reads

$$E = \mathcal{E} + \hbar\omega\left(n + \frac{1}{2} \pm \frac{1 + \delta}{2}\right) . \tag{10.3.12}$$

We observe a near degeneracy: in the approximation $\delta \simeq 0$, for a given $\mathcal{E}$ and $k_y$ we have one state with energy $\mathcal{E}$, and two states each with energies $\mathcal{E} + \hbar\omega$, $\mathcal{E} + 2\hbar\omega$, etc.

Because the energies (10.3.12) do not depend on $k_y$, these energy levels exhibit a very large further degree of degeneracy. Suppose the electrons are confined in a square slab, with $-L_x/2 \leq x \leq L_x/2$ and $-L_y/2 \leq y \leq L_y/2$. The harmonic oscillator wave functions (2.5.13) extend around $x_0$ in the $x$-direction over a microscopic distance $\simeq (\hbar/m_e\omega)^{1/2}$, which we assume to be very much less than $L_x$, so $x_0$ in Eq. (10.3.8) must have $|x_0| < L_x/2$, which according to Eq. (10.3.9) gives $|k_y| < eB_zL_x/2\hbar c$. As in Eq. (1.1.1), the wave number $k_y$ can only take values $2\pi n_y/L_y$, where $n_y$ is a positive or negative integer, so the number of states with a given $n$, $\mathcal{E}$, and $s_z$, satisfying the condition that $|k_y|$ is less than $eB_zL_x/2\hbar c$, is the number of positive or negative integers with magnitude less than $(eB_zL_x/2\hbar c)(L_y/2\pi)$, which is

$$\mathcal{N}_y = \frac{eB_zA}{2\pi\hbar c} , \tag{10.3.13}$$

where $A = L_xL_y$ is the area of the slab.

To go further, we need to make some assumption about the term $\mathcal{H}$ in the Hamiltonian that governs the $z$-dependence of the wave function, given by Eq. (10.3.4). We will concentrate on the simplest case, assuming that we are dealing with a slab of metal so thin in the $z$-direction that the eigenvalues $\mathcal{E}$ of $\mathcal{H}$ are very far apart, so that we can assume that all conduction electrons are in the eigenstate of $\mathcal{H}$ with lowest energy $\mathcal{E}_0$.

If we assume that all of the harmonic oscillator states are occupied by electrons up to a maximum energy $\mathcal{E}_F$ (the Fermi energy less $\mathcal{E}_0$), then the total number of conduction electrons will be

$$N = 2\left(\frac{\mathcal{E}_F}{\hbar\omega}\right)\mathcal{N}_y = \frac{\mathcal{E}_Fm_eA}{\pi\hbar^2} . \tag{10.3.14}$$

Without a magnetic field, we would have just the same relation between the Fermi energy and the number $N/A$ of electrons per area:

$$N = 2\left(\frac{L_x}{2\pi}\right)\left(\frac{L_y}{2\pi}\right)\int_0^{\sqrt{2m_e\mathcal{E}_F}/\hbar} 2\pi k \, dk = \frac{\mathcal{E}_Fm_eA}{\pi\hbar^2} .$$

Where the magnetic field makes a difference is in the quantization of the energy levels. According to Eq. (10.3.12) (with $\delta = 0$), if all the energy levels (10.3.12) up to some maximum energy are completely filled, then the partial Fermi energy $\mathcal{E}_F$ must be a whole-number multiple of $\hbar\omega$, which is not necessarily true of the value of $\mathcal{E}_F$ given according to Eq. (10.3.14) for a particular number per unit area $N/A$ of conduction electrons. When the partial Fermi energy $\mathcal{E}_F$ is not a whole-number multiple of $\hbar\omega$, the highest of the harmonic oscillator energy levels is not completely filled. Specifically, if $[\mathcal{E}_F/\hbar\omega]$ is the largest integer less than or equal to $\mathcal{E}_F/\hbar\omega$, then all of the energy levels up to $\hbar\omega[\mathcal{E}_F/\hbar\omega]$ will be fully occupied, and the fraction $f$ of the next highest energy level that is occupied will be given by the condition that

$$\left(\left[\frac{\mathcal{E}_F}{\hbar\omega}\right] + f\right)\hbar\omega = \mathcal{E}_F \, ,$$

or in other words

$$f = \frac{\mathcal{E}_F}{\hbar\omega} - \left[\frac{\mathcal{E}_F}{\hbar\omega}\right] . \tag{10.3.15}$$

As the magnetic field increases, the ratio $\mathcal{E}_F/\hbar\omega$ decreases as $1/B_z$, so $f$ decreases until $\mathcal{E}_F/\hbar\omega$ is an integer, where $f = 0$. With a continued increase in $B_z$, the occupancy $f$ will jump up from zero to nearly one, and then decrease to zero again when $\mathcal{E}_F/\hbar\omega$ equals the next lowest integer, and so on. Many properties of the metal therefore show a periodicity in $1/B_z$, with a period equal to the decrease in $1/B_z$ required for $\mathcal{E}_F/\hbar\omega$ to decrease by one unit:

$$\Delta\left(\frac{1}{B_z}\right) = \frac{\hbar e}{m_e c \mathcal{E}_F} . \tag{10.3.16}$$

The observed periodicities in electrical resistivity and magnetic susceptibility are known as the Shubnikov–de Haas effect and the de Haas–van Alphen effect, respectively. By measuring such periodicities for various magnetic field orientations, it is possible to determine the relation between electron energies and momenta in a crystal.

Similar periodicities are also seen in slabs with a finite thickness in the $z$-direction, in which many different eigenstates of $\mathcal{H}$ are occupied. Here the eigenvalues $\mathcal{E}$ are functions of the $z$-component $k_z$ of the Bloch wave number, and the oscillations are associated with maxima or minima in $\mathcal{E}(k_z)$.

## 10.4  The Aharonov–Bohm Effect

As emphasized in Section 10.1, even though in classical physics the introduction of vector and scalar potentials is a mere mathematical convenience, in quantum mechanics it is essential. This is vividly demonstrated by the existence of an effect predicted by Aharonov and Bohm,[3] in which the vector potential can have measurable effects on a charged particle, even though the magnetic field vanishes everywhere along the particle's path.

First let's consider how to calculate the wave function of an electron (ignoring spin effects) of energy $E$ in a static electromagnetic field, in a case where the scale of length over which the field varies appreciably is large compared with the electron wavelength. In this case we can use the eikonal approximation described in Section 7.10, with a Hamiltonian given by Eq. (10.1.9) for charge $-e$ and with no non-electromagnetic potential $\mathcal{V}$:

---

[3] Y. Aharonov and D. Bohm, *Phys. Rev.* **115**, 485 (1959).

$$H(\mathbf{x}, \mathbf{p}) = \frac{1}{2m_e} \left[ \mathbf{p} + \frac{e}{c} \mathbf{A}(\mathbf{x}) \right]^2 - e\phi(\mathbf{x}) . \tag{10.4.1}$$

We write the wave function as

$$\psi(\mathbf{x}) = N(\mathbf{x}) \exp(iS(\mathbf{x})/\hbar) \tag{10.4.2}$$

with $N$ and $S$ real, and we make the approximation that the phase $S(\mathbf{x})/\hbar$ varies much more rapidly with position than does the amplitude $N(\mathbf{x})$. As described in Section 7.10, to find $S$ we must construct ray paths, defined by the Hamiltonian equations (7.10.4), which for the Hamiltonian (10.4.1) read

$$\frac{dx_i}{d\tau} = \frac{1}{m_e} \left[ p_i + \frac{e}{c} A_i(\mathbf{x}) \right] , \tag{10.4.3}$$

$$\frac{dp_i}{d\tau} = -\frac{e}{m_e c} \sum_j \left[ p_j + \frac{e}{c} A_j(\mathbf{x}) \right] \frac{\partial A_j(\mathbf{x})}{\partial x_i} + e \frac{\partial \phi(\mathbf{x})}{\partial x_i} , \tag{10.4.4}$$

where $\tau$ parameterizes the path through phase space. Boundary conditions on the wave function are specified on an initial surface, on which to leading order the phase of the wave function is a constant, which we can take as zero, on which $d\mathbf{x}/d\tau$ is normal to this surface, and on which the Hamiltonian $H$ equals the electron energy $E$. (For instance, if the potentials vanish for $z$ large and negative, and the wave function in this case is proportional to $\exp(ikz)$, then we can take the initial surface to be any plane at large negative $z$ normal to the $z$-axis.) Equations (10.4.3) and (10.4.4) then give $H = E$ along any path. For any point $\mathbf{x}$ in at least a neighborhood of the initial surface there will be some point $\mathbf{X}(\mathbf{x})$ on the initial surface such that the path starting from $\mathbf{X}(\mathbf{x})$ at $\tau = 0$ and obeying the Hamiltonian equations (10.4.3) and (10.4.4) will eventually reach $\mathbf{x}$, at some value $\tau = \tau_\mathbf{x}$ of the path parameter. The phase $S(\mathbf{x})/\hbar$ is then given by the general formula

$$S(\mathbf{x}) = \int_0^{\tau_\mathbf{x}} \mathbf{p}(\tau) \cdot \frac{d\mathbf{x}(\tau)}{d\tau} \, d\tau . \tag{10.4.5}$$

As shown in Section 7.10, this has the consequence that

$$\mathbf{p}(\tau_\mathbf{x}) = \nabla S(\mathbf{x}) , \tag{10.4.6}$$

with it understood here that $\mathbf{p}(\tau)$ is the solution of Eqs. (10.4.3) and (10.4.4) for the ray path that runs from the initial surface to $\mathbf{x}$. (This ensures that $H(\nabla S, \mathbf{x}) = E$, which is the Schrödinger equation in the approximation that gradients of $N$ are neglected.) In our case, using Eq. (10.4.3) and setting the Hamiltonian (10.4.1) equal to $E$, Eq. (10.4.5) gives

$$S(\mathbf{x}) = \int_0^{\tau_\mathbf{x}} \left[ -\frac{e}{c} \mathbf{A}(\mathbf{x}(\tau)) \cdot \frac{d\mathbf{x}(\tau)}{d\tau} + 2\Big( E + e\phi(\mathbf{x}(\tau)) \Big) \right] d\tau . \tag{10.4.7}$$

To calculate the amplitude $N(\mathbf{x})$, we use the probability conservation law (10.1.5). Since the wave function here is time-independent, this gives

$$\nabla \cdot \boldsymbol{\mathcal{J}} = 0 \qquad (10.4.8)$$

with the current $\boldsymbol{\mathcal{J}}$ given by Eq. (10.1.6). Again neglecting gradients of $N$ in the eikonal approximation, this current is

$$\boldsymbol{\mathcal{J}} = \frac{1}{m} N^2 \left( \nabla S + \frac{e}{c} \mathbf{A} \right) . \qquad (10.4.9)$$

Following the argument of Section 7.10, consider all the ray paths that reach a small patch of area $\delta a$ around $\mathbf{x}$, normal to these paths. These paths will have started on the initial surface in a small patch of area $\delta A$ around $\mathbf{X}(\mathbf{x})$. We can draw a thin tube, whose ends are these two patches, and whose sides are formed from ray paths that go from the edges of the patch on the initial surface to the edges of the patch around $\mathbf{x}$. Equations (10.4.8) and (10.4.9) and Gauss's theorem tell us that the integral over this surface of the component of $N^2 (\nabla S + (e/c)\mathbf{A})$ in the direction of the outward normal to the surface of the tube vanishes. According to Eqs. (10.4.3) and (10.4.6), the combination $S + (e/c)\mathbf{A}$ is just proportional to $d\mathbf{x}/d\tau$, and hence points in the direction of the ray path, so the normal component of $N^2(\nabla S + (e/c)\mathbf{A})$ vanishes on the sides of the tube, which are in the direction of the ray path. The vector $N^2(\nabla S + (e/c)\mathbf{A})$ on the patch at $\mathbf{x}$ is in the direction of the outward normal to this patch, while on the corresponding patch on the initial surface it is in the direction of the inward normal to this surface, so Gauss's theorem tells us that

$$N^2(\mathbf{x}) \left| \left( \frac{d\mathbf{x}(\tau)}{d\tau} \right)_{\tau = \tau_\mathbf{x}} \right| \delta a - N^2(\mathbf{X}(\mathbf{x})) \left| \left( \frac{d\mathbf{x}(\tau)}{d\tau} \right)_{\tau = 0} \right| \delta A = 0 , \qquad (10.4.10)$$

it being understood that $d\mathbf{x}(\tau)/d\tau$ is here to be calculated for the ray path that goes to $\mathbf{x}$ from the corresponding point $X(\mathbf{x})$ on the initial surface. The only feature of Eq. (10.4.10) that will be needed below is that the ratio of $N^2$ at $\mathbf{x}$ to its value at the corresponding point $\mathbf{X}(\mathbf{x})$ on the initial surface depends only on the energy $E$ and on the field strengths $\mathbf{B}$ and $\mathbf{E}$ acting on the electron, but not on the vector potential except as it affects these fields. This is because it follows from Eqs. (10.4.3) and (10.4.4) that $\mathbf{x}(\tau)$ obeys an equation of motion analogous to Eq. (10.1.1):

$$m_e \ddot{\mathbf{x}}(t) = -e \left[ \mathbf{E} \big( \mathbf{x}(t), t \big) + \frac{1}{c} \dot{\mathbf{x}}(t) \times \mathbf{B} \big( \mathbf{x}(t), t \big) \right] , \qquad (10.4.11)$$

while according to Eqs. (10.4.1) and (10.4.3) the value of $d\mathbf{x}/d\tau$ on the initial surface depends only on $E$ and $\phi$. The ray paths $\mathbf{x}(\tau)$ therefore do not depend on the vector potential, except as it affects the magnetic field, and the same is then true of the path expansion ratio $\delta a/\delta A$ and the ratio

$$\left| \left(\frac{d\mathbf{x}(\tau)}{d\tau}\right)_{\tau=\tau_\mathbf{x}} \middle/ \left(\frac{d\mathbf{x}(\tau)}{d\tau}\right)_{\tau=0} \right| ,$$

so according to Eq. (10.4.10) it is also true of the ratio of $N^2$ at $\mathbf{x}$ to its value at the corresponding point on the initial surface.

Now suppose that by some arrangement of fields, screens, and/or beam splitters, a single coherent beam of electrons is split into two parts, so that there are two ray paths to a detector at $\mathbf{x}$. The wave function at $\mathbf{x}$ will take the form

$$\psi(\mathbf{x}) = N_1(\mathbf{x}) \exp\left(i S_1(\mathbf{x})/\hbar\right) + N_2(\mathbf{x}) \exp\left(i S_2(\mathbf{x})/\hbar\right), \qquad (10.4.12)$$

where the subscripts 1 and 2 denote the two paths to the detector. The probability density at $\mathbf{x}$ then depends on the difference of the phases:

$$|\psi(\mathbf{x})|^2 = N_1^2(\mathbf{x}) + N_2^2(\mathbf{x}) + 2N_1(\mathbf{x})N_2(\mathbf{x}) \cos\left([S_1(\mathbf{x}) - S_2(\mathbf{x})]/\hbar\right). \quad (10.4.13)$$

According to Eq. (10.4.7), the phase difference appearing here may be written as an integral over a curve that goes from the point $X_1(\mathbf{x})$ on the initial surface along path 1 to $\mathbf{x}$, and then back along path 2 to the point $\mathbf{X}_2(\mathbf{x})$ on the initial surface. But by definition the phase $S$ is constant on the initial surface, so the integral can just as well be taken over the closed curve $C_{12}$ that goes from $X_1(\mathbf{x})$ to $\mathbf{x}$ on path 1, then from $\mathbf{x}$ to $X_2(\mathbf{x})$ backward on path 2, and then on the initial surface from $X_2(\mathbf{x})$ to $X_1(\mathbf{x})$:

$$\frac{1}{\hbar}\left[S_1(\mathbf{x}) - S_2(\mathbf{x})\right] = \frac{1}{\hbar}\oint_{C_{12}}\left[-\frac{e}{c}\mathbf{A}(\tau)\cdot\frac{d\mathbf{x}(\tau)}{d\tau} + 2\left(E + e\phi(\mathbf{x}(\tau))\right)\right]d\tau .$$
$$(10.4.14)$$

According to the Stokes theorem, the first term in the phase difference is proportional to the magnetic flux through the surface $\mathcal{A}_{12}$ bounded by $C_{12}$:

$$-\frac{e}{\hbar c}\oint_{C_{12}}\mathbf{A}(\tau)\cdot\frac{d\mathbf{x}(\tau)}{d\tau}\,d\tau = -\frac{e}{\hbar c}\Phi , \qquad (10.4.15)$$

where the flux is

$$\Phi = \int_{\mathcal{A}_{12}}\mathbf{B}\cdot\hat{n}\,d\mathcal{A}, \qquad (10.4.16)$$

where $\hat{n}$ is the unit vector normal to the surface $\mathcal{A}_{12}$. Thus the phase difference (10.4.14) and hence the intensity (10.4.13) depend on the values of the magnetic field in places in the interior of the curve $C_{12}$, where the electron does not go.

In the particular case considered by Aharonov and Bohm, a magnetic solenoid is inserted between paths 1 and 2, carrying a magnetic flux $\Phi$ that is entirely contained within the solenoid. As we have seen, the ray paths and the values of $N^2$ are only affected by the electric and magnetic fields along the paths, and so are unaffected by the solenoid. But the vector potential of the solenoid does extend outside it, and this contributes a term $-e\Phi/\hbar c$ to the phase difference, even though the magnetic field of the solenoid vanishes along both ray paths. There are other contributions to the phase difference (10.4.14), but the contribution of the solenoid can be observed by changing its flux $\Phi$, while making no other change to the system. As shown by Eqs. (10.4.13)–(10.4.15), the electron probability density at the detector will be periodic in $\Phi$, with a period $2\pi\hbar c/e = 4.14 \times 10^{-7}$ Gauss cm$^2$. This effect has been observed in a long series of experiments.[4]

The Aharonov–Bohm effect has been described here in a time-independent context, but we can also consider it to be the effect of the changing magnetic field seen in the rest frame of the electron. In this sense, we can regard Eq. (10.4.15) as an example of the Berry phase discussed in Section 6.7.

## Problems

1. Consider a system in an external electromagnetic field. Suppose that the part of the Lagrangian that depends on the scalar potential $\phi$ and vector potential $\mathbf{A}$ takes the form

$$L_{\text{int}}(t) = \int d^3x \; [-\rho(\mathbf{x}, t)\phi(\mathbf{x}, t) + \mathbf{J}(\mathbf{x}, t) \cdot \mathbf{A}(\mathbf{x}, t)] \;,$$

   where $\rho$ and $J$ depend on the matter variables but not on $\phi$ or $\mathbf{A}$. What condition must be satisfied by $\rho$ and $\mathbf{J}$ for the action to be gauge-invariant?

2. Consider a homogeneous rectangular slab of metal, with edges $L_x$, $L_y$, and $L_z$. Assume that the electric potential $\phi$ vanishes within the slab, and that the wave functions of conduction electrons in the slab satisfy periodic boundary conditions at the slab faces. Suppose that the slab is in a constant magnetic field in the $z$-direction that is strong enough that the cyclotron frequency $\omega$ is very much larger than $\hbar/m_e L_z^2$. Suppose that there are $n_e$ conduction electrons per unit volume in the slab. Calculate the maximum energy of individual conduction electrons, in the limit $\omega m_e L_z^2/\hbar \to \infty$.

3. Consider a non-relativistic electron in an external electromagnetic field. Calculate the commutators of different components of its velocity.

---

[4] R. G. Chambers, *Phys. Rev. Lett.* **5**, 3 (1960); H. A. Fowler, L. Marton, J. A. Simpson, and J. A. Suddeth, *J. Appl. Phys.* **22**, 1153 (1961); H. Boersch, H. Hamisch, K. Grohmann, and D. Wohlleben, *Z. Phys.* **165**, 79 (1961); G. Möllenstedt and W. Bayh, *Phys. Blätter* **18**, 299 (1962); A. Tomomura, T. Matsuda, R. Suzuki, A. Fukuhara, N. Osakabe, H. Umezaki, J. Endo, K. Shinagawa, Y. Sagita, and H. Fujiwara, *Phys. Rev. Lett.* **48**, 1443 (1982).

# 11

# The Quantum Theory of Radiation

We now come back to the problem that gave rise to quantum theory at the beginning of the twentieth century – the nature of electromagnetic radiation.

## 11.1 The Euler–Lagrange Equations

In order to quantize the electromagnetic field, we will work with a Lagrangian that leads to Maxwell's equations. But before introducing this Lagrangian, it will be helpful first to explain in general terms how in field theories the field equations can be derived from a Lagrangian.

The canonical variables $q_N(t)$ in general field theories are fields $\psi_n(\mathbf{x}, t)$, for which $N$ is a compound index, including a discrete label $n$ indicating the type of field and a spatial coordinate $\mathbf{x}$. Correspondingly, the Lagrangian $L(t)$ is a *functional* of $\psi_n(\mathbf{x}, t)$ and $\dot{\psi}_n(\mathbf{x}, t)$, depending on the form of all of the functions $\psi_n(\mathbf{x}, t)$ and $\dot{\psi}_n(\mathbf{x}, t)$ for all $\mathbf{x}$, but at a fixed time $t$. In consequence, the partial derivatives with respect to $q_N$ and $\dot{q}_N$ in the equations of motion must be interpreted as functional derivatives with respect to $\psi_n(\mathbf{x}, t)$ and $\dot{\psi}_n(\mathbf{x}, t)$, so that these equations read

$$\frac{\partial}{\partial t}\left(\frac{\delta L(t)}{\delta \dot{\psi}_n(\mathbf{x}, t)}\right) = \frac{\delta L(t)}{\delta \psi_n(\mathbf{x}, t)} , \tag{11.1.1}$$

where the functional derivatives $\delta L/\delta \dot{\psi}_n$ and $\delta L/\delta \psi_n$ are defined so that the change in the Lagrangian produced by independent infinitesimal changes $\delta \psi_n(\mathbf{x}, t)$ and $\delta \dot{\psi}_n(\mathbf{x}, t)$ in $\psi_n(\mathbf{x}, t)$ and $\dot{\psi}_n(\mathbf{x}, t)$ at a fixed time $t$ is

$$\delta L(t) = \sum_n \int d^3x \left[ \frac{\delta L(t)}{\delta \psi_n(\mathbf{x}, t)} \delta \psi_n(\mathbf{x}, t) + \frac{\delta L(t)}{\delta \dot{\psi}_n(\mathbf{x}, t)} \delta \dot{\psi}_n(\mathbf{x}, t) \right] . \tag{11.1.2}$$

Likewise, the canonical conjugate to $\psi_n(\mathbf{x}, t)$ is

$$\pi_n(\mathbf{x}, t) = \frac{\delta L(t)}{\delta \dot{\psi}_n(\mathbf{x}, t)} , \tag{11.1.3}$$

361

and in a theory with no constraints, the canonical commutation relations are

$$[\psi_n(\mathbf{x}, t), \pi_m(\mathbf{y}, t)] = i\hbar\delta_{nm}\delta^3(\mathbf{x} - \mathbf{y}) , \tag{11.1.4}$$

$$[\psi_n(\mathbf{x}, t), \psi_m(\mathbf{y}, t)] = [\pi_n(\mathbf{x}, t), \pi_m(\mathbf{y}, t)] = 0 . \tag{11.1.5}$$

Typically (though not always), the Lagrangian in a field theory will be an integral of a local *Lagrangian density* $\mathcal{L}$:

$$L(t) = \int d^3x \; \mathcal{L}\Big(\psi(\mathbf{x}, t), \nabla\psi(\mathbf{x}, t), \dot{\psi}(\mathbf{x}, t)\Big) . \tag{11.1.6}$$

The variation of the Lagrangian action due to infinitesimal changes in the $\psi_n$ and their space and time derivatives that vanish for $|\mathbf{x}| \to \infty$ is

$$\delta L(t) = \int d^3x \; \sum_n \left[ \frac{\partial\mathcal{L}}{\partial\psi_n}\delta\psi_n + \sum_i \frac{\partial\mathcal{L}}{\partial(\partial_i\psi_n)}\frac{\partial}{\partial x_i}\delta\psi_n + \frac{\partial\mathcal{L}}{\partial\dot{\psi}_n}\frac{\partial}{\partial t}\delta\psi_n \right] .$$

Integrating by parts, this is

$$\delta L(t) = \int d^3x \; \sum_n \left[ \left(\frac{\partial\mathcal{L}}{\partial\psi_n} - \sum_i \frac{\partial}{\partial x_i}\frac{\partial\mathcal{L}}{\partial(\partial_i\psi_n)}\right)\delta\psi_n + \frac{\partial\mathcal{L}}{\partial\dot{\psi}_n}\frac{\partial}{\partial t}\delta\psi_n \right] .$$

This may be expressed as formulas for the variational derivatives of the Lagrangian

$$\frac{\delta L}{\delta\psi_n} = \frac{\partial\mathcal{L}}{\partial\psi_n} - \sum_i \frac{\partial}{\partial x_i}\frac{\partial\mathcal{L}}{\partial(\partial_i\psi_n)}, \tag{11.1.7}$$

$$\frac{\delta L}{\delta\dot{\psi}_n} = \frac{\partial\mathcal{L}}{\partial\dot{\psi}_n} . \tag{11.1.8}$$

The equations of motion (11.1.1) then take the form of the *Euler–Lagrange field equations*

$$\frac{\partial\mathcal{L}}{\partial\psi_n} - \sum_i \frac{\partial}{\partial x_i}\frac{\partial\mathcal{L}}{\partial(\partial_i\psi_n)} = \frac{\partial}{\partial t}\frac{\partial\mathcal{L}}{\partial\dot{\psi}_n} . \tag{11.1.9}$$

(In relativistically invariant theories it is convenient to write this as

$$\frac{\partial\mathcal{L}}{\partial\psi_n} = \sum_\mu \frac{\partial}{\partial x^\mu}\frac{\partial\mathcal{L}}{\partial(\partial_\mu\psi_n)} . \tag{11.1.10}$$

Here $\mu$ is a four-component index, summed over the values $i = 1, 2, 3$, and 0, with $x^i = x_i$ and $x^0 = ct$.) Similarly, in theories with a local Lagrangian density, the field variable (11.1.3) that is canonically conjugate to $\psi_n(\mathbf{x}, t)$ is

$$\pi_n = \frac{\delta L}{\delta\dot{\psi}_n} = \frac{\partial\mathcal{L}}{\partial\dot{\psi}_n} . \tag{11.1.11}$$

## 11.2 The Lagrangian for Electrodynamics

The electric field $\mathbf{E}(\mathbf{x}, t)$ and magnetic field $\mathbf{B}(\mathbf{x}, t)$ are governed by the inhomogeneous Maxwell equations:[1]

$$\nabla \times \mathbf{B} - \frac{1}{c}\frac{\partial \mathbf{E}}{\partial t} = \frac{4\pi}{c}\mathbf{J}, \quad \nabla \cdot \mathbf{E} = 4\pi\rho, \tag{11.2.1}$$

as well as the homogeneous Maxwell equations, already encountered in Section 10.1:

$$\nabla \times \mathbf{E} + \frac{1}{c}\frac{\partial \mathbf{B}}{\partial t} = 0, \quad \nabla \cdot \mathbf{B} = 0. \tag{11.2.2}$$

Here $\rho(\mathbf{x}, t)$ is the electric charge density, defined so that the electric charge within any volume is the integral of $\rho$ over that volume, and $\mathbf{J}(\mathbf{x}, t)$ is the electric current density, defined so that the charge per second passing through a small area is the component of $\mathbf{J}$ normal to the area, times the area. They satisfy the charge conservation condition

$$\frac{\partial \rho}{\partial t} + \nabla \cdot \mathbf{J} = 0, \tag{11.2.3}$$

which is needed for the consistency of Eqs. (11.2.1). For instance, for a set of non-relativistic point particles with charges $e_n$ and coordinate vectors $\mathbf{x}_n(t)$, the charge and current densities are

$$\rho(\mathbf{x}, t) = \sum_n e_n \delta^3\Big(\mathbf{x} - \mathbf{x}_n(t)\Big), \quad \mathbf{J}(\mathbf{x}, t) = \sum_n e_n \dot{\mathbf{x}}_n(t)\delta^3\Big(\mathbf{x} - \mathbf{x}_n(t)\Big). \tag{11.2.4}$$

It is easy to see that these satisfy the conservation condition (11.2.3), by use of the relation

$$\frac{\partial}{\partial t}\delta^3\Big(\mathbf{x} - \mathbf{x}_n(t)\Big) = -\dot{\mathbf{x}}_n(t) \cdot \nabla\delta^3\Big(\mathbf{x} - \mathbf{x}_n(t)\Big).$$

As in Section 10.1, to construct a Lagrangian for electromagnetism, we need to express the electric and magnetic fields in terms of a vector potential $\mathbf{A}(\mathbf{x}, t)$ and a scalar potential $\phi(\mathbf{x}, t)$:

$$\mathbf{E} = -\frac{1}{c}\dot{\mathbf{A}} - \nabla\phi, \quad \mathbf{B} = \nabla \times \mathbf{A}, \tag{11.2.5}$$

so that the homogeneous Maxwell equations (11.2.2) are automatically satisfied. We saw in Eq. (10.1.3) that the term in the Lagrangian for the interaction of a set of non-relativistic particles with an electromagnetic field is

---

[1] The factor $4\pi$ appears here because in this book we are using unrationalized units for electric charges and currents, so that the electric field produced by a charge $e$ at a distance $r$ is $e/r^2$ rather than $e/4\pi r^2$. These are sometimes called Gaussian units.

$$L_{\text{int}}(t) = \sum_n \left[ - e_n \phi\left(\mathbf{x}_n(t), t\right) + \frac{e_n}{c} \dot{\mathbf{x}}_n(t) \cdot \mathbf{A}\left(\mathbf{x}_n(t), t\right) \right] .$$

This can be expressed as the integral of a local density

$$L_{\text{int}}(t) = \int d^3 x \, \mathcal{L}_{\text{int}}(\mathbf{x}, t) , \tag{11.2.6}$$

where

$$\mathcal{L}_{\text{int}}(\mathbf{x}, t) = -\rho(\mathbf{x}, t)\phi(\mathbf{x}, t) + \frac{1}{c}\mathbf{J}(\mathbf{x}, t) \cdot \mathbf{A}(\mathbf{x}, t) . \tag{11.2.7}$$

We will take this as the interaction Lagrangian density for any sort of charges and currents.

To (11.2.7), we must add a Lagrangian density $\mathcal{L}_0$ for the electromagnetic fields themselves, so that the part of the Lagrangian that involves electromagnetic fields is the integral of the density

$$\mathcal{L}_{\text{em}} = \mathcal{L}_0 + \mathcal{L}_{\text{int}} . \tag{11.2.8}$$

As we will now see, the electromagnetic field Lagrangian that yields the correct Maxwell equations is

$$\mathcal{L}_0 = \frac{1}{8\pi} \left[ \mathbf{E}^2 - \mathbf{B}^2 \right] , \tag{11.2.9}$$

with $\mathbf{E}$ and $\mathbf{B}$ expressed in terms of $\mathbf{A}$ and $\phi$ by means of Eq. (11.2.5). The total Lagrangian for the system is

$$L(t) = \int d^3 x \, \mathcal{L}_{\text{em}}(\mathbf{x}, t) + L_{\text{mat}}(t) , \tag{11.2.10}$$

where $L_{\text{mat}}(t)$ depends only on the matter coordinates and their rates of change, but not on the electromagnetic potentials, and therefore plays no role in determining the electromagnetic field equations.

The derivatives of the Lagrangian density with respect to the potentials and their derivatives are then

$$\frac{\partial \mathcal{L}_{\text{em}}}{\partial(\partial_j A_i)} = -\frac{1}{4\pi} \sum_k \epsilon_{kji} B_k , \quad \frac{\partial \mathcal{L}_{\text{em}}}{\partial \dot{A}_i} = -\frac{1}{4\pi c} E_i , \quad \frac{\partial \mathcal{L}_{\text{em}}}{\partial A_i} = \frac{1}{c} J_i , \tag{11.2.11}$$

$$\frac{\partial \mathcal{L}_{\text{em}}}{\partial(\partial_i \phi)} = -\frac{1}{4\pi} E_i , \quad \frac{\partial \mathcal{L}_{\text{em}}}{\partial \dot{\phi}} = 0 , \quad \frac{\partial \mathcal{L}_{\text{em}}}{\partial \phi} = -\rho , \tag{11.2.12}$$

where $i, j, k$ run over the three coordinate axes $1, 2, 3$, and as before $\epsilon_{kji}$ is the totally antisymmetric quantity with $\epsilon_{123} = +1$. It is then easy to see that the inhomogeneous Maxwell equations (11.2.1) are the same as the Euler–Lagrange equations (11.1.9) for $A_i$ and $\phi$:

$$\frac{\partial \mathcal{L}_{em}}{\partial A_i} - \sum_j \frac{\partial}{\partial x_j} \frac{\partial \mathcal{L}_{em}}{\partial (\partial_j A_i)} = \frac{d}{dt} \frac{\partial \mathcal{L}_{em}}{\partial \dot{A}_i} , \quad \frac{\partial \mathcal{L}_{em}}{\partial \phi} - \sum_i \frac{\partial}{\partial x_i} \frac{\partial \mathcal{L}_{em}}{\partial (\partial_i \phi)} = \frac{d}{dt} \frac{\partial \mathcal{L}_{em}}{\partial \dot{\phi}} .$$

(11.2.13)

So $\mathcal{L}_{em}$ can indeed be taken as the Lagrangian density for the electromagnetic fields. Of course, we could multiply the whole Lagrangian $L$ for matter and radiation with an arbitrary constant factor, and still get the same electromagnetic field equations and particle equations of motion. As we will see, the normalization here of $L$ is chosen to give sensible results for the energies of photons and charged particles.

## 11.3 Commutation Relations for Electrodynamics

From Eqs. (11.2.12) and (11.2.11), we see that the canonical conjugates to $A_i$ and $\phi$ are[2]

$$\Pi_\phi \equiv \frac{\partial \mathcal{L}}{\partial \dot{\phi}} = 0 , \tag{11.3.1}$$

$$\Pi_i \equiv \frac{\partial \mathcal{L}}{\partial \dot{A}_i} = -\frac{1}{4\pi c} E_i = \frac{1}{4\pi c} \left[ \frac{1}{c} \dot{\mathbf{A}} + \nabla \phi \right]_i . \tag{11.3.2}$$

The constraint (11.3.1) is clearly inconsistent with the usual commutation rule $[\phi(\mathbf{x}, t), \Pi_\phi(\mathbf{y}, t)] = i\hbar \delta^3(\mathbf{x} - \mathbf{y})$. Also, the field equation for $\mathbf{E}$ tells us that $\Pi_i$ is subject to a further constraint,

$$\nabla \cdot \mathbf{\Pi} = -\rho/c . \tag{11.3.3}$$

Equation (11.3.3) is inconsistent with the usual canonical commutation relations, which would require that $[A_i(\mathbf{x}, t), \Pi_j(\mathbf{y}, t)] = i\hbar \delta_{ij} \delta^3(\mathbf{x} - \mathbf{y})$, and that $A_i(\mathbf{x}, t)$ commutes with $\rho(\mathbf{y}, t)$.

In the language of Dirac described in Section 9.5, the constraints (11.3.1) and (11.3.3) are "first class," because the Poisson bracket of $\Pi_\phi$ and $\nabla \cdot \mathbf{\Pi} + \rho/c$ vanishes. On the other hand (and not unrelated to the presence of first-class constraints), gauge invariance gives us a freedom to impose additional conditions on the dynamical variables. There are various possibilities, but the most common choice is *Coulomb gauge*, in which we impose the condition that the vector potential is solenoidal:

$$\nabla \cdot \mathbf{A} = 0 . \tag{11.3.4}$$

---

[2] I am using an upper case letter for the canonical conjugate to $A_i$, in order to distinguish the Heisenberg-picture operators $A_i$ and $\Pi_i$ from their counterparts in the interaction picture, which in Section 11.5 will be denoted $a_i$ and $\pi_i$.

(Note that this can always be done, because if $\mathbf{V} \cdot \mathbf{A}$ does not vanish, then it can be made to vanish by a gauge transformation (10.2.1), (10.2.2):

$$\mathbf{A} \mapsto \mathbf{A}' = \mathbf{A} + \mathbf{V}\alpha , \quad \phi \mapsto \phi' = \phi - \dot{\alpha}/c ,$$

with $\nabla^2 \alpha = -\mathbf{V} \cdot \mathbf{A}$, which makes $\mathbf{V} \cdot \mathbf{A}' = 0$.) With the gauge choice (11.3.4), the field equation $\mathbf{V} \cdot \mathbf{E} = 4\pi\rho$ gives $\nabla^2 \phi = -4\pi\rho$, so $\phi$ is not an independent field variable, but a function of $\mathbf{x}$ and of the matter coordinates at the same time:[3]

$$\phi(\mathbf{x}, t) = \int d^3 y \, \frac{\rho(\mathbf{y}, t)}{|\mathbf{x} - \mathbf{y}|} = \sum_n \frac{e_n}{|\mathbf{x} - \mathbf{x}_n(t)|} . \tag{11.3.5}$$

So now we don't need to worry about the vanishing of the $\Pi_\phi$. We do still have two constraints, (11.3.3) and (11.3.4), which in line with the notation of Section 9.5, we will write as $\chi_1 = \chi_2 = 0$, where

$$\chi_1 = \mathbf{V} \cdot \mathbf{A} , \quad \chi_2 = \mathbf{V} \cdot \mathbf{\Pi} + \rho/c . \tag{11.3.6}$$

As in Section 9.5, we define a matrix

$$C_{r\mathbf{x},s\mathbf{y}} \equiv [\chi_r(\mathbf{x}), \chi_s(\mathbf{y})]_P , \tag{11.3.7}$$

where $[\cdot, \cdot]_P$ denotes the Poisson bracket (9.4.19), and $r$ and $s$ run over the values 1 and 2. (Recall that the Poisson bracket is what the commutators would be, aside from a factor $i\hbar$, if the canonical commutation relations applied here.) This "matrix" has elements

$$C_{1\mathbf{x},2\mathbf{y}} = -C_{2\mathbf{y},1\mathbf{x}} = \sum_{ij} \delta_{ij} \frac{\partial^2}{\partial x_i \partial y_j} \delta^3(\mathbf{x} - \mathbf{y}) = -\nabla^2 \delta^3(\mathbf{x} - \mathbf{y}) , \tag{11.3.8}$$

$$C_{1\mathbf{x},1\mathbf{y}} = C_{2\mathbf{x},2\mathbf{y}} = 0 . \tag{11.3.9}$$

This has a matrix inverse

$$C_{1\mathbf{x},2\mathbf{y}}^{-1} = -C_{2\mathbf{y},1\mathbf{x}}^{-1} = -\frac{1}{4\pi |\mathbf{x} - \mathbf{y}|} , \tag{11.3.10}$$

$$C_{1\mathbf{x},1\mathbf{y}}^{-1} = C_{2\mathbf{x},2\mathbf{y}}^{-1} = 0 , \tag{11.3.11}$$

in the sense that

$$\int d^3 y \begin{pmatrix} 0 & C_{1\mathbf{x},2\mathbf{y}} \\ C_{2\mathbf{x},1\mathbf{y}} & 0 \end{pmatrix} \begin{pmatrix} 0 & C_{1\mathbf{y},2\mathbf{z}}^{-1} \\ C_{2\mathbf{y},1\mathbf{z}}^{-1} & 0 \end{pmatrix}$$
$$= \begin{pmatrix} \delta^3(\mathbf{x} - \mathbf{z}) & 0 \\ 0 & \delta^3(\mathbf{x} - \mathbf{z}) \end{pmatrix} . \tag{11.3.12}$$

---

[3] Here we are using the relation $\nabla_\mathbf{y}^2 |\mathbf{y} - \mathbf{z}|^{-1} = -4\pi \delta^3(\mathbf{y} - \mathbf{z})$. It is easy to check that this quantity vanishes for $\mathbf{y} \neq \mathbf{z}$, because $d/dr(r^2 d/dr(1/r)) = 0$. But Gauss's theorem tells us that its integral over a ball centered on $\mathbf{z}$ equals the integral of $(d/dr)(1/r)$ over the surface of the ball, which is $-4\pi$.

That is,

$$\int d^3y \, C_{1\mathbf{x},2\mathbf{y}} C_{2\mathbf{y},1\mathbf{z}}^{-1} = \int d^3y \, [-\nabla^2 \delta^3(\mathbf{x} - \mathbf{y})] \left[ \frac{1}{4\pi |\mathbf{y} - \mathbf{z}|} \right]$$

$$= \int d^3y \, [\delta^3(\mathbf{x} - \mathbf{y})] \left[ -\nabla^2 \frac{1}{4\pi |\mathbf{y} - \mathbf{z}|} \right]$$

$$= \delta^3(\mathbf{x} - \mathbf{z}) \,,$$

and likewise for $\int d^3y \, C_{2\mathbf{x},1\mathbf{y}} C_{1\mathbf{y},2\mathbf{z}}^{-1}$. We also note the Poisson brackets

$$[A_i(\mathbf{x}, t), \chi_{2\mathbf{x}'}(t)]_{\mathrm{P}} = \frac{\partial}{\partial x_i'} \delta^3(\mathbf{x} - \mathbf{x}') \,, \quad [A_i(\mathbf{x}, t), \chi_{1\mathbf{x}'}(t)]_{\mathrm{P}} = 0 \,,$$

$$[\chi_{1\mathbf{y}'}(t), \Pi_j(\mathbf{y}, t)]_{\mathrm{P}} = \frac{\partial}{\partial y_j'} \delta^3(\mathbf{y}' - \mathbf{y}) \,, \quad [\chi_{2\mathbf{y}'}(t), \Pi_j(\mathbf{y}, t)]_{\mathrm{P}} = 0 \,.$$

Then according to Eqs. (9.5.17)–(9.5.19), the commutators of the canonical variables are

$$[A_i(\mathbf{x}, t), \Pi_j(\mathbf{y}, t)] = i\hbar \Bigg[ \delta_{ij} \delta^3(\mathbf{x} - \mathbf{y}) - \int d^3x' \int d^3y' \, [A_i(\mathbf{x}, t), \chi_{2\mathbf{x}'}(t)]_{\mathrm{P}}$$

$$\times C_{2\mathbf{x}',1\mathbf{y}'}^{-1} [\chi_{1\mathbf{y}'}(t), \Pi_j(\mathbf{y}, t)]_{\mathrm{P}} \Bigg]$$

$$= i\hbar \Bigg[ \delta_{ij} \delta^3(\mathbf{x} - \mathbf{y}) - \int d^3x' \int d^3y' \left[ \frac{\partial}{\partial x_i'} \delta^3(\mathbf{x} - \mathbf{x}') \right]$$

$$\times \left[ \frac{1}{4\pi |\mathbf{x}' - \mathbf{y}'|} \right] \left[ \frac{\partial}{\partial y_j'} \delta^3(\mathbf{y} - \mathbf{y}') \right] \Bigg]$$

$$= i\hbar \left[ \delta_{ij} \delta^3(\mathbf{x} - \mathbf{y}) - \frac{\partial^2}{\partial x_i \, \partial y_j} \frac{1}{4\pi |\mathbf{x} - \mathbf{y}|} \right] \,, \qquad (11.3.13)$$

$$[A_i(\mathbf{x}, t), A_j(\mathbf{y}, t)] = [\Pi_i(\mathbf{x}, t), \Pi_j(\mathbf{y}, t)] = 0 \,. \qquad (11.3.14)$$

There is an awkward feature about the canonical commutation relations in Coulomb gauge, that we have not yet uncovered. Although the commutators of the particle coordinates $x_{nj}$ with $A_i$ and $\Pi_i$ all vanish, the particle momenta $p_{nj}$ have non-vanishing commutators with $\Pi_i$. According to the Dirac prescription and Eqs. (11.3.8)–(11.3.11), this commutator is

$$[\Pi_i(\mathbf{x}, t), p_{nj}(t)] = -i\hbar \int d^3y \int d^3z \, [\Pi_i(\mathbf{x}, t), \chi_{1\mathbf{y}}(t)]_P$$

$$\times \, C^{-1}_{1\mathbf{y}, 2\mathbf{z}}[\chi_{2\mathbf{z}}(t), p_{nj}(t)]_P$$

$$= -i\hbar \int d^3y \int d^3z \left[ -\frac{\partial}{\partial y_i} \delta^3(\mathbf{x} - \mathbf{y}) \right] \left[ \frac{-1}{4\pi |\mathbf{y} - \mathbf{z}|} \right]$$

$$\times \left[ \frac{1}{c} \frac{\partial}{\partial y_{nj}} \rho(\mathbf{z}) \right]$$

$$= \frac{i\hbar e_n}{4\pi c} \frac{\partial^2}{\partial x_i \, \partial x_{nj}} \frac{1}{|\mathbf{x} - \mathbf{x}_n(t)|} \, . \tag{11.3.15}$$

We can avoid this complication by introducing as a replacement for $\mathbf{\Pi}$ its solenoidal part

$$\mathbf{\Pi}^\perp \equiv \mathbf{\Pi} - \frac{1}{4\pi c} \nabla\phi = \frac{1}{4\pi c^2} \dot{\mathbf{A}} \, , \tag{11.3.16}$$

for which in Coulomb gauge

$$\nabla \cdot \mathbf{\Pi}^\perp = 0 \, . \tag{11.3.17}$$

The Dirac bracket of the term $-\nabla\phi/4\pi c$ with $p_{nj}$ is just the Poisson bracket, so

$$\left[ \frac{\partial}{\partial x_i} \phi(\mathbf{x}, t), p_{nj}(t) \right] = i\hbar e_n \frac{\partial^2}{\partial x_i \, \partial x_{nj}} \frac{1}{|\mathbf{x} - \mathbf{x}_n(t)|} \, . \tag{11.3.18}$$

So we see that

$$[\mathbf{\Pi}^\perp(\mathbf{x}, t), p_{nj}(t)] = 0 \, . \tag{11.3.19}$$

Also, since $\phi$ has vanishing Poisson brackets with $\chi_1$ and $\chi_2$, it has vanishing commutators with $\mathbf{A}$ and $\mathbf{\Pi}$, and so the commutators of the components of $\mathbf{\Pi}^\perp$ with each other and with $\mathbf{A}$ are the same as for $\mathbf{\Pi}$:

$$[A_i(\mathbf{x}, t), \Pi_j^\perp(\mathbf{y}, t)] = i\hbar \left[ \delta_{ij} \delta^3(\mathbf{x} - \mathbf{y}) - \frac{\partial^2}{\partial x_i \, \partial y_j} \frac{1}{4\pi |\mathbf{x} - \mathbf{y}|} \right], \tag{11.3.20}$$

$$[A_i(\mathbf{x}, t), A_j(\mathbf{y}, t)] = [\Pi_i^\perp(\mathbf{x}, t), \Pi_j^\perp(\mathbf{y}, t)] = 0 \, . \tag{11.3.21}$$

Note that these commutation relations are consistent with the vanishing of the divergences of both $\mathbf{A}$ and $\mathbf{\Pi}^\perp$.

## 11.4 The Hamiltonian for Electrodynamics

Now let us construct the Hamiltonian for this theory. In Coulomb gauge, because $\phi$ is no longer an independent physical variable, the total Hamiltonian is

$$H = \int d^3x \left[ \mathbf{\Pi} \cdot \dot{\mathbf{A}} - \mathcal{L}_0 \right] + H_{\text{mat}}, \tag{11.4.1}$$

where $\mathcal{L}_0$ is the purely electromagnetic Lagrangian density (11.2.9), and $H_{mat}$ is the Hamiltonian for matter, now including its interaction with electromagnetism. Because $\nabla \cdot \mathbf{A} = 0$, we can replace $\mathbf{\Pi}$ in the first term with $\mathbf{\Pi}^\perp$, and then use Eq. (11.3.16) to replace $\dot{\mathbf{A}}$ with $4\pi c^2 \mathbf{\Pi}^\perp$. We can also use Eqs. (11.3.16) and (11.2.5) to replace $\mathbf{E}$ in $\mathcal{L}_0$ with $-4\pi c \mathbf{\Pi}$:

$$ H = \int d^3x \left[ 4\pi c^2 [\mathbf{\Pi}^\perp]^2 - \frac{1}{8\pi} [4\pi c \mathbf{\Pi}^\perp + \nabla \phi]^2 + \frac{1}{8\pi} (\nabla \times \mathbf{A})^2 \right] + H_{mat} . $$

Integrating by parts gives $\int d^3x \, \mathbf{\Pi}^\perp \cdot \nabla \phi = 0$ and

$$ -\frac{1}{8\pi} \int d^3x \, (\nabla \phi)^2 = \frac{1}{8\pi} \int d^3x \, \phi \nabla^2 \phi = -\frac{1}{2} \int d^3x \, \rho \phi . $$

The Hamiltonian is then

$$ H = \int d^3x \left[ 2\pi c^2 [\mathbf{\Pi}^\perp]^2 + \frac{1}{8\pi} (\nabla \times \mathbf{A})^2 \right] + H'_{mat} , \tag{11.4.2} $$

where

$$ H'_{mat} = H_{mat} - \frac{1}{2} \int d^3x \, \rho \phi . \tag{11.4.3} $$

For instance, in the case where the matter consists of non-relativistic charged point particles in a general local potential $\mathcal{V}$, Eq. (10.1.9) gives

$$ H_{mat} = \sum_n \frac{1}{2m_n} \left[ \mathbf{p}_n - \frac{e_n}{c} \mathbf{A}(\mathbf{x}_n, t) \right]^2 + \sum_n e_n \phi \left( \mathbf{x}_n, t \right) + \mathcal{V}(\mathbf{x}) . $$

and furthermore, here[4]

$$ \phi(\mathbf{x}, t) = \sum_m \frac{e_m}{|\mathbf{x} - \mathbf{x}_m(t)|} , \qquad \int d^3x \, \rho(\mathbf{x}, t) \phi(\mathbf{x}, t) = \sum_{n \neq m} \frac{e_n e_m}{|\mathbf{x}_n - \mathbf{x}_m(t)|} . $$

Hence,

$$ H'_{mat} = \sum_n \frac{1}{2m_n} \left[ \mathbf{p}_n - \frac{e_n}{c} \mathbf{A}(\mathbf{x}_n) \right]^2 + \frac{1}{2} \sum_{n \neq m} \frac{e_n e_m}{|\mathbf{x}_n - \mathbf{x}_m|} + \mathcal{V}(\mathbf{x}) . \tag{11.4.4} $$

(Time arguments are suppressed here.) We recognize the second term as the usual Coulomb energy of a set of charged point particles. The factor $1/2$ in this term arises from the combination of a term $\int d^3x \, \rho\phi$ in $H_{mat}$ and the term $-(1/2) \int d^3x \, \rho\phi$ in Eq. (11.4.3). This factor serves to eliminate double counting; for instance, for two particles, the sum over $n$ and $m$ includes both a term with $n = 1, m = 2$, and an equal term with $n = 2, m = 1$.

---

[4] In imposing the restriction $n \neq m$ on the sum over $n$ and $m$, we are dropping an infinite c-number term in the Hamiltonian, which only shifts all energies by the same amount, and has no effect on rates of change derived from the Hamiltonian.

Let's check that we recover Maxwell's equations from this Hamiltonian. Using the commutators (11.3.20) and (11.3.21) and Eq. (11.3.17), the Hamiltonian equations of motion for $\mathbf{A}$ and $\mathbf{\Pi}$ are

$$\dot{A}_i = \frac{i}{\hbar}[H, A_i] = 4\pi c^2 \Pi_i^\perp \,, \tag{11.4.5}$$

$$\begin{aligned}
\dot{\Pi}_i^\perp &= \frac{i}{\hbar}[H, \Pi_i^\perp] \\
&= -\frac{1}{4\pi}(\nabla \times \nabla \times \mathbf{A})_i \\
&\quad + \sum_{nj} \frac{e_n}{m_n c}\left(p_{nj} - \frac{e_n}{c}A_j(\mathbf{x}_n)\right) \\
&\quad \times \left[\delta^3(\mathbf{x} - \mathbf{x}_n)\delta_{ij} - \frac{\partial^2}{\partial x_i \, \partial x_{nj}}\frac{1}{4\pi|\mathbf{x} - \mathbf{x}_n|}\right].
\end{aligned} \tag{11.4.6}$$

(The expression in the last factor of the last term in Eq. (11.4.6) arises from the commutator (11.3.20). In Eq. (11.4.5) and in the first term of Eq. (11.4.6) we do not need to keep the second term in this commutator, because $\mathbf{\Pi}^\perp$ and $\nabla \times \mathbf{A}$ both have zero divergence.) To make contact with Maxwell's equations, we recall that, according to Eq. (10.1.8), we have $\mathbf{p}_n - e_n\mathbf{A}(\mathbf{x}_n)/c = m_n\dot{\mathbf{x}}_n$. Hence Eqs. (11.4.5) and (11.4.6) give

$$\ddot{\mathbf{A}} = -c^2 \, \nabla \times \mathbf{B} + 4\pi c\mathbf{J} - c \, \nabla\dot{\phi} \,,$$

or in other words,

$$\dot{\mathbf{E}} = c \, \nabla \times \mathbf{B} - 4\pi\mathbf{J} \,,$$

which is the same as the first of the inhomogeneous Maxwell equations (11.2.1). In Coulomb gauge the other inhomogeneous Maxwell equation $\nabla \cdot \mathbf{E} = 4\pi\rho$ just follows directly from the formula (11.2.5) for $\mathbf{E}$ in terms of $\dot{\mathbf{A}}$ and $\nabla\phi$, together with the constraint (11.3.4) and Eq. (11.3.5) for $\phi$. The two homogeneous Maxwell equations (11.2.2) follow directly from the definition (11.2.5) for the fields in terms of the potentials. So the Hamiltonian (11.4.2) together with the commutation relations (11.3.20) and (11.3.21) does indeed complete the set of Maxwell equations.

## 11.5 Interaction Picture

In order to use the time-dependent perturbation theory described in Section 8.7, it is necessary to split the Hamiltonian $H$ into a term $H_0$ that will be treated to all orders, plus a term $V$ in which we expand:

$$H = H_0 + V \,. \tag{11.5.1}$$

In order to calculate the rates for radiative transitions between otherwise stable states of atoms or molecules, we split the Hamiltonian $H$ given by Eqs. (11.4.2) and (11.4.4) into

$$H_0 = H_{0\gamma} + H_{0\,\mathrm{mat}}, \tag{11.5.2}$$

$$H_{0\gamma} = \int d^3x \left[ 2\pi c^2 [\mathbf{\Pi}^\perp]^2 + \frac{1}{8\pi} (\nabla \times \mathbf{A})^2 \right], \tag{11.5.3}$$

$$H_{0\,\mathrm{mat}} = \sum_n \frac{\mathbf{p}_n^2}{2m_n} + \frac{1}{2} \sum_{n \neq m} \frac{e_n e_m}{|\mathbf{x}_n - \mathbf{x}_m|} + \mathcal{V}(\mathbf{x}), \tag{11.5.4}$$

plus a term $V$ consisting of the terms in (11.4.4) involving the vector potential:

$$V = -\sum_n \frac{e_n}{m_n c} \mathbf{A}(\mathbf{x}_n) \cdot \mathbf{p}_n + \sum_n \frac{e_n^2}{2m_n c^2} \mathbf{A}^2(\mathbf{x}_n). \tag{11.5.5}$$

In the first term in $V$ we have replaced $\mathbf{A}(\mathbf{x}_n) \cdot \mathbf{p}_n + \mathbf{p}_n \cdot \mathbf{A}(\mathbf{x}_n)$ with $2\mathbf{A}(\mathbf{x}_n) \cdot \mathbf{p}_n$, which is allowed because, in Coulomb gauge,

$$\mathbf{A}(\mathbf{x}_n) \cdot \mathbf{p}_n - \mathbf{p}_n \cdot \mathbf{A}(\mathbf{x}_n) = i\hbar \nabla \cdot \mathbf{A}(\mathbf{x}_n) = 0 \,.$$

We also need to introduce interaction-picture operators, whose time-dependence is governed by $H_0$ instead of $H$. For the interaction-picture vector potential $\mathbf{a}$ and the solenoidal part $\pi^\perp$ of its canonical conjugate, the time-dependence can be found in the interaction picture by calculating their commutators with $H_{0\gamma}$, in the same way as we did for the Heisenberg picture operators in the previous section. The results will obviously be the same, except that now there is no contribution from the interaction $V$, and so we find just Eqs. (11.4.5) and (11.4.6), but with all terms involving the charges $e_n$ dropped:

$$\dot{\mathbf{a}} = 4\pi c^2 \pi^\perp, \tag{11.5.6}$$

$$\dot{\pi}^\perp = -\frac{1}{4\pi} \nabla \times \nabla \times \mathbf{a}. \tag{11.5.7}$$

The interaction-picture operators are related to the corresponding Heisenberg-picture operators at $t = 0$ by a unitary transformation

$$\mathbf{a}(\mathbf{x}, t) = e^{iH_0 t/\hbar} \mathbf{A}(\mathbf{x}, 0) e^{-iH_0 t/\hbar}, \quad \pi^\perp(\mathbf{x}, t) = e^{iH_0 t/\hbar} \mathbf{\Pi}^\perp(\mathbf{x}, 0) e^{-iH_0 t/\hbar}, \tag{11.5.8}$$

so these operators satisfy the same time-independent conditions as the Heisenberg-picture operators:

$$\nabla \cdot \mathbf{a} = \nabla \cdot \pi^\perp = 0. \tag{11.5.9}$$

In consequence, $\nabla \times \nabla \times \mathbf{a} = -\nabla^2 \mathbf{a}$. By eliminating $\pi^\perp$ from Eqs. (11.5.6) and (11.5.7), we find a wave equation for $\mathbf{a}$:

$$\ddot{\mathbf{a}} = c^2 \nabla^2 \mathbf{a}. \tag{11.5.10}$$

The general Hermitian solution of Eqs. (11.5.9) and (11.5.10) may be expressed as a Fourier integral

$$\mathbf{a}(\mathbf{x}, t) = \int d^3k \left[ e^{i\mathbf{k}\cdot\mathbf{x}} e^{-i|\mathbf{k}|ct} \boldsymbol{\alpha}(\mathbf{k}) + e^{-i\mathbf{k}\cdot\mathbf{x}} e^{i|\mathbf{k}|ct} \boldsymbol{\alpha}^\dagger(\mathbf{k}) \right] , \qquad (11.5.11)$$

where the operator $\boldsymbol{\alpha}(\mathbf{k})$ is subject to the condition

$$\mathbf{k} \cdot \boldsymbol{\alpha}(\mathbf{k}) = 0 . \qquad (11.5.12)$$

Equation (11.5.6) then gives the solenoidal part of the canonical conjugate to $\mathbf{a}$ as

$$\boldsymbol{\pi}^\perp(\mathbf{x}, t) = -\frac{i}{4\pi c} \int |\mathbf{k}| \, d^3k \left[ e^{i\mathbf{k}\cdot\mathbf{x}} e^{-i|\mathbf{k}|ct} \boldsymbol{\alpha}(\mathbf{k}) - e^{-i\mathbf{k}\cdot\mathbf{x}} e^{i|\mathbf{k}|ct} \boldsymbol{\alpha}^\dagger(\mathbf{k}) \right] . \qquad (11.5.13)$$

We need to work out the commutators of the operators $\boldsymbol{\alpha}(\mathbf{k})$ and their Hermitian adjoints. Again, since the interaction-picture operators are related to the corresponding Heisenberg-picture operators at $t = 0$ by a unitary transformation, they must satisfy the same equal-time commutation relations (11.3.20), (11.3.21) as the Heisenberg-picture operators:

$$[a_i(\mathbf{x}, t), \pi_j^\perp(\mathbf{y}, t)] = i\hbar \left[ \delta_{ij} \delta^3(\mathbf{x} - \mathbf{y}) - \frac{\partial^2}{\partial x_i \, \partial y_j} \frac{1}{4\pi|\mathbf{x} - \mathbf{y}|} \right], \qquad (11.5.14)$$

$$[a_i(\mathbf{x}, t), a_j(\mathbf{y}, t)] = [\pi_i^\perp(\mathbf{x}, t), \pi_j^\perp(\mathbf{y}, t)] = 0 , \qquad (11.5.15)$$

and both $\mathbf{a}$ and $\boldsymbol{\pi}^\perp$ commute with all matter coordinates and momenta. From Eqs. (11.5.11) and (11.5.13), we find the commutator of $a_i(\mathbf{x}, t)$ and $\pi_j^\perp(\mathbf{y}, t)$:

$$[a_i(\mathbf{x}, t), \pi_j^\perp(\mathbf{y}, t)] = \frac{i}{4\pi c} \int d^3k \int d^3k' \, |\mathbf{k}'|$$
$$\times \left[ e^{i(\mathbf{k}\cdot\mathbf{x} - \mathbf{k}'\cdot\mathbf{y})} e^{ict(-|\mathbf{k}| + |\mathbf{k}'|)} [\alpha_i(\mathbf{k}), \alpha_j^\dagger(\mathbf{k}')] \right.$$
$$- e^{i(-\mathbf{k}\cdot\mathbf{x} + \mathbf{k}'\cdot\mathbf{y})} e^{ict(|\mathbf{k}| - |\mathbf{k}'|)} [\alpha_i^\dagger(\mathbf{k}), \alpha_j(\mathbf{k}')]$$
$$- e^{i(\mathbf{k}\cdot\mathbf{x} + \mathbf{k}'\cdot\mathbf{y})} e^{ict(-|\mathbf{k}| - |\mathbf{k}'|)} [\alpha_i(\mathbf{k}), \alpha_j(\mathbf{k}')]$$
$$\left. + e^{i(-\mathbf{k}\cdot\mathbf{x} - \mathbf{k}'\cdot\mathbf{y})} e^{ict(|\mathbf{k}| + |\mathbf{k}'|)} [\alpha_i^\dagger(\mathbf{k}), \alpha_j^\dagger(\mathbf{k}')] \right] . \qquad (11.5.16)$$

Equation (11.5.14) shows that this must be time-independent, so the terms with positive-definite or negative-definite frequency must both vanish, and therefore

$$[\alpha_i(\mathbf{k}), \alpha_j(\mathbf{k}')] = [\alpha_i^\dagger(\mathbf{k}), \alpha_j^\dagger(\mathbf{k}')] = 0 . \qquad (11.5.17)$$

To calculate the remaining commutators, we use the Fourier transforms

$$\delta^3(\mathbf{x} - \mathbf{y}) = \int \frac{d^3k}{(2\pi)^3} e^{i\mathbf{k}\cdot(\mathbf{x}-\mathbf{y})} , \qquad \frac{1}{4\pi|\mathbf{x} - \mathbf{y}|} = \int \frac{d^3k}{(2\pi)^3 |\mathbf{k}|^2} e^{i\mathbf{k}\cdot(\mathbf{x}-\mathbf{y})} ,$$

and rewrite Eq. (11.5.14) as

$$[a_i(\mathbf{x}, t), \pi_j^\perp(\mathbf{y}, t)] = i\hbar \int \frac{d^3k}{(2\pi)^3} e^{i\mathbf{k}\cdot(\mathbf{x}-\mathbf{y})} \left[ \delta_{ij} - \frac{k_i k_j}{|\mathbf{k}|^2} \right] . \tag{11.5.18}$$

Comparing this with the first two terms in Eq. (11.5.16), we see that

$$[\alpha_i(\mathbf{k}), \alpha_j^\dagger(\mathbf{k}')] = \frac{4\pi c\hbar}{2|\mathbf{k}|(2\pi)^3} \delta^3(\mathbf{k} - \mathbf{k}') \left[ \delta_{ij} - \frac{k_i k_j}{|\mathbf{k}|^2} \right] . \tag{11.5.19}$$

The commutation relations (11.5.15) then follow automatically.

Like any vector perpendicular to a given $\mathbf{k}$, the operator $\boldsymbol{\alpha}(\mathbf{k})$ may be expressed as a linear combination of any two independent vectors $\mathbf{e}(\hat{k}, \pm 1)$ perpendicular to $\mathbf{k}$:

$$\boldsymbol{\alpha}(\mathbf{k}) = \sqrt{\frac{4\pi c\hbar}{2|\mathbf{k}|(2\pi)^3}} \sum_\pm \mathbf{e}(\hat{k}, \pm 1) a(\mathbf{k}, \pm 1) , \tag{11.5.20}$$

with the factor $\sqrt{4\pi c\hbar/2|\mathbf{k}|(2\pi)^3}$ inserted to simplify the commutation relations of the operators $a(\mathbf{k}, \pm 1)$ that will be found. For instance, for $\mathbf{k}$ in the $z$-direction, we can take

$$\mathbf{e}(\hat{z}, \pm 1) = \frac{1}{\sqrt{2}} \left( 1, \pm i, 0 \right) \tag{11.5.21}$$

and for $\mathbf{k}$ in any other direction, we take $e_i(\hat{k}, \pm) = \sum_j R_{ij}(\hat{z}) e_j(\hat{z}, \pm 1)$, where $R_{ij}(\hat{k})$ is the rotation matrix that takes the $z$-direction into the direction of $\mathbf{k}$. It follows that for any $\mathbf{k}$, we have

$$\mathbf{k} \cdot \mathbf{e}(\hat{k}, \sigma) = 0 , \quad \mathbf{e}(\hat{k}, \sigma) \cdot \mathbf{e}^*(\hat{k}, \sigma') = \delta_{\sigma\sigma'} . \tag{11.5.22}$$

Also,

$$\sum_\sigma e_i(\hat{k}, \sigma) e_j^*(\hat{k}, \sigma) = \delta_{ij} - \hat{k}_i \hat{k}_j . \tag{11.5.23}$$

(It is easiest to prove Eqs. (11.5.22) and (11.5.23) by direct calculation in the case where $\hat{k}$ is in the $z$-direction, and then note that these equations preserve their form under rotations.) The commutation relations (11.5.19) are then satisfied if

$$[a(\mathbf{k}, \sigma), a^\dagger(\mathbf{k}', \sigma')] = \delta_{\sigma'\sigma} \delta^3(\mathbf{k} - \mathbf{k}'). \tag{11.5.24}$$

Also, the commutation relations (11.5.17) are satisfied if

$$[a(\mathbf{k}, \sigma), a(\mathbf{k}', \sigma')] = [a^\dagger(\mathbf{k}, \sigma), a^\dagger(\mathbf{k}', \sigma')] = 0 . \tag{11.5.25}$$

We recognize Eqs. (11.5.24) and (11.5.25) as the commutation relations (2.5.8) and (2.5.9) for the raising and lowering operators of a harmonic oscillator, but with the 3-component indices $i$ and $j$ replaced here with the compound indices $\mathbf{k}, \sigma$ and $\mathbf{k}', \sigma'$.

The Hamiltonian $H_{0\gamma}$ for the free electromagnetic field can be calculated in the interaction picture by setting $t = 0$ in Eq. (11.5.3), and then applying the unitary transformation (11.5.8), which gives a Hamiltonian of the same form:

$$H_{0\gamma} = \int d^3x \left[ 2\pi c^2 [\boldsymbol{\pi}^\perp]^2 + \frac{1}{8\pi} (\nabla \times \mathbf{a})^2 \right]. \qquad (11.5.26)$$

We can uncover the physical significance of the operators $a(\mathbf{k}, \sigma)$ and $a^\dagger(\mathbf{k}, \sigma)$ by expressing the free-field Hamiltonian $H_{0\gamma}$ in terms of these operators. They appear in the formulas for $\mathbf{a}(\mathbf{x}, t)$ and $\boldsymbol{\pi}^\perp(\mathbf{x}, t)$:

$$\mathbf{a}(\mathbf{x}, t) = \sqrt{4\pi c\hbar} \sum_\sigma \int \frac{d^3k}{\sqrt{2k(2\pi)^3}} \left[ e^{i\mathbf{k}\cdot\mathbf{x}} e^{-ictk} \mathbf{e}(\mathbf{k}, \sigma) a(\mathbf{k}, \sigma) + \text{H.c.} \right],$$

$$(11.5.27)$$

$$\boldsymbol{\pi}^\perp(\mathbf{x}, t) = -i \frac{\sqrt{4\pi c\hbar}}{4\pi c} \sum_\sigma \int \frac{k\, d^3k}{\sqrt{2k(2\pi)^3}} \left[ e^{i\mathbf{k}\cdot\mathbf{x}} e^{-ictk} \mathbf{e}(\mathbf{k}, \sigma) a(\mathbf{k}, \sigma) - \text{H.c.} \right],$$

$$(11.5.28)$$

where $k \equiv |\mathbf{k}|$, and "H.c." denotes the Hermitian conjugate of the preceding term. The integral over $\mathbf{x}$ in Eq. (11.5.26) gives delta functions for the wave numbers times $(2\pi)^3$. We then have

$$\int d^3x\, (\nabla \times \mathbf{a})^2$$

$$= 2\pi c\hbar \sum_{\sigma'\sigma} \int k\, d^3k \left[ \mathbf{e}^*(\hat{k}, \sigma) \cdot \mathbf{e}(\hat{k}, \sigma') a^\dagger(\mathbf{k}, \sigma) a(\mathbf{k}, \sigma') \right.$$

$$+ \mathbf{e}^*(\hat{k}, \sigma') \cdot \mathbf{e}(\hat{k}, \sigma) a(\mathbf{k}, \sigma) a^\dagger(\mathbf{k}, \sigma')$$

$$+ \mathbf{e}(\hat{k}, \sigma) \cdot \mathbf{e}(-\hat{k}, \sigma') a(\mathbf{k}, \sigma) a(-\mathbf{k}, \sigma') e^{-2ickt}$$

$$\left. + \mathbf{e}^*(\hat{k}, \sigma) \cdot \mathbf{e}^*(-\hat{k}, \sigma') a^\dagger(\mathbf{k}, \sigma) a^\dagger(-\mathbf{k}, \sigma') e^{2ickt} \right],$$

$$\int d^3x\, (\boldsymbol{\pi}^\perp)^2$$

$$= -\frac{\hbar}{8\pi c} \sum_{\sigma'\sigma} \int k\, d^3k \left[ -\mathbf{e}^*(\hat{k}, \sigma) \cdot \mathbf{e}(\hat{k}, \sigma') a^\dagger(\mathbf{k}, \sigma) a(\mathbf{k}, \sigma') \right.$$

$$- \mathbf{e}^*(\hat{k}, \sigma') \cdot \mathbf{e}(\hat{k}, \sigma) a(\mathbf{k}, \sigma) a^\dagger(\mathbf{k}, \sigma')$$

$$+ \mathbf{e}(\hat{k}, \sigma) \cdot \mathbf{e}(-\hat{k}, \sigma') a(\mathbf{k}, \sigma) a(-\mathbf{k}, \sigma') e^{-2ickt}$$

$$\left. + \mathbf{e}^*(\hat{k}, \sigma) \cdot \mathbf{e}^*(-\hat{k}, \sigma') a^\dagger(\mathbf{k}, \sigma) a^\dagger(-\mathbf{k}, \sigma') e^{2ickt} \right].$$

When we add the two terms in Eq. (11.5.26), we see that the time-dependent terms cancel (as they must, since $H_{0\gamma}$ commutes with itself). This is just as well, since $\mathbf{e}(\hat{k}, \sigma) \cdot \mathbf{e}(-\hat{k}, \sigma)$ depends on how we choose the rotations that take $\hat{z}$ into

$\hat{k}$ and $-\hat{k}$. On the other hand, the two terms in Eq. (11.5.26) make equal contributions to the time-independent terms. These remaining terms can be evaluated using Eq. (11.5.22), which gives $\mathbf{e}^*(\hat{k}, \sigma) \cdot \mathbf{e}(\hat{k}, \sigma') = \delta_{\sigma'\sigma}$, and we find

$$H_{0\gamma} = \frac{1}{2} \sum_{\sigma} \int d^3k \, \hbar c k \left[ a^\dagger(\mathbf{k}, \sigma) a(\mathbf{k}, \sigma) + a(\mathbf{k}, \sigma) a^\dagger(\mathbf{k}, \sigma) \right] . \quad (11.5.29)$$

The physical interpretation of this result is described in the next section.

## 11.6 Photons

According to the commutation relations (11.5.24) and (11.5.25), the commutators of the unperturbed electromagnetic Hamiltonian (11.5.29) with the operators $a^\dagger(\mathbf{k}, \sigma)$ and $a(\mathbf{k}, \sigma)$ are

$$[H_{0\gamma}, a^\dagger(\mathbf{k}, \sigma)] = \hbar c k a^\dagger(\mathbf{k}, \sigma) , \quad (11.6.1)$$

$$[H_{0\gamma}, a(\mathbf{k}, \sigma)] = -\hbar c k a(\mathbf{k}, \sigma) . \quad (11.6.2)$$

Hence $a^\dagger(\mathbf{k}, \sigma)$ and $a(\mathbf{k}, \sigma)$ are raising and lowering operators for the energy. That is, if $\Psi$ is an eigenstate of $H_{0\gamma}$ with eigenvalue $E$, then $a^\dagger(\mathbf{k}, \sigma)\Psi$ is an eigenstate with energy $E + \hbar c k$, and $a(\mathbf{k}, \sigma)\Psi$ is an eigenstate with energy $E - \hbar c k$.

Although not compelled by the formalism of quantum mechanics, we are led by the stability of matter to assume that there is a state $\Psi_0$ of lowest energy. The only way to avoid having a state $a(\mathbf{k}, \sigma)\Psi_0$ of energy that is lower by an amount $\hbar c k$ is to suppose that

$$a(\mathbf{k}, \sigma)\Psi_0 = 0 . \quad (11.6.3)$$

We can find the energy of the state $\Psi_0$ by using the commutation relations (11.5.24) to write Eq. (11.5.29) as

$$H_{0\gamma} = \sum_{\sigma} \int d^3k \, \hbar c k a^\dagger(\mathbf{k}, \sigma) a(\mathbf{k}, \sigma) + E_0 , \quad (11.6.4)$$

where $E_0$ is the infinite constant

$$E_0 = \sum_{\sigma} \int d^3k \, \frac{\hbar c k}{2} \delta^3(\mathbf{k} - \mathbf{k}) . \quad (11.6.5)$$

We can give this a meaning of sorts by putting the system in a box of volume $\Omega$. Then $\delta^3(\mathbf{k} - \mathbf{k})$ becomes $\Omega/(2\pi)^3$, so we have an energy per volume

$$E_0/\Omega = (2\pi)^{-3} \int d^3k \, \hbar c k . \quad (11.6.6)$$

This energy may be attributed to the unavoidable quantum fluctuations in the electromagnetic field. As shown by Eqs. (11.5.18) and (11.5.6), it is not possible for the vector potential at any point in space to vanish (or take any definite fixed value) for a finite time interval; if the field vanishes at one moment, then its rate of change at that moment cannot take any definite value, including zero. The energy density (11.6.6) has no effect in ordinary laboratory experiments, as it inheres in space itself, and space cannot normally be created or destroyed, but it does affect gravitation, and hence influences the expansion of the universe and the formation of large bodies like galaxy clusters. Needless to say, an infinite result is not allowed by observation. Even if we cut off the integral at the highest wave number probed in laboratory experiments, say $10^{15}$ cm$^{-1}$, the result is larger than allowed by observation by a factor of roughly $10^{56}$. The energy due to fluctuations in the electromagnetic field and other bosonic fields can be canceled by the negative energy of fluctuations in fermionic fields, but we know of no reason why this cancellation should be exact, or even precise enough to bring the vacuum energy down to a value in line with observation. Since $E_0/\Omega$ was known to be vastly smaller that the value estimated from vacuum fluctuations at accessible scales, for decades most physicists who thought at all about this problem simply assumed that some fundamental principle would be discovered that imposes on any theory the condition that makes $E_0/\Omega$ vanish. This possibility was ruled out by the discovery[5] in 1998 that the expansion of the universe is accelerating, in a way that indicates a value of $E_0/\Omega$ about three times larger than the energy density in matter. This remains a fundamental problem for modern physics,[6] but it can be ignored as long as we do not deal with effects of gravitation.

We can now construct states spanning what is called *Fock space*:

$$\Psi_{\mathbf{k}_1,\sigma_1;\mathbf{k}_2,\sigma_2;\ldots;\mathbf{k}_n,\sigma_n} \propto a^\dagger(\mathbf{k}_1,\sigma_1)a^\dagger(\mathbf{k}_2,\sigma_2)\ldots a^\dagger(\mathbf{k}_n,\sigma_n)\Psi_0 , \qquad (11.6.7)$$

which according to Eq. (11.6.1) (and dropping the term $E_0$) has the energy

$$\hbar c k_1 + \hbar c k_2 + \cdots + \hbar c k_n .$$

We interpret this as a state of $n$ photons, with energies $\hbar c k_1$, $\hbar c k_2$, ..., $\hbar c k_n$.

To work out the momentum of these states, we note that according to the general results of Section 9.4, the operator that generates the infinitesimal translation $a_i(\mathbf{x},t) \mapsto a_i(\mathbf{x}-\boldsymbol{\epsilon},t)$ is given by Eq. (9.4.4) as

$$\boldsymbol{\epsilon}\cdot\mathbf{P}_\gamma = -\sum_i \int d^3x\,\pi_i^\perp(\mathbf{x},t)(\boldsymbol{\epsilon}\cdot\nabla)a_i(\mathbf{x},t) . \qquad (11.6.8)$$

(That is, the sum over $N$ in Eq. (9.4.4) is replaced with a sum over the vector index $i$ and an integral over the argument $\mathbf{x}$ of the field.) Using the commutation relations (11.5.14) and (11.5.15), we have

---

[5] This is the independent result of two teams: The Supernova Cosmology Project [S. Perlmutter *et al.*, *Astrophys. J.* **517**, 565 (1999); also see S. Perlmutter *et al.*, *Nature* **391**, 51 (1998).] and the High-$z$ Supernova Search Team [A. G. Riess *et al.*, *Astron. J.* **116**, 1009 (1998); also see B. Schmidt *et al.*, *Astrophys. J.* **507**, 46 (1998).]

[6] For a review, see S. Weinberg, *Rev. Mod. Phys.* **61**, 1 (1989).

$$[\mathbf{P}_\gamma, a_i(\mathbf{x}, t)] = i\hbar \nabla a_i(\mathbf{x}, t) , \quad [\mathbf{P}_\gamma, \pi_i^\perp(\mathbf{x}, t)] = i\hbar \nabla \pi_i^\perp(\mathbf{x}, t) . \quad (11.6.9)$$

(The second term in square brackets in Eq. (11.5.14) does not contribute because $\nabla \cdot \mathbf{a} = 0$ and $\nabla \cdot \boldsymbol{\pi}^\perp = 0$.) Then $\mathbf{P}_\gamma$ commutes with $H_{0\,\gamma}$ as it does with the integral over $\mathbf{x}$ of any function of $a_i(\mathbf{x}, t)$ and $\pi_i^\perp(\mathbf{x}, t)$ and their gradients. Inserting Eqs. (11.5.11) and (11.5.13) in Eq. (11.6.9) gives

$$[\mathbf{P}_\gamma, a(\mathbf{k}, \sigma)] = -\hbar k a(\mathbf{k}, \sigma) , \quad [\mathbf{P}_\gamma, a^\dagger(\mathbf{k}, \sigma)] = \hbar k a^\dagger(\mathbf{k}, \sigma) . \quad (11.6.10)$$

Assuming that the state $\Psi_0$ is translation-invariant, this tells us that the states (11.6.7) have momentum

$$\hbar \mathbf{k}_1 + \hbar \mathbf{k}_2 + \cdots + \hbar \mathbf{k}_n .$$

So we can interpret these states as consisting of $n$ photons, each with a momentum $\hbar \mathbf{k}$ and an energy $\hbar ck$. Because the energy $E$ of a photon is related to its momentum $\mathbf{p}$ by $E = c|\mathbf{p}|$, the photon is a particle of mass zero.

By using the commutation relations (11.5.24), we see that the operators $a(\mathbf{k}, \sigma)$ and $a^\dagger(\mathbf{k}, \sigma)$ acting on the states (11.6.7) have the effect

$$a(\mathbf{k}, \sigma)\Psi_{\mathbf{k}_1,\sigma_1;\mathbf{k}_2,\sigma_2;...;\mathbf{k}_n,\sigma_n} \propto \sum_{r=1}^{n} \delta^3(\mathbf{k} - \mathbf{k}_r)\delta_{\sigma\sigma_r}$$

$$\times \Psi_{\mathbf{k}_1,\sigma_1;\mathbf{k}_2,\sigma_2;...\mathbf{k}_{r-1},\sigma_{r-1};\mathbf{k}_{r+1},\sigma_{r+1};...;\mathbf{k}_n,\sigma_n} ,$$
$$(11.6.11)$$

$$a^\dagger(\mathbf{k}, \sigma)\Psi_{\mathbf{k}_1,\sigma_1;\mathbf{k}_2,\sigma_2;...;\mathbf{k}_n,\sigma_n} \propto \Psi_{\mathbf{k},\sigma;\mathbf{k}_1,\sigma_1;\mathbf{k}_2,\sigma_2;...;\mathbf{k}_n,\sigma_n}. \quad (11.6.12)$$

Thus $a(\mathbf{k}, \sigma)$ and $a^\dagger(\mathbf{k}, \sigma)$ respectively annihilate and create a photon of momentum $\hbar \mathbf{k}$ and spin index $\sigma$.

Now we must consider the physical significance of the $\sigma$ label carried by each photon. For this purpose, we need to work out the properties of the operators $a(\mathbf{k}, \sigma)$ under rotations. Let us consider a wave vector $\mathbf{k}$ in the $z$-direction $\hat{z}$, and limit ourselves to rotations that leave $\hat{z}$ invariant. According to Eq. (4.1.4), under a rotation represented by an orthogonal matrix $R_{ij}$, a vector like $\boldsymbol{\alpha}(k\hat{z})$ undergoes the transformation

$$U^{-1}(R)\alpha_i(k\hat{z})U(R) = \sum_j R_{ij}\alpha_j(k\hat{z}) . \quad (11.6.13)$$

Inserting the decomposition (11.5.20), this gives

$$\sum_\sigma e_i(\hat{z}, \sigma)U^{-1}(R)a(k\hat{z}, \sigma)U(R) = \sum_\sigma \sum_j R_{ij}e_j(\hat{z}, \sigma)a(k\hat{z}, \sigma) .$$

The rotations that leave $\hat{z}$ invariant have the form

$$R_{ij}(\theta) = \begin{pmatrix} \cos\theta & -\sin\theta & 0 \\ \sin\theta & \cos\theta & 0 \\ 0 & 0 & 1 \end{pmatrix} .$$

A simple calculation shows that

$$\sum_j R_{ij}(\theta) e_j(\hat{z}, \sigma) = e^{-i\sigma\theta} e_i(\hat{z}, \sigma) , \qquad (11.6.14)$$

so by equating the coefficients of $e_i(\hat{z}, \sigma)$, we have

$$U^{-1}(R) a(k\hat{z}, \sigma) U(R) = e^{-i\sigma\theta} a(k\hat{z}, \sigma) . \qquad (11.6.15)$$

Now, for infinitesimal $\theta$, $R_{ij} = \delta_{ij} + \omega_{ij}$, where the non-vanishing elements of $\omega_{ij}$ are $\omega_{xy} = -\omega_{yx} = -\theta$, so according to Eqs. (4.1.7) and (4.1.11),

$$U(\theta) \to 1 - (i/\hbar)\theta J_z ,$$

and Eq. (11.6.15) becomes

$$(i/\hbar)[J_z, a(k\hat{z}, \sigma)] = -i\sigma a(k\hat{z}, \sigma) .$$

Taking the adjoint gives

$$[J_z, a^\dagger(k\hat{z}, \sigma)] = \hbar\sigma a^\dagger(k\hat{z}, \sigma) .$$

Assuming that the no-photon state $\Psi_0$ is rotationally invariant, the one-photon state $\Psi_{k\hat{z},\sigma} \equiv a^\dagger(k\hat{z}, \sigma)\Psi_0$ satisfies

$$J_z \Psi_{k\hat{z},\sigma} = \hbar\sigma \Psi_{k\hat{z},\sigma} . \qquad (11.6.16)$$

There is nothing special about the $z$-direction, so we can conclude that a general one-photon state $\Psi_{\mathbf{k},\sigma}$ has a value $\hbar\sigma$ for the helicity, the angular momentum $\mathbf{J} \cdot \hat{k}$ in the direction of motion. For this reason, the photon is said to be a particle of spin one, but it is a peculiarity of massless particles that the state with $\mathbf{J} \cdot \hat{k} = 0$ is missing. In classical terms, photons with helicity $\pm 1$ make up a beam of left- or right-circularly polarized light.

Of course, photons do not have to be circularly polarized. In the general case, a photon of momentum $\hbar\mathbf{k}$ is a superposition

$$\Psi_{\mathbf{k},\xi} \equiv \left(\xi_+ a^\dagger(\mathbf{k}, +) + \xi_- a^\dagger(\mathbf{k}, -)\right)\Psi_{0\gamma} , \qquad (11.6.17)$$

where $\xi_\pm$ are a pair of generally complex numbers. According to Eq. (11.5.24), the scalar products of these states are

$$\left(\Psi_{\mathbf{k}',\xi'} , \Psi_{\mathbf{k},\xi}\right) = \delta^3(\mathbf{k}' - \mathbf{k}) \left(\xi_+'^* \xi_+ + \xi_-'^* \xi_-\right) , \qquad (11.6.18)$$

so in particular these one-photon states are properly normalized if $|\xi_+|^2 + |\xi_-|^2 = 1$. Such a state is associated with a polarization vector

$$e_i(\hat{k}, \xi) \equiv \xi_+ e_i(\hat{k}, +) + \xi_- e_i(\hat{k}, -) , \qquad (11.6.19)$$

in the sense that

$$\left(\Psi_{0\gamma}, \mathbf{a}(\mathbf{x}, t)\Psi_{\mathbf{k},\xi}\right) = \frac{\sqrt{4\pi c\hbar}}{(2\pi)^{3/2}\sqrt{2k}} e^{i\mathbf{k}\cdot\mathbf{x}} e^{-ickt} \mathbf{e}(\hat{k}, \xi) . \qquad (11.6.20)$$

Circular polarization is the extreme case where either $\xi_+$ or $\xi_-$ vanishes, and the photon has definite helicity. In the opposite extreme case, $|\xi_-| = |\xi_+| = 1/\sqrt{2}$, the polarization vector is real up to an overall phase, and we have the case of linear polarization. For instance, with $\mathbf{k}$ in the $z$-direction, we have a polarization vector

$$e(\hat{z}, \xi) = (\cos \zeta, \sin \zeta, 0) \qquad (11.6.21)$$

if we take

$$\xi_\pm = e^{\mp i \zeta}/\sqrt{2} . \qquad (11.6.22)$$

(Since there is no physical difference between the state vectors $\Psi_{\mathbf{k},\xi}$ and $-\Psi_{\mathbf{k},\xi}$, there is no physical difference between a polarization vector and its negative, or between the polarization angles $\zeta$ and $\zeta + \pi$.) One consequence of Eqs. (11.6.18) and (11.6.22) that we will need in Section 11.8 is that if an observer finds a photon to have linear polarization in a direction $\zeta$, and then re-sets an analyzer to tell if the photon has polarization direction $\zeta'$, the probability of a polarization in this direction is

$$P(\xi \mapsto \xi') = \left| \xi'^*_+ \xi_+ + \xi'^*_- \xi_- \right|^2 = \cos^2(\zeta - \zeta'). \qquad (11.6.23)$$

A complete orthonormal basis is provided by polarizations in directions $\zeta$ and $\zeta + \pi/2$ for any $\zeta$.

The intermediate case in which $|\xi_+|$ and $|\xi_-|$ are unequal but neither vanishes is the case of elliptical polarization.

It is characteristic of massless particles that they come in only two states, with helicity $\pm \hbar j$, where $j$ can be an integer or half-integer. We have seen that $j = 1$ for photons; the quantization of the gravitational field shows that for gravitons, $j = 2$.

Because $a(\mathbf{k}, \sigma)$ and $a^\dagger(\mathbf{k}, \sigma)$ do not commute, it is not possible to find eigenstates of both operators. But the $a(\mathbf{k}, \sigma)$ commute with each other for all $\mathbf{k}$ and $\sigma$, so we can find states $\Phi_{\mathcal{A}}$ that are eigenstates of all these annihilation operators:

$$a(\mathbf{k}, \sigma) \Phi_{\mathcal{A}} = \mathcal{A}(\mathbf{k}, \sigma) \Phi_{\mathcal{A}} , \qquad (11.6.24)$$

with $\mathcal{A}$ an arbitrary complex function of $\mathbf{k}$ and $\sigma$. These are called *coherent states*. In a coherent state, the expectation value of the electromagnetic field (11.5.11) is

$$\frac{\left( \Phi_{\mathcal{A}}, \mathbf{a}(\mathbf{x}, t) \Phi_{\mathcal{A}} \right)}{\left( \Phi_{\mathcal{A}}, \Phi_{\mathcal{A}} \right)} = \int d^3 k \sum_\sigma \sqrt{\frac{4\pi c \hbar}{2|\mathbf{k}|(2\pi)^3}}$$
$$\times \left[ e^{i\mathbf{k}\cdot\mathbf{x}} e^{-ic|\mathbf{k}|t} \mathbf{e}(\mathbf{k}, \sigma) \mathcal{A}(\mathbf{k}, \sigma) \right.$$
$$\left. + e^{-i\mathbf{k}\cdot\mathbf{x}} e^{ic|\mathbf{k}|t} \mathbf{e}^*(\mathbf{k}, \sigma) \mathcal{A}^*(\mathbf{k}, \sigma) \right]. \qquad (11.6.25)$$

(We have here used the defining property of the adjoint, that $(\Phi, a^{\dagger}\Phi) = (a\Phi, \Phi)$.) The coherent state $\Phi_A$ appears classically as if the electromagnetic vector potential has the value (11.6.25). This coherent state contains an unlimited number of photons, for if $\Phi_A$ were a superposition of states (11.6.7) with some maximum number $N$ of photons, then $a(\mathbf{k}, \sigma)\Phi_A$ would be a superposition of states with a maximum number $N - 1$ of photons, and could not possibly be proportional to $\Phi_A$.

## 11.7 Radiative Transition Rates

We now want to calculate the rate of atomic or molecular transitions $a \rightarrow b + \gamma$, where $\Psi_a$ and $\Psi_b$ are eigenstates of the matter Hamiltonian (11.5.4):

$$H_{0\,\text{mat}}\Psi_a = E_a\Psi_a , \qquad H_{0\,\text{mat}}\Psi_b = E_b\Psi_b . \qquad (11.7.1)$$

Both $\Psi_a$ and $\Psi_b$ are zero-photon states, with

$$a(\mathbf{k}, \sigma)\Psi_a = a(\mathbf{k}, \sigma)\Psi_b = 0 , \qquad (11.7.2)$$

for any photon wave number $\mathbf{k}$ and helicity $\sigma$. Hence the final state of the radiative decay process, containing a photon $\gamma$ with a particular wave number $\mathbf{k}$ and helicity $\sigma$, may be expressed as

$$\Psi_{b,\gamma} = \hbar^{-3/2}a^{\dagger}(\mathbf{k}, \sigma)\Psi_b . \qquad (11.7.3)$$

The factor $\hbar^{-3/2}$ is inserted here so that the scalar product of these states involves a delta function for momenta rather than wave numbers; that is, using Eqs. (11.7.2), (11.7.3), and (11.5.24)

$$\left(\Psi_{b',\gamma'}, \Psi_{b,\gamma}\right) = \hbar^{-3}\delta^3(\mathbf{k}' - \mathbf{k})\left(\Psi_{b'}, \Psi_b\right) = \delta^3(\hbar\mathbf{k}' - \hbar\mathbf{k})\left(\Psi_{b'}, \Psi_b\right) .$$

The S-matrix element for the transition $a \rightarrow b + \gamma$ is given to first order in the interaction $V$ by Eq. (8.6.2) [or by Eq. (8.7.14), using $(\Psi_{b\gamma}, V(\tau)\Psi_a) = \exp(-i(E_a - E_b - \hbar ck)\tau/\hbar)(\Psi_{b\gamma}, V(0)\Psi_a)$], as

$$S_{b\gamma,a} = -2\pi i\delta(E_a - E_b - \hbar ck)\left(\Psi_{b\gamma}, V(0)\Psi_a\right)$$

$$= -2\pi i\hbar^{-3/2}\delta(E_a - E_b - \hbar ck)\left(\Psi_b, a(\mathbf{k}, \sigma)V(0)\Psi_a\right) . \qquad (11.7.4)$$

The interaction $V$ at $\tau = 0$ is given by Eq. (11.5.5), which can be written in terms of interaction-picture operators since they are the same as Heisenberg-picture operators at $\tau = 0$:

$$V = -\sum_n \frac{e_n}{m_n c}\mathbf{a}(\mathbf{x}_n) \cdot \mathbf{p}_n + \sum_n \frac{e_n^2}{2m_n c^2}\mathbf{a}^2(\mathbf{x}_n) . \qquad (11.7.5)$$

(We now are dropping the time argument $\tau = 0$.) The $\mathbf{a}^2$ term in Eq. (11.7.5) can only create or destroy two photons, or leave the number of photons unchanged, so it can be dropped here, leaving us with

$$S_{b\gamma,a} = 2\pi i\hbar^{-3/2}\delta(E_a - E_b - \hbar ck)\sum_n \frac{e_n}{m_n c}\left(\Psi_b, a(\mathbf{k}, \sigma)\mathbf{a}(\mathbf{x}_n) \cdot \mathbf{p}_n\Psi_a\right).$$

We insert Eq. (11.5.27) and use the commutation relations (11.5.24) and (11.5.25) to write this as

$$S_{b\gamma,a} = \frac{2\pi i\sqrt{4\pi c\hbar}}{\sqrt{2k(2\pi\hbar)^3}}\delta(E_a - E_b - \hbar ck)\mathbf{e}^*(\hat{k}, \sigma) \cdot \sum_n \frac{e_n}{m_n c}\left(\Psi_b, e^{-i\mathbf{k}\cdot\mathbf{x}_n}\mathbf{p}_n\Psi_a\right).$$

$$(11.7.6)$$

Of course, momentum as well as energy is conserved in the decay process. To see how this works, and for reasons that will become clear later, let us define relative particle coordinates $\bar{\mathbf{x}}_n$ as

$$\bar{\mathbf{x}}_n \equiv \mathbf{x}_n - \mathbf{X}, \qquad (11.7.7)$$

where $\mathbf{X}$ is the center-of-mass coordinate, and $M$ is the total mass

$$\mathbf{X} \equiv \sum_n m_n\mathbf{x}_n/M, \qquad M \equiv \sum_n m_n. \qquad (11.7.8)$$

(Of course, the $\bar{\mathbf{x}}_n$ are not independent, but are subject to a constraint $\sum_n m_n\bar{\mathbf{x}}_n = 0$.) Thus the matrix element in Eq. (11.7.6) may be written as

$$\left(\Psi_b, e^{-i\mathbf{k}\cdot\mathbf{x}_n}\mathbf{p}_n\Psi_a\right) = \left(\Psi_{\bar{b}}, e^{-i\mathbf{k}\cdot\bar{\mathbf{x}}_n}\mathbf{p}_n\Psi_a\right), \qquad (11.7.9)$$

where

$$\Psi_{\bar{b}} \equiv e^{i\mathbf{k}\cdot\mathbf{X}}\Psi_b. \qquad (11.7.10)$$

Note that $[\mathbf{P}, e^{i\mathbf{k}\cdot\mathbf{X}}] = \hbar\mathbf{k}e^{i\mathbf{k}\cdot\mathbf{X}}$, so the operator $e^{i\mathbf{k}\cdot\mathbf{X}}$ just has the effect of a Galilean transformation of the state, that shifts its momentum by $\hbar\mathbf{k}$:

$$\mathbf{P}\Psi_{\bar{b}} = (\mathbf{p}_b + \hbar\mathbf{k})\Psi_{\bar{b}}. \qquad (11.7.11)$$

The operator $\mathbf{P}$ commutes with $\bar{\mathbf{x}}_n$ and with $\mathbf{p}_n$, so the matrix element (11.7.9) vanishes unless $\mathbf{p}_b + \hbar\mathbf{k} = \mathbf{p}_a$, and can therefore be written

$$\left(\Psi_{\bar{b}}, e^{-i\mathbf{k}\cdot\bar{\mathbf{x}}_n}\mathbf{p}_n\Psi_a\right) = \delta^3(\mathbf{p}_b + \hbar\mathbf{k} - \mathbf{p}_a)\mathbf{D}_{n\,ba}(\hat{k}), \qquad (11.7.12)$$

with $\mathbf{D}_{n\,ba}(\hat{k})$ free of delta functions. (We write $\mathbf{D}_{n\,ba}(\hat{k})$ as a function of $\hat{k}$ rather than of $\mathbf{k}$, because the value of $k = |\mathbf{k}|$ is fixed by energy conservation.)

To see how the calculation of this function works in practice, note that in coordinate space the wave functions representing the states $\Psi_a$ and $\Psi_b$ take the form $(2\pi\hbar)^{-3/2}\exp(i\mathbf{p}_a\cdot\mathbf{X}/\hbar)\psi_a(\overline{\mathbf{x}})$ and $(2\pi\hbar)^{-3/2}\exp(i\mathbf{p}_b\cdot\mathbf{X}/\hbar)\psi_b(\overline{\mathbf{x}})$, so the matrix element is

$$\left(\Psi_b, e^{-i\mathbf{k}\cdot\mathbf{x}_n}\mathbf{p}_n\Psi_a\right)$$

$$= (2\pi\hbar)^{-3}\int d^3X \int \left(\prod_m d^3\overline{x}_m\right)\delta^3\left(\sum_m m_m\overline{\mathbf{x}}_m/M\right)$$

$$\times \exp(-i\mathbf{p}_b\cdot\mathbf{X}/\hbar)\,\psi_b^*(\overline{\mathbf{x}})$$

$$\times \exp(-i\mathbf{k}\cdot\overline{\mathbf{x}}_n)\exp(-i\mathbf{k}\cdot\mathbf{X})\,(-i\hbar\,\nabla_n)\exp(i\mathbf{p}_a\cdot\mathbf{X}/\hbar)\psi_a(\overline{\mathbf{x}})\ .$$

We will work in the center-of-mass frame, so $\mathbf{p}_a = 0$, and the $\mathbf{X}$-dependent factors can be combined into a single exponential. The integral over $\mathbf{X}$ then gives

$$\left(\Psi_b, e^{-i\mathbf{k}\cdot\mathbf{x}_n}\mathbf{p}_n\Psi_a\right) = \delta^3(\mathbf{p}_b + \hbar\mathbf{k})\int\left(\prod_m d^3\overline{x}_m\right)\delta^3\left(\sum_m m_m\overline{\mathbf{x}}_m/M\right)$$

$$\times \psi_b^*(\overline{\mathbf{x}})e^{-i\mathbf{k}\cdot\overline{\mathbf{x}}_n}(-i\hbar\,\nabla_n)\psi_a(\overline{\mathbf{x}})\ .$$

Comparing this with Eq. (11.7.12) for $\mathbf{p}_a = 0$, we have

$$\mathbf{D}_{n\,ba}(\hat{k}) = \int\left(\prod_m d^3\overline{x}_m\right)\delta^3\left(\sum_m m_m\overline{\mathbf{x}}_m/M\right)$$

$$\times \psi_b^*(\overline{\mathbf{x}})e^{-i\mathbf{k}\cdot\overline{\mathbf{x}}_n}(-i\hbar\,\nabla_n)\psi_a(\overline{\mathbf{x}})\ . \tag{11.7.13}$$

Returning now to the calculation of the S-matrix element, we can put together Eqs. (11.7.6), (11.7.9), and (11.7.12), and find

$$S_{b\gamma,a} = \delta(E_a - E_b - \hbar ck)\delta^3(\mathbf{p}_a - \mathbf{p}_b - \hbar\mathbf{k})M_{b\gamma,a}\ , \tag{11.7.14}$$

where

$$M_{b\gamma,a} = \frac{2\pi i\sqrt{4\pi c\hbar}}{\sqrt{2k(2\pi\hbar)^3}}\mathbf{e}^*(\hat{k},\sigma)\cdot\sum_n\frac{e_n}{m_n c}\mathbf{D}_{n\,ba}(\hat{k})\ . \tag{11.7.15}$$

The rate for the decay $a \to b + \gamma$ in the center-of-mass frame (where $\mathbf{p}_a = 0$ and $\mathbf{p}_b = -\hbar\mathbf{k}$), with $\hat{k}$ in an infinitesimal solid angle $d\Omega$, is then given by Eq. (8.2.13) as

$$d\Gamma = \frac{1}{2\pi\hbar}|M_{\beta\alpha}|^2\mu\hbar k\,d\Omega\ , \tag{11.7.16}$$

where $\mu$ is given by Eq. (8.2.11), which in the usual case where $E_b \approx Mc^2 \gg \hbar ck$ gives

$$\mu \equiv \frac{E_b\hbar ck}{c^2(E_b + \hbar ck)} \simeq \frac{\hbar k}{c}. \tag{11.7.17}$$

Using Eqs. (11.7.15) and (11.7.17) in Eq. (11.7.16) then gives

$$d\Gamma(\hat{k}, \sigma) = \frac{k}{2\pi\hbar} \left| \mathbf{e}^*(\hat{k}, \sigma) \cdot \sum_n \frac{e_n}{m_n c} \mathbf{D}_{n\,ba}(\hat{k}) \right|^2 d\Omega. \tag{11.7.18}$$

When photon polarization is not measured, the transition rate is the sum of this over $\sigma$. Using Eq. (11.5.23), this is

$$d\Gamma(\hat{k}) \equiv \sum_\sigma d\Gamma(\hat{k}, \sigma)$$

$$= \frac{k}{2\pi\hbar} \sum_{nmij} \frac{e_n e_m}{m_n m_m c^2} \mathbf{D}_{nabi}(\hat{k}) \mathbf{D}^*_{mabj}(\hat{k}) \left[ \delta_{ij} - \hat{k}_i \hat{k}_j \right] d\Omega. \tag{11.7.19}$$

It is frequently possible to make a great simplification in these results. A typical value of the energy $\hbar ck$ emitted in the transition is $\approx e^2/r$, where $r$ is a typical separation of particles from the center-of-mass. Hence the argument of the exponential $\exp(-i\mathbf{k} \cdot \bar{\mathbf{x}}_n)$ in Eqs. (11.7.12) and (11.7.13) is of the order $kr \approx e^2/\hbar c \simeq 1/137$. Since this is small, as long as $\mathbf{D}_{n\,ba}(\hat{k})$ does not vanish, it is a good approximation to set the argument of the exponential $\exp(-i\mathbf{k} \cdot \bar{\mathbf{x}}_n)$ in Eq. (11.7.13) equal to zero, so that here

$$\mathbf{D}_{n\,ab}(\hat{k}) = (b|\mathbf{p}_n|a) \tag{11.7.20}$$

with the reduced matrix element $(b|\mathbf{p}_n|a)$ defined by Eq. (11.7.12) as just the matrix element of $\mathbf{p}_n$ without the delta function:

$$\left( \Psi_{\bar{b}}, \mathbf{p}_n \Psi_a \right) = \delta^3(\mathbf{p}_a - \mathbf{p}_b - \hbar\mathbf{k})(b|\mathbf{p}_n|a) . \tag{11.7.21}$$

In coordinate-space calculations, we have

$$(b|\mathbf{p}_n|a) = \int \left( \prod_m d^3\bar{x}_m \right) \delta^3 \left( \sum_m m_m \bar{\mathbf{x}}_m/M \right) \psi_b^*(\bar{\mathbf{x}})(-i\hbar\nabla_n)\psi_a(\bar{\mathbf{x}}) . \tag{11.7.22}$$

Because the reduced matrix element is now independent of the direction of $\hat{k}$, Eq. (11.7.19) gives the angular dependence of the transition rate explicitly:

$$d\Gamma(\hat{k}) = \frac{k}{2\pi\hbar} \sum_{nmij} \frac{e_n e_m}{m_n m_m c^2} (b|p_{ni}|a)(b|p_{mj}|a)^* \left[ \delta_{ij} - \hat{k}_i \hat{k}_j \right] d\Omega . \tag{11.7.23}$$

We can therefore integrate Eq. (11.7.19) over the directions $\hat{k}$, and find the total radiative decay rate

$$\Gamma = \frac{4k}{3\hbar} \left| \sum_n \frac{e_n}{m_n c}(b|\mathbf{p}_n|a) \right|^2 . \tag{11.7.24}$$

We have seen this formula before, though in a somewhat different form, involving matrix elements of coordinates rather than momenta. To see the connection, note that

$$[H_{0\text{ mat}}, \bar{\mathbf{x}}_n] = -i\hbar \left[ \frac{\mathbf{p}_n}{m_n} - \frac{\mathbf{P}}{M} \right] .$$

Because we are in the center-of-mass frame, with $\mathbf{P}\Psi_a = 0$, we can drop the second term in the square brackets, and write the matrix element in Eq. (11.7.22) as

$$\left( \Psi_{\bar{b}}, \mathbf{p}_n \Psi_a \right) = \frac{im_n}{\hbar} \left( \Psi_{\bar{b}}, [H_{0\text{ mat}}, \bar{\mathbf{x}}_n]\Psi_a \right) = \frac{im_n}{\hbar}(E_{\bar{b}} - E_a)\left( \Psi_{\bar{b}}\bar{\mathbf{x}}_n\Psi_a \right) .$$

Because the state $\Psi_{\bar{b}}$ has momentum $\mathbf{p}_b + \hbar\mathbf{k} = \mathbf{p}_a = 0$, its energy $E_{\bar{b}}$ is not precisely equal to $E_b$, but rather to $E_b$ minus the actual recoil kinetic energy $(\hbar k)^2/2M$. In any non-relativistic system, this recoil energy will be very small compared with the energy splitting $E_b - E_a = \hbar ck$, because $E_a - E_b \ll Mc^2$. Hence we can take $E_{\bar{b}} - E_a \simeq \hbar ck$, so that

$$\left( \Psi_{\bar{b}}, \mathbf{p}_n \Psi_a \right) = ickm_n\left( \Psi_{\bar{b}}, \bar{\mathbf{x}}_n\Psi_a \right) . \tag{11.7.25}$$

Of course, momentum is still conserved here, so we can write

$$\left( \Psi_{\bar{b}}, \bar{\mathbf{x}}_n\Psi_a \right) = \delta^3(\mathbf{p}_b + \hbar c\mathbf{k})(b|\bar{\mathbf{x}}_n|a) \tag{11.7.26}$$

and by the same argument as that which led to Eq. (11.7.22)

$$(b|\bar{\mathbf{x}}_n|a) = \int \left( \prod_m d^3\bar{x}_m \right) \delta^3 \left( \sum_m m_m\bar{\mathbf{x}}_m/M \right) \psi_b^*(\bar{\mathbf{x}})\bar{\mathbf{x}}_n\psi_a(\bar{\mathbf{x}}) . \tag{11.7.27}$$

So Eq. (11.7.24) may be written

$$\Gamma = \frac{4\omega^3}{3c^3\hbar} \left| \sum_n e_n(b|\bar{\mathbf{x}}_n|a) \right|^2 , \tag{11.7.28}$$

where $\omega \equiv ck$. The operator $\sum_n e_n\bar{\mathbf{x}}_n$ is the electric-dipole operator, so as mentioned in Section 4.4, this is called an E1 or *electric-dipole* radiation.

This formula is a slight generalization of Eq. (1.4.5), which was derived in 1925 by Heisenberg on the basis of an analogy with radiation by a classical charged oscillator. As discussed in Section 6.5, the same result was re-derived by Dirac in 1926 on the basis of the calculation of stimulated emission in a classical light wave, together with the Einstein relation (1.2.16) between the rates of stimulated and spontaneous emission. The derivation given here, due originally to Dirac in 1927,[7] was the first that showed how photons are created through the interaction of a quantized electromagnetic field with a material system.

---

[7] P. A. M. Dirac, *Proc. Roy. Soc.* A **114**, 710 (1927).

The operators $\mathbf{p}_n$ and $\bar{\mathbf{x}}_n$ are spatial vectors, and therefore as shown in Eq. (4.4.6) behave under rotations like operators with $j = 1$. According to the rules for addition of angular momentum described in Section 4.3, such operators have zero matrix elements between the states $\Psi_a$ and $\Psi_{\bar{b}}$ unless the angular momenta $\hbar j_a$ and $\hbar j_b$ of these states satisfy $|j_a - j_b| \leq 1 \leq j_a + j_b$. Also, these operators change sign under a reflection of space coordinates, so these matrix elements vanish unless the states $a$ and $b$ have opposite parity. As already mentioned, transitions satisfying the selection rules that $|j_a - j_b| \leq 1 \leq j_a + j_b$ and that $a$ and $b$ have opposite parity are called *electric-dipole*, or E1, transitions. Thus for instance, aside from small effects involving electron spin, the formula (11.7.28) can be used to calculate the rate of single-photon emission in transitions in hydrogen such as the E1 Lyman-$\alpha$ transition $2p \rightarrow 1s$, but not $3d \rightarrow 1s$ or $3p \rightarrow 2p$.

To calculate the rates for single-photon emission in transitions that do not satisfy the electric-dipole selection rules, we must include higher terms in the expansion of the exponential in Eq. (11.7.13). Suppose we have a transition in which the matrix elements $(\Psi_{\bar{b}}, \mathbf{p}_n \Psi_a)$ and $(\Psi_{\bar{b}}, \bar{\mathbf{x}}_n \Psi_a)$ all vanish. In this case we can try to calculate the transition rate by including the first-order term in the expansion of the exponential in Eq. (11.7.13), so that in place of Eq. (11.7.20) we have

$$D_{nabi}(\hat{k}) = -i \sum_j k_j (b|\bar{x}_{nj} p_{ni}|a) , \qquad (11.7.29)$$

with the reduced matrix element of any operator $\mathcal{O}$ that commutes with the total particle momentum defined by

$$(\Psi_{\bar{b}}, \mathcal{O}\Psi_a) = \delta^3(\mathbf{p}_b + \hbar \mathbf{k} - \mathbf{p}_a)(b|\mathcal{O}|a) . \qquad (11.7.30)$$

The differential decay rate (11.7.19) can then be written

$$d\Gamma(\hat{k}) = \frac{k^3}{2\pi\hbar} \sum_{nmijkl} \frac{e_n e_m}{m_n m_m c^2} (b|\bar{x}_{nk} p_{ni}|a)(b|\bar{x}_{ml} p_{mj}|a)^* \hat{k}_k \hat{k}_l \left[ \delta_{ij} - \hat{k}_i \hat{k}_j \right] d\Omega .$$

$$(11.7.31)$$

To integrate over the directions of $\hat{k}$, we now need the formula[8]

$$\int d\Omega \, \hat{k}_i \hat{k}_j \hat{k}_k \hat{k}_l = \frac{4\pi}{15} \left[ \delta_{ij}\delta_{kl} + \delta_{ik}\delta_{jl} + \delta_{il}\delta_{jk} \right] ,$$

as well as the previously used formula

$$\int d\Omega \, \hat{k}_k \hat{k}_l = \frac{4\pi}{3} \delta_{kl} .$$

---

[8] The right-hand sides of these formulas are, up to a constant factor, the unique combinations of Kronecker deltas that are symmetric in the indices. The numerical coefficients can be calculated by noting that if we contract all pairs of indices, the integral must equal $4\pi$.

The decay rate is then

$$\Gamma = \frac{2k^3}{15\hbar} \sum_{nmijkl} \frac{e_n e_m}{m_n m_m c^2} (b|\overline{x}_{nk} p_{ni}|a)(b|\overline{x}_{ml} p_{mj}|a)^* \left[4\delta_{ij}\delta_{kl} - \delta_{ik}\delta_{jl} - \delta_{jk}\delta_{il}\right] \,.$$

(11.7.32)

It is helpful to decompose the final factor into a term symmetric in $i$ and $k$ and in $j$ and $l$, and a term antisymmetric in $i$ and $k$ and in $j$ and $l$:

$$4\delta_{ij}\delta_{kl} - \delta_{ik}\delta_{jl} - \delta_{jk}\delta_{il} = \frac{3}{2}\left(\delta_{ij}\delta_{kl} + \delta_{kj}\delta_{il} - \frac{2}{3}\delta_{ik}\delta_{jl}\right) + \frac{5}{2}\left(\delta_{ij}\delta_{kl} - \delta_{kj}\delta_{il}\right) \,.$$

(11.7.33)

Correspondingly, the rate (11.7.32) may be expressed as

$$\Gamma = \frac{2k^3}{15\hbar c^2} \sum_{ij} \left[\frac{3}{4}|(b|Q_{ij}|a)|^2 + \frac{5}{4}|(b|M_{ij}|a)|^2\right] \,,$$

(11.7.34)

where

$$(b|Q_{ij}|a) \equiv \sum_n \frac{e_n}{m_n} \left[(b|\overline{x}_{ni} p_{nj}|a) + (b|\overline{x}_{nj} p_{ni}|a) - \frac{2}{3}\delta_{ij}\sum_l (b|\overline{x}_{nl} p_{nl}|a)\right] \,,$$

(11.7.35)

$$(b|M_{ij}|a) \equiv \sum_n \frac{e_n}{m_n} \left[(b|\overline{x}_{ni} p_{nj}|a) - (b|\overline{x}_{nj} p_{ni}|a)\right] \,.$$

(11.7.36)

The reduced matrix elements $(b|Q_{ij}|a)$ and $(b|M_{ij}|a)$ are known as the *electric-quadrupole* (E2) and *magnetic-dipole* (M1) matrix elements. The operators involved transform under rotations as operators with $j = 2$ and $j = 1$, so these matrix elements vanish unless the following selection rules are satisfied:

$$\text{E2:} \quad |j_a - j_b| \le 2 \le j_a + j_b \,, \qquad \text{M1:} \quad |j_a - j_b| \le 1 \le j_a + j_b \,. \qquad (11.7.37)$$

Also, unlike the electric-dipole case, these matrix elements vanish unless the states $a$ and $b$ have the *same* parity. Thus for instance, in hydrogen the transitions $3d \to 2s$ and $3d \to 1s$ are dominated by the electric-quadrupole matrix element, while the transition $3p \to 2p$ receives contributions from both the electric-quadrupole and the magnetic-dipole matrix elements.

The formulas (11.7.35) and (11.7.36) for the E2 and M1 matrix elements can be put in a more useful form. In the same way that we derived Eq. (11.7.25), it is easy to show that the E2 matrix element is

$$(b|Q_{ij}|a) = ick \sum_n e_n \left[(b|\overline{x}_{ni}\overline{x}_{nj}|a) - \frac{1}{3}(b|\mathbf{x}_n^2|a)\right] \,.$$

(11.7.38)

We cannot use this trick for the M1 matrix element, but we note instead that

$$(b|M_{ij}|a) = \sum_k \epsilon_{ijk} \sum_n \frac{e_n}{m_n} (b|L_{nk}|a) \,,$$

(11.7.39)

where $\mathbf{L}_n$ is the orbital angular momentum $\mathbf{x}_n \times \mathbf{p}_n$ of the $n$th particle.

So far, we have ignored any spin of the charged particles, but, to the accuracy of this calculation, we now need also to include the effects of magnetic moments. As noted in Eq. (10.3.1), the effect of magnetic moments is to add to the interaction a term

$$\Delta V = - \sum_n \boldsymbol{\mu}_n \cdot \left( \nabla \times \mathbf{a}(\mathbf{x}_n) \right) , \qquad (11.7.40)$$

where (for any spin) $\boldsymbol{\mu}_n = \mu_n \mathbf{S}_n / s_n$, with $\mathbf{S}_n$ the spin operator of the $n$th particle and $\mu_n$ the quantity known as the $n$th particle's magnetic moment. Following the same analysis that led to Eq. (11.7.34), we find that the effect of this addition of Eq. (11.7.40) is to replace Eq. (11.7.39) with

$$(b|M_{ij}|a) = \sum_k \epsilon_{ijk} \sum_n \frac{e_n}{m_n} (b|L_{nk} + g_n S_{nk}|a) , \qquad (11.7.41)$$

where $g_n$ is the gyromagnetic ratio, a dimensionless constant generally of order unity, defined by $\mu_n = e_n g_n s_n / 2m_n$, or in other words, $\boldsymbol{\mu}_n = e_n g_n \mathbf{S}_n / 2m_n$. (For electrons, $g = 2.002322 \ldots$.) For instance, in the important transition of the $1s$ state of the hydrogen atom with total (electron plus nucleon) spin equal to one into the $1s$ state with total spin zero, which produces photons with a wavelength of 21 cm, the rate is dominated by the M1 matrix element, arising entirely from the second term in Eq. (11.7.41).

This analysis can be continued. The matrix element for a transition that does not satisfy the selection rules for electric-dipole, electric-quadrupole, or magnetic-dipole moments can be calculated by including terms in the exponential in Eq. (11.7.12) or (11.7.13) of higher than first order in $\mathbf{k} \cdot \bar{\mathbf{x}}_n$. But there is one kind of transition that is forbidden to all orders in $\mathbf{k} \cdot \bar{\mathbf{x}}_n$ – single-photon transitions between states with $j_a = j_b = 0$. This rule follows immediately from the conservation of the component of angular momentum along the direction $\hat{k}$. Where $j_a = j_b = 0$, the states $a$ and $b$ necessarily have value zero for this component (or any component) of angular momentum, while the photon can only have a value $\hbar$ or $-\hbar$ for this component. Thus, for instance, the decay of the charged spinless meson $K^+$ into the charged spinless meson $\pi^+$ and a single photon is absolutely forbidden.

## 11.8 Quantum Key Distribution

Since ancient times people have attempted to send messages that cannot be understood by anyone but the designated recipient, even when the message is intercepted by an eavesdropper. Any message can be regarded as a whole number $m$, for instance by interpreting the dots and dashes of Morse code as ones and zeros, and treating the resulting string of ones and zeros as the binary expression of a number. An encryption is a function, agreed on between the sender (Alice) and the designated recipient (Bob) but unknown to a possible eavesdropper

(Eve), that takes the message number $m$ into some other whole number $f(m)$. If the same encryption is used many times, Eve can usually deduce the nature of the encryption and read the messages by frequency analysis – for instance, for English-language messages interpreting the most commonly encountered sequence of ones and zeros as the letter $e$. For this reason, it is common to let the encryption depend on a frequently changed key, which can be regarded as another whole number $k$, so that a message $m$ is sent as the number $f(m, k)$ depending on the key. One simple common method is to take $f(m, k)$ as the product $km$. Knowing $k$, it is trivial for Bob to retrieve the message $m$ from the encrypted signal $km$ by just dividing by $k$, but if Eve does not know $k$, then it is necessary for her to try all the possible factorizations of the signal $km$ into a product of whole numbers, which takes a time that grows with $km$ faster than any power. But if the key is to be frequently changed, Alice and Bob must frequently exchange messages that establish new keys, and these messages too may be intercepted by Eve. Quantum key distribution defeats Eve's attempt to learn the key by exploiting the feature of quantum mechanics that it is not possible to measure any quantity without changing the state vector to one in which that quantity has some definite value.

In the widely used BB84 protocol,[9] Alice sends the key to Bob as a sequence of linearly polarized photons, say with momentum along the 3-direction, and so with polarization vectors of the form

$$\mathbf{e} = (\cos \zeta, \sin \zeta, 0) \,,$$

where the $\zeta$ are various angles. Alice represents ones and zeros by values of the angle $\zeta$ in either one of two modes, which she chooses at random for each successive photon. In mode I, a zero and a one are represented respectively by the orthogonal polarization vectors with $\zeta = 0$ and $\zeta = \pi/2$, while in mode II a zero and a one are represented respectively by two different orthogonal polarization vectors, with $\zeta = \pi/4$ and $\zeta = 3\pi/4$. (This is summarized in Table 11.1.) Receiving the photon, Bob at random makes a choice between two modes of setting his polarization analyzer: in mode I he measures whether $\zeta = 0$ or $\zeta = \pi/2$ (for instance by setting his analyzer so that all photons with $\zeta = 0$ go through, and all photons with $\zeta = \pi/2$ are blocked) while in mode II Bob measures whether $\zeta = \pi/4$ or $\zeta = 3\pi/4$. If Alice sends a photon in some mode and Bob analyzes its polarization using the same mode, then Bob finds the value of $\zeta$ used by Alice, and records the same one or zero as intended by Alice. But if Alice uses mode I and Bob happens to use mode II, then he observes a polarization angle $\zeta = \pi/4$ or $3\pi/4$ with probabilities each given by Eq. (11.6.23) as 50%, so he records a one or a zero that has just a 50% chance of being what Alice intended. The outcomes are the same if Alice chooses a polarization according

9    C. H. Bennett and G. Brassard, in *Proceedings of the IEEE International Conference on Computers, Systems, and Signal Processing, Bangalore, India, 1984* (IEEE, New York, 1984), pp. 175–179.

Table 11.1   The BB84 protocol

| Mode | Binary digit | $\zeta$ |
|------|--------------|---------|
|      | 0            | 0       |
| I    |              |         |
|      | 1            | $\pi/2$ |
|      | 0            | $\pi/4$ |
| II   |              |         |
|      | 1            | $3\pi/4$ |

to mode II and Bob measures the polarization using mode I. For each photon, there is a 50% chance that Alice and Bob will be using different modes, and if they are then there is a 50% chance that Bob will record the same one or zero as sent by Alice, so 25% of the binary digits recorded by Bob will be wrong. To weed these out, after all the photons have been sent and observed, Bob and Alice compare notes about the modes they both used (using messages that can be sent back and forth *en clair*, without encryption), and they discard the 50% of binary digits for which Alice and Bob happened to have used different modes of choosing and analyzing the photon polarization. The resulting string of binary digits, which are the same for Alice and Bob, is the new key.

By intercepting the photons sent from Alice to Bob, Eve can prevent this key distribution, but what Eve really wants is that Alice and Bob *should* establish a key, but one that Eve knows, so that she can secretly read the messages sent from Alice to Bob. Unfortunately for Eve, even though she may know all about the BB84 protocol, her eavesdropping inevitably destroys the key, and this will become known to Alice and Bob.[10] The only way that Eve can eavesdrop is by intercepting the photons sent by Alice, measuring their polarizations, and then sending substitute photons with these polarizations on to Bob. But while this is going on, Eve like Bob does not know the mode that Alice is using in choosing each photon polarization. If for some photon Eve sets her polarization analyzers in a mode different from the mode used by Alice to send the photon, then there is only a 50% chance that the substitute photon sent by Eve to Bob will have the same polarization that it had when it was sent by Alice. For instance, if Alice using mode I sends a photon with $\zeta = \pi/2$, representing a one, and Eve happens to set her analyzers in mode II, then she will find either $\zeta = \pi/4$ or $\zeta = 3\pi/4$, each with 50% probability. Whichever of these polarizations Eve

---

[10] The security of the BB84 protocol was rigorously proved by P. W. Shor and J. Preskill, *Phys. Rev. Lett.* **85**, 441 (2000).

chooses for the photon she sends on to Bob, he will record either a zero or a one with equal probability irrespective of whether he sets his analyzer in mode I or mode II. After all this is over, when Alice and Bob compare notes, they identify the photons that had been sent when Alice and Bob had by chance been using the same modes, and Eve too may learn this information, but by then it is too late. Even when Alice and Bob had been using the same mode for a given photon, there is only a 50% chance that Eve had used the same mode that they had used, and if she had not then there is only a 50% chance that Bob would have observed the same polarization that had been sent by Alice, so 25% of the binary digits in the key that Bob had received from Eve would not match the corresponding digits in the key understood by Alice. When Alice and Bob try to communicate using their respective keys, the keys generally will not work. For instance, if Alice encrypts a message represented by a number $m$ by multiplying by a key represented by a number $k$, and Bob tries to decrypt this signal by dividing by the number $k'$ representing what he thinks is the key, the result $mk/k'$ will typically not be a whole number. Even if it is, and even if this number represents what could have been a possible message, Alice and Bob can detect Eve's intervention by comparing a part of the key, and observing that 25% of the digits don't match. Eve will have succeeded only in preventing the construction of a key, not in secretly learning a key that will be used by Alice and Bob.

## Problems

1. Calculate the rate for emission of a photon in the transition $3d \to 2p$, and in the transition $2p \to 1s$ in hydrogen. Give formulas and numerical values. You can use the facts that the proton is much heavier than the electron, and the wavelength of the photon emitted in these processes is much larger than the atomic size, and neglect electron spin.

2. What powers of the photon wave number appear in the rates for single-photon emission in the decays of the $4f$ state of hydrogen into the $3s$, $3p$, and $3d$ states?

3. Consider the theory of a real scalar field $\varphi(\mathbf{x}, t)$, interacting with a set of particles with coordinates $\mathbf{x}_n(t)$. Take the Lagrangian as

$$
L(t) = \frac{1}{2} \int d^3x \left[ \left( \frac{\partial \varphi(\mathbf{x}, t)}{\partial t} \right)^2 - c^2 \left( \nabla \varphi(\mathbf{x}, t) \right)^2 - \mu^2 \varphi^2(\mathbf{x}, t) \right]
$$
$$
- \sum_n g_n \varphi(\mathbf{x}_n(t), t) + \sum_n \frac{m_n}{2} \left( \dot{\mathbf{x}}_n(t) \right)^2 - V\left( \mathbf{x}(t) \right),
$$

where $\mu$, $m_n$, and $g_n$ are real parameters, and $V$ is a real local function of the differences of the particle coordinates.

(a) Find the field equations and commutation rules for $\varphi$.

(b) Find the Hamiltonian for the whole system.

(c) Express $\varphi$ in the interaction picture in terms of operators that create and destroy the quanta of the scalar field.

(d) Calculate the energy and momentum of these quanta.

(e) Give a general formula for the rate of emission per solid angle of a single $\varphi$ quantum in a transition between eigenstates of the matter part of the Hamiltonian (that is, the part of the Hamiltonian involving only the coordinates $\mathbf{x}_n$ and their canonical conjugates).

(f) Integrate this formula over solid angles in the case where the wavelength of the emitted quanta is much larger than the size of the initial and final particle system. What are the selection rules for these transitions?

4. Express the coherent state $\Phi_A$ as a superposition of states (11.6.7) with definite numbers of photons.

# 12

# Entanglement

There is a troubling weirdness about quantum mechanics. Perhaps its weirdest feature is entanglement, the need to describe even systems that extend over macroscopic distances in ways that are inconsistent with classical ideas.

## 12.1 Paradoxes of Entanglement

Einstein had from the beginning resisted the idea that quantum mechanics could provide a complete description of reality. His reservations were crystallized in a 1935 paper[1] with Boris Podolsky (1896–1966) and Nathan Rosen (1909–1995). They considered an experiment in which two particles that move along the $x$-axis with coordinates $x_1$ and $x_2$ and momenta $p_1$ and $p_2$ were somehow produced in an eigenstate of the observables $x_1 - x_2$ and $p_1 + p_2$: specifically, $p_1 + p_2$ has an eigenvalue zero, and $x_2 - x_1 = x_0$, where $x_0$ is some length that is taken to be macroscopically large, much too large for particles 1 and 2 to exert any influence on each other. Quantum mechanics itself presents no obstacle to this, for these two observables commute. Indeed, we can easily write the wave function for such a state:

$$\psi(x_1, x_2) = \int_{-\infty}^{\infty} dk \, \exp[ik(x_1 - x_2 + x_0)] = 2\pi \delta(x_1 - x_2 + x_0) . \quad (12.1.1)$$

Of course, this wave function is not normalizable, but this is just the usual problem with continuum wave functions; the wave function (12.1.1) can be approximated arbitrarily closely with a normalizable wave function, such as

$$\exp(-\kappa(x_1 + x_2)^2) \int_{-\infty}^{\infty} dk \, \exp[ik(x_1 - x_2 + x_0)] \exp\left(-L^2(k - k_0)^2\right) ,$$

with $L$ and $\kappa$ both very small.

Einstein *et al.* imagined that an observer who studies particle 1 measures its momentum, and finds a value $\hbar k_1$. The momentum of particle 2 is then known

---

[1] A. Einstein, B. Podolsky, and N. Rosen, *Phys. Rev.* **47**, 777 (1935).

to be $-\hbar k_1$, up to an arbitrarily small uncertainty. But suppose that the observer instead measures the position of particle 1, finding a position $x_1$, in which case the position of particle 2 would have to be $x_1 + x_0$. We understand that the measurement of the position of particle 1 can interfere with its momentum, and vice versa, so that whichever measurement is done last would interfere with the result of earlier measurement. But how can these measurements interfere with the properties of particle 2, if the particles are far apart? And if they do not, then must we not conclude that particle 2 has both a definite momentum $-k_1$ and a definite position $x_1 + x_0$, contradicting the fact that these observables do not commute?

Einstein *et al.* did not spell out how to construct such a state, but one can imagine that two particles that are originally bound in some sort of unstable molecule at rest fly apart freely in opposite directions, with equal and opposite momenta, until their separation becomes macroscopically large. If they have the initial separation $x_1^{\text{init}} - x_2^{\text{init}}$, then (assuming that the particles have equal mass $m$), after a time $t$ their separation will be

$$x_1 - x_2 = x_1^{\text{init}} - x_2^{\text{init}} + (p_1 - p_2)t/m \ .$$

We cannot actually take the initial separation $x_1^{\text{init}} - x_2^{\text{init}}$ to be precisely known, because then the relative momentum $p_1 - p_2$ would be entirely uncertain, making the separation $x_1 - x_2$ soon also uncertain. If we take the initial separation to be known within an uncertainty $\Delta|x_1^{\text{init}} - x_2^{\text{init}}| = L$, then the uncertainty in the relative momentum will be at least of order $\hbar/L$, and after a time $t$ the uncertainty in the separation will be at least of order $L + \hbar t/mL$. This has a minimum when $L = \sqrt{\hbar t/m}$, at which the uncertainty in $x_1 - x_2$ is also of order $\sqrt{\hbar t/m}$. But this does not obviate the Einstein–Podolsky–Rosen paradox, because we can measure $k_2$ as accurately as we like, and we can measure $x_2$ to an accuracy of about $\sqrt{\hbar t/m}$, so the product of these uncertainties can be as small as we like, contradicting the uncertainty principle.

The problem posed by Einstein, Rosen, and Podolsky was made sharper by David Bohm[2] (1917–1992). A system of zero total angular momentum decays into two particles, each with spin 1/2. Using the Clebsch–Gordan coefficients for combining spin 1/2 and spin 1/2 to make spin zero, the spin state vector is then

$$\Psi = \frac{1}{\sqrt{2}}\big[\Psi_{\uparrow\downarrow} - \Psi_{\downarrow\uparrow}\big] \ , \tag{12.1.2}$$

where the two arrows indicate the signs of the $z$-component of the two particles' spins. After a long time, the particles are far apart, and then measurements are made of the spin components of particle 1. If the $z$-component of the spin

---

[2] D. Bohm, *Quantum Theory* (Prentice-Hall, Inc., New York, 1951), Chapter XXII. Also see D. Bohm and Y. Aharonov, *Phys. Rev.* **108**, 1070 (1957).

of particle 1 is measured, it must have a value $\hbar/2$ or $-\hbar/2$, and then the $z$-component of the spin of particle 2 must correspondingly have a value $-\hbar/2$ or $+\hbar/2$, respectively. Bohm reasoned that since the two particles are so far apart, the measurement of the spin of particle 1 could not have influenced the spin of particle 2, so it must have had that $z$-component all along. But the observer could have measured the $x$-component of the spin of particle 1 instead of its $z$-component, and by the same reasoning, if a value $\hbar/2$ or $-\hbar/2$ were found for the $x$-component of the spin of particle 1 then also the $x$-component of the spin of particle 2 must have been $-\hbar/2$ or $\hbar/2$ all along. Likewise for the spin $y$-components. So according to this reasoning, all three components of the spin of particle 2 have definite values, which is impossible since these spin components do not commute.

Bohm was led to suppose that either the content or the interpretation of quantum mechanics needs modification. Most physicists today would instead respond to both the Einstein–Podolsky–Rosen paradox and the Bohm paradox by accepting that no matter how far apart the two particles are, the measurement of the properties of one of them does affect the wave function of the other. Though the particles are far apart, their properties remain entangled.

The existence of entanglement in quantum mechanics naturally raises the question whether a measurement of one isolated part of an entangled system can be used to send messages to another isolated part instantaneously, with no limitation set by the finite speed of light. No, it can't. In the Einstein–Podolsky–Rosen case, there is no way that an observer of particle 2 can tell that it does or does not have a definite momentum – if she measures the momentum she gets some value, but she does not know whether there is any other value she could have gotten. Even if this experiment is repeated many times, the observer of particle 2 cannot tell what measurements have been made on particle 1. She may find various different values for the momentum of particle 2, but she can't know whether this is because the position of particle 1 was measured, or whether particle 1 was in a superposition of momentum eigenstates to begin with.

This can be put in very general terms, described most simply for systems like those considered by Bohm, in which the measured quantities take only discrete values. As described in Section 3.7, both the deterministic unitary evolution of states in quantum mechanics and the probabilistic change produced in a measurement, or in any combination of unitary evolution and measurement, will produce a linear transformation $\rho \mapsto \rho'$ of the density matrix, which takes the general form

$$\rho'_{M'N'} = \sum_{MN} K_{M'M,N'N}\, \rho_{MN} \,, \tag{12.1.3}$$

where $K$ is some c-number kernel independent of $\rho$. In order for $\rho'$ to have unit trace for an arbitrary $\rho$ with unit trace, it is necessary and sufficient that

$$\sum_{M'} K_{M'M,M'N} = \delta_{MN} . \qquad (12.1.4)$$

Suppose that a system consists of two isolated parts, subsystems *I* and *II*, and replace the indices $M$, $N$, etc. with compound indices $ma$, $nb$, etc., with the first letter labeling the states of subsystem *I* and the second the states of subsystem *II*. The possibility of entanglement does not in general allow the density matrix to factor into a product $\rho_{mn}^{(I)}\rho_{ab}^{(II)}$ of density matrices for the two subsystems, but if the subsystems are isolated (with no physical influence *or* information flowing between them) then the kernel in Eq. (12.1.3) does factorize:

$$K_{m'a'ma,n'b'nb} = K_{m'm,n'n}^{(I)} \, K_{a'a,b'b}^{(II)} , \qquad (12.1.5)$$

where $K^{(I)}$ and $K^{(II)}$ are the kernels that would describe the transformation of the density matrix in subsystems *I* and *II* if the other subsystem did not exist. For instance, if we make a measurement of some physical quantities in subsystem *I* that take definite values in a complete orthonormal set of states $\Phi_\mu^{(I)}$ and also make a measurement of some physical quantities in subsystem *II* that take definite values in a complete orthonormal set of states $\Phi_\alpha^{(II)}$, then this puts the whole system in a state with projection operator

$$[\Lambda_{\mu\alpha}]_{m'a',ma} = [\Lambda_\mu^{(I)}]_{m'm}[\Lambda_\alpha^{(II)}]_{a'a} ,$$

where $\Lambda_\mu^{(I)}$ and $\Lambda_\alpha^{(II)}$ are the projection operators onto the states $\Phi_\mu^{(I)}$ and $\Phi_\alpha^{(II)}$, respectively. According to Eq. (3.7.2) the effect of the joint measurement is a mapping with kernel

$$K_{m'a'ma,n'b'nb} = \sum_{\mu\alpha}[\Lambda_{\mu\alpha}]_{m'a',ma}[\Lambda_{\mu\alpha}]_{nb,n'b'}$$

$$= \left(\sum_\mu [\Lambda_\mu^{(I)}]_{m'm}[\Lambda_\mu^{(I)}]_{nn'}\right)\left(\sum_\alpha [\Lambda_\alpha^{(II)}]_{a'a}[\Lambda_\alpha^{(II)}]_{bb'}\right) . \qquad (12.1.6)$$

In the case of the ordinary unitary evolution of state vectors, the factorization of the kernel follows as a consequence of the property of isolated systems that

$$H_{ma,nb} = H_{mn}^{(I)}\delta_{ab} + H_{ab}^{(II)}\delta_{mn} .$$

Since the two terms in each exponential in Eq. (3.7.3) commute, the exponential of the sum is the product of the exponentials, and so here Eq. (3.7.3) gives

$$K_{m'a'ma,n'b'nb} = \left[[\exp(-iH^{(I)}(t'-t)/\hbar)]_{m'm}[\exp(+iH^{(I)}(t'-t)/\hbar)]_{nn'}\right]$$

$$\times \left[[\exp(-iH^{(II)}(t'-t)/\hbar)]_{a'a}[\exp(+iH^{(I)}(t'-t)/\hbar)]_{bb'}\right] . \qquad (12.1.7)$$

Equations (12.1.6) and (12.1.7) exhibit the factorization (12.1.5) characteristic of isolated subsystems. The same factorization applies for any combination of measurements interspersed with ordinary unitary evolution.

Now, since both $K^{(I)}$ and $K^{(II)}$ are possible physical kernels, they each satisfy the analog of Eq. (12.1.4):

$$\sum_{m'} K^{(I)}_{m'm,m'n} = \delta_{mn} \;, \qquad \sum_{a'} K^{(II)}_{a'a,a'b} = \delta_{ab} \;. \qquad (12.1.8)$$

In the absence of any information about subsystem *II*, the density matrix of subsystem *I* is

$$\rho^{(I)}_{mn} = \sum_a \rho_{ma,na} \;. \qquad (12.1.9)$$

As mentioned in Section 3.3, this follows from the requirement that the mean value $\mathrm{Tr}(\rho A)$ of any physical quantity represented by an operator of the form $A_{ma,nb} = A^{(I)}_{mn}\delta_{ab}$, which acts non-trivially only on subsystem *I*, should be equal to $\mathrm{Tr}(\rho^{(I)}A^{(I)})$. According to Eqs. (12.1.3), (12.1.5), and (12.1.9) its evolution is given by

$$\rho^{(I)}_{m'n'} \mapsto \rho'^{(I)}_{m'n'} = \sum_{a'} \sum_{mnab} K^{(I)}_{m'm,n'n} \, K^{(II)}_{a'a,a'b} \rho_{ma,nb} \;.$$

Using Eq. (12.1.8) for $K^{(II)}$ and Eq. (12.1.9), this is

$$\rho'^{(I)}_{m'n'} = \sum_{mn} K^{(I)}_{m'm,n'n} \rho^{(I)}_{mn} \;, \qquad (12.1.10)$$

so the evolution of $\rho^{(I)}$ is independent of $\rho^{(II)}$. Therefore, even though in entangled states it is possible to modify the state vector of subsystem *I* by making measurements in subsystem *II* or by modifying its Hamiltonian, this cannot change the density matrix of subsystem *I*. The subsequent evolution of the density matrix of subsystem *I* and the results of any measurements in this subsystem depend only on the density matrix, so entanglement does not create any possibilities of instantaneous communication at a distance.

But this is a special feature of quantum mechanics, arising from the fact that both measurement and the Hamiltonian evolution of the state vector produce a mapping of the density matrix into a linear function only of the density matrix, not depending on the state vector. Any attempt to generalize quantum mechanics by allowing small non-linearities in the evolution of state vectors risks the introduction of instantaneous communication between separated observers.[3]

Of course, according to present ideas a measurement in one subsystem does change the state vector for a distant isolated subsystem – it just doesn't change the density matrix. If it were possible to probe state vectors, other than by making measurements, then faster-than-light communication could be possible. As mentioned in Section 3.7, the phenomenon of entanglement thus poses an obstacle to any interpretation of quantum mechanics that attributes to the wave

---

[3]  N. Gisin, *Helv. Phys. Acta* **62**, 363 (1989); J. Polchinski, *Phys. Rev. Lett.* **66**, 397 (1991).

function or the state vector any physical significance other than as a means of predicting the results of measurements.

$$* \; * \; * \; * \; *$$

Section 3.3 described a quantity, the von Neumann entropy:

$$S \equiv -k_{\mathrm{B}} \mathrm{Tr} \, (\rho, \ln \rho) = -k_{\mathrm{B}} \sum_{N} \lambda_N \ln \lambda_N \, , \qquad (12.1.11)$$

where the sum runs over all eigenvalues $\lambda_N$ of the density matrix. This vanishes for a pure state for which $\rho$ is a projection operator with a single unit eigenvalue and all other eigenvalues zero, and is positive-definite in all other cases.

Entropy defined in this way is a useful quantity because, as shown in Section 3.3, in the absence of entanglement it is an extensive quantity. Matters are very different for two isolated systems when they are entangled. In particular, in a pure state of the whole system the von Neumann entropy vanishes, but the entropies of the individual subsystems do not vanish, but are in fact both positive and equal. In a pure state $\Psi$ the density matrix has the components

$$\rho_{ma,nb} = \psi_{ma} \psi_{nb}^* \, , \qquad (12.1.12)$$

where $\psi_{ma}$ are the components of the normalized state $\Psi$ along a complete orthonormal set of state vectors with $m$ and $a$ labeling the states of subsystems *I* and *II*, respectively. (This is of course not of the form (3.3.42) unless the wave function itself factorizes, i.e., unless $\psi_{ma}$ is a function of $m$ times a function of $a$, which is the case of no entanglement.) According to Eq. (12.1.9), the density matrix of subsystem *I* is

$$\rho_{mn}^{(I)} = \sum_{a} \rho_{ma,na} = (\psi \psi^\dagger)_{mn} \, , \qquad (12.1.13)$$

where $\psi$ is here the matrix with components $\psi_{ma}$. The eigenvalues of $\rho^{(I)}$ are thus the eigenvalues of $\psi \psi^\dagger$, which are positive-definite or zero. Similarly, the density matrix of subsystem *II* is

$$\rho_{ab}^{(II)} = \sum_{m} \rho_{ma,mb} = (\psi^\dagger \psi)_{ba} \, , \qquad (12.1.14)$$

so its eigenvalues are the eigenvalues of the matrix $\psi^\dagger \psi$, and also positive definite or zero. These matrices have the same non-zero eigenvalues, because if $\psi \psi^\dagger u = \lambda u$ then, multiplying with $\psi^\dagger$, we find $(\psi^\dagger \psi)(\psi^\dagger u) = \lambda(\psi^\dagger u)$, and $\psi^\dagger u$ cannot vanish if $\lambda \neq 0$, so every non-zero eigenvalue of $\psi \psi^\dagger$ is an eigenvalue of $\psi^\dagger \psi$. In the same way, if $\psi^\dagger \psi v = \lambda' v$ and $\lambda' \neq 0$ then $(\psi \psi^\dagger)(\psi v) = \lambda'(\psi v)$, so every non-zero eigenvalue of $\psi^\dagger \psi$ is an eigenvalue of $\psi \psi^\dagger$. Since the non-zero eigenvalues of $\rho^{(I)}$ and $\rho^{(II)}$ are the same, their entropies are the same. This common value is known as the *entanglement entropy* of the system.

## 12.2 The Bell Inequalities

It might be supposed that the weird entanglement encountered in quantum mechanics could be avoided by a modification of quantum mechanics, based on the introduction of local hidden variables. Suppose that in the situation described by Bohm, the two-electron state is not (12.1.2), but instead is an ensemble of possible states, characterized by some parameter or set of parameters collectively called $\lambda$, such that the value of the component of the first particle's spin in any direction $\hat{a}$ is a definite function $(\hbar/2)S(\hat{a}, \lambda)$, where $S(\hat{a}, \lambda)$ can only take the values $\pm 1$. Both experience and the conservation of angular momentum then tell us that the component of the second particle's spin in the same direction will be $-(\hbar/2)S(\hat{a}, \lambda)$. The parameter $\lambda$ is fixed before the two particles separate from each other, so no non-locality is involved, but in order to imitate the probabilistic features of quantum mechanics, the value of $\lambda$ is taken to be random, with some probability density $\rho(\lambda)$, about which it is only necessary to assume that $\rho(\lambda) \geq 0$ and $\int \rho(\lambda) \, d\lambda = 1$. The correlation between the spins of the two particles can be expressed as the average value of the product of the $\hat{a}$ component of the spin of particle 1 and the $\hat{b}$ component of the spin of particle 2:

$$\left\langle (\mathbf{s}_1 \cdot \hat{a}) \, (\mathbf{s}_2 \cdot \hat{b}) \right\rangle = -\frac{\hbar^2}{4} \int d\lambda \, \rho(\lambda) S(\hat{a}, \lambda) S(\hat{b}, \lambda) \,, \qquad (12.2.1)$$

where $\hat{a}$ and $\hat{b}$ are any two unit vectors. In quantum mechanics, the spin of particle 1 is an operator satisfying[4]

$$(\mathbf{s}_1 \cdot \hat{a}) \, (\mathbf{s}_1 \cdot \hat{b}) = \frac{\hbar^2}{4} \hat{a} \cdot \hat{b} + i \frac{\hbar}{2} \left( \hat{a} \times \hat{b} \right) \cdot \mathbf{s}_1 \,, \qquad (12.2.2)$$

so in the state (12.1.2), in which $\mathbf{s}_2 = -\mathbf{s}_1$ and $\mathbf{s}_1$ has zero expectation value, the average of the product of spin components is

$$\left\langle (\mathbf{s}_1 \cdot \hat{a}) \, (\mathbf{s}_2 \cdot \hat{b}) \right\rangle_{\mathrm{QM}} = -\frac{\hbar^2}{4} \hat{a} \cdot \hat{b}. \qquad (12.2.3)$$

There is no obstacle to constructing a function $S$ and a probability density $\rho$ for which (12.2.1) and (12.2.3) are equal for any single pair of directions $\hat{a}$ and $\hat{b}$. So it is not possible experimentally to distinguish between local hidden-variable theories and quantum mechanics by studying spin components in just two directions. But in a 1964 paper[5] John Bell (1928–1990) was able to show

---

[4]  The easiest way to see this is to recall that the spin operator **s** for spin 1/2 may be represented as $(\hbar/2)\boldsymbol{\sigma}$, where the components of $\boldsymbol{\sigma}$ are the Pauli matrices (4.2.18). Direct calculation shows that these matrices satisfy the multiplication rule $\sigma_i \sigma_j = \delta_{ij} \mathbf{1} + i \sum_k \epsilon_{ijk} \sigma_k$, from which Eq. (12.2.2) immediately follows.

[5]  J. S. Bell, *Physics* **1**, 195 (1964). This journal is no longer published; the article by Bell can be found in the collection *Quantum Theory and Measurement*, eds. J. A. Wheeler and W. Zurek (Princeton University Press, Princeton, NJ, 1983). For a review, see N. Brunner, D. Cavalcanti, S. Pironio, V. Scarani, and S. Wehner, *Rev. Mod. Phys.* **86**, 419 (2014).

that such a conflict does exist when one considers spin components for three different directions $\hat{a}$, $\hat{b}$, and $\hat{c}$. In this case, the correlation functions (12.2.1) satisfy inequalities that are not in general satisfied by the quantum-mechanical expectation values (12.2.3).

To see this, we note that according to the general properties of local hidden-variable theories assumed above,

$$\left\langle (s_1 \cdot \hat{a})(s_2 \cdot \hat{b}) \right\rangle - \left\langle (s_1 \cdot \hat{a})(s_2 \cdot \hat{c}) \right\rangle = -\frac{\hbar^2}{4} \int \rho(\lambda)\, d\lambda \left[ S(\hat{a}, \lambda) S(\hat{b}, \lambda) \right.$$
$$\left. - S(\hat{a}, \lambda) S(\hat{c}, \lambda) \right]. \tag{12.2.4}$$

Since $S^2(\hat{b}, \lambda) = 1$, this can be written

$$\left\langle (s_1 \cdot \hat{a})(s_2 \cdot \hat{b}) \right\rangle - \left\langle (s_1 \cdot \hat{a})\, (s_2 \cdot \hat{c}) \right\rangle = -\frac{\hbar^2}{4} \int \rho(\lambda)\, d\lambda \; S(\hat{a}, \lambda) S(\hat{b}, \lambda)$$
$$\times \left[ 1 - S(\hat{b}, \lambda) S(\hat{c}, \lambda) \right]. \tag{12.2.5}$$

The absolute value of an integral is at most equal to the integral of the absolute value, so

$$\left| \left\langle (s_1 \cdot \hat{a})\, (s_2 \cdot \hat{b}) \right\rangle - \left\langle (s_1 \cdot \hat{a})(s_2 \cdot \hat{c}) \right\rangle \right| \leq \frac{\hbar^2}{4} \int \rho(\lambda)\, d\lambda \left[ 1 - S(\hat{b}, \lambda) S(\hat{c}, \lambda) \right]$$

and therefore

$$\left| \left\langle (s_1 \cdot \hat{a})(s_2 \cdot \hat{b}) \right\rangle - \left\langle (s_1 \cdot \hat{a})(s_2 \cdot \hat{c}) \right\rangle \right| \leq \frac{\hbar^2}{4} + \left\langle (s_1 \cdot \hat{b})(s_2 \cdot \hat{c}) \right\rangle. \tag{12.2.6}$$

This is the original Bell inequality.

The important thing is that, at least for some choices of the directions $\hat{a}$, $\hat{b}$, and $\hat{c}$, this inequality is *not* satisfied by the quantum-mechanical correlation function (12.2.3). For instance, suppose we take

$$\hat{b} \cdot \hat{a} = 0, \qquad \hat{c} = [\hat{a} + \hat{b}]/\sqrt{2}. \tag{12.2.7}$$

Then for the quantum-mechanical correlation function (12.2.3), the left-hand side of the inequality (12.2.6) is

$$\left| \left\langle (s_1 \cdot \hat{a})(s_2 \cdot \hat{b}) \right\rangle_{QM} - \left\langle (s_1 \cdot \hat{a})(s_2 \cdot \hat{c}) \right\rangle_{QM} \right| = \frac{\hbar^2}{4\sqrt{2}}, \tag{12.2.8}$$

while the right-hand side is

$$\frac{\hbar^2}{4} + \left\langle (s_1 \cdot \hat{b})(s_2 \cdot \hat{c}) \right\rangle_{QM} = \frac{\hbar^2}{4} - \frac{\hbar^2}{4\sqrt{2}}. \tag{12.2.9}$$

Needless to say, the quantity (12.2.8) is greater, not less, than the quantity (12.2.9). So measurement of the correlation functions $\left\langle (s_1 \cdot \hat{a})(s_2 \cdot \hat{b}) \right\rangle$,

$\langle(\mathbf{s}_1 \cdot \hat{a})(\mathbf{s}_2 \cdot \hat{c})\rangle$, and $\langle(\mathbf{s}_1 \cdot \hat{b})(\mathbf{s}_2 \cdot \hat{c})\rangle$ can provide a clear verdict between the predictions of quantum mechanics and those of any local hidden-variable theory.

Not only can experiment deliver such a verdict; it has done so. The experiments, carried out by Alain Aspect and his collaborators,[6] actually tested a generalization of the original Bell inequality. Consider any quantity $S_n(\hat{a})$ for a particle $n$ that (like the electron spin component $\hat{a} \cdot \mathbf{s}_n$ in units of $\hbar/2$) can only take the values $\pm 1$. In a local hidden-variable theory the measured value of $S_n(\hat{a})$ will be a definite function $S_n(\hat{a}, \lambda)$ of some parameter or set of parameters $\lambda$ whose value is fixed before the particles separate, with a probability $\rho(\lambda) d\lambda$ of getting a value between $\lambda$ and $\lambda + d\lambda$. The correlation between the value of $S_1(\hat{a})$ for particle 1 and the value of $S_2(\hat{b})$ for particle 2 is the average of the product:

$$\langle S_1(\hat{a}) S_2(\hat{b})\rangle = \int d\lambda\, \rho(\lambda) S_1(\hat{a}, \lambda) S_2(\hat{b}, \lambda) . \qquad (12.2.10)$$

Consider the quantity

$$\langle S_1(\hat{a}) S_2(\hat{b})\rangle - \langle S_1(\hat{a}) S_2(\hat{b}')\rangle + \langle S_1(\hat{a}') S_2(\hat{b})\rangle + \langle S_1(\hat{a}') S_2(\hat{b}')\rangle$$
$$= \int d\lambda\, \rho(\lambda) \Big[ S_1(\hat{a}, \lambda) S_2(\hat{b}, \lambda) - S_1(\hat{a}, \lambda) S_2(\hat{b}', \lambda)$$
$$+ S_1(\hat{a}', \lambda) S_2(\hat{b}, \lambda) + S_1(\hat{a}', \lambda) S_2(\hat{b}', \lambda) \Big]$$

for four different directions, $\hat{a}$, $\hat{b}$, $\hat{a}'$, and $\hat{b}'$. For any given $\lambda$, each product $S_1 S_2$ in the square brackets can only have the value $\pm 1$, so the sum can only have the value[7] $0$, $+2$, or $-2$. The average must therefore satisfy the inequality

$$\left| \langle S_1(\hat{a}) S_2(\hat{b})\rangle - \langle S_1(\hat{a}) S_2(\hat{b}')\rangle + \langle S_1(\hat{a}') S_2(\hat{b})\rangle + \langle S_1(\hat{a}') S_2(\hat{b}')\rangle \right| \le 2 .$$
$$(12.2.11)$$

Note that this inequality holds for a wider class of theories than the original Bell inequality (12.2.6), because in its derivation we did not need to use the previous assumption that $S_2(\hat{a}, \lambda) = -S_1(\hat{a}, \lambda)$ for all directions $\hat{a}$.

For the inequality (12.2.11) to be of use in distinguishing hidden-variable theories from quantum mechanics, the value of the left-hand side given by quantum mechanics must violate the inequality. To calculate this value, we need of course

---

[6] A. Aspect, P. Grangier, and G. Roger, *Phys. Rev. Lett.* **47**, 460 (1981); **49**, 91 (1982); A. Aspect, J. Dalibard, and G. Roger, *Phys. Rev. Lett.* **49**, 1804 (1982). The discussion here mostly follows the second of these papers.

[7] It is not possible for the sum in the integrand to have the value $+4$ for any $\lambda$, because in order for the first three terms to have the value $+1$ it would be necessary to have $S_1(\hat{a}, \lambda) = S_2(\hat{b}, \lambda) = -S_2(\hat{b}', \lambda) = S_1(\hat{a}', \lambda)$, which would make the fourth term equal to $-1$, and the sum equal to $+2$ rather than $+4$. Similarly, it is not possible for the sum to have the value $-4$ for any $\lambda$, because in order for the first three terms to have the value $-1$ it would be necessary to have $S_1(\hat{a}, \lambda) = -S_2(\hat{b}, \lambda) = S_2(\hat{b}', \lambda) = S_1(\hat{a}', \lambda)$, which would make the fourth term equal to $+1$, and the sum equal to $-2$ rather than $-4$.

to specify a particular experimental arrangement. Following earlier experiments of Clauser *et al.*,[8] Aspect *et al.* measured photon polarization correlations in a two-photon transition that had been previously studied by Kocher and Commins.[9] The two photons are emitted in a cascade decay in calcium atoms, the first from a state with $j = 0$ and even parity to a short-lived intermediate state with $j = 1$ and odd parity, and the second from that state to another state with $j = 0$ and even parity. These photons are directed into polarizers. One polarizer sends photon 1 into one photomultiplier if it has linear polarization along a direction $\hat{a}$ (orthogonal to the photon direction $\hat{k}$), in which case a value $S_1(\hat{a}) = +1$ is recorded, and into a different photomultiplier if it is linearly polarized along a direction orthogonal to both $\hat{a}$ and $\hat{k}$, in which case a value $S_1(\hat{a}) = -1$ is recorded. Similarly, the other polarizer sends photon 2 into one photomultiplier if it has linear polarization along a direction $\hat{b}$ (orthogonal to the photon direction $-\hat{k}$), in which case a value $S_2(\hat{b}) = +1$ is recorded, and into a different photomultiplier if it is linearly polarized along a direction orthogonal to both $\hat{b}$ and $-\hat{k}$, in which case a value $S_2(\hat{b}) = -1$ is recorded. The polarizers can be rotated so that either $\hat{a}$ is replaced with $\hat{a}'$ or $\hat{b}$ is replaced with $\hat{b}'$, or both. Since the two-photon transition is between atomic states with $j = 0$, the amplitude for the transition must be a scalar function of the two polarizations, and since the initial and final atomic states have even parity the scalar $\hat{k} \cdot (\mathbf{e}_1 \times \mathbf{e}_2)$ is ruled out, so the amplitude must be proportional to $\mathbf{e}_1 \cdot \mathbf{e}_2$, and the probability of particle 1 having polarization in the direction $\hat{a}$ and particle 2 having polarization in the direction $\hat{b}$ is therefore $(\hat{a} \cdot \hat{b})^2/2$. (The factor $1/2$ is fixed by the condition that the sum over two orthogonal directions of $\hat{a}$ and of $\hat{b}$ must be unity.) By adding $S_1(\hat{a}) S_2(\hat{b})$ for the four possibilities $S_1(\hat{a}) = \pm 1$, $S_2(\hat{b}) = \pm 1$ weighted with these probabilities, we see that the quantum-mechanical expectation value of $S_1(\hat{a})$ times $S_2(\hat{b})$ is

$$\Big\langle S_1(\hat{a}) S_2(\hat{b}) \Big\rangle_{\text{QM}} = \frac{1}{2} \left( \cos^2 \theta_{ab} - \sin^2 \theta_{ab} - \sin^2 \theta_{ab} + \cos^2 \theta_{ab} \right) = \cos 2\theta_{ab} ,$$

$$(12.2.12)$$

where $\theta_{ab}$ is the angle between $\hat{a}$ and $\hat{b}$. Thus in quantum mechanics, the left-hand side of Eq. (12.2.11) is

$$\Big\langle S_1(\hat{a}) S_2(\hat{b}) \Big\rangle_{\text{QM}} - \Big\langle S_1(\hat{a}) S_2(\hat{b}') \Big\rangle_{\text{QM}} + \Big\langle S_1(\hat{a}') S_2(\hat{b}) \Big\rangle_{\text{QM}} + \Big\langle S_1(\hat{a}') S_2(\hat{b}') \Big\rangle_{\text{QM}}$$

$$= \cos 2\theta_{ab} - \cos 2\theta_{ab'} + \cos 2\theta_{a'b} + \cos 2\theta_{a'b'} .$$

$$(12.2.13)$$

---

[8] J. F. Clauser, M. A. Horne, A. Shimony, and R. A. Holt, *Phys. Rev. Lett.* **23**, 880 (1969). For a review of various versions of Bell inequalities and their experimental tests, see J. F. Clauser and A. Shimony, *Rep. Prog. Phys.* **41**, 1881 (1978).

[9] C. A. Kocher and E. D. Commins, *Phys. Rev. Lett.* **18**, 575 (1967).

This is a maximum[10] if $\theta_{ab} = \theta_{a'b} = \theta_{a'b'} = 22.5°$ and $\theta_{ab'} = 67.5°$, in which case

$$\left\langle S_1(\hat{a})S_2(\hat{b})\right\rangle_{\text{QM}} - \left\langle S_1(\hat{a})S_2(\hat{b}')\right\rangle_{\text{QM}} + \left\langle S_1(\hat{a}')S_2(\hat{b})\right\rangle_{\text{QM}} + \left\langle S_1(\hat{a}')S_2(\hat{b}')\right\rangle_{\text{QM}}$$
$$= 2\sqrt{2} = 2.828 .$$

Because the polarizers in this experiment were not perfectly efficient, the expected value was only $2.70 \pm 0.05$. The experimental result for the left-hand side of Eq. (12.2.11) was $2.697 \pm 0.0515$, in good agreement with quantum mechanics, and in clear disagreement with the inequality (12.2.11) satisfied by all local hidden-variable theories.

## 12.3 Quantum Computation

In recent years much attention has been given to the opportunities provided for computation by quantum mechanics.[11] This section will give only a brief glimpse of the capabilities of quantum computers, and their limitations.

It is the existence of entanglement in quantum mechanics that provides a possibility of calculations with quantum computers that in a classical computer would require exponentially greater resources. The working memory of a quantum computer may be considered to consist of $n$ *qbits*, elements like atoms of total angular momentum 1/2 or electric currents in superconducting loops, for which some physical quantity, such as the $z$-component of the angular momentum or the direction of the current, can only take two values. We will label these two values with an index $s$, that only takes the values 0 and 1, and define $\Psi_{s_1 s_2 ... s_n}$ as the normalized state vector in which the qbits take values $s_1, s_2, ..., s_n$. The general state of the memory is then

$$\Psi = \sum_{s_1 s_2 ... s_n} \psi_{s_1 s_2 ... s_n} \Psi_{s_1 s_2 ... s_n} , \qquad (12.3.1)$$

where the $\psi_{s_1 s_2 ... s_n}$ are complex numbers, subject to the normalization condition

$$\sum_{s_1 s_2 ... s_n} \left| \psi_{s_1 s_2 ... s_n} \right|^2 = 1 . \qquad (12.3.2)$$

Since the moduli of the $\psi_{s_1 s_2 ... s_n}$ are subject to this condition, and the over-all phase of $\psi_{s_1 s_2 ... s_n}$ is irrelevant, there are $2^n - 1$ independent coefficients, which

---

[10] All the directions $\hat{a}, \hat{b}, \hat{a}'$, and $\hat{b}'$ are normal to $\hat{k}$, so they all lie in the same plane. The maximum value of (12.2.13) is achieved by putting them in an order such that $\theta_{ab'} = \theta_{ab} + \theta_{a'b} + \theta_{a'b'}$, and then setting the derivatives of the expression (12.2.13) with respect to $\theta_{ab}$ and $\theta_{a'b}$ and $\theta_{a'b'}$ all equal to zero.

[11] See, e.g., N. D. Mermin, *Quantum Computer Science – An Introduction* (Cambridge University Press, Cambridge, 2007). For an on-line review of quantum computation, see J. Preskill, http://www.theory.caltech.edu/people/preskill/ph229/#lecture.

can be taken as the ratios of the $\psi_{s_1 s_2 \ldots s_n}$. Hence a quantum computer with $n$ qbits has a memory that can contain $2^n - 1$ independent complex numbers, in the sense that this is the information on which the computer can act during calculations. (As we shall see, this information is not in general available to be read out from the memory.)

This may be compared with a classical digital computer. The state of a classical memory containing $n$ bits is just a string of $n$ zeros and ones, which can be regarded as the binary expression of a single integer taking a value between 0 and $2^n - 1$. It is the comparison of a quantum memory containing $2^n - 1$ unconstrained complex numbers and a classical memory containing a single integer between 0 and $2^n - 1$ that makes the difference between quantum and classical computers. A classical digital computer can do anything a quantum computer can do, but at the cost of needing an exponentially larger memory.

As with a classical computer, we can think of the indices $s_1, s_2, \ldots, s_n$ on $\psi$ and $\Psi$ as a string of zeros and ones, and replace them with a single integer $\nu$ between zero and $2^n - 1$ whose binary expansion is $s_1 s_2 \ldots, s_n$. (For instance, in the case $n = 2$, we would define $\Psi_0 \equiv \Psi_{00}$, $\Psi_1 \equiv \Psi_{01}$, $\Psi_2 \equiv \Psi_{10}$, and $\Psi_3 \equiv \Psi_{11}$.) We can thus write Eq. (12.3.1) as

$$\Psi = \sum_{\nu=0}^{2^n-1} \psi(\nu)\Psi_\nu \, , \tag{12.3.3}$$

and think of $\psi(\nu)$ as a single complex-valued function of the integer $\nu$.

By exposing the $n$ qbits to various external influences, it is possible in principle to act on their state vector with an operator of the form $\exp(-iHt/\hbar)$, where $H$ is any sort of Hermitian operator, and in this way subject the state vector to any unitary transformation $\Psi \to U\Psi$ we like. The effect on the wave function will be

$$\psi(\nu) \mapsto \sum_{\mu=0}^{2^n-1} U_{\mu\nu}\psi(\mu), \tag{12.3.4}$$

where $U_{\nu\mu}$ is some more-or-less arbitrary unitary matrix. In this way, a quantum computer can convert functions into other functions. For example, the construction of an algorithm for finding the prime factors of large integers[12] makes use of a unitary transformation with

$$U_{\mu\nu} = 2^{-n/2} \exp\left(2i\pi\mu\nu/2^n\right) , \tag{12.3.5}$$

by which $\psi(\nu)$ is converted to its Fourier transform:

---

[12] P. W. Shor, *J. Sci. Statist. Comput.* **26**, 1484 (1997). The use of such factorization in cryptography is briefly described in Section 11.8.

$$\psi(\nu) \mapsto 2^{-n/2} \sum_{\mu=0}^{2^n-1} \exp\left(2i\pi\mu\nu/2^n\right) \psi(\mu). \qquad (12.3.6)$$

This is unitary, because for $\mu$ and $\mu'$ integers between 0 and $2^n - 1$, we have

$$\sum_{\nu=0}^{2^n-1} U_{\mu\nu} U_{\mu'\nu}^* = 2^{-n} \sum_{\nu=0}^{2^n-1} \exp\left(2i\pi(\mu - \mu')\nu/2^n\right) = \delta_{\mu\mu'}.$$

In order not to lose the advantages of quantum computers, it is necessary to build up such useful unitary transformations out of "gates" – unitary transformations that act on no more than a fixed number of qbits at a time. For instance, the reference cited in footnote 12 shows that it is possible to construct the unitary transformation (12.3.5) by using gates of just two kinds: a gate $R_j$ that acts on the two states of the $j$th qbit with a unitary matrix

$$R_j: \quad \frac{1}{\sqrt{2}} \begin{pmatrix} 1 & 1 \\ 1 & -1 \end{pmatrix},$$

and a gate $S_{ij}$ that acts on the four states of the $j$th and $k$th qbits (with $j < k$):

$$S_{j,k}: \quad \begin{pmatrix} 1 & 0 & 0 & 0 \\ 0 & 1 & 0 & 0 \\ 0 & 0 & 1 & 0 \\ 0 & 0 & 0 & \exp(i\pi 2^{j-k}) \end{pmatrix},$$

in which the rows and columns correspond to the two-qbit states with indices 00, 01, 10, and 11, in that order.

Quantum computation is subject to limitations, both intrinsic and extrinsic. It faces intrinsic limitations in reading out the contents of the memory of a quantum computer. For a memory in a general state (12.3.3) with unknown coefficients $\psi(\nu)$, no single measurement of the state of each qbit can by itself tell us anything precise about the values of these coefficients. Even if we repeat identical computations many times and measure the state of each qbit each time, we only learn the values of the moduli $|\psi(\nu)|$. On the other hand, if we know that a computation has put the memory into one of the basis states $\Psi_\nu$, then we can find the integer $\nu$ by measuring the state of each qbit. In particular, in factoring large numbers into products of primes, the output is a set of numbers, represented by states $\Psi_\nu$, and there is no problem in finding these numbers by a measurement of the state of each qbit.

More general measurements are also possible. If we know that a quantum computation has put the memory in a state for which

$$\sum_{\nu=0}^{2^n-1} A_{\mu\nu}^r \psi(\nu) = a^r \psi(\mu)$$

with some set of Hermitian matrices $A^r$, then by appropriate measurements we can find the eigenvalues $a^r$. (The previously mentioned example, where a computation leaves the memory in a state $\Psi_\nu$, is just the case where these matrices are $A^\nu_{\mu'\mu} = \nu\delta_{\nu\mu'}\delta_{\nu\mu}$.)

Another intrinsic limitation: because of the linearity of the operations $U$ that can be carried out on the contents of a memory register, there are some things that can be done easily with a classical computer that cannot be done with a quantum computer. One of them is copying the contents of one memory register into another register.[13] The state of two independent registers can be represented as a direct product, $\Psi \otimes \Phi$, where $\Psi$ and $\Phi$ are the states of the two registers. (That is, if $\Psi = \sum_\nu \psi(\nu)\Psi_\nu$ and $\Phi = \sum_\mu \phi(\mu)\Phi_\mu$, then $\Psi \otimes \Phi = \sum_{\nu\mu} \psi(\nu)\phi(\mu)\Psi_{\nu\mu}$.) A copying operator $U$ would be one with the property that

$$U(\Psi \otimes \Phi_0) = \Psi \otimes \Psi , \qquad (12.3.7)$$

where $\Psi$ is an arbitrary state of the first register and $\Phi_0$ is some fixed "empty" state of the second register. If this is true for any $\Psi$, it must be true when $\Psi$ is a sum $\Psi_A + \Psi_B$, so

$$U\Big((\Psi_A + \Psi_B) \otimes \Phi_0\Big) = (\Psi_A + \Psi_B) \otimes (\Psi_A + \Psi_B)$$
$$= \Psi_A \otimes \Psi_A + \Psi_A \otimes \Psi_B + \Psi_B \otimes \Psi_A + \Psi_B \otimes \Psi_B . \qquad (12.3.8)$$

But if $U$ is linear, then

$$U\Big((\Psi_A+\Psi_B) \otimes \Phi_0\Big) = U\Big(\Psi_A \otimes \Phi_0\Big) + U\Big(\Psi_B \otimes \Phi_0\Big) = \Psi_A \otimes \Psi_A + \Psi_B \otimes \Psi_B , \qquad (12.3.9)$$

in contradiction with Eq. (12.3.8).

The extrinsic limitation on quantum computation is the necessity of counteracting errors, which if not dealt with will accumulate during extended calculations, making such calculations useless. One sort of error is a change of phase, in which interaction with its environment changes the state of some qbit from $\psi_0\Psi_0 + \psi_1\Psi_1$ to $e^{i\alpha_0}\psi_0\Psi_0 + e^{i\alpha_1}\psi_1\Psi_1$. Even if the phases $\alpha_i$ are very small this amounts to a change of the complex number $\psi_1/\psi_0$ represented by this qbit. For large uncontrolled phase changes the entanglement between this qbit and other qbits is destroyed. A disentangled state, in which $\psi_{s_1...s_n}$ is effectively a product of functions of the indices, can contain only $n - 1$ rather than $2^n - 1$ independent complex numbers, so that the advantage of quantum computers over classical computers is lost. Another sort of error is a bit flip; the state $\Psi_1$ of some qbit changes to $\Psi_0$, or vice versa.

---

[13] W. R. Wooters and W. H. Zurek, *Nature* **299**, 802 (1982); D. Dicks, *Phys. Lett.* A **92**, 271 (1982).

It is possible to give a quantum computer the ability to detect and correct such errors by writing programs in terms of synthetic qbits, which are assembled from a number of real qbits.[14] In one popular scheme,[15] nine real qbits are joined into three triplets, forming a single synthetic qbit. Its general state is

$$\psi_0 \Big( \Psi_{000} + \Psi_{111} \Big) \otimes \Big( \Psi_{000} + \Psi_{111} \Big) \otimes \Big( \Psi_{000} + \Psi_{111} \Big)$$
$$+ \psi_1 \Big( \Psi_{000} - \Psi_{111} \Big) \otimes \Big( \Psi_{000} - \Psi_{111} \Big) \otimes \Big( \Psi_{000} - \Psi_{111} \Big) , \qquad (12.3.10)$$

in which the direct products symbolized by $\otimes$ should be understood in the sense that, for instance, $\Psi_{000} \otimes \Psi_{111} \otimes \Psi_{000}$ is the nine-qbit state $\Psi_{000111000}$. This allows errors affecting a single real qbit to be detected and corrected by majority rule. (The details of the procedure are described in the references cited in footnotes 14 and 15.) A phase change of any one real qbit that alters the state of one of the triplets of qbits from $\Psi_{000} + \Psi_{111}$ or $\Psi_{000} - \Psi_{111}$ to any other linear combination (perhaps $\Psi_{000} - \Psi_{111}$ or $\Psi_{000} + \Psi_{111}$, respectively) can be corrected by changing its state to the state of the other two triplets. A bit flip, which converts one of the triplet states into an illegal state in which one qbit is in the 0 state and two are in the 1 state, can be corrected by converting this triplet into the legal state $\Psi_{111}$, while a bit flip that converts a triplet state into an illegal state in which one qbit is in the 1 state and two are in the 0 state can be corrected by replacing this triplet with the other legal state, $\Psi_{000}$.

Phase changes and bit flips do not act directly on synthetic qbits, but only on the real qbits from which they are formed. Hence, if errors affecting real qbits are corrected by methods like those described above, no errors will disturb the coefficients $\psi_0$ or $\psi_1$ in the synthetic qbit state (12.3.10), or similar coefficients in entangled states formed from the assemblage of many such synthetic qbits. The development of error-correcting codes of this sort, together with impressive progress in the physical performance of individual qbits,[16] leaves the problems of combining hundreds of qbits in a useful quantum computer, and of writing programs for such computers.

---

[14] For reviews, see J. Preskill, http://www.theory.caltech.edu/people/preskill/ph229/#lecture, Chapter 7; D. Gottesman, in *Quantum Computation: A Grand Mathematical Challenge for the Twenty-First Century and the Millennium*, ed. S. J. Lononaco, Jr. (American Mathematical Society, Providence, RI, 2002), pp. 221–235.

[15] P. W. Shor, *Phys. Rev.* A **52**, 2493 (1995).

[16] For example, see T. P. Harty, D. T. Allcock, C. J. Ballance, L. Guidoni, H. A. Janacek, N. M. Linke, D. N. Stacey, and D. M. Lucas, *Phys. Rev. Lett.* **113**, 220501 (2014) [arXiv:1403.1524].

# Author Index

# Subject Index

Printed in the United States
by Baker & Taylor Publisher Services